Advances in Intelligent Systems and Computing

Volume 520

Series editor

Janusz Kacprzyk, Polish Academy of Sciences, Warsaw, Poland
e-mail: kacprzyk@ibspan.waw.pl

About this Series

The series "Advances in Intelligent Systems and Computing" contains publications on theory, applications, and design methods of Intelligent Systems and Intelligent Computing. Virtually all disciplines such as engineering, natural sciences, computer and information science, ICT, economics, business, e-commerce, environment, healthcare, life science are covered. The list of topics spans all the areas of modern intelligent systems and computing.

The publications within "Advances in Intelligent Systems and Computing" are primarily textbooks and proceedings of important conferences, symposia and congresses. They cover significant recent developments in the field, both of a foundational and applicable character. An important characteristic feature of the series is the short publication time and world-wide distribution. This permits a rapid and broad dissemination of research results.

More information about this series at http://www.springer.com/series/11156

Álvaro Rocha · Mohammed Serrhini
Carlos Felgueiras
Editors

Europe and MENA Cooperation Advances in Information and Communication Technologies

Editors
Álvaro Rocha
Faculdade de Ciências e Tecnologia,
Departamento de Engenharia Informática
Universidade de Coimbra
Coimbra
Portugal

Carlos Felgueiras
School of Engineering Polytechnic of Porto
CIETI/ISEP
Porto
Portugal

Mohammed Serrhini
University Mohammed First Oujda
Oujda
Morocco

ISSN 2194-5357 ISSN 2194-5365 (electronic)
Advances in Intelligent Systems and Computing
ISBN 978-3-319-46567-8 ISBN 978-3-319-46568-5 (eBook)
DOI 10.1007/978-3-319-46568-5

Library of Congress Control Number: 2016951698

Printed on acid-free paper

This Springer imprint is published by Springer Nature
The registered company is Springer International Publishing AG
The registered company address is: Gewerbestrasse 11, 6330 Cham, Switzerland

Preface

This book contains a selection of papers accepted for presentation and discussion at the Europe, Middle East and North Africa Conference on Technology and Security to Support Learning 2016 (EMENA-TSSL'16). This conference had the support of the University Mohamed First Oujda, Morocco, AISTI (Iberian Association for Information Systems and Technologies/Associação Ibérica de Sistemas e Tecnologias de Informação), and School of Engineering Polytechnic of Porto, Portugal. It took place at in city of Saïdia, Oujda, Morocco during October 3–5, 2016.

EMENA-TSSL'16 conference has two aims; first it provides the ideal opportunity to bring together professors, researchers and high education students of different disciplines, discuss new issues, and discover the most recent developments, researches and trends on information and communication technologies, emerging technologies and security to support in learning. The second goal is focusing on to boost future collaboration and cooperation between researchers and academicians from Europe Middle East and North Africa universities (EMENA).

The Program Committee of EMENA-TSSL'16 was composed of a multidisciplinary group of experts and those who are intimately concerned with information and communication, e-learning, and security. They have had the responsibility for evaluating, in a 'blind-review' process, the papers received for each of the main themes proposed for the conference: (A) Online Education; (B) Emerging Technologies in Education; (C) Artificial Intelligence in Education; (D) Gamification and Serious games; (E) Network & Web Technologies Applications; (F) Online experimentation and Virtual Laboratories; (G) Multimedia Systems and Applications; (H) Security and Privacy; (I) Multimedia, Computer Vision and Image Processing; (J) Cloud, Big Data Analytics and Applications; (K) Human-Computer Interaction; (L) Software Systems, Architectures, Applications and Tools; (M) Online Languages and Natural Language Processing (N) E-content Development, Assessment and Plagiarism; (O) Secure E-Learning Development and Auditing; (P) Internet of Things and Wireless Sensor Networks.

EMENA-TSSL'16 received contributions from 37 countries around the world. The papers accepted for presentation and discussion at the conference are published by Springer (this book) and by EMENA-TSSL'16 (another e-book) and will be submitted for indexing by ISI, EI-Compendex, SCOPUS, DBLP and/or Scholar Google, among others. Extended versions of selected best papers will be published in relevant journals, including SCI/SSCI and Scopus indexed journals.

We acknowledge all who contributed to the staging of EMENA-TSSL'16 (authors, committees, and sponsors); their involvement and support is very much appreciated.

Saïdia, Oujda
October 2016

Álvaro Rocha
Mohammed Serrhini
Carlos Felgueiras

Organization

General Chairs

Mohammed Serrhini, University Mohammed First Oujda, MA

Conference Co-chair

Álvaro Rocha, University of Coimbra, PT
Carlos Felgueiras, School of Engineering, Polytechnic of Porto, PT

Local Chairs

Abdelaziz Ait Moussa, University Mohamed First Oujda, MA
El Bekkay Mermri, University Mohamed First Oujda, MA
Abdelillah Monir, University Mohamed First Oujda, MA
Ahmed Tahiri, University Mohamed First Oujda, MA
Mohcine Kodad, University Mohamed First Oujda, MA
Fouad Mehdaoui, University Mohamed First Oujda, MA

Advisory Committee

El-Mostafa Daoudi, University Mohamed First Oujda, MA
Abdelmajid Dargham, University Mohamed First Oujda, MA
El Miloud Jaara, University Mohamed First Oujda, MA

Program Committee

Gustavo Alves, School of Engineering, Polytechnic of Porto, Portugal
Mohsine Eleuldj, Mohammadia School of Engineering, Morocco
Ounsa Roudies, Mohammadia School of Engineering, Morocco

Álvaro Rocha, University of Coimbra, Portugal
Ernest Cachia, Dean Faculty of ICT University of Malta, Malta
Roza Dumbraveanu, University, Chisinau, Moldova
Raúl Cordeiro Correia, Instituto Politécnico de Setúbal, Portugal
Ronan Champagnat Universite de La Rochelle, France
Rita Francese, University of Salerno, Italy
Naceur Ben Hadj Braiek, Polytechnic School of Tunis, Tunisia
Fernando Moreira, Oporto Global University, Portugal
Maytham Hassan Safar, Kuwait University, Kuwait
Maria José Angélico Gonçalves, ISCAP/Polytechnic Institute of Porto, Portugal
Hamid Harroud, Al Akhawayn University in Ifrane, Morocco
Maria José Sousa, Universidade Europeia de Lisboa, Portugal
James Uhomoibhi, University of Ulster, UK
Jarno Limnéll, Aalto University, Finland
Esteban Vázquez Cano, Universidad Nacional de Educación a Distancia, Spain
Juarez Bento Silva, Universidade Federal de Santa Catarina, Brasil
Anouar Belahcen, Aalto University, Finland
Peter Mikulecky, University of Hradec Kralove, Czech
Katherine Maillet, Institut Mines-Télécom Paris, France
Rafael Valencia-Garcia, Universidad de Murcia, Spain
Luis Anido Rifon, Universidade de Vigo, Spain
Mraoui Hamid, Faculty of Sciences Oujda, Morocco
Rolou Lyn Rodriguez Maata, Faculty of Computing Sciences—Gulf College Oman, Oman
Ali Shaqour, Najah National University, Palestine
Abdullah Al-Hamdani of Sultan Qaboos University Muscat, Oman
Muzafer Saracevic, International University of Novi Pazar, Serbia
Maouche Mourad, Philadelphia University, Jordan
Manuel Caeiro Rodríguez, Universidade de Vigo, Spain
Chikh Mohammed Amine, University of Tlemcen, Algieria
Rafik Zitouni, Ecole d'ingénieur généraliste et high-tech à Paris, France
Utku ZKose, Usak University, Turkey
Tajullah Sky-Lark, Sustainable Knowledge Global Solutions, United States
Otmane Ait Mohamed, Concordia University, Canada
Mohammad Hamdan, Yarmouk University, Jordan
Yaser Khamayseh, Jordan University of Science and Technology, Jordan
Wail Mardini, Jordan University of Science and Technology, Jordan
Francesca Pozzi, Istituto Tecnologie Didattiche—CNR, Italy
Filipe Cardoso, Polytechnic Institute of Setubal, Portugal
Abdel-Badeeh Salem, Ain Shams University, Egypt
Mohammad Al-Smadi, Jordan University of Science and Technology, Jordan
Mohamad Badra, Zayed University, United Arab Emirate
Amal Zouaq, Royal Military College of Canada, Canada
Pedro Guerreiro, Universidade do Algarve, Portugal
Martin Llamas-Nistal, University of Vigo, Spain

Camille Salinesi, CRI, Université de Paris1 Panthéon-Sorbonne, France
Jorge Pires, Polytechnic Institute of Castelo Branco, Portugal
Ali Jaoua, Qatar University, Qatar
Osama Shata, Qatar University, Qatar
Abdelkarim Erradi, Qatar University, Qatar
Osama Halabi, Qatar University, Qatar
Rachid Nourine, Oran University, Algieria
Abdelhafid Bessaid, University of Tlemcen, Algieria
Lehsaini Mohamed, University of Tlemcen, Algieria
Nissrine Souissi, ENIM, Morocco
John Sahalos, University of Nicosia, Cyprus
Lebbah Yahia, University of Oran 1, Algieria
Kashif Saleem, King Saud University, Saudi Arabia
Amjad Gawanmeh, Khalifa University, United Arab Emirate
Abdulmalik Al-Salman, King Saud University, Saudi Arabia
Olivier Markowitch, Université Libre de Bruxelles, Belgium
Ladan Mohammad, Rafik Hariri University, Lebanon
Alaoui Talibi, University Mohamed First Oujda, Morocco
Tolga Ensari, Istanbul University, Turkey
David Baneres, Universitat Oberta de Catalunya, Spain
Yahya Tashtoush, Jordan University of Science and Technology, Jordan
Alberto Cardoso, University of Coimbra, Portugal
Sjouke Mauw, FSTC Université du Luxembourg, Luxembourg
StephanieTeufel, University of Fribourg, Switzerland
HatemBen-Sta, Insight Centre for Data Analytics, Ireland
Associate Dean Majida Ali Abed Alasady, Tikrit University, Iraq
Pierre Manneback, Faculté Polytechnique de Mons, Belgium
Mohammed Benjelloun, Faculté Polytechnique de Mons, Belgium

Contents

Part I Educational Technology

Part III Network & Web Technologies Applications

Part I
Educational Technology

Cell Phone for Classroom Learning: Challenges and Recommendations in a Developing Country Context

Maria Miiro Kafuko, Fatuma Namisango and Gorretti Byomire

Abstract The study was based on cell phone and classroom learning experiences in Uganda. The wide spread use and advancement in functionalities of a cell phone has not rolled out to include classroom learning capabilities in Uganda. A number of views are composed and generally referred to as challenges to m-learning at large. Cell phone use in Uganda is a very silent issue that has not been extensively discussed or presented for lucrative implementation. The embryonic challenges to this question have not been explicitly explored. A survey on mobile learning (m-learning) among learners in two universities in Uganda was conducted. The findings presented in this paper provide a review of the use of cell phones to teach and learn by learners and tutors in institutions of higher learning and discusses challenges encountered and applicable recommendations for successful cell phone use for teaching and learning.

Keywords Cell phone · M-learning · Developing country

1 Introduction

As mobile learning (m-learning) has significantly set off in developed countries like in the U.S and Europe, it is still in the infant stage in developing countries like Uganda. For example, [1] brings to light the fact that the use of cell phones in

M.M. Kafuko (✉) · F. Namisango · G. Byomire
Makerere University Business School, P.O Box 1337, Kampala, Uganda
e-mail: mariamiiro@gmail.com

F. Namisango
e-mail: fnamisango@gmail.com

G. Byomire
e-mail: bgorretti@gmail.com

Á. Rocha et al. (eds.), *Europe and MENA Cooperation Advances in Information and Communication Technologies*, Advances in Intelligent Systems and Computing 520, DOI 10.1007/978-3-319-46568-5_1

classroom learning environment was prohibited by the Ministry of Education and Sports (MOES) in Uganda. The latter is explained by the fact that cell phones are largely seen as destructive devices whose usefulness exists outside the classroom. The real challenges of cell phone use and necessary approaches to the successful usage in class room learning have been ignored hence categorizing them as merely distractive devices that cannot add value to learning. According to [2] the number of Telecom subscribers and Penetration Data is increasing annually, implying that the use of cell phone for online related functionalities and benefits is expanding to encompass all possibilities such as learning. Cell phones provide functionalities (like voice, text, browsing, downloading etc.) that would fully apply into the learning context, however, the benefits have not been attained since cell phones are not applied in learning in Uganda. Learning at Higher Education Institutions (HEIs) in developing countries has remained largely conventional amidst the ever growing telecom subscription and penetration data. Many young people have adopted smart technologies more than the elderly and are almost willing to use these technologies in all aspects of their lives, however, cell phone usage for learning has not been quite common practice. There is considerable concern about the doom that is yet to come onto the education system in Uganda because it is failing to keep up with the young digital natives and society. There is much needed drive for m-learning in the entire education sector so as to avoid the impending disconnect between the way digital natives operate in their day-to-day lives and the way existent educational institutions interact with them [3]. Anecdotal evidence shows that portable technology tools such as cell phones engage learners and promote learning [4].

Learning using mobile media like cell phones has achieved cultural significance in educational institutions because it is viewed as culturally situated making-meaning inside and outside of school [3]. The limited use of cell phones may not only depend on facts from the learners' perspective but also the tutors'. On the same, [5] explain that there is little evidence that teachers are adopting innovative approaches to teaching and learning with the devices. Reference [5] notes that there several challenges that impact negatively on the uptake and use of cell phones in innovative teaching. Makerere University Business School is one institution that has an m-learning system that has not quite caught on. It is tedious as students phone numbers (partial to a particular mobile network-MTN) have to be registered before using the m-learning portal. Secondly registration is costly as one needs to have airtime (money for making calls). Additionally, it only serves as a mobile notice board, each exchange is paid for and the visual impact is not appealing to either tutor or learner. Reference [6] reveal that the future of mobile learning depends largely on the level of social acceptance it receives. It may be imperative to understand the issues that may hinder its successful usage in countries like Uganda. Herein are objectives of this study, the methodology used and the findings thereof.

2 Objectives

The study sought to (1) establish the challenges encountered with cell phone use for learning that affect its use in Higher Education Institutions in developing countries; and (2) provide recommendations to the notable challenges to use of cell phones for learning in Higher Education Institutions in developing countries.

3 Methodology

A survey on learners at university education level was conducted using self-administered questionnaires. It was appropriate because the sample was a large group and it necessitated the responses gathered to be standardized. The questions were semi-structured so that they could answer the inquiries regarding challenges and recommendations of cell phone based learning. In addition to that, open-ended questions were included to allow respondents a chance to express and clarify their views. In addition to that, open-ended questions were included to allow respondents a chance to express and clarify their views. A population of respondents from two universities in Uganda was used to derive the study sample. A simple formula by [7] was used to derive the samples indicated in Table 1.

4 Technology Description/A Global Move Towards Student-Centred Learning

4.1 Cell Phone as a Device for M-Learning

Of interest to this study is the fact that a cell phone prevails as a handheld device that can be used to foster learning among peers and teachers hence its appropriateness in classroom learning. Although a number of other devices such as tablets and laptops can be used in m-learning, a cell phone would be more recognizable due to the fact that it has achieved an enormous subscription rate in developing countries. For the case of Africa to be more specific [2, 8] in the case of Uganda can testify to this. Reference [8] confirms the observance of cell phone based activities as it shows that developing countries are home to more than three quarters of all

Table 1 Sample distribution

Institution name	Status	Population	Sample size
Makerere University Business School	Public	10731	385
Kampala International University	Private	13938	388
Total sample size			**773**

mobile-cellular subscriptions and that mobile-broadband penetration in Africa reaches close to 20 % in 2014, up from 2 % in 2010 specifying that Africa leads in mobile broadband growth. Henceforth focusing on cell phone use for classroom learning is quite justified as they have gained a lot of popularity now.

A cell phone being the most popular device that is used by almost every individual in Uganda, it is therefore of specific interest for this review. Studies conducted by [9] on different m-learning projects in Africa, a number of premises for m-learning in Africa were established. These premises were summarized as follows;

(a) M-learning is a supportive mode of education and not a primary mode of education
(b) M-learning provides flexibilities for various learning- and life-styles
(c) The most appropriate mobile device for learners in Africa is a mobile phone
(d) Possibilities and latest developments in mobile technologies must be tested against practicality, usability and cost-effectiveness
(e) The use of multimedia on mobile phones must be tested against the envisaged leaning outcomes
(f) The major focus of m-learning should be more on communication and interaction than on content

4.2 M-Learning in a Developing Country Context

In the case of still developing countries, environment reliance to available and affordable technologies is the most realistic way forward [10]. Such technologies may include mobile technologies which support m-learning. According to [11] m-learning represents a way to address a number of educational problems even for developing countries. Additionally [11] shows that devices such as smart phones and tablets used in conjunction with 4G/3G wireless connectivity enable innovation and help learners, teachers, and parents gain access to digital content and personalized assessment vital for a post-industrial world and are essential tools to improve learning for learners. References [12, 13] inform this study that in the world's developing regions (Africa in particular), there is vast prospect to improve access to education through the use of mobile technology. Mobile technology makes it possible to connect learners in remote parts of the world where the deployment of wire line infrastructure is not cost effective.

Reference [13] further shows that there is awareness amongst mobile network operators (MNOs) that Africa could host their next billion subscribers, that these subscribers might represent a market for educational content and connectivity. The implication is that the cell phone will emerge as a highly used and important device and its impact on the education system cannot be under estimated. However, for the tertiary education sector of a developing country like Uganda, there is no formal Information and Communication Technology (ICT) policies or initiatives taken on to harness the ever-increasing benefits of these technologies and mobile technologies in particular [14]. The possibility of efforts geared towards m-learning with cell phones has not manifested anywhere living with many educational stakeholders wondering.

5 Emerging Issues and Challenges

Although m-learning has taken a brighter path over the years, a number of technical, socio-cultural and educational issues have emerged as challenging factors to its further acceptance and use. More issues within the same context have surfaced during m-learning implementation which include; non-delivery or late delivery of messages which is attributed to either the bulk SMS gateway provider or service glitches at certain telecommunication providers [15]. Reference [15] explain that in situations where a message is not received at the targeted time, the Mobile Learning team generally waits until the following day to see if it is a permanent failure. In such cases, the same message will be reset for the following day and the subsequent SMS is also rescheduled if it was scheduled for the day after the failed SMS. More technical issues such as synchronizing the device with a PC, or laptop, navigation and file storage, short battery life and device breakages were problematic, similarly, paucity of appropriate software, together with uncertainty as to how the devices might best be used to enhance teaching and learning, inhibited the development of innovative models of teaching and learning [10]. The challenges experienced are divergently presented ranging from technical, social and educational challenges.

6 Results

6.1 Challenges to Cell Phone Use in Classroom Learning

The study intended to structure an ideal scale at which m-learning can start by establishing the challenges that prevail as presented in Table 2. We hence provide an insight on the boundaries at which m-learning in the classroom may not exceed for initial success.

It was noted that learners feel that cell phones cause distraction in the classroom as they create interruption of normal class proceedings. Over-reliance on cell phones leading to laziness and failure to think out of technology assisted environment as [16] also noted. The emergence of negative social implications such as pornography in the classroom impacts negatively on morally acceptable behaviour of learners. Additionally, without uniformity of the cell phone in use, some learners can feel inferior because some have cheap but internet enabled phones yet others

Table 2 Challenges of using cell phone for class learning

Theme	Prevailing challenge
Distracting class	Too much focus on cell phone hence interruption of class
	Misuse of cell phones for social networking e.g. Facebook and twitter while viewing related data
	Noise from cell phones i.e. ringtones, notifications and message tones.
	May lead to learners' mistrusting the information from tutors
Social implication	Brings pornography to the classroom
	Low self-esteem as a result of lack of expression of ideas
	Learners showing off expensive phones
	Leads to laziness and failure to use one's brain.
	Cell phone will replace the need to go to class
	Can result into health risks such as eye sight problems
Mobile phone technology concerns	Some cell phones do not have enough storage capacity
	Some cell phones do not have Bluetooth
	Not all phones can access the internet as are not GPRS enabled.
	Poor battery life/shutting down of phones due to power
Internet technology concerns	Disruption due malware such as viruses
	Inexperience about the internet (does not cater for ignorant learners)
	Poor and unstable network systems
	Cost of acquiring internet bundles
	No wireless internet within classroom
	Lack of thorough explanations on the internet
Other issues	Expensive to maintain cell phones and internet access
	Cell phones are sometimes prohibited in the classroom
	Mixed attitudes or misconception about cell phone use in classroom by stakeholders

might pull off poses with expensive phones. This may present the issue of inequality, the mobile divide, feeling of inclusion or exclusion within a class as [16, 17] also confirmed.

The theme of inadequacy of mobile phone technology for learning takes the third place with issues such as screen size, poor storage capacity and poor battery life. Reference [16] states that a cell phone may be portable but it is poor at recording videos and needs detailed co-operation from the parties involved to get output that is usable which causes negative interference.

Being a developing country, challenges of internet technology accessibility were almost an obvious challenge. Learners expresses concerns on the issue of unstable network, limited network coverage, lack of wireless internet (WIFI) in classrooms and the high cost of purchasing alternative to WIFI internet connection more specifically known as "bundles". Without stable internet connection an inverted classroom or m-learning enabled environment becomes almost impossible.

6.2 *Recommendations to Challenges Faced in Using Cell Phones in Classroom Learning*

The analysis of findings indicates that sustainability of cell phone use in classroom learning would (1) entail preparing tutors and learners on how to use cell phones for the purpose of learning, (2) require restrictive cell phone use policy, (3) Free or relatively cheap internet platform. Learners concurred to cell phone use in classroom learning with such terms and conditions observed. Concerns of using cell phones to access class material implied that learners would have the tutor teaching normally but only use the cell phone for other purposes such as getting class notes and searching for more information to enhance class discussions. The internet access gap is a major challenge to cell phone use for m-learning in the classroom.

A learning centred goal is imperative as a solution to the negative social impacts. Subsequently, explanations from the tutor as well as sharing information among learners and personal review of learning material are expected. This implies that learning can be enhanced in classrooms if tutors can further explain on what learners discover on the internet. The ability to exchange notes and have access to

Table 3 Recommendations to challenges faced in using cell phones for classroom learning

Theme	Possible recommendations
Restrictive use of phone policy as a solution to distracting the class	Only using phones to access class material
	Putting phones in silent mode during lectures
	Receiving calls from outside the classroom
	Not using the loudspeaker option on cell phones
	Limiting the time learners use their phones for
Provision of the internet and m-learning enablers as a solution to inadequacy of mobile phone technology for learning	Availing a free internet platform
	Having sockets in class to ensure charged phones
	Providing wireless internet in the classroom
Having a learning-focused goal as a solution to negative social implications of cell phones in class	Tutors explaining what learners have found out
	Listening more to the tutors than reading what is on the cell phones
	Emphasizing what learners should use phones for
	Encouraging learners to read on their own
	Sharing information with learners who do not have internet enabled phones

them is clearly shown here. Reference [18] suggest that amidst the learner autonomy in m-learning, there is a need to ensure that learners have appropriate skills in locating and evaluating resources and the ability to reflect on their own learning. The solutions in can be promoted through the use of cheap Short Message Service (SMS), easy sharing of numbers by tutors and also sharing information, easy upload and download of lecture podcasts through a free WIFI connectivity. This gives learners the freedom of asking questions and sending messages through SMS and support m-learning in the classroom. Reference [5] also envisioned this as a solution for the shy learners (Table 3).

7 Conclusions

Cell phone use for m-learning in Uganda has not quite taken effect in the classroom as these devices are taken as unacceptable in classroom. However, its use outside the classroom for learning is to a small extent evident under certain circumstances as learners indicated a list of activities for which a cell phone has been applied. In addition to the prevailing activities undertaken, learners feel they have a lot more educational activities that they would accomplish with a cell phone if it was fully accepted. A number of challenges manifest when the field of cell phone use in classroom learning is projected. Class distraction, social inequality and internet connection are paramount challenges raised. We appreciate that strategies on effective implementation of cell phones for m-learning in classroom environment would apply to avert foreseen failure. Restrictive cell phone use policy is imperative to lead way for workable cell phone use in classroom adding to the fact that establishing stable internet connections that are relatively cheap or even better free provides a strong staring point in this matter.

References

1. Ouga, S.: Uganda: ministry of education bans mobile phones in schools. In: The New Vision. http://allafrica.com/stories/201309101261.html. Accessed 9 Sept 2013
2. UCC: Facts and Figures. http://www.ucc.co.ug/data/qmenu/3/Facts-and-Figures.html (2014)
3. Pachler, N.: The socio-cultural ecological approach to mobile learning: an overview. Medienbildung in neuen Kulturräumen: Die deutschprachige und britische Diskussion, 153–167 (2010)
4. Hlodan, O.: Mobile learning anytime, anywhere. Bioscience **60**(9), 682–682 (2010)
5. Facer, K., Faux, F., McFarlane, A.: Challenges and opportunities: making mobile learning a reality in schools. In: Mobile Technology: The future of Learning in Your Hands, pp. 53–56 (2005)
6. Mehdipour, Y., Zerehkafi, M.: Mobile learning for education: benefits and challenges. Int. J. Comput. Eng. Res. **03**(6), 93–101 (2013)
7. Yamane, T.: Statistics, An Introductory Analysis, 2nd edn. Harper and Row, New York (1967)

8. ITU: ITU releases 2014 ICT figures. Mobile-broadband penetration approaching 32 per cent; Three billion Internet users by end of this year. Newsroom Press Release. https://www.itu.int/net/pressoffice/press_releases/2014/23.aspx (2014)
9. Brown, T.: The role of m-learning in the future of e-learning in Africa. In: 21st ICDE World Conference, vol. 110. http://www.tml.tkk.fi/Opinnot. June 2003
10. Trucano, M.: 10 principles to consider when introducing ICTs into remote, low-income educational environments. World Bank Edutech Blog. http://blogs.worldbank.org/edutech/10-principles-consider-when-introducing-icts-remote-low-income-educational-environments (2013)
11. West, D.: Mobile Learning: Transforming Education, Engaging Students, and Improving Outcomes. http://www.insidepolitics.org/brookingsreports/MobileLearning.pdf (2013)
12. Jacobs, I.M.: Modernizing Education and Preparing Tomorrow's Workforce through Mobile Technology, Innovation for Jobs Summit 2013. QUALCOMM, Menlo Park, USA (2013)
13. Traxler, J.: Potential of learning with mobiles in Africa. In: World Innovations Summit for Education. University of Wolverhampton. http://www.wise-qatar.org/content/prof-john-traxler-potential-learning-mobiles-africa (2012)
14. Farrell, G.: Survey of ICT and Education in Africa, ICT in Education in Uganda, Uganda Country Report. http://www.infodev.org/infodev-files/resource/InfodevDocuments_435.pdf (2007)
15. Lim, T., Fadzil, M., Mansor, N.: Mobile learning via SMS at open university Malaysia: equitable, effective, and sustainable. Int. Rev. Res. Open Distrib. Learn. **12**(2), 122–137 (2011)
16. Traxler, J.: Defining mobile learning. In IADIS International Conference Mobile Learning, pp. 261–266. June, 2005
17. Brown, M., Diaz, V.: Mobile learning: contexts and prospects. A report on the ELI focus session. EDUCAUSE learning initiative (2010)
18. McFarlene, A., Roche, N., Triggs, P.: Mobile Learning: research Findings. A report to Becta. University of Bristol. http://www.becta.org.uk (2007)

E-learning Foresight for Renewable Energy Technology in Higher Education in Morocco

Amina Laaroussi, Souad Ajana, Soumia Bakkali, Kenza Faraj and Omar Cherkaoui

Abstract Nowadays, education strategy aims to introduce some potential approaches into teaching and learning taking into account sustainable development. Renewable energy sources represent the future utilization of energy. The use of this technology do not harm the environment because it is clean and environmentally friendly. Consequently, using E-learning platform in renewable energy education is a novel and a creative method to empower teaching for students and researchers to take action for sustainable development. This paper presents a survey of some existing E-learning platforms in energy education. As a result, it proposes to implement an E-learning platform as a beneficial and advantageous solution for developing countries such Morocco to train the needed competencies.

Keywords E-learning · Renewable energy · Green energy · Renewable energy education · Sustainable development · Moodle

1 Introduction

Development in Information and Communication Technologies (ICT) is increasingly transforming and promoting the way of learning. The new ICTs permit to optimize the modern society's requirements, so a lot of educational institutions throughout the world have developed new educational methods based on electronic, so called virtual, teaching. Nowadays, E-leaning is considered as a new approach to enhance learning and training relying on ICT as a practical means of delivering and presenting educational contents. It has become of paramount importance as it covers a very large number of various educational fields by offering new and

A. Laaroussi (✉) · S. Ajana · S. Bakkali · K. Faraj
ENSEM, Hassan II University in Casablanca, Casablanca, Morocco
e-mail: laaroussi.amina5@gmail.com

O. Cherkaoui
ESITH, Casablanca, Morocco

Á. Rocha et al. (eds.), *Europe and MENA Cooperation Advances in Information and Communication Technologies*, Advances in Intelligent Systems and Computing 520, DOI 10.1007/978-3-319-46568-5_2

convivial tools helping to incorporate constructive learning strategies. E-learning provides also the possibility to personalize learning anywhere (classroom, home etc) and anytime. It meets the requirements of students according to their age and previous knowledge. Various definitions of E-learning are given in literature, [1, 2] present e-leaning as communication and learning activities via electronic means and through computer and networks. E-learning is defined in [3] as self-learning through information technology. Collaborative learning is also one of the E-learning applications.

In the field of energy engineering, ICTs have been abundantly integrated practically into education programs for many years, for instance by simulation software for energy systems.

Today we primarily use fossil fuels (coal, oil, and natural gas) to meet our energy needs like heating and powering homes and also fuelling cars. However, we have limited reserves of these fuels on the Earth and we are utilizing them much more rapidly. In this context, new alternative energy sources are developed to be efficient and environmentally friendly. Renewable energy is the use of new energy sources like solar, wind, geothermal, hydro, ocean and biomass to deliver power and heat to the end-user. Using renewable energies is better for the environment because they are less pollutant. This is why renewable energy technologies are often called clean or green. Therefore, green energy means clean energy included in sustainable development plans for environment protection and free pollution [4].

In regard of the importance given to renewable energy in the world, universities are integrating and developing educational programs taking into account renewables. Renewable energy education (REE) is currently of paramount importance as it is one of the key elements of sustainable development [5]. In the meantime, advancements in ICTs have allowed the use of didactical tools to present some internet based courses on different topics such as renewable energy. The rapid advancements in these technologies have put more emphasis on this educational discipline which requires further changes. In this respect, various E-learning platforms are proposed to fulfill the educational requirements with novel methods. Experiences in other domains of study suggest that the knowledge obtained through experiments in laboratories can be gained almost equally through suitable E-learning platforms. These platforms can be usefully used to conduct some experiments in laboratory by students and can also be very helpful for teachers. In the domain of renewable energy where the founding of advanced laboratories in the developing countries may be unwieldy, a set of E-leaning platforms for renewable energy increasingly facilitate high quality REE worldwide.

2 Renewable Energy in Morocco

Unarguably, the glory value of energy, as a fundamental ingredient in our daily life, has become a key element of debate on E-learning, social-media, social, economic and environmental dimensions of sustainable development.

Different kinds exist in energy like fossil-based energy, which commonly contains coal, fossil fuel and natural gas and the big part of energy is produced with this last.

The fossil fuel has given birth to some major human living system problems and human health diseases, because of their expanded use in several industrial and non-industrial sectors. Such problems are detailed in [6]. Hence, when we discuss around green energy we intend by this non-exhaustible resources produced from renewable energy sources l solar, hydro, biomass, wind, geothermal and ocean [7].

In Morocco, the majority of resources data currently in existence are used and collected as follows for an eco-friendly and economical over the long-term.

2.1 Solar Energy

According to the two Moroccan agencies, the Research Institute for Solar Energy and New Energies (IRESEN) and the Moroccan Agency for Solar Energy in Morocco (MASEN), solar energy is emerging as a major trumping in the growth and the development of the country. We don't just provide green and clean energy for our country but also the possibility of exporting energy to neighboring markets for the future.

Since 2009, His Majesty King Mohamed VI announced the launch of a National Solar initiative spotted with a budget of $9bn in order to sustain Morocco's ambition of reaching 42 % of renewable in its energy mix by 2020. The following convention between the projects DESERTEC and NAREVA for an extra 500 MW plant is to be implemented. In addition to this, another worldwide complex solar power called NOOR is in construction in OUARZAZAT in the central south of Morocco. It is considered as the world's largest solar power plant. The whole project is divided into three parts, the gigantic one namely NOOR-I was launched in February 2016 producing 160 MW and using 5000 panels, while the two others are now under construction. NOOR is developed on 3000 hectares and is expected to produce 580 MW in the horizon of 2018.

2.2 Hydro Energy

Morocco is a country, which contains an immense and a wide renewable energy potential. This potential has been exploited early on in the shape of hydro. To give an exhaustive vision, Morocco as an agrarian economy, hydro energy sought to confer itself with an energy power to whip up its development and growth. In the early years after its independence, there was a reasonable and rationalistic option as it would avail generating energy as a byproduct and water security.

King Hassan the Second made it a national policy to construct one dam per year. As outcome nowadays, Morocco disposes of 26 hydro power stations totaling 1 360 MW in capacity. Future expansion and evolution are actually centered around micro dams and 200 potential sites have formally been defined. Actually, Al Wahda, the second largest dam in Africa is Moroccan According to IRESEN.

2.3 *Wind Energy*

As part of the energy strategy utilized to sustain the development of renewable energy and energy efficiency in the country, Morocco has undertaken a wide wind energy program. According to the National Agency for the Development of Renewable Energy and Energy Efficiency (ADEREE) and Ministry of Energy, Mines, Water and Environment, the total investment concerning the Moroccan integrated wind energy project over a period of 10 years was estimated at 31.5 billion dirhams. This investment will authorize the country to bring capacity based on wind energy from 280 MW in 2010 to 2000 MW in 2020.

3 Renewable Energy Education

Education plays a vital role in the sustainable development in societies. It is the cornerstone in any development strategy. It aims primarily to perform a powerful social change by raising awareness in all fields, and secondly, it also provides training for professionals namely researchers who will develop the next generation systems. In energy domain, renewable energy is evolving rapidly as an advanced academic field. Generally, REE is a relatively new area and previously it was not a major part of engineering courses. Presently, REE must be integrated into educational programs as it requires special techniques that are not normally met in other disciplines. Studying a limited number of units on renewable energy that are added into traditional science and technique courses doesn't seem probable to provide enough knowledge and skills to the graduates in order to use renewables efficiently. REE programs propose integrated packages that include various disciplines and skills such as the study of the resources, technology, design of the systems, industry, economics and policies. This gives the ability to the graduates to analyze and design systems with a range of available options and be particularly aware of industrial environment.

According to [8], engineering education and training in renewable energy have both long-term and short-term goals, and both are important. The aim of the long term is to form educated specialists at all levels in all fields of engineering and generalists whose education includes an appreciation of how renewable energy will figure in their fields. In the short term, the goal is to re-educate specialists already in the work-force. These goals need various kinds of education effort.

4 E-learning Platforms in Renewable Energy Education

New Information and Communication Technologies allow learning without limitations of place, time, occupation or age of the students. E-learning platforms offer courses that are educational concepts with technical, didactical and administrative materials to transfer the contents of any subject of knowledge. An E-learning platform is a set of tools managing the interaction between system users (students, tutors and administrators) and the distant server. Among different E-learning platforms concerning energy education we cite and present the following ones.

4.1 RegEn-M E-learning Platform

RegEn-M (Renewable Energy Multimedia) system is developed in the Otto-von-Guericke University of Magdeburg in Germany. The major aim of the realization of this system is to introduce the technology and implementation of the dispersed energy resources (DER) into the power system closer to the students [9]. RegEn-M platform is a web system designed to offer a didactic tool on DER technologies and to stimulate the interest of students in this field for the future of the power system.

The teaching system RegEn-M is a server-client platform; the E-learning courses remain open systems implement new materials and information in the future. The study area can be adapted to the user and the teaching contents are presented in a tree structure, which facilitates the orientation of the student during the studying. Petri Networks have been used to control and guide the students through the teaching contents. The access to the project user is possible by simply using web browsers such as Mozilla and Explorer etc.

RegEn-M platform has been conceived as a set of modules. Each module introduces a special type of DER. Different types of animations and introductory videos are realized to introduce the teaching concepts. The modules are categorized into three parts: basic principles, system technology and power network connection. Basic knowledge is presented as small text blocks, illustrated with figures, tables, graphics, animations, acoustic and non-acoustic and also simulations in MATLAB. In-depth knowledge includes additional contents of teaching, internal and also Internet links. Teaching materials are allowed to be downloaded. Each teaching part finishes with a test module that consists of a set of multiple choice questions and calculation tasks to assess the reached knowledge. Six teaching modules are available. They consist of the following topics:

Basic principles of energy production; Wind as an energy source; Photovoltaic energy production; Small water power plants; Energy storage devices; Fuel cell system.

4.2 E-learning Website Energy University by Schneider Electric

Schneider Electric, a global specialist in energy management, has launched an online educational community, Energy University. It offers the information and professional training on energy efficiency concepts. The E-learning courses aim to provide help to the specialists in this domain. Apply safe, reliable and cost-effective measures and take care of efficiency issues are the main objectives of the imparted knowledge. Typically less than 1 h to complete each course. More than 21000 users have been registered since the E-learning platform launch in June 2009.

4.3 Energy Power Lab E-learning Platform

Energy Power Lab proposes an e-leaning platform focusing on the renewable energy industry including international finance, logistics, project management, etc. Besides the courses of renewable energy like solar, wind and biofuels, the fundamentals of renewable energy are taught such as thermal transfer, fluid mechanical engineering and economics of renewable energy.

Energy Power Lab can provide its E-learning service in one or a combination of the following ways: online, with support (video chat regularly with experts/ instructors) and face-to-face, together with online. The proposed online courses are:

- The fundamentals of thermal transfer, fluid mechanical engineering, principles of
- electricity
- The fundamentals of the solar energy systems, photovoltaic cells and systems
- The fundamentals of the wind source, the fundamentals of biomass energy
- Grid integration and renewable energy integration
- Solar energy devices and solar energy systems
- Heat Transfer
- Bio energy
- Wind energy technology
- Energy efficiency and storage, economics of renewable energy sources.

4.4 Solar Energy International Online Courses

Solar Energy International (SEI) was founded in 1991 in United States of America as a nonprofit educational organization. It aims to offer industry technical training and teaching and also an expertise in renewable energy to empower people, communities, and businesses worldwide. SEI proposes various online courses related to the renewable energy topics:

Why Renewable Energy?; Conservation and Efficiency; Basics of Electricity; Solar Thermal; Wind Power; Micro-Hydro; Other Renewable Energy Technologies; Appropriate Technology for the Developing World; Economics of Renewable Energy.

4.5 *Distance Learning Program on Renewable Energy by TERI University*

TERI University in New Delhi, India provides online programs on renewable energy education. It offers three kinds of certificates; each certificate has its related courses. The proposed diplomas are:

- Advanced PG Diploma in Renewable Energy, prepared during 2 years.
- PG Diploma in Renewable Energy prepared during 1 years.
- Certificate courses during 20 weeks.

4.6 *Comparison of Platforms*

In this section, we present a qualitative comparison of four E-learning platforms already cited, knowing that we do not have the external access to the RegEn-M E-learning platform to put it in comparison (Table 1).

In summary of a few qualitative points of comparison, we can say that all of these four platforms studied use either dynamic websites manually developed or CMS (Content Management System). Nowadays there are other more effective and adapted ways to learn in a distance-learning environment.

Facing the development of E-learning, standards has been created to guide its practices. Thus, the content used as part of the E-learning generally meet one or more of the AICC, IMS, SCORM, LOM and QTI. These standards have several aims, the main ones:

- Standardize the indexing of various contents
- Allow to share pedagogical content between different environments
- Ensure interoperability of content between themselves and with the E-learning platforms
- Allow the combination of basic training modules to create a customized training.

Table 1 Qualitative comparison between E-learning platforms

E-learning platforms	Used technologies (Must important)	Strong points (Must important)	Websites of platforms
Website energy university by Schneider electric	PHP 5.3.3 Apache 2.2.15.	Security, flexibility, more stable.	http://www.schneider-electric.fr/sites/france/fr/produits-services/formations/formations-energy-university.page-05.04.2016
Energy power lab e-learning platform	PHP 5.3.29 Apache	Security, flexibility.	http://www.energypowerlab.com/31-renewabletraining.htm-05.04.2016
Solar energy international online courses	CMS WordPress 4.3.1 PHP 5.4.43	Content Management System (CMS) made simple, easy update, most used, easy customization.	http://www.solarenergy.org/online/-05.04.2016
TERI university	CMS Joomla PHP 5.3.3 Apache 2.2.15	CMS made simple, easy update.	http://www.teriuniversity.ac.in/index.php?option=com_program&task=program&sno=20-05.04.2016

5 Proposed E-learning Energy Platform

Morocco has known a great development in the field of renewable energy especially in solar and wind energy. With this strategy aiming to improve green education, E-learning platform will be a suitable solution for renewable energy education programs in Morocco. In this work, we will propose an E-learning solution that will be implemented and documented in our future works. Until now, the courses of the platform are developed in collaboration with experts in the field of renewable energy.

5.1 The Goals of the Proposed Platform

- Self-study that allows the student to progress according to his own capacity and evaluate his progress at any time. It puts at disposal the course material and also practical activities.
- The system allows the interaction between the course members so that they can share information and discuss the proposed topics. Also, a part of knowledge will be constructed by the group.

- Students are supported and guided by the tutor during the learning and teaching process.
- The student becomes an active element in the teaching and learning process.
- This model consists of how students learn and not how teachers teach.
- The targeted educational levels are at university level.
- The training must help students to understand green energy concepts.
- The students must be responsible in the learning process, especially in time management.

5.2 Courses Organization

The list of topics will be developed by specialists in the green energy technologies. Each of topics is presented with a didactical manner to attract students' attention during the learning process. The common structure of the course is: introduction, contents, case studies, extra sections and bibliography.

Indeed, the topic presented is based on different learning activities (forums, videos, presentations, animations, audio, simulations, e-books etc). During the learning process, some practical examples together with some exercises will be available to help the student well understand the topic and also to make him/her familiar with green energy concept. At the end of each course, students are evaluated through tests to measure their knowledge and study pace. Moreover, the tutors will guide the students via e-mails, forums and conversation (chat, interactive talks or, video). The platform provides different units: communication unit (forums, discussion groups, e-mails etc), information services (dictionaries, e-books, e-journals, important web sites etc) and assessment unit (tests, exercises, auto-evaluations, monitoring tools etc). Thus, several tools allow content management of platform to exist, but the most adapted tool to our platform is the LMS that has huge features.

5.3 Selected Platform

There is a diversity of various open source Learning Management Systems (LMS), such as ATutor, Moodle, Dokeos, Ganesha, Eliademy, Claroline, Sakaï, Chamilo, etc. LMS is defined in [10] as a system used to ease the process of communication between students and instructors. For this purpose, a detailed selection was made among different open-source platforms currently used in large universities and institutions structures to pick and choose the best suited tool to our training environment and our vision. As a result, the choice was made on MOODLE (Modular Object-Oriented Dynamic Learning Environment). Several definitions of MOODLE are presented in literature, [11] present MOODLE as a technology platform which

Table 2 Essential advantages of moodle

Level advantages	Advantages	Category
1st-level	• Open source software that any educational organization can acquire it for free, without worrying neither about the purchase costs nor about rights to use license	Financial
2nd-level	• Offers more interactivity between learner and teacher • Works by the social-constructivist approach • Proposes planned activities to allow learners to create content, such as (forums, wikis, glossaries, messaging, etc.) to construct their own knowledge	Pedagogical
3rd-level	• Offers more usability to creator or designer of online course • Provides opportunities for unlimited configurations and learning becomes more personalized	Creation process
	• Application is constantly revised and updated for free by hundreds of programmers	

offers educational institutions and training organizations and gives the capacity and the capability to create courses online and E-learning websites. Thus, it allows the organization offering online courses to arrange content manage learning, facilitating interactivity between learner and teacher and assess the learner.

This technology is easily integrated into the technological infrastructure of the university. Furthermore, it is most adapted to the interests of programmers and operators of education (learners, teachers and administrators) because it presents some benefits that other learning software do not offer. The following table shows essential advantages collected of Moodle (Table 2).

6 Conclusion

In this paper, we have reported E-learning platforms as an attractive and new method for teaching and learning especially in the case of energy education since Morocco is developing its renewable energy infrastructure.

E-learning platforms such as RegEn-M, Energy University and Power Energy Lab E-learning and others are presented and a qualitative comparison between platforms was made. Generally, they use either dynamic websites manually developed or they use CMS. They provide a variety of teaching materials, although, there exist other more effective and adapted ways to learn in a distance-learning environment.

For developing countries such Morocco, this kind of solution has clearly shown its importance as it offers a huge number of tools to enrich and empower the competencies of engineers and researchers during their curriculum to fulfill the requirements of green energy industry. This paper also highlights our vision about the E-learning platform that we will implement, based on the studies concerning renewable energy in Morocco. In this context, a detailed selection was made among different open-source platforms to choose the best suited to our training environment and our vision. As a result, the choice was made on MOODLE platform.

References

1. Moore, J.L., Dickson-Deane, C., Galyen, K.: E-Learning, online learning, and distance learning environments: are they the same? Internet High. Educ. **14**, 129–135 (2011)
2. Mahenge, M.P.J., Sanga, C.: ICT for E-learning in three higher education institutions in Tanzania. J. Knowl. Manag. E-learn. **8**, 200–212 (2016)
3. Chen, H., Moore, J.L., Chen, W.: Understand and analyzing learning objects: A foundation for long-term substantiality and use for elearning. J. Knowl. Manag. E-learn. **7**, 280–296 (2015)
4. Lu, S.M., et al.: Development strategy of green energy industry for Taipei-a modern medium sized city. Energy Policy **62**, 484–492 (2013)
5. Garg, H.P.: Trends and issues in renewable energy education. In: World Renewable Energy Congress. Florence (2006)
6. Ibrahim, D.: Renewable energy and sustainable development: a crucial review. Renew. Sustain. Energy Rev. **4**, 157–175 (2000)
7. Adnan, M., Ibrahim, D., Murat, A.: A green energy strategies for sustainable development. Energy Policy **34**, 3623–3633 (2006)
8. Boris, B., Charles, M.G.: Strengthening human resources for new and renewable energy technologies of the 21st Century, UNESCO engineering education and training programme. Renew. Energy **10**, 441–450 (1997)
9. Hadzi-Kostova, B., Styczynski, Z.: Teaching renewable energy using multimedia. In: Power Systems Conference and Exposition, pp. 843–847. IEEE PES (2004)
10. Azmi, M.A., Singh, D.: Schoolcube: gamification for learning management system through microsoft sharepoint. Int. J. Comput. Games Technol. 2015 (2015)
11. Oproiu, G.C.: A Study about using E-learning platform (Moodle) in university teaching process. In: 6th International Conference Edu World, pp. 426–432 (2015)

An Approach to Improving Student Retention in a Programming Course that Is Constructively Aligned Around Automatic Online Assessment

Heikki Hyyrö

Abstract The principle of constructive alignment formulated by Biggs [Enhancing teaching through constructive alignment, Higher Education, 1996] has received considerable attention both among education practitioners and within the broader field of educational psychology. The principle states that the learning activities and the final assessment of a course should be designed in such manner that they directly correspond to the learning goals of the course. In this paper we report on our experiences from introducing automatic online assessment into a traditional C++ programming course in an increasingly constructively aligned manner. Initially the student retention rate dropped significantly: from 63 to 43 %. In reaction to this, we introduced "learning by example"-style lab sessions to the course. As a result, the retention was raised back to 63 % while still achieving, according to our subjective evaluation, better learning results than without automated assessment.

1 Introduction

Automatic assessment tools have been used for decades in programming courses in order to improve student learning and reduce teaching staff workload (see e.g. [1–3] for fairly comprehensive surveys). The most fundamental form of automatic assessment concerns automated compiling and testing of program code that students have submitted as their solutions to course exercise questions. This usually implies the need to formulate the exercise questions in such manner that a correct solution produces a unique output that can easily be checked for correctness. In addition to programming courses, this type of automated assessment is widely used also in various types of programming contests, such as the ACM International Collegiate Programming Contest (ICPC) [4] and the International Olympiad in Informatics (IOI) [5]. The first automated assessment tools were typically scripts or programs that the teachers used in order to automate testing of students' solutions. Currently,

---

H. Hyyrö (✉)
School of Information Sciences, University of Tampere, Tampere, Finland
e-mail: heikki.hyyro@uta.fi

Á. Rocha et al. (eds.), *Europe and MENA Cooperation Advances in Information and Communication Technologies*, Advances in Intelligent Systems and Computing 520, DOI 10.1007/978-3-319-46568-5_3

as internet-based learning platforms systems have become ubiquitous in programming courses (and teaching in general), automated assessment is typically integrated directly to the same internet-based system that students use for submitting their exercise solutions. This type of systems enable students to receive immediate feedback about the correctness of their solutions.

In addition to checking program output for correctness, automated assessment tools can be augmented with various other types of features, such as static code analysis (e.g. check program style, try to identify bad coding practices) and plagiarism detection [1].

There exists a comprehensive body of literature on the results of using automated assessment. In a very recent work, Pettit et al. [3] reviewed over 100 papers on the topic and concluded that the existing empirical results support the view that the use of automated assessment indeed does improve student learning in programming courses. Hence it seems clear that the use of automated assessment is one very important aspect to consider when designing a programming course. In addition to this type of assistive technologies, it is naturally important to also pay attention to general pedagogical considerations. Recently the model of constructive alignment proposed by Biggs [6] has become one of the most popular frameworks for education design. The model has roots in constructivist learning theory and views learning as a result of students' own actions/activities. The concept "constructive alignment" refers to the idea that all aspects of teaching should, to as large extent as possible, direct the students towards activities that support achieving the learning goals. One additional key assumption is that students tend to put emphasis on activities that they expect to be evaluated (i.e. that affect grade). The process of designing a constructively aligned course may be described by the following steps (which may be considered in an iterative manner) [7]:

1. Define the learning goals.
2. Formulate the assessment criteria: how will learning be evaluated?
   - The assessment criteria should correlate strongly with reaching the learning goals.
   - The criteria will be made known to the students.
3. Design course activities that are directed towards fulfilling the assessment criteria.

The present paper concentrates on our experience with teaching and redesigning an intermediate-level C++ programming course from 2011 until 2015. The students who attend the course have already studied elementary programming using the Java programming language. The aim of the course is to both teach the fundamentals of the C++ programming language as well as to further develop the students' overall programming skills. The course is an optional element in the students' curriculum. We describe our experience from redesigning the course by inducing constructive alignment around automated assessment into the course: we made automated assessment a central aspect of all student activities in the course. This at first resulted in a significant drop in student retention: from 63 to 43 %. Student retention climbed back to 63 % after we introduced extensive "learning by example"-style guidance sessions into the course. To our best knowledge, this is a novel aspect in the literature. For example Kumar [8] presents an overview on the effect of using different types of lab sessions in programming courses, but none of the covered lab types seem to be directly comparable with our approach.

2 Initial Setting

The C++ course was taught in 2011 and 2012 in a rather traditional manner, using the following components:

- Lectures.
  - Attendance is neither mandatory nor rewarded in course grading.
- Weekly exercises.
  - Programming tasks ("open labs").
    Students work on the exercise questions on their own and submit their solutions via a web-based system.
    Neither automated nor manual assessment!
    Students need to do at least 40 % of the exercises.
  - Exercise review sessions.
    Students must submit their exercise solutions before the corresponding review session.
    Students need to attend the review session in order to receive points for their solutions.
    Randomly selected students present (and explain) their solutions to all attendants. This serves as the (only) control point for the correctness of students' solutions.
    If deemed necessary (e.g. the student's solutions have clear deficiencies), the teacher presents also a model solution.
- A programming project (as an "open lab").
  - A little bit more extensive programming task.
  - Students work on their own and also create a document describing their solution.
    Students receive instructions on how to test their solutions before submitting them.
  - All solutions are assessed in a semi-automatic manner.
    Correct behavior is checked with an automated script.
    The teacher inspects the code manually to check matters related to good programming style (e.g. adequate commenting).

- Final exam.
  - Done using pen and paper in a regular examination hall. The examination time is 4 h.
  - Both essay and programming questions.

When teaching the course in the traditional manner, we quickly became concerned about the following aspects:

1. Ensuring students' learning: open labs provide a weak guarantee about how much of the work has been done by each respective student.
2. The exam: a paper-based exam has very little in common with the actual work students do during the course.

These concerns led us to start redesigning the course.

2.1 Redesigning the Course

There were two background factors that guided the redesign of the course. One is that we had coached our home country's national team in the International Olympiad in Informatics and hence were familiar with automated assessment tools. A second factor is that we had attended an educational psychology course that our university offers to its teaching staff, and hence we were aware of Biggs' model of constructive alignment. As a result, we decided to redesign the C++ course by (1) strengthening the constructive alignment of the course and (2) introducing automated assessment. These steps are described in the following subsections. Due to practical limitations, such as the time it takes to develop new tools and teaching procedures, the eventual course development was carried out gradually.

Figure 1 shows some key statistics from the course in the years 2011–2015. We will further discuss these numbers in the following sections. One thing to note is that we do not report the number of students that have registered into the course: this is because the course does not have a strict registration policy. We instead measure the number of students in the course as the number of students who submitted an

	Submitted at least once	**Passed the exam**	**Exam mean**	**Exam median**
2011	61 students	42 students (69%)	13.1 points	13 points
2012	46 students	29 students (63%)	14.4 points	15 points
2013	40 students	17 students (43%)	15.8 points	16 points
2014	56 students	27 students (43%)	16 points	15 points
2015	71 students	45 students (63%)	13.4 points	14 points

Fig. 1 Statistics from the C++ course from 2011 to 2015. The maximum points for all exams have been scaled to be 20. The mean and median have been calculated among passing students. The retention rates, calculated as the percentage of students that passed the final exam among those that submitted at least once, are shown inside the parentheses in the *middle column*

answer to at least one exercise question. Student retention rate is then measured as the percentage of these students who eventually pass the final exam.

2.2 Strengthening Constructive Alignment

The first step was to define the learning goals and corresponding aligned assessment criteria. We settled with a straight-forward solution: the goal of the course is to learn to program non-trivial and correctly working programs in C++, and the assessment criteria is a practical and controlled examination of how well the students are able to produce such programs. The emphasis was moved completely to program correctness: other aspects, such as coding style, were discarded as being of secondary importance to the present course.

2.3 Phase 1 (2013): Automatically Assessed Weekly Exercises

The first development step was to adopt automated assessment for all weekly exercise questions and the programming project. At the same time about half of the weekly exercise questions were made mandatory: previous rules stated that students need to complete at least a certain number of questions, but now the set of mandatory questions became fixed. This was done in the spirit of constructive alignment: the set of mandatory questions was designed in such manner that they cover the basic learning goals of the course.

The automated assessment system accepted only those solutions that produced a correct result. We anticipated that this might result in many students to soon drop out of the course as a result of failing to submit working answers in time to all mandatory tasks of some weekly exercises. As a countermeasure against this, we adopted a rule that students were allowed to return their answers even a little bit after the review session. The idea was that having the opportunity to see the descriptions (and code) of working solutions in the review sessions should help students fix their own code.

The final exam remained unchanged: it was still done on pen and paper and contained both essay and programming questions.

Results As shown in Fig. 1, the student retention rate dropped from 63 to 43 % after mandatory automatically assessed exercise questions were introduced into the course. On the other hand the exam scores improved. A likely explanation is that the mandatory automatically assessed questions created a sort of survival-of-the-fittest type rule on who survives the course until the final exam: a significant number of students dropped out from the course early after having failed to return accepted solutions to some week's mandatory questions. The precaution of allowing submission even after the review session was surprisingly ineffective in preventing this.

2.4 Phase 2 (2014): Automatically Assessed Electronic Exam

The second development step was to move from a pen-and-paper exam to an electronic examination where the students are presented with similar questions as in the weekly exercises and their answers are assessed automatically during the exam. One consequence is that the students know in real time during the exam how many points they have amassed so far (and whether they have reached enough points to pass the exam). The students were not allowed to bring any own material to the exam. The only tools they had available were a simple editor with syntax highlighting. For example a debugger was not available. If a particular question required remembering some potentially tricky details e.g. about the C++ standard library, such information was provided as part of the problem statement for that particular question.

The examination time of the electronic examination was kept as the same 4 h as in the pen-an-paper exam, and there were 5 programming tasks. Because this was our first experiment with an electronic and automatically assessed exam, the students were asked to answer only 3 of the 5 questions, and they were free to choose which 3 to answer. The idea was to provide some flexibility regarding possible "dead ends", where a student becomes completely stuck with a question due to some difficult-to-find bug etc. Passing the exam required at least one completely and one partially solved question. Giving partial points was also one precaution against the possibility that grading based on fully automated assessment would prove to strict. But even in this case partial points were rewarded only for solutions that were at least almost correct. This made the manual assessment take relatively little time: only those answers where the code compiled but produced only slightly wrong results needed to be inspected, as well as answers where the compiler message already hinted that there is only a very small syntactic error.

The other parts of the course remained same as in 2013.

Results As seen in Fig. 1, both the student retention rate and the mean and median of the exam points were virtually the same as in 2013. There, however, was one significant improvement in comparison to 2013: moving from manually graded paper-and-pen exam to a mostly automatically assessed electronic examination resulted in a significant reduction in the work required to arrange and grade an exam. Although automatically assessed questions take more time to prepare, the workload was still much less than the time required for grading pen-and-paper exams manually.

Another improvement is that, at least according to our own subjective view, the exam was now a better indicator of the students' C++ programming skills. In part due to the greater flexibility of using a text editor, and in part due to the immediate feedback received from automatic assessment (e.g. regarding compiler errors), the programming questions in an electronic examination could be made considerably more complicated (and realistic) than what is feasible in a paper-and-pen exam. For example one of the harder questions in the electronic exam asked the student to implement a template-based general linked list from scratch. In the era of paper-and-pen exams, a corresponding hard question might have asked the student to write code for a function template that sorts an array. The latter question puts more emphasis

	Percentage of students that solved all mandatory questions							
	Week 1	Week 2	Week 3	Week 4	Week 5	Week 6	Week 7	Week 8
2013	70%	60%	50%	45%	45%	45%	45%	45%
2014	96%	68%	64%	54%	52%	52%	48%	48%
2015	86%	79%	77%	73%	72%	69%	66%	-

Fig. 2 The percentages of students who solved all mandatory questions in each individual weekly exercises. The course spanned 8 weeks in 2013 and 2014 and 7 in 2015

on remembering the template syntax by heart, whereas the former question requires several times more lines of code and also considerable care e.g. in dynamic memory handling.

Another positive note about the electronic exam was that 10 students out of the 24 who passed the exam scored full points: they solved 3 questions out of the 5 available. This was an encouraging sign as we feel that the overall level of the questions was harder than in the previous paper-and-pen exams.

One point we would like to emphasis here is that mandatory automatically assessed exercises are the main culprit behind the fairly low retention rate. A large majority of those who dropped out did so already well before the final exam. This is evident from Fig. 2 that shows how many percent of students who submitted at least one answer to any question managed to solve all mandatory questions in each weekly exercises. This data is shown for the years 2013–2015 and the trend is similar in each year: the percentage of students who solve all mandatory exercises in the last week is quite close to the percentage of students that pass the final exam, and the majority of students who dropped out did so already during the first half of the course.

2.5 *Phase 3 (2015): Allowing Students to "Learn by Example"*

As far as the course arrangements are concerned, phase 2 reached the basic goals we initially defined for redesigning the course. But the considerable drop in the retention rate after 2012 prompted us to seek further measures to prevent students from dropping out due to failure to complete mandatory weekly exercises.

Our approach was to change the weekly exercise review session into a "learn by example"-style session where the teacher guides the students step-by-step to working solutions for all mandatory questions of that particular week. The teacher coded each presented solution from scratch, and explained each step in a fairly great detail, also answering students' questions. The students could use either their own laptops or computers present in the classroom, and they also had access to the exercise solution submission system during the session.

A second change in comparison to 2014 was that now optionality was removed from the electronic examination: there were still 5 questions, but now all of them were a fixed part of the exam. This change was done as a response to the high proportion of students who received full points in the previous year's exams. At the same time the exam grading was made completely automatic: partial points were no longer rewarded even for "almost, but not quite" correct answers. The points rewarded by the automated assessment system were the final exam points.

Results We can see from Fig. 2, that switching to a "learning by example"-style constructive walkthrough of all mandatory questions seems to have had a very significant positive improvement on the percentage of students who eventually submitted accepted solutions to all mandatory tasks. The same effect also carries over to the exam, as shown in Fig. 1.

Changing the exam to consist of 5 mandatory questions proved to be an appropriate choice. This time 4 out of the 45 students that passed the exam received full points, that is, solved all 5 questions. As many as 18 students solved 4 questions. The students were required to solve at least 2 questions in order to pass the exam. We received no protests from students about the automated all-or-nothing grading. This is probably because the students are already accustomed to this type of grading: exactly the same rules are used in grading the weekly exercises.

3 Concluding Remarks

It is in some sense curious that the impact of using "learning by example"-style in the exercise sessions seemed to be so significant. We could of course think that this is a result of the students simply copying the readily given model answers completely as such, but this option was already possible to the students in the previous years: also the sessions in 2013 and 2014 involved showing the source code of working solutions to the students, and the students were allowed to submit answers even after the exercise session in each of the years 2013–2015.

We would still like to emphasise that the current results are very much preliminary: there might e.g. be some other hidden factors, such as qualitative differences between the students who attended the course in different years, that could explain some of our reported results regarding student retention. We may properly assess the effects of "learning by example"-style exercise sessions only after a longitudinal study carried out over several years. In any case, at least the early results seem promising.

As far as learning outcomes are concerned, one might be tempted to think that the "learning by example"-style sessions lead the students to simply copy the readily given answers without actually learning much. We would, however, counter this claim by referring to how the higher level of returning solutions to mandatory questions carried over also to the electronical exam. In our opinion even the more easier questions in the electronic exams were complicated enough that the students should not be able to solve them in the limited exam environment without having at least some real C++ programming skills. As mentioned before, the exam environment for example provides neither detailed C++ documentation nor an interactive debugger.

To give a more concrete idea of this last claim, we describe the easiest question from the final exam of the 2015 course. The question asked the student to define a function template `vector<T> sorted(const vector<T> &t)` that returns a vector that contains the same values as the parameter vector `t`, but in sorted order. Answering this question succesfully requires that the student (1) knows the syntax of defining a function template and (2) knows how to use a vector and (3) either knows how to use the sorting function provided by the C++ standard library or alternatively is able to implement sorting on her own.

We also note that the question that we just described as being among *the easiest* in the electronic exam is actually virtually identical to what we described in Sect. 2.4 as an example of *the most difficult* questions in the earlier pen-and-paper exams. Therefore we believe that the redesigned course, that incorporates automated assessment, electronic exams and "learning by example"-style weekly exercise sessions, leads into significantly better learning outcomes than the previous traditional course from 2012 and before. Student retention seems to be at a similar level but the teacher workload especially regarding manual inspection of student code has been greatly reduced. Therefore the overall effect of redesigning the course seems to have been positive.

References

1. Ala-Mutka, K.M.: A survey of automated assessment approaches for programming assignments. Comput. Sci. Educ. **15**(2), 83–102 (2005)
2. Douce, C., Livingstone, D., Orwell, J.: Automatic test-based assessment of programming: a review. J. Educ. Resour. Comput. (JERIC) **5**(3) (2005)
3. Pettit, R.S., Homer, J.D., Holcomb, K.M., Simone, N., Mengel, S.A.: Are automated assessment tools helpful in programming courses? In: 2015 ASEE Annual Conference and Exposition (2015)
4. Baylor University: Acm international collegiate programming contest. Accessed 30 Apr 2016
5. IOI: International olympiad in informatics. Accessed 30 Apr 2016
6. Biggs, J.: Enhancing teaching through constructive alignment. High. Educ. **32**(3), 347–364 (1996)
7. University College Dublin: Using biggs' model of constructive alignment in curriculum design. Accessed 30 Apr 2016
8. Kumar, A.N.: Closed labs in computer science I revisited in the context of online testing. In: Proceedings of the 41st ACM Technical Symposium on Computer Science Education (SIGCSE 2010), pp. 539–543 (2010)

Cooperative m-Learning Based on EXPROLM Protocol

Karima Aissaoui, El Hassane Ettifouri and Mostafa Azizi

Abstract Using mobile devices in our daily activities has become elementary. Starting from this, we were interested on the field of mobile learning and how mobile apps can contribute together in order to collaborate with each other and to offer collaborative mobile apps without the need to unify the use of just one application. This paper presents a solution for cooperative m-learning systems based on a new protocol that we have created for this goal. The proposed architecture is composed by three layers. For every m-learning system, a special plug-in is easily created in order to communicate with the global m-learning application. This methodology contributes to create collaborative mobile system and helps all actors in a learning process to increase their concentration on the pedagogical side.

Keywords Mobile · m-learning · Architecture · Cooperative m-learning

1 Introduction

With the great development that knows the mobile world, the utilization of mobile devices in our life has become primary. The integration of this technology in different fields makes us more and more attached to mobile devices. One of the fields that were touched by this technology is learning. This integration has led to a mobile form of learning and has created the mobile learning (m-learning). During the last years, many applications of m-learning systems were developed and researchers are more and more interested on how and when this type of learning is more suitable

K. Aissaoui (✉) · M. Azizi
MATSI Laboratory, ESTO, Mohammed First University, Oujda, Morocco
e-mail: aissaoui.karima@gmail.com

M. Azizi
e-mail: azizi.mos@gmail.com

E.H. Ettifouri
LSEII Laboratory, ENSAO, Mohammed First University, Oujda, Morocco
e-mail: h.ettifouri@gmail.com

Á. Rocha et al. (eds.), *Europe and MENA Cooperation Advances in Information and Communication Technologies*, Advances in Intelligent Systems and Computing 520, DOI 10.1007/978-3-319-46568-5_4

to apply. Mobile operating systems are several, we find Apple OS, Google Android, Blackberry OS and windows mobile phone. Mobile applications are also dependent on these operating systems which makes them several and special for each one of those systems.

The choice of a mobile learning application in institutions is personal. Every teacher can choose hit preferred platform according to his needs. As a consequence, the heterogeneity in systems used in the same institution becomes bigger. This situation constitutes an obstacle for students behind the learning using mobile devices, which discourage them and make them less motivated. In order to address this problem, we propose in this paper a new methodology based on a new architecture and a new protocol called EXPROLM (EXchange PROtocol for M-learning) in order to obtain a cooperative and interoperable system where the student has not to be connected independently on all mobile learning platforms used by his teachers. As a result, the problem of heterogeneity of platforms in institutions will be resolved and students will be more motivated and encouraged to use mobile apps in learning, and consequently to improve the efficiency of this way of learning.

This paper is organized as follows: the first section is dedicated to definitions and issues of mobile learning, and then we will present the context and the state of the art of using this type of applications. The third section will introduce our proposed architecture and the details of every element composing the solution.

2 Mobile Learning

Many definitions have been given to m-learning in different communities; however, the characteristic of mobility is common between all these definitions.

According to [1], m-learning is learning that happens when using mobile devices regardless of location in time or space. In [2], m-learning is defined as a kind of learning that happens when the learner is not bound to a fixed or predetermined location while taking advantage of mobile technologies for learning. In [3], we find that the process of m-learning requires the use of handheld devices for learning. In [4] the concept of D-learning (distance learning) was introduced in e-learning (electronic learning) while considering the mobility with m-learning and ubiquity with ubiquitous learning (U-learning, ubiquitous learning).

For us, we define m-learning as each act of learning that happens when the learner is using mobile devices and is independent from a fixed location. By this, not just student are interested by this type of learning, but also employees in professional contexts. In the next section, we will define the state of art of m-learning and we will present some contexts where it can be used and when the learner can take advantage from m-learning.

3 State of the Art

Nowadays, and according to different studies, the great majority of students prefer using mobile applications than using mobile browsers. This type of applications is characterized by its independence. That means that each application is used independently and has its own database, specifications and users. In a case where a student or a learner must use more than one mobile learning system, he/she must be sign on different applications in order to be connected on those applications (This case is present when teachers of the same student are using different mobile learning system). Also, learners must be familiarized with all those mobile learning systems. In the other side, if two teachers do not use the same platform, they will not have a collaborative space where they can exchange their courses and groups of students.

Moreover, we know that mobile learning is used even in professional contexts, when the user wants to learn using a mobile learning system, he should not waste time on connecting independently and looking for the learning content on all applications.

Actually, all mobile learning apps are used independently. That means that even in institutions where more than one system is used, students must be connected on all platforms and they do not have a collaborative space where they can find all what they need even for two teachers who use two different systems. This situation makes the learning task more difficult, thing which discourages the students to profit from the advantages of mobile devices in the learning process.

In a previous work [5], we have proposed a new methodology to allow users of e-learning systems to have a unified access to all platforms used by their teachers. Our objective is to facilitate the learning task and to offer a comfortable way to all actors in e-learning and m-learning. As we know, the principal aim of m-learning is to improve the efficiency of the learning process. In order to help m-learning users to achieve this goal, we propose in this paper our idea that comes as a real solution for constraints faced by actors in such system. We suggest a unified system that offers to its user all what he/she needs according the technique of Single sign on. Due to our system, the learner is connected just one time on GMLA (Global Mobile Learning Application) to be connected on all platforms installed and configured in the system. In the next sections, we will present this system, its components and how it can be used.

4 Our Proposed Architecture

4.1 General View of the System

As explained above, our system aims to offer a collaborative architecture that can be used in order to make m-learning context more efficient, more practice, and more intuitive. The next Fig. 1 illustrates this architecture.

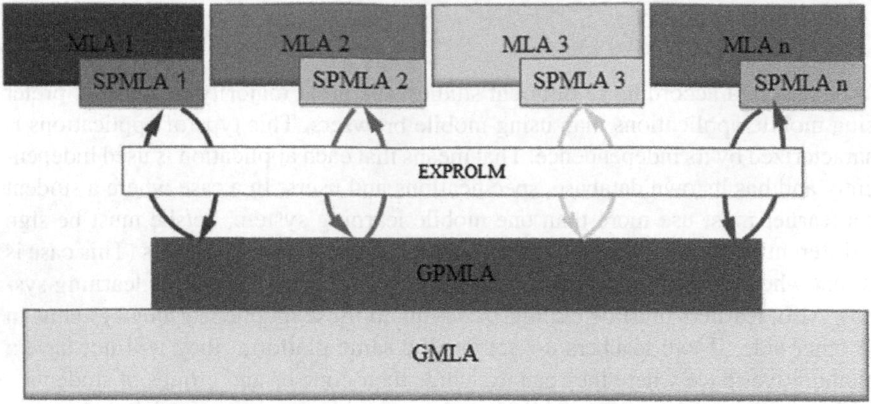

Fig. 1 General view of the proposed system

It is composed by three layers; each layer is characterized by its elements which are composed as follows:

- MLA (M-Learning Application) is used for m-learning application;
- SPMLA is used for the Special Plug-in of the M-learning Application MLA;
- GPMLA is used for the Global Plug-in of the M-learning Application MLA;
- GMLA is used for the Global M-learning Application;
- EXPROLM is used for the EXchange PROtocol for M-learning Applications.

In the next sub sections, we will explain each element and how it can be used. We will start with the GML Application.

4.2 GMLA

Our proposed system is based on a Global Mobile Learning Application (GMLA) that covers the great majority of specifications of a mobile learning system. This application was designed after studies of some existing systems. It is compatible with the most operating systems used (Android, iOS, Windows phone).

The first step was to describe the most essential actors in such system. Those actors are the common users between all existing mobile learning systems. Also, they must be defined in order to give specific roles of every user, what will make management of access rules of users easier. Thus, we found the following actors:

- Student/Learner;
- Teacher;
- Administration;
- Guest;

The second step was to find the essential functionalities that need each type of actors. We found diversity in those functionalities from a platform to another. Consequently, we focused our study on the primordial functionalities that must exist in a mobile learning system.

For students and teachers, the most essential functionalities are:

- News: For example when a new course or exercise is posted, the student receives a notification on his smart phone.
- Calendar/Agenda: events must be reordered in a calendar, which will help students and teachers to organize coming events in order to avoid any forget of interesting events.
- Courses/Exercises/Exams: Students and teachers are in most time in interaction with each other. The functionality of ensuring this interaction constitutes one of the elements forming the core of a mobile learning system.
- Synchronous/Asynchronous communication: the application must ensure this functionality because one of objectives of our system is to make the task of learning easier and to offer to all users our system a comfortable way to be involved in the learning process. This functionality can be used by the administration too.
- Download/Upload files: for example when a teacher posts a course, he/she needs this functionality in order to improve the quality of the posted course. This functionality can be used by the administration too.

For the administration, basically some data concerning the use of the m-learning application must be taken. For example:

- Users: the administration should know the rate of use of the m-learning application. Those data will be helpful in statistics for example.
- Updates: when the m-learning application needs some updates (for example case of mobile operating systems).

As shows Fig. 2, GMLA can be used by all m-learning system actors. Each actor is oriented to his domain according to his profile.

4.3 SPMLA: The Special Plug-In

The Special Plug-in SPMLAi is specifically dedicated to the m-learning application MLAi. This plug-in is composed from two parts:

The specific part which is special for the platform i (This part is specific for each m-learning application and depends on its architecture and the targeted language of the application). It communicates with MLAi in order to obtain data. Here, we find two cases:

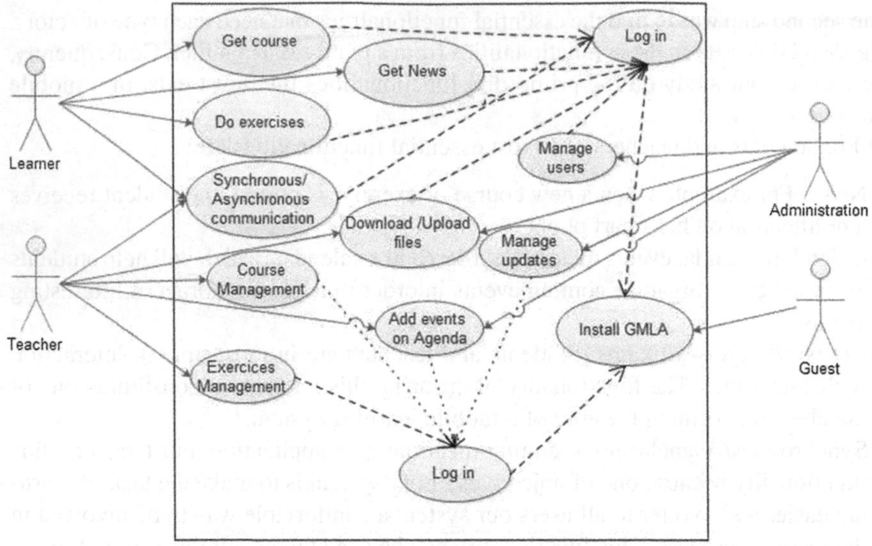

Fig. 2 GMLA use cases

Fig. 3 SPMLAs structure

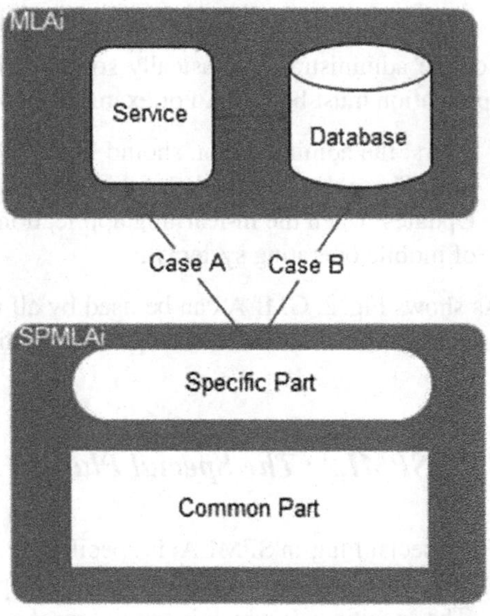

- Case A: The MLAi communicates those data using services or web links.
- Case B: The MLAi has only the database, so the specific part interrogates it in order to obtain data.

The common part is the front interface. It is common between all plug-ins of all platforms. The Fig. 3 illustrates the structure of the SPMLA.

4.4 Communication Between GPMLA and SPMLA

Communication between the common part (GPMLA) and the specific one (SPMLA) is done according to request/response architecture and using our new protocol EXPROLM. Common part or the front interface sends requests to the specific part through functions that are already defined. They could be functions requesting details of users, learning resources

In order to obtain the interoperability needed in mobile learning field, our system was designed to be compatible with most operating systems used (in our case Android and iOs). Syntax of functions depends on the targeted operating system of the device. For example if the learner uses a device with Android OS, the specific part is implemented according this system specifications and respecting its syntax.

In the next sub section, we will explain the EXPROLM protocol, its specifications and its architecture.

4.5 EXPROLM: Protocol of Exchange

Cooperation between different m-learning systems is done due to communication between the common and the specific parts. EXPROLM protocol is responsible of success of this communication. In this paper, we have created this protocol based on JSON (JavaScript Object Notation) [6] for data representation, REST (Representational State Transfer) [7] and HTTP (Hypertext Transfer Protocol) [8] for transfer of data. The next figure illustrates the architecture of EXPROLM protocol (Fig. 4).

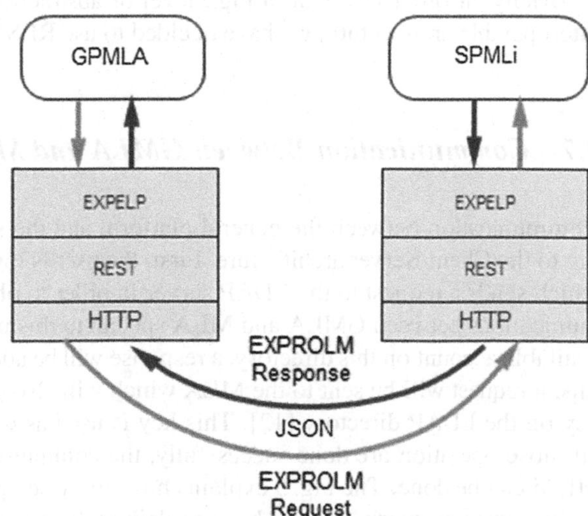

Fig. 4 EXPROLM architecture

4.6 Tools Used in the Protocol Implementation

4.6.1 JSON

Exchanging data in a standard format supported by the most existing operating system was our goal. This specification let us thinking on two formats: XML (Extensible Markup Language) and JSON. In order to know which is the more suitable for our case, we have looked for a comparison between these two tools. According to [9], XML [10] is used primary in Remote Procedure Calls (RPC) [11] and object serialization for transfer of data between applications. In our case, we have chosen JSON because of the multiple requests and data that are exchanged between GPMLA and SPMLi which makes of the speed of data exchange a primordial factor. Data transformation between the platform and JSON format is done in a bidirectional way.

4.6.2 REST Protocol

Representational State Transfer [7] is an architecture style for designing networked applications. It can be considered as a lightweight alternative to Web Services and RPC. Applications based on this architecture use HTTP requests [8] in order to do CRUD operations (Create, Read, Update and Delete). We have chosen this tool for several reasons. First, REST is Platform independent. That means that there is no care if the client and the server do not run on the same platform. Also, it is independent of the language which allows two different systems developed using different languages to communicate without any problem. In addition, REST uses HTTP which makes this tool based on this standard and gives it a powerful character.

Briefly, in order to obtain a high level of abstraction, and consequently a very interoperable architecture, we have decided to use REST tool.

4.7 Communication Between GMLA and MLAi

Communication between the general platform and the specific one is done according to the Client/Server architecture. First, the user is logged using GMLA interface which sends a request to the LDAP server in order to obtain a key for all next communications between GMLA and MLA special to this user. Then, if the user has an available account on this directory, a response will be addressed to the GMLA. After this, a request will be sent to the MLA which will also check the availability of this key on the LDAP directory [12]. This key is used as a security mechanism. When all those operation are done successfully, the communication between GMLA and MLAi can be done. The Fig. 5 explains how all those operations are done.

In a previous work [13], we have explained the advantages and disadvantages of unified access to e-learning platforms using LDAP directory.

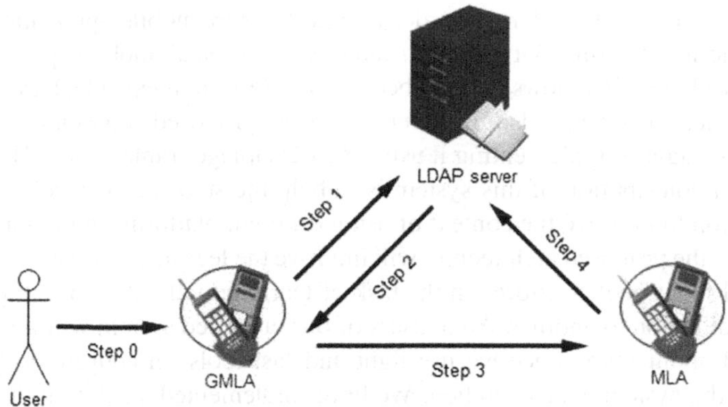

Fig. 5 Communication between GMLAi and MLAi

5 Adding New m-Learning Systems to Our Proposed Solution

In order to add new m-learning systems to the proposed solution, two simple steps must be followed to add the new platform to our proposed system. First, the specific part of the SPMLA must be created. As said above, the SPMLA plug-in is composed by two parts: public and specific. The public one is common and does not need any modification since it communicates with GPMLA. Consequently, the administrator of the platform should implement only the specific part of the SPMLA plug-in which will be in direct communication with this new platform.

The second step is to add the URL of the new implemented SPMLA in the general plug-in in order to have a channel of communication between both plugs-in.

6 Discussion

Using mobile devices in learning is going to be more and more paramount. This new way of learning makes students and teachers independent from the traditional classrooms and formal contexts. Developers of m-learning applications are interested to offer a system compatible with all operating systems and suitable for the majority of users. In this article, our aim is to help institutions to have a cooperative system connecting all m-learning systems used.

Using our proposed architecture, students will not be forced to access independently to all m-learning systems used by their teachers (for example one student needs to access Moodbile used by teacher 1 and Claroline-mobile used by teacher 2). He/she (the student) must log in only one time using the GMLA interface. Also, teachers will be able to collaborate between them without any need to use only one platform. In the side of administration, there is no doubt if a new platform is coming.

Two simple steps must be done in order to add this new mobile application to the system. In addition to this, our approach can be applied on all mobile operating systems (Android, iOS, Windows phone,) because it offers an independent architecture of any technical platform. Also, in order to run the proposed solution on different supports, we suggest implementing it using ZeroCouplage framework [14].

The great contribution of this system is to help the student, the teacher and the administration to improve the context of using different platforms in the same institution. Thus, the proposed architecture will improve the learning process because the student will focus his/her efforts on the task of learning without the need to switch from an application to another. Also, users of our proposed system will not feel the background architecture since we use light and fast tools. In addition to this, the security of the system is also studied. We have implemented a solution that uses a security key in order to exchange data.

On the pedagogical side, our proposed architecture constitutes a real contribution since it helps students to be more motivated when using mobile devices for learning. Moreover, teachers will be free to choose the suitable platform for them without thinking on how to convince the administration or students on using the chosen platform. As a result, we offer throw this paper a technical and pedagogical solution for possible problems that can occur when mobile systems are numerous and different in the same institution.

All these characteristics make our solution the perfect one to face problems of heterogeneity of using numerous m-learning applications in the same institution.

7 Conclusion

Characterized by its efficiency and its simplicity in use, our proposed system offers a global application where the most important functionalities are installed. This application communicates with other existing mobile learning applications using EXPROLM which is responsible of making a communication channel between the specific part of the mobile learning application and the global one. Thus, student has not to be connected independently on all existing mobile learning applications used by his/her teachers.

In order to improve the learning process, researchers are looking for new pedagogical and technical methods to make the task of learning easier for the learner and the teacher. In the same context, our proposed methodology in this paper aims to offer a cooperative system for all users of mobile learning applications and consequently to make the task of learning easier and to offer to all users of our system a comfortable way to be involved in the learning process.

In this paper, we have presented a new architecture dedicated for m-learning apps. A plug-in called SPMLA can be easily created for every platform. Composed by two parts, this plug-in is used to obtain data from the mobile app and communicate it to the GPMLA plug-in throw the EXPROLM protocol.

Our proposed architecture is a real solution for institutions, teachers, students and all actors in the learning process. Both the technical and the pedagogical constraints were studied and resolved. Also, our architecture aims to offer an interoperable system able to improve the quality of learning and consequently to obtain better results in this field, where we really need the best quality.

References

1. Wu, W.-H., Wu, Y.-C.J., Chen, C.-Y., et al.: Review of trends from mobile learning studies: a meta-analysis. Comput. Educ. **59**(2), 817–827 (2012)
2. O'Malley, G., et al.: Exploring the usability of a mobile app for adolescent obesity management. JMIR mHealth uHealth **2**(2), e29 (2014)
3. Keegan, D.: The incorporation of mobile learning into mainstream education and training. In: World Conference on Mobile Learning, p. 11, Cape Town (2005)
4. David, B.T., Yin, C., Chalon, R.: Contextual mobile learning for repairing industrial machines: system architecture and development process. iJAC **1**(2), 9–14 (2008)
5. Aissaoui, K., Ettifouri. E.H., Azizi, M.: EXPEL protocol based architecture for cooperative cooperative E-learning. Int. J. Emerg. Technol. Learn. (jET) (2016)
6. Shin, S.: Introduction to JSON (javascript object notation). Presentation (2010). https://www.javapassion.com
7. Windley, P.J.: REST: representational state transfer, 11,pp. 237–261
8. Fielding, R., Reschke, J.: Hypertext transfer protocol (HTTP/1.1): message syntax and routing (2014)
9. Nurseitov, N., Paulson, M., Reynolds, R.,: Comparison of JSON and XML data interchange formats: a case study, vol. 9, pp. 157–162. Caine (2009)
10. W3schools, Introduction to XML. Accessed Jul 2016 from http://www.w3schools.com/xml/xml_whatis.asp
11. Corbin, J.R.: The Art of Distributed Applications: Programming Techniques for Remote Procedure Calls. Springer (2012)
12. Joshi, C.G., Shah, R.J.: Portable lightweight LDAP directory server and database. U.S. Patent No. 9,032,193, 12 May 2015
13. Aissaoui, K., Azizi, M.: Taxonomy and Unified Access of E-Learning Platforms. MedICT, vol. 2 (2015). ISBN: 978-3-319-30296-6
14. Ettifouri, E.H., Rhouati, A., Dahhane, W., Bouchentouf, T.: ZeroCouplage Framework: A Framework for Multi-supports Applications (Web, Mobile and Desktop). MedICT, vol. 2 (2015). ISBN: 978-3-319-30296-6

Toward a New Approach: Extending a Game-Based Learning Authoring Tool eAdventure to Multiple Mobile Devices

Lamyae Bennis and Said Benhlima

Abstract Over the last decade the mobile revolution has done the most fundamental changes to education, and as the fact that learning games are gradually having an increasing impact in Technology Enhanced Learning, we have suggested to combine the both approaches learning game "LG" and mobile learning called: "Mobile Game Based Learning" (MGBL). However there is a huge lack of authoring tool that facilitates the generation of an adaptive LG to multiple mobile devices, at the time when the features of mobile devices are increasingly evolved and the development cost are so expensive. To this end in this paper we present an extension of the chosen LG authoring tool also known as eAdventure, which aims to support an automatic adaptation of MGBL to multiple mobile devices platforms.

Keywords Learning game · Mobile game based learning · Authoring tool · Ergonomic criteria

1 Introduction

Initial research has revealed a pervasive use of mobile phones across the population and specially the new generation. Mobiles have known a specifically high approval among teenagers and young adults. Additionally various studies have evinced that Mobile telephony has overrun across cultural groups, economics layer, and age units [1]. Mobile devices can also produce more lively learning involvements that Develop student commitment, education, and lesson retention [2]. Furthermore recent finding in this researches work have shown that using mobile devices in learning activities have a good influence on learners apprenticeship in different courses for example

L. Bennis (✉) · S. Benhlima
Faculty of Science, Department of Mathematics and Computer Science,
Moulay Ismail University, Meknes, Morocco
e-mail: lamyaebennis@gmail.com

S. Benhlima
e-mail: saidbenhlima@yahoo.fr

Á. Rocha et al. (eds.), *Europe and MENA Cooperation Advances in Information and Communication Technologies*, Advances in Intelligent Systems and Computing 520, DOI 10.1007/978-3-319-46568-5_5

social science [3] and language courses [4, 5]. For this reason a learning environment called as mobile learning "m-learning" has been done by education world. The m-learning is a type of learning model allowing learners to obtain learning materials anywhere and anytime using mobile technologies and the internet [6]. During the last few years an emergent interest has been also raised concerning Learning Games "LG" or what we named Game-Based Learning "GBL", this is one of LG approaches revealed in this study: "Games-Based Learning offer covenant learning and completely personalized schedules, plus the motivational improvements of the learning process" [7–9]. The interaction with this mobile devices and Game Based learning for education purposes extend the old-style learning Model into a new phenomenon so-called Mobile Game Based Learning (MGBL), and as the big task for educators and designers round the world however, is one of Comprehending and exploring how better we might utilize these resources to sustain Game Based Learning. The key target of this paper is to adapt the chosen preexisting educational adventure games authoring tool "eAdventure" to multiple mobile devices permitting to Facilitate the sharing and dissemination of generated LG into different platforms such us Android, iOS, Mac OS and Windows, reducing costs and development times and the most important thing is that allowing the good integration of the Gamification with Education which is a difficult task [10]. This paper is structured as follows: Related Work Section shows a literature review about evaluations methods of learning game authoring tool. Case study Section discusses the chosen Learning Games authoring tools to make a comparison and provide the best way to meet all the game requirements, final point shows the application of evaluation methods on our chosen authoring tools aiming to extract a final analysis of the comparison in order to choose the one which is more suitable to generate a Mobile Game Based Learning which is presented in results and discussion section followed by our contribution and finally there is the conclusion.

2 Background

Estimate a Game Based Learning (GBL) authoring tools is a stiff task, and as we see the evaluation techniques existing are numerous and diverse and each one has its advantages and disadvantages and as a result section 1 decides to apply the following evaluation methods: List of evaluation elements made by Tom Murraya and The Ergonomic Criteria of Bastien & Scapin.

2.1 List of Evaluation Elements Made by Tom Murray:

Many studies show that an Authoring tool attains its goals utilizing a number of methods or features. In this order I illustrate eight methods in detail, authoring tools should support [11]:

- ☐ Interoperability: Can the system work with other systems? That's mean one application can ask another for specific types of information. A higher level of interoperability includes "record-ability", in which an application is diffusing information. A further level of interoperability contains "script-ability" which permits one application to control another.
- ☐ Manageability: Can the overall system pursue information about the learner and learning session?.
- ☐ Reusability: Can learning objects be re-utilized by developers and applications?.
- ☐ Durability and scalability: Will the item stay usable as technology and standards evolve? Designing software so that it is extensible is one approach.
- ☐ Accessibility: Can the learner or content developer find and download the material in a reasonable time?

2.2 The Ergonomic Criteria of Bastien & Scapin:

The Ergonomic Criteria presented in the next section contains eight key criteria. Some of these measures are divided into sub-criteria: there are 18 requisite criteria [12]. Ergonomic Criteria have been extended to the Web [8] and interactions with virtual environments [13] (Fig. 1).

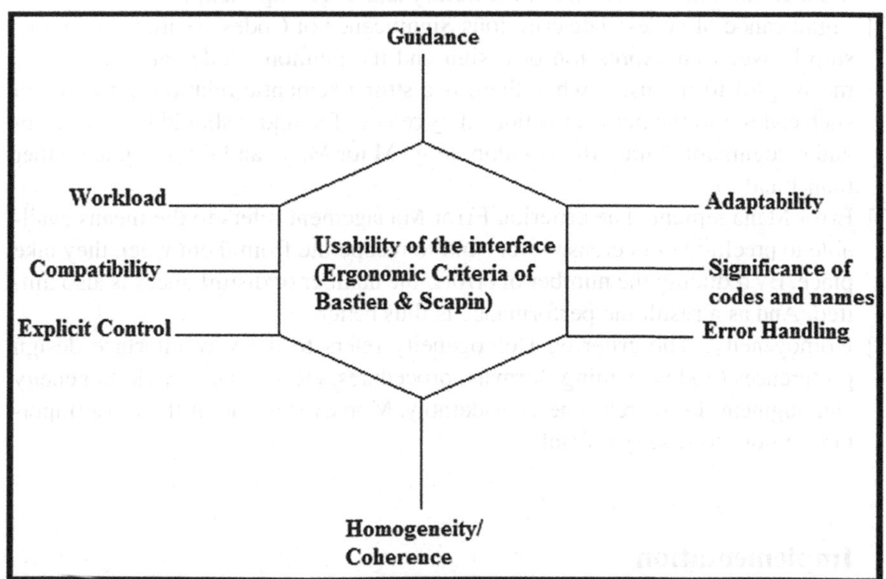

Fig. 1 The ergonomic criteria of Bastien & Scapin

- ☐ Guidance: User Guidance denotes to the process offered to guide, advice, and train the users throughout their communications with a computer (messages, alarms, labels, etc.). The standard Guidance contains four criteria: Encouragement, Assemblage/Dissimilarity of Items, Instant Feedback, and legibility.
- ☐ Workload: The measure Workload concerns all interface mechanisms that contribute in the diminution of the users' cognitive load, and in the growth of the dialogue efficacy. The criterion Workload is subdivided into two criteria: Shortness (which includes Concision and Minimal Actions), and Data Density. The probability of committing errors is elevated if the workload is elevated. Furthermore the fewer users are distracted by superfluous information, the more they will be capable to achieve their task efficiently.
- ☐ Compatibility: The criterion Compatibility debates the relative among operators' characters (memorial, observations, abilities, etc.) psychological characteristics; and task characteristics on the other hand. It refers also to the coherence between environments and between applications and as result the efficiency is increased and the performances are better.
- ☐ Explicit Control: The criterion Explicit Control relays to both the system treating of specific user actions, and the control users have on the treating of their actions by the system. The criterion Explicit Control is subdivided into two criteria: Explicit User Action, and User Control.
- ☐ Adaptability: The adaptability of a system refers to its capacity to act contextually and relating to the users' needs and tendencies. The criterion Adaptability is subdivided into two criteria: Flexibility and User Experience.
- ☐ Significance of codes: The criterions Significance of Codes assure the relationship between an expression or a sign and its mention. Codes and names are meaningful to the users when there is a strong semantic relationship between such codes and the items or actions they refer to for, and it should be significant and recognizable rather than random (e.g., M for Male, and F for Female rather than 1 and 2)
- ☐ Error Management: The criterion Error Management refers to the means available to preclude or decrease errors and to recuperate from them when they take place. By reducing the number of errors, the number of disturbances is also limited. And as a result the performance is thus better.
- ☐ Homogeneity: The criterion Homogeneity refers to the way interface design preferences (codes, naming, formats, procedures, etc.). A lack of Homogeneity can augment the search time considerably. Moreover is one of the most important reasons for user's refusal.

3 Implementation

The following methodology allowing us to explore and look for other tools to study how they are used and what kinds of GBL they are able to design.

3.1 Scratch 2

Scratch is a free introduction to programming launched in 2007 that allows the creation of serious games, interactive stories, animations, or greeting cards. Programming is based on graphical blocks which allow its handling by a very young audience. In addition, Scratch has the advantage of integrating the dimension of sharing in the software. Scratch allows the teacher to develop his learning game by an almost playful interactivity? It was also designed to faster collaborative work. Scratch is used in over 150 different countries and available in over 40 languages. Scratch is a project of the Lifelong Kindergarten Group at the MIT Media Lab, it'is written in Squeak and is an Open Source project.

3.2 GameMaker 8

Game Maker is a programming tool that facilitates the creation of 2D and 3D games and knowledgeable game development professionals equally, allowing them to create in a short time and at a fraction of the cost of ordinary tools! It is paid software, its platform is written in English, but the demo version insufficient to begin. This software was launched in 1999 by the Professor Mark Overmars. Game Maker uses an object-oriented programming. Behavior patterns are defined in the software without the need to program them, unlike Scratch; you have to do it by yourself.

3.3 eAdventure

eAdventure is an open source tool developed by Universidad Complutense de Madrid, it's written in java. These include a graphical interface that assists author's scenarios and directly produces a coded program in the specific language of e-Game executable by the engine of the game, this GUI even offers integrated access to all graphical editor and motor interpreter language dedicate. There are many advantages in using eAdventure, this tool permits to build a serious game point and click. It is able to produce Third player games with point-and-click features and first player games with static images; moreover player-learner can move and interact either with Non-Player Characters (NPCs) or with objects. NPCs are able to have conversations with the Third Person Games. The player can have actions on objects such as taking, using, watch, etc. These objects also have states (on, open, visited, etc.) that the player can change and gives the dynamics of the game, these are implemented directly, they are designed to be easily written and read by novices users, it is supported by all universal operational systems, like MAC OS, Windows and Linux.

Table 1 Comparison between the four GBL authorings tools through the evaluation elements of (Tom Murray 2003)

Software name	Evaluation methods				
	Interoperability	Manageability	Reusability	Scalability	Accessibility
Scratch 2			✓		GPL
eAdventure	✓	✓	✓	✓	LGPL
Game Maker 8			✓		Gratuit/Payant (Professional Version)

4 Result and Discussion

We presented below each educational authoring tool along with the result of its assessment and the synthesis of the study of these chosen GBL authoring tools (Scratch 2, eAdventure, Game Maker 8) following the criteria used in the implementation of the diverse evaluation methodology (List of evaluation elements made by Tom Murray, The Ergonomic Criteria of Bastien & Scapin, Educational Criteria) as shown in the Tables 1 and 2. With this thoroughly evaluation, we revealed that the authoring tools tested are founded on an event-driven programming. The quality of this GUI programming and the careful equilibrium of education and play and the easy use of a platform by novice are central to the quality of an author tool set. We also notice that these authoring tools are characterized by the absence of mechanism to generate an adaptable Game Based Learning to multiple mobile devices for this reason the platform which is better to develop an effective GBL and which one provide us to extend it to Mobile Game Based Learning is the eAdventure platform.

5 Our Approach Based on Extending E-Adventure for Multiple Mobile Devices

5.1 The Diverse Multi Platforms Mobile Technologies and the Dominant Solutions

The market for phones and applications is fragmented. If we wish to distribute an application to many people we need to deploy it on multiple operating systems. When we say platform we mean by that all the devices that share the same operating system and which are very compatible with each other as regards applications. Another important point in the development of mobile applications concerns the development technology is that each platform indeed requires different development tools, so if we want to deploy an application in our case is (Mobile Game Based Learning)

Table 2 Comparison between the four GBL authorings tools through the the ergonomic criteria of Bastien & Scapin

Software name	Good-guidance	Workload	Explicit Control	Adaptability	Error-management	Consistency	Codes-significance	Compatibility
Scratch 2	✓					✓	✓	
eAdventure	✓	✓	✓	✓	✓	✓	✓	✓
Game Maker 8			✓		✓	✓	✓	✓

	Operating Systems and Their Editors					
	IOs	Android	Blackberry Os	Windows Mobile	Widonws Phone 7	Symbian
The Solutions	Apple	Google	RIM	Microsoft	Microsoft	Nokia
PHONEGAP	×	×	×			×
TITANIUM	×	×	×			
FLEX	×	×	×			
MOBL	×	×	×			
OPENPLUG	×	×		×		×
CORONA	×	×				
APPMOBI	×	×				
QUICKCONNECT	×	×	×			×
WORKLIGHT	×	×	×	×		

Fig. 2 Comparison of the different development solutions of mobile multiplatform

on various platforms, it seems necessary to develop as many times we want these applications contact platforms, for this reason there are diverse solutions allowing to develop the application only once, then to deploy it on other platforms. Thanks to those methods we can reduce cost and time developing. The table below presents the diverse solutions existing and the difference between them, each with its advantages and drawbacks (See Fig. 2).

5.2 The New Architecture of eAdventure

In this section we briefly introduce our approach concentrating especially on the use of eAdventure to generate an adaptable Game Based Learning for different Mobile devices platform, as shown in this article many studies made in this domain approve that the usage of mobile devices is very varied and unlocks an extensive choice of possibilities in the scholastic field and as This authoring tool was established to decrease development charges for GBL [14, 15]. The eAdventure game editor is designed for instructors, minimizing the difficulty use of GBL authoring tool platform by novice and containing diverse structures that ease the integration of educational features for this purposes we have extended the old architecture of eAdventure to this new one allowing an automatic adaptation to diverse mobile devices (See Fig. 3).

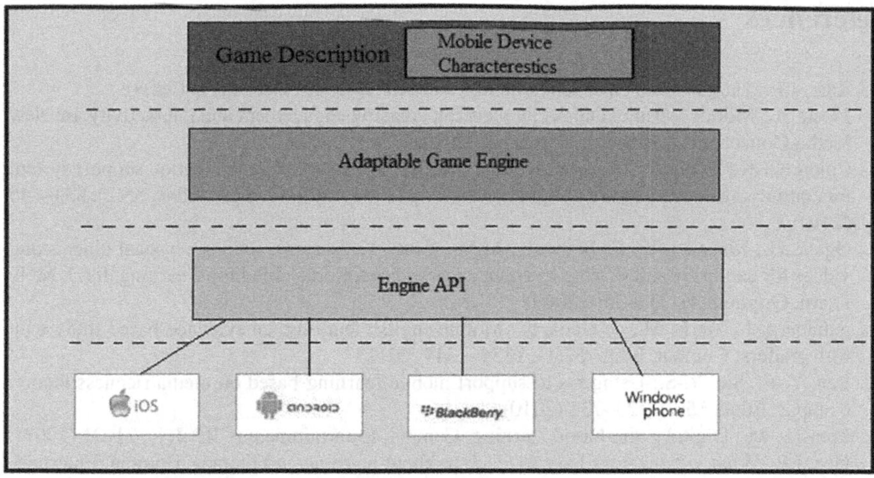

Fig. 3 The new architecture of the eAdventure

6 Conclusion

Understanding and exploring an educational authoring tool necessitates the application of several evaluation methods. In this paper we have presented the new architecture of eAdventure that will permit the development of reachable 2D Game Based Learning for multiple mobile device meeting the devices criteria already seized by user. The main advantage of this approach is that eAdventure could easily support automatically extensions to welcome several types of mobile device platforms. This article has only mention the technical difficulties and solutions bound to the GBL accessibility. On the other hand our future Works involves the development and evaluation of accessible Mobile Game Based learning using the New architecture of the eAdventure and the next steps will be the adaptation of this generated Mobile Game Based Learning for players profile such as: cognitive abilities, learning style (Figurative, symbolic and semantic), and these methods give us the possibility to adapt the generate MGBL according to his preferences and to his level and as result we will provide an effective learning [16].

Acknowledgments This research paper is made possible through the help and support from everyone, including:parents, teachers, family, friends, and in essence, all sentient beings. Especially, please allow me to dedicate my acknowledgment of gratitude toward the following significant advisors and contributors: First and foremost, I would like to thank Pr. Said Benhlima for his most support and encouragement. He kindly read my paper and offered invaluable detailed advices on grammar, organization, and the theme of the paper. Finally, I would like to thank Pr. M Ali bekri to read my paper and to provide valuable advices.

References

1. Katz, J.E.: Handbook of Mobile Communication Studies. The MIT Press (2008)
2. Jooste, T.: Mobile learning and social media: increasing engagement and interactivity. In: New Media Consortium Conference, pp. 9–12 (2010)
3. Chiou, C.-K., Tseng, J.C.R., Hwang, G.-J., Heller, S.: An adaptive navigation support system for conducting context-aware ubiquitous learning in museums. Comput. Educ. **55**(2), 834–845 (2010)
4. Ogata, H., Matsuka, Y., El-Bishouty, M.M., Yano, Y.: Lorams: linking physical objects and videos for capturing and sharing learning experiences towards ubiquitous learning. Int. J. Mob. Learn. Organ. **3**(4), 337–350 (2009)
5. Sandberg, J., Maris, M., de Geus, K.: Mobile english learning: an evidence-based study with fifth graders. Comput. Educ. **57**(1), 1334–1347 (2011)
6. Lan, Y.-F., Sie, Y.-S.: Using rss to support mobile learning based on media richness theory. Comput. Educ. **55**(2), 723–732 (2010)
7. Prensky, M.: Digital game-based learning. Comput. Entertainment (CIE) **1**(1), 21–21 (2003)
8. Gee, J.P.: What video games have to teach us about learning and literacy. Comput. Entertainment (CIE) **1**(1), 20–20 (2003)
9. Burgos, D., Tattersall, C., Koper, R.: Re-purposing existing generic games and simulations for e-learning. Comput. Hum. Behav. **23**(6), 2656–2667 (2007)
10. Bennis, L., Benhlima, S.: Comparative study of the process model of serious game design through the generic model dice. In: Intelligent Systems and Computer Vision (ISCV), 2015, pp. 1–5. IEEE (2015)
11. Murray, T.: An overview of intelligent tutoring system authoring tools: updated analysis of the state of the art. In: Authoring Tools for Advanced Technology Learning Environments, pp. 491–544. Springer (2003)
12. Bastien, J.M.C., Scapin, D.L.: Ergonomic criteria for the evaluation of human-computer interfaces (1993)
13. Bach, C., Scapin, D.L.: Adaptation des critères ergonomiques aux interactions hommeenvironnements virtuels. IHM'2003-15ème Conférences Francophone sur l'Interaction Homme-Machine (2003)
14. Bennis, L., Benhlima, S., Bekri, M.A.: An authoring tool for generating an adaptive mobile game based learning
15. Bennis, L., Benhlima, S.: Building an Adaptive Game-Based Mobile Learning Using The Felder-Silverman Learning Style Model (FSLSM) Approach (2016)
16. Bennis, L., Benhlima, S.: A new approach to design an attractive game based learning in various domains. J. Theor. Appl. Inf. Technol. **85**(3) (2016)

Best Practices in Distance Education: An Investigation of a Hybrid Faculty Development Program

Ebony Terrell Shockley and Izolda Fotiyeva

Abstract Guided by program goals and faculty feedback, this paper aims to examine best practices for a hybrid faculty development certificate program. Researchers evaluated outcome measures and report faculty responses in an effort to add to the literature on this topic and to provide a model for institutions of higher learning.

Keywords Distance education · Online learning · Online courses · Faculty development · Hybrid courses · Professional development · Best practices

1 Introduction

Online teaching is becoming increasingly prevalent in higher education as many institutions are increasing their online and hybrid offerings in an effort to reach more students [1–3]. Universities are forming programs to respond to this growing need for online educators, and seeking information on how to reach additional faculty in meaningful ways. In one faculty development program, referred to as PAL, Professional Academy of Learning, participants receive professional development on building online courses in a hybrid setting. As participants of PAL, professors convert an existing face-to-face course to an online course while acquiring pedagogical knowledge and instructional technology practices.

E.T. Shockley (✉) · I. Fotiyeva
University of Maryland, College Park, USA
e-mail: pcarreteiro@gmail.com

E.T. Shockley · I. Fotiyeva
Howard University, Washington, DC, USA

Á. Rocha et al. (eds.), *Europe and MENA Cooperation Advances in Information and Communication Technologies*, Advances in Intelligent Systems and Computing 520, DOI 10.1007/978-3-319-46568-5_6

1.1 Theoretical Framework

Using connectivism as a framework, an investigation ensued of best practices for higher education faculty seeking online certification. Connectivism is an appropriate lens for the study as it considers other learning theories (e.g. behaviorism, constructivism, cognitivism), but also considers their limits, given that the theories emerged prior to the digital age [4]. As such, connectivism is a "successor" to these theories. Connectivism supports the ability to realize the relationships between current knowledge, various disciplines, and diversity of ideas. The consideration of connectivism as a learning theory is critical, so that the resources and decisions for faculty continue to evolve. As best practices in the development of online and hybrid faculty development programs are explored, this framework provides an appropriate angle for this work.

2 Review of Literature

As online and hybrid programs for educators continue to abound throughout professional organizations, colleges, and universities, there are commonalities that emerged. To determine best practices for faculty growth and retention, it was important to examine the models implemented by available programs. For instance, Public Broadcast Service, a nationally televised service with educational media and programming, has offered more than 130 online courses with more than 100 instructors [5]. They implement a model of Select, Prepare, Support and Feedback. They *select* individuals who fit their criteria, (Master's degree earned, a previous online learner, and knowledge of content). PBS *prepares* faculty by having them participate in two courses, including one that teaches techniques for online communication, evaluating web resources, preparing online discussion, delivering effective online assessment, planning course facilitation and policies, enhancing pedagogy of online learning, and supporting online communities. Faculty are *supported* with a mentor to teach their first online course, and they *reflect* on their teaching and review course evaluations. In the last stage, they participate in a cycle of annual *feedback*. The feedback stage is a data point and is the last, but ongoing part of the professional development stage.

Abdous [2] suggests a process-oriented framework that includes a preparation, planning, design, facilitation, and reflection phase. In the *preparation* phase, faculty are encouraged to participate in development activities to develop an understanding of online teaching pedagogy, adequate knowledge about copyright and intellectual property, knowledge of teaching and learning technology and their limitations, and an understanding of the students based on demographics. The next phase, which is the *planning* phase, is an opportunity for faculty to develop a syllabus that is structured and student-centered. During this phase faculty should brainstorm options for technology tools for various learning activities. In the last two phases of

Abdous' pre-course work (*design and facilitation*), it is suggested that faculty align syllabi with a menu of learning activities and that they should *design* appropriate assessments. In the last phase, faculty should direct students to a LMS orientation and assist with *facilitating* and welcoming learners to the platform and discuss school policies. This is the first part of Abdous' process-oriented framework, and it is the portion that parallels the faculty development program. As instructors move through the cycle of the framework, they are directed to *reflect*.

Similar to the above research, reflection logs and journals were a primary resource for data collection throughout the literature [2, 5]. Some programs incorporated self-reflections from faculty regarding their progress in their faculty development program and others reflected on their teaching after exiting the program [5]. In addition to reflection logs, methods for collecting data included, surveys and rating scales, some distributed prior to faculty joining an online faculty development program, that required information regarding prior preparation and development [1, 6]. Researchers also used focus groups to evaluate their faculty development programs [7]. In general, the instruments used for data collection were limited in the research. Additionally, the research was limited regarding the type of faculty (e.g. K-12 or higher education), learning management system utilized, and the time span of the faculty development program. The absence of this information is challenging for institutions that may be interested in including some of the efforts as appropriate for their faculty, or providing results to institution leaders who are considering online faculty development programs. It is also arduous for institutions with programs that maybe interested in collaborating and drawing conclusions based on the findings of multiple institutions. The articles that were selected for this article included programs as short as 3 days and as long as 3 months [1, 6–8]. A review of the literature revealed that, barriers to online faculty development include: course structure, compensation, increased workloads, faculty dispositions, organizational change, and trepidation of technology use [2, 6, 8].

3 Research Context and Questions

The literature divulged the following best practices for online faculty development programs: developing faculty learning communities, incorporating the principles of andragogy, creating a data-driven culture, incorporating a variety of professional development techniques, using topics in faculty development that are relevant to subject area or job description, involving faculty in the planning and evaluation of their development, pre and post assessment using the same instrument, target experienced online instructors for further training and assessment [1, 3, 5–7].

3.1 Research Questions

Based on the literature and our data, we will identify our best practices and lessons learned from PAL. We will provide the explicit process of our data collection to add to the limited research, and share a matrix of how we assessed our program. This is summarized by the following research questions:

1. In what ways are we evaluating our faculty certification program for online/hybrid teaching?
2. What outcome measures are we using to assess our faculty certification program for online/hybrid teaching?
3. What are the best practices of our faculty certification program for online/hybrid teaching?
4. How did we determine the best practices of our faculty certification program for online/hybrid teaching (including our evaluation process)?

4 Methodology

In this study, we focused on faculty participants in two versions of professional development programs: a 6-week face-to-face PAL offered during the summers of 2012, 2013, and 2014; and the two accelerated 9—week hybrid PAL offered in fall 2013 and spring 2014. The differences between these offerings are that the 6-week face-to-face model is a summer course with presentations of course instructors, and the 9-week offerings are hybrid (i.e. Friday face-to-face sessions) as participants cannot meet daily during the regular semester. The content is the same for both. The overarching goals for faculty in both programs were to increase knowledge about the principles of online teaching and the corresponding instructional practices; to design and develop their own courses, and to gain experience and confidence in teaching those courses in upcoming semesters. The expectation of both programs was that by providing knowledge on instructional design, pedagogy, best practices, and educational technologies, faculty would reflect this knowledge and experience in their subsequent online instruction. Using connectivism as a lens, we analyzed participant data, resulting in an analysis of next steps for course instructors.

4.1 Participants

A new team of faculty participated in each PAL offering. Each team involved faculty from different departments. During each offering, the participant faculty designed all instructional units of the future courses that included learning objectives, instructional materials, assessment of student learning that was aligned with

Table 1 PAL participant numbers and disciplines

Summer 2011	n = 7	Biology (2), English (2), Mass Media (1), Music (1), Nursing (1)
Summer 2012	n = 7	Business (2), Education (2), English (2), Nutrition (1)
Summer 2013	n = 8	Business (2), Education (2), English (3), Law (1)
Summer 2014	n = 8	Academic Advising (1), Business (4), Mass Media (2), Social Work
Fall 2013	n = 14	Business (5), Education (2), Mass Media (1), Political Science (2), Social Work (1), Speech Pathology (1), Urban Planning (1)
Spring 2014	n = 8	Business (2), Education (3), Mathematics (2)
Total participants	52	Biology (2), Business (11), Education (11), English (4), Law (1)

those objectives; and active, learner-centered teaching strategies, such as cooperative learning, and implementation of educational technologies. All faculty were invited to participate in this professional development, and we included faculty who responded by submitting the PAL application and were recommended by their Deans. The majority of study participants had doctoral degrees in various disciplines. The remaining of the participants had the law and masters' degrees. The program had an overall total of 52 participants. Participants include faculty from assistant, associate, and full professor faculty rank. Table 1 displays the participants per offering and their corresponding disciplines.

4.2 Instrumentation and Procedures

Two sources of data were used in this study: surveys (self-reports) from the PAL participants using SurveyMonkey® and the Blackboard site which was the course LMS. The pre-assessment survey serves, to some extent, as a baseline survey, in which faculty were asked to report their knowledge of and experience with online teaching and was administered electronically via survey monkey to participants prior to the start of each PAL. It contained a combination of "check all that apply" questions such as: "In the courses I teach, I present information in the following ways…" to assess participants teaching experiences, and "I am familiar with the following intermediate/advanced features in the Blackboard Learning Management System" to assess participant competencies.

The post-assessment survey, administered on the final day of each PAL offering, collected information ranging from the competencies participants gained during the PAL experience to overall program evaluation. In this final survey, faculty responded to a variety of questions, including those concerning their knowledge of and experience with online teaching, their confidence about their preparation as a an online professor, the level support that they received from the PAL instructors, and the challenges that they encountered. The Likert scale questions of 1–3 were used to gain their level of agreement with the course units being helpful (3—helpful, 2—somewhat helpful, and 0—not helpful). Scale questions (1—lowest to 3—highest), agreement questions (don't agree, strongly agree and agree), and open-ended response were also used to allow participants to rate content areas, activities and strategies.

The Blackboard site that hosted both, the PAL course and participants' own newly-developed courses, served a dual role. During the PAL course a mid-program reflection survey was included and faculty were asked to list the reforms that they implemented in their courses based on PD (including assessments, instructional strategies, and teaching materials). Faculty's own courses provided opportunity for the PAL instructors to observe the actual implementation of what was learned and experienced during the PAL program.

5 Results

Qualitative analysis was used for the data collected from the above-identified sources. The data were imputed verbatim and summarized first per corresponding survey question, and finally, categorized by the appropriate research question.

There were multiple ways we used to evaluate our faculty certification program for online/hybrid teaching (research question 1). Direct faculty evaluations and opinions about the effectiveness of PAL certification program, on expertize and helpfulness of PAL instructors, and on overall satisfaction with this training experience, were extracted from pre-PAL survey assessment, mid—semester reflection, and post-PAL assessments assessment. Another way to evaluate our faculty certification program for online/hybrid teaching included the faculty success rates finishing the PAL offerings (be present for all face-to-face sessions, actively participate in all in-class and online activities, and fulfill all assignments). Finally, the number of faculty who designed and developed their own online or hybrid courses (along with these courses' effectiveness and quality) was another important evaluating factor. Those courses were subsequently peer-reviewed and recommended to be included into the official university course catalog. This evaluation factor is very important since it is aligned with the expectation of PAL certification program that anticipates the participating faculty to reflect their knowledge and experience gained from this certification program in their subsequent online instruction.

The types of evaluation that we were using to evaluate our faculty certification program for online/hybrid teaching involved specific outcome measures (research question 2) that are summarized in Table 2. In this study, we focused on the following outcome measures: (1) usefulness of program offering, (2) relevance of program offering, (3) developed skill or competency during program, (4) improvement of teaching, (5) the number of faculty who successfully finished the PAL course, (6) and the numbers of courses that were designed and developed by participating faculty, with the subsequent peer-review process and recommendation to include those course to the University course catalog. As depicted in Table 2, PAL program's usefulness was evaluated by the pre-PAL assessment, the mid-semester reflection, and the post-PAL assessment. The relevance of the program was only evaluated by the pre PAL and post-PAL assessments. Developed skills or competency was addressed by the mid-semester evaluation, post-PAL assessment. The improvement of teaching was assessed by the pre/post PAL assessments and the mid semester reflection survey. The Blackboard site and the PAL website helped to gained data for outcome measures 5 and 6.

The overall purpose of the PAL is to expose participating faculty to the online teaching and learning environment as both students and instructors. To achieve its goals PAL covers six areas (1) An Overview of Online Teaching, (2) Online Course Design and Organization, (3) Online Course Assessment, (4) Enhancing Collaborative Learning, (5) Fair Use and Accessibility, and (6) Integrating Technology.

The practices that we used in PAL courses are to include but are not limited to: creating the course assessments that are aligned with its objectives and instructional materials; offering a variety of activities/instructions, including the customized digital lessons and instructor-made videos; extensive instructor support and couching, and an ongoing evaluation of the PAL course by the participating faculty. Participating faculty overwhelmingly approved the best practices used, and this approval was reflected in their responses to the PAL course evaluation and reflective feedback on the course (research question 3). The particular faculty opinions on best practices of our faculty certification program for online/hybrid teaching were drawn from participants' surveys and mid-semester reflection responses (research question 4). The responses on best practices that emerged from the original PAL course evaluation, the Blackboard site, and reflective feedback on the course were categorized into the following themes: (a) use of Quality Matters, (b) the course assessments/their alignment with learning objectives (c) teaching support. In the following part of the result section, we will discuss the identified best practices in turn.

Table 2 Types of PAL evaluation and outcome measures

	Pre PAL assessment	Mid semester reflection	Post PAL assessment	Blackboard/ PAL website
Outcome measure				
Usefulness of program offering	x	x	x	
Relevance of program offering	x		x	
Developed skill or competency during program		x	x	
Improvement of teaching	x	x	x	
Number of completers				x
Number of courses developed				x

5.1 Rubrics and Standards

The application of the Quality Matters Rubric, a leader in quality assurance for online education, was acknowledged to be one of the best practices of the PAL certification program. The Quality Matters Rubric received national recognition for its peer-based approach to continuous improvement in online education and student learning. All PAL participants were required to have completed the Applying the Quality Matters Rubric workshop training as a prerequisite to apply for PAL certification course. Therefore, all participants had a baseline familiarity with the rubric and process before they started the certification training. During the PAL course, all participants extensively used the rubric and process as a guiding tool while they were designing and constructing their own courses. In addition, they used the QM rubric to peer review their colleagues (other PAL participants) courses and continuously received the instructors' feedback on the quality of their reviews. One PAL participant stated, "Using the QM rubric as a guiding tool enhances the course content and clearly identifies the learning objectives of the course along with the reading assignments and activities." This is echoed by an additional remark that "The Quality Matters Rubric [is a] great roadmap for developing a good class… [it] definitely force [d] me to think about and implement all of the important facets of the class… obtaining QM certification is a strong incentive." The collective response of the participants to the use of Quality Matters is captured by the following statement, "…implementing the QM rubric—The annotated sections were most helpful in understanding the scope of each rubric as are the peer review activities."

5.2 Alignment of Assessments with Learning Objectives

The alignment of the PAL course assessments with the learning objectives was found to be another best practice and strong feature of our certification program, and the one that many PAL participating faculty wanted to adopt for their own courses. All PAL participants (100 %, n = 52) responded positively to the following survey question, "How well do you feel the course content/activities/assessments aligned with units and course learning objectives?" One participant remarked, "I feel they [course assessments] aligned great with the content/activities/assessments. I walked away each week with the feeling that I accomplished the objectives for that week." Analyzing participants' responses, it became clear that many faculty members have not previously implemented a consistent and comprehensive assessment process to measure the extent to which students have mastered the course objectives. This is highlighted by the statement of one of the PAL participants: "I have used examinations, quizzes, case studies, term paper, etc. but have not tied them back to the course objectives. Being in the PAL has facilitated my further developing course objectives for each unit. I have a clearer understanding of assessment modalities and how varied they can be. And I am more cognizant of the importance of closing the objectives-assessment loop."

5.3 Teaching Support

Many participating faculty members were new to the online environment and did not a have a full picture of what is required in terms of both knowledge and time to develop a good online or hybrid course. They also lacked familiarity with andragogy, the Quality Matters rubric, educational technologies, and many other essential components. Moreover, many were intimidated by the fact that they have to teach online and design and develop their own courses. Therefore, the PAL instructors anticipated to scaffold support through the whole training process and be prepared to support and advise the faculty on every step of this certification program. As the result of this realization, the faculty support component was built into the PAL course as one of the major feature of this certification program. Some examples of this support system are the following: holding mentoring sessions every face-to-face meeting (one-to-one coaching labs), assigning personal couches to each PAL participant for additional training and advising, inclusion of the instructor-made videos (PAL instructors) on challenging topics, providing constant email/discussion board Q&A. These examples mirror most closely with the outcome of the Public Broadcast System model in the literature review (see Sect. 2).

According to the survey responses, the participating faculty valued and praised this support system and considered it one of the best practices. They viewed it as an essential factor in their successful completion of the PAL certification program. One participant noted, "The supportive and friendly instructors of PAL made the daunting thought of becoming an online instructor quite tangible!" Versions of this comment were shared by other colleagues who viewed teaching online as "intimidating". Teacher support during the program was appreciated by the faculty on a number of levels. Highlighting this, another participant commented, "This was a profound and eye-opening experience. I'm grateful to have been offered the opportunity to grow in this professional capacity. The coaching team is phenomenal and patient. The environment was extremely supportive. It was helpful to know that they and my colleagues were with me EVERY step of the way. I look forward to this new chapter in my professional journey…!" Other participants also expressed how helpful it was to have "skilled professionals" to help throughout PAL. One participant found "Having hands-on activities and lab time with instructors present" to be the most valuable part of the entire program."

6 Next Steps

Three more PAL iterations were offered since the data for this article was collected –during Spring 2015, Summer 2015, and Spring 2016, with more than 30 participants overall. Lessons learned from this study allowed the PAL faculty to redesign and improve this professional development course to make it more effective and relevant to the participating faculty. We worked with the Deans and Department Chairpersons and they helped to incentivize the program, at the same time making it more rigorous and accountable. At the time of the study, the completion of the PAL course ensured the certification award for participating faculty, and the complete course development and its successful QM peer review were a separate process for the faculty. Presently, those two stages are tied together, and the complete course development and its successful QM peer review are necessary requirements for PAL certification. As the result of these modifications we anticipate a greater rate of new approved online and hybrid courses that will be offered at the university course catalog after each PAL offering.

To meet the needs of our faculty and accommodate their schedules, we are working on the development of a completely online version of our professional development program. We will begin to gather data from the students of PAL completers. We plan to utilize the latest trends and practices in online and face to face pedagogy, using novel technology, including but not limited to digitally augmented physical spaces, mobile video analysis, multimedia applications, interactive tools, and transmedia storytelling,

References

1. Horvitz, B., Beach, S.: Professional development to support online learning. J. Fac. Dev. **25**, 24–32 (2011)
2. Abdou, M.: A process-oriented framework for acquiring online teaching competencies. J. Comput. High. Educ. **23**, 60–77 (2011)
3. Barker, A.: Faculty development for teaching online: educational and technological issues. J. Contin. Educ. Nurs. **34**(6) (2003)
4. Bell, F.: Connectivism: It's place in theory-reformed research and innovation in technology-enabled learning. Int. Rev. Res. **2**(3) (2011)
5. Storandt, B., Dossin, L., Lacher, A.: Toward an understanding of what works in professional development for online instructors. J. Asynchronous Learn. Netw. **16**(2), 121–162 (2012)
6. Koepke, K., O'Brien, A.: Advancing pedagogy: evidence for the role of online instructor training in improved pedagogical practices. J. Asynchronous Learn. Netw. **16**(2), 73–83 (2012)
7. Lee, D., Paulus, T., Loboda, I., Phipps, G., Wyatt, T., Myers, C., Mixer, S.: A faculty development program for nurse educators learning to teach online. Tech. Trends **54**(6), 20–28 (2010)
8. Johnson, T., Wisniewski, M., Kuhlemeyer, G., Isaacs, G., Krzykowski, J.: Technology adoption in higher education: overcoming anxiety through faculty bootcamp. J. Asynchronous Learn. Netw. **16**(2), 63–72 (2012)

Setting up an Intelligent IDS Based on Markov Chains Theory

Noureddine Rahmoun, Yassine Ayachi, El Hassane Ettifouri, Jamal Berrich and Bouchentouf Toumi

Abstract The tremendous growth of the web-based applications has increased information security vulnerabilities over the Internet. Security administrators use Intrusion-Detection System (IDS) to monitor network traffic and host activities to detect attacks against hosts and network resources. The solutions proposed in the literature actually achieved good results for the detection rate, while there is still-room for reducing the false positive rate or even predict beforehand attack according to a website visitor behavior. For this purpose we propose a probabilistic approach applied for an "intelligent IDS" whose main role is to predict attacks on a website according to a real time probabilistic calculation based on Markov chains.

Keywords WEB attacks · Intrusion detection system (IDS) · Markov chain · False positive

N. Rahmoun (✉) · Y. Ayachi · E.H. Ettifouri · J. Berrich · B. Toumi
Laboratory LSE2I, National School of Applied Sciences,
First Mohammed University, Oujda, Morocco
e-mail: rahmoun.noureddine@gmail.com
URL: http://wwwensa.ump.ma

Y. Ayachi
e-mail: ayachi.yassine@gmail.com

E.H. Ettifouri
e-mail: h.ettifouri@gmail.com

J. Berrich
e-mail: jberrich@gmail.com

B. Toumi
e-mail: tbouchentouf@gmail.com

Á. Rocha et al. (eds.), *Europe and MENA Cooperation Advances in Information and Communication Technologies*, Advances in Intelligent Systems and Computing 520, DOI 10.1007/978-3-319-46568-5_7

1 Introduction

Several works that focused on anomaly-based high-speed classification, proposed the use of simple statistics on the application-layer payload to characterize the normal behavior of Web applications [1–7]. We share the definition of "normal behavior" provided by [8]: the term normal behavior generally refers to a set of characteristics (e.g., the distribution of the characters of string parameters, the mean and standard deviation of the values of integer parameters) extracted from HTTP messages that are observed during normal operation. Initially, the main obstacle to the large-scale deployment of anomaly based IDS solutions has been the too high false positive rate, as not all the detected anomalies are actually related to attack attempts. The payload is the data portion of a network packet that is the portion of the network packet, which carries the HTTP message. According to RFC 2616, the HTTP payload contains the Request-Line plus the Request-Header fields used by the client to pass additional information to the server (RFC 2616). In last year's, several statistical models for IDS based on solutions anomaly; among them, there is the work [9] on IDS, NaÏve Bayes classifier is used. This technique identifies the most important HTTP traffic features that can be used to detect attacks.

The highlight is to enhance IDS performance through preparing the training dataset allowing detecting malicious connections that exploit the HTTP service. The weakness of this work has not decreased the false positive rate.

In this work, we are interested in the analysis of the HTTP service and trying to improve security by simply analyzing the network traffic incoming to the Web server, without a detailed knowledge of how the Web application is implemented and works. This can be done by implementing an "intelligent" IDS solution based on Markov chains; the main idea revolves around the calculation of the occurrence probability of a web attack after n transition (between the website pages), depending on the current state and the matrix of all possible transitions (the state space) [10]. This probability calculation is based on visit history statistics present on a real dataset assembled over a long period of time (more than 7 years), the dataset include information on visitors, visited pages, request content, attack attempts.

This article is organized as follows: In Sect. 2 we present the existing work on the assessment of IDS, in Sect. 3 we present the dataset that will be used by the IDS, finally in Sect. 4 we present the modeling of the problem using Markov chains and his application to build an intelligent IDS.

2 Related Work

Given particular IDS, how can we determine if it behaves as expected? We can test it with various inputs, but we can never test it with every possible input or every possible interleaving of all possible inputs. This problem has already been addressed in the field of software testing, which defined the concept of equivalence classes to reduce the number of inputs against which a software should be tested. The idea is that test cases belong to the same equivalence class if they stimulate (activate) the same parts of the software in the same conditions and thus produce equivalent results. The idea is thus to apply this principle to IDS evaluation by defining a classification for elementary attack actions. By classification, we mean a "systematic arrangement, in groups or categories, according to established criteria".

In [9] they proposed IDS containing classifier Naive Bayes by using the database NSL-KDD [11] which contains data in the form of features with the goal of improvement of the performance of time and the detection of malicious connections which exploits service HTTP as long as the rate of positive forgery reached approximately 6.6 %. Another proposal for IDS containing Markov Chains hidden in [12]. This system with the same purpose described in [9] but more efficient because it is characterized analysis of the abnormal bytes for the protocols which use service HTTP (TCP), like attacks Web detected XSS, SQL-injection, SHELL-code and the attacks of CLET with a rate of 0.1 %.

3 Intrusion Detection Evaluation Dataset

3.1 Existing Datasets and Scenario Based Detection

A dataset is a practical solution in terms of evaluating the performance of IDS; but in the realm of network intrusion detection (IDS), the scarcity of adequate datasets is a well-known issue. Another problem adds to it: most existing datasets [we tested OSVDB [13], National Vulnerability Database (NVD) [14], NSL_KDD [11], DARPA are classifying attacks at the frame level, which ignores the attack scenarios as a whole and can't predict if a sequence of frame leads to an attack,

From this perspective, we created our own dataset, which include an over 7 years visits history of a specific web site, this dataset contains real attacks scenarios on which we based our analysis and study.

3.2 Building a Dataset

The limitations of existing dataset prompted us to study the possibility of proposing a dataset that can be adapted to ours needs, specifically the complete scenario of an attack.

For this purpose, we used a combination of personalised access log to a specific website and build a CSV file containing nearly 359710 lines divided between normal flow and attacks,

In this dataset, we marked all attack scenarios from the first connection until the occurrence of the attack; several sort criteria are taken into account in order to diversify our studies perspective.

3.3 Presentation of Our Idea

To use the Markov chains approach, we must adapt our context and build a model that represent the possible transitions (navigations) as a graph and then as a matrix of probabilities.

Date and time based analysis

The access history contain a huge amount of information which we sorted by IP address (to identify users), access date and time and content of the requests,

Categorization of the web pages

All visits are following a logic that can lead to a normal visit or an attack. On this base, we divided our pages to categories as follow, a user sends a request (POST, GET…) a URL is returned to his with a specific content, which can be shown in a web page, that web pages was classified into categories (Fig. 1) as follow:

Category 1: (Connection): represented by all pages leading to a connection to the website.
Category 2 (Logout): Page shown after clicking the button logout.
Category 3 (The 404 error): or not found pages error.
Category 4 (The 50X errors): or access privileges errors.
Category 5 (Normal pages): Shown content matches requested content.
Category 6: Pages with false obligatory id.
Category 7: Pages with false pagination id.
Category 8 (Attacks): Correspond to SQL script or commands, with the objective to retrieve information or to interrupt service.

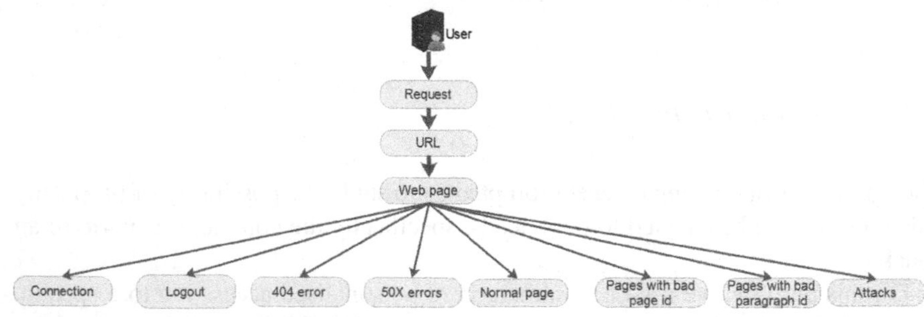

Fig. 1 Classification in categories pages

Table 1 The website visit logs

idReq	*idC*	*dateReq*	*Time*	*Page_name*	*id_categ*
368	36	26/02/2015	17:24:10	contact.php	3
369	36	26/02/2015	17:24:10	contact.php	3
370	36	26/02/2015	17:24:10	observation.php	3
371	36	26/02/2015	17:24:10	observation.php	3
372	36	26/02/2015	17:24:10	observation.php	3
373	36	26/02/2015	17:24:10	observation.php	3
374	36	26/02/2015	17:24:10	observation.php	3
375	36	26/02/2015	17:24:11	observation.php	3
376	36	26/02/2015	17:24:27	contact.php	1
377	36	26/02/2015	17:24:32	index.php	3
378	36	26/02/2015	17:24:33	index.php	5
379	36	26/02/2015	17:24:33	index.php	3

Potential navigation between categories

For a specific user we can follow all transitions between our website categories according to the following log entries (Table 1):

With the following column notation:

IdReq: The request identification.
IdC: The client identification.
DateReq, Time: The request date and time of occurrence.
Page_name: The visited page during this request.
Id_categ: The page categories according to the above classification.

A complete inventory of all navigation possibilities can lead us to a node graph representation (Fig. 3) as follows:

4 Problem Modeling

4.1 Markov Chains

In machine learning, Markov chain is a family of simple probabilistic. A stochastic process X(t), t ϵ T is a function of time whose value at every moment depends on the outcome of a random experiment. At each time t ϵ T, X(t) is a random variable. If we consider a discrete-time rating is then X_n, n ϵ N a stochastic process in discrete time. If we finally assume that the random variables X_n can take only discrete set of values, we refer to process discrete time and discrete state space. X_n, n ϵ N is a Markov chain in discrete (1) time if and only if:

$$\mathbf{P(X_n = j | X_{n-1} = x_{n-1}, y | X_{n-2} = x_{n-2}, \ldots, X_0 = x_0) = P(X_n = j | X_{n-1} = x_{n-1})} \quad (1)$$

The probability that the chain is in a certain state to the nime of the process depends only on the state of the process in the previous step (the nime) and not states in which it was to earlier stages. Is defined as a homogeneous Markov chain when this probability does not depend on n. We can then define the probability of transition (2) from state i to state j denoted pij:

$$\mathbf{p_{ij} = P(X_n = j | X_{n-1} = i), \forall n \in N} \quad (2)$$

By introducing the set of possible states (3) denoted E, we have:

$$\sum_{ij} p_{ij} = 1 \quad (3)$$

We then define the transition matrix (4):

$$\begin{pmatrix} p_{11} & p_{12} & p_{13} & \cdots & \cdots \\ p_{21} & p_{22} & p_{23} & \cdots & \cdots \\ \vdots & p_{32} & \vdots & \cdots & \vdots \\ \vdots & \cdots & \vdots & \cdots & \vdots \end{pmatrix} \quad (4)$$

4.2 Problem Modelling with Markov Chain

Our goal is to set up a new approach permitting to anticipate a visitor attack by calculating the probability of occurrence of these attacks basing on category scheme (cf §c.iii) and the browsing between different graph nodes.

This steered us to applicate Markov chains to calculate the probability to transit from a category C_i to another category C_j after n transition:

Consider the following notation:

n: The total website category number,
C_i: The number i category,
C_j: The number j category,
C_k: The current category,
Applying the Markov chains we can write this definition: $\mathbf{P(n) = P(0).P^n}$
P(n): The probability to transit from a C_k category to C_k' after n transition,
P(0): The probability to be on C_k category at initial time (*t = 0*),
P: The **Pij** matrix representing the transition between graph categories, with i, j [*1, category number*], each Pij represent the probability to transit from a **Ci** to **Cj** category,

$$\mathbf{P(0)} = \begin{pmatrix} 0.8 \\ 0 \\ 0.025 \\ 0 \\ 0.125 \\ 0.025 \\ 0.025 \\ 0 \end{pmatrix}$$

$$\mathbf{P} = \begin{pmatrix} 0.966 & 0 & 0.012 & 0 & 0.021 & 0.001 & 0 & 0 \\ 0 & 0 & 0 & 0 & 1 & 0 & 0 & 0 \\ 0.064 & 0 & 0.924 & 0 & 0.013 & 0 & 0 & 0 \\ 0.286 & 0 & 0 & 0 & 0.714 & 0 & 0 & 0 \\ 0.081 & 0.012 & 0.012 & 0.041 & 0.826 & 0.029 & 0 & 0 \\ 0 & 0 & 0 & 0 & 1 & 0 & 0 & 0 \\ 0 & 0 & 0 & 0 & 0 & 0 & 0 & 0 \\ 0 & 0 & 0 & 0 & 0 & 0 & 0 & 0 \end{pmatrix}$$

An example of calculation of a probability P_{14} will be as follows:

$$\mathbf{P} = \begin{pmatrix} 0.966 & 0 & 0.012 & 0 & 0.021 & 0.001 & 0 & 0 \\ 0 & 0 & 0 & 0 & 1 & 0 & 0 & 0 \\ 0.064 & 0 & 0.924 & 0 & 0.013 & 0 & 0 & 0 \\ 0.286 & 0 & 0 & 0 & 0.714 & 0 & 0 & 0 \\ 0.081 & 0.012 & 0.012 & 0.041 & 0.826 & 0.029 & 0 & 0 \\ 0 & 0 & 0 & 0 & 1 & 0 & 0 & 0 \\ 0 & 0 & 0 & 0 & 0 & 0 & 0 & 0 \\ 0 & 0 & 0 & 0 & 0 & 0 & 0 & 0 \end{pmatrix}$$

Being in a page of category 4, what is the probability to come from a page of category 1? (Fig. 3).

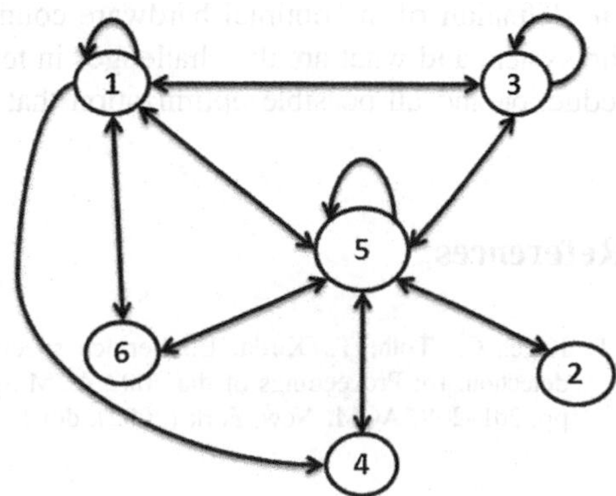

Fig. 2 Possible transitions between categories

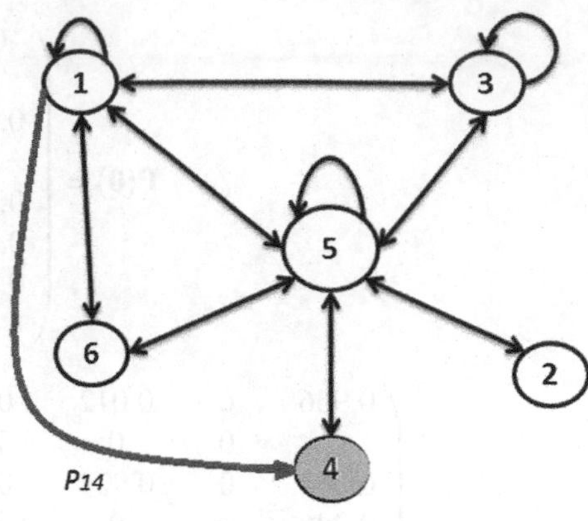

Fig. 3 Possible transitions from category 1 to category 4

Using the website visit log (Fig. 2) it's possible to calculate the occurrences of visits incoming from pages of categories 1 and divide it by all incoming visits to category 4 (Fig. 3).

For example: $P_{14} = 2/7 = 0.286$.

5 Conclusion and Future Works

The implementation of such a solution puts us forward in the resolutions of several security issues among which we mention:

- To establish after what number of transitions between pages of a web site, and for what navigation path the probability of an attack happening is highest and proceed to prevention of it.
- To enable IDS to learn automatically when an attack is most likely to recur, according to his access history, and prevent the recurrence of new attacks.
- Set the most vulnerable pages in a website and strengthen their security.

Other research areas can be addressed, following the establishment of our IDS as the definition of the optimal hardware configuration for the proper functioning of the system and what are the challenges in terms of performance and false positives reduction and all possible optimization that may be made to this system.

References

1. Krgel, C., Toth, T., Kirda, E.: Service specific anomaly detection for network intrusion detection. In: Proceedings of the 2002 ACM Symposium on Applied Computing (SAC'02), pp. 201–208. ACM, New York (2002). doi:10.1145/508791.508835

2. Mahoney, M.V.: Network traffic anomaly detection based on packet bytes. In: Proceedings of the 2003 ACM Symposium on Applied Computing (SAC'03), pp. 346–350. ACM, New York (2003). doi:10.1145/952532.952601
3. Wang, K., Stolfo, S.J.: Anomalous payload-based network intrusion detection. In: Recent Advances in Intrusion Detection, pp. 203–222. Springer, Berlin (2004). doi:10.1007/978-3-540-30143-1 11
4. Wang, K., Cretu, G., Stolfo, S.J.: Anomalous payload based worm detection and signature generation. In: Valdes, A., Zamboni, D. (eds.) Proceedings of the 8th International Conference on Recent Advances in Intrusion Detection (RAID'05), pp. 227–246. Springer, Berlin (2005). doi:10.1007/11663812 12
5. Wang, K., Parekh, J.J., Stolfo, S.J.: Anagram: a content anomaly detector resistant to mimicry attack. In: Zamboni, D., Kruegel, C. (eds.) Proceedings of the 9th International Conference on Recent Advances in Intrusion Detection (RAID'06), pp. 226–248. Springer, Berlin (2006). doi:10.1007/11856214 12
6. Perdisci, R., Ariu, D., Fogla, P., Giacinto, G., Lee, W.: McPAD: a multiple classifier system for accurate payload based anomaly detection. Comput. Netw. **53**(6), 864–881 (2009). doi:10.1016/j.comnet.2008.11.011
7. Tartakovsky, A.G., Rozovskii, B.L., Blaek, R.B., Kim, H.: Detection of intrusions in information systems by sequential change-point methods. Stat. Methodol. **3**(3), 252–293 (2006). doi:10.1016/j.stamet.2005.05.003. ISSN 1572-3127
8. Maggi, F., Robertson, W., Kruegel, C., Vigna, G.: Protecting a moving target: addressing web application concept drift. In: Kirda, E., Jha, S., Balzarotti, D. (eds.) Proceedings of the 12th International Symposium on Recent Advances in Intrusion Detection (RAID'09), pp. 21–40. Springer, Berlin (2009). doi:10.1007/978-3-642-04342-02
9. Abd-Eldayem, M.M.: A proposed HTTP service based IDS. Egypt. Inf. J. **15**(1), 13–24 (2014). doi:10.1016/j.eij.2014.01.001. ISSN 1110-8665
10. Norris, J.R.: Markov Chains. Cambridge University Press (1998). https://en.wikipedia.org/wiki/Markov_chain
11. The NSL-KDD data set. http://nsl.cs.unb.ca/NSL-KDD/. Accessed 1 Oct 13
12. Ariu, D., Tronci, R., Giacinto, G.: HMMPayl: an intrusion detection system based on hidden markov models **30**(4), 221–241 (2011). doi:10.1016/j.cose.2010.12.004
13. Open source vulnerability databse-OSVDB. http://osvdb.org/
14. Mitre's common vulnerability and exposure (CVE) (2009). http://cve.mitre.org/

Serious Game to Enhance and Promote Youth Entrepreneurship

Rachid Lamrani, El Hassan Abdelwahed, Souad Chraibi, Sara Qassimi, Meriem Hafidi and Abdelaziz El Amrani

Abstract Entrepreneurship constitutes a key driver providing jobs and wealth creation. Besides, promoting entrepreneurship can play significant role to ensure the prosperity of a society. In this respect our article aims to foster entrepreneurship education, through focusing on the effectiveness of serious games as educational tools, whose goal supporting young people to actively being entrepreneurs. In this paper, we unveil the entrepreneurial context notably at the national level, where we highlight the important role of entrepreneurship potential by presenting some initiatives. Moreover, we introduce the concept of serious games, as well as we outline some of their applications, but also their relation to entrepreneurship through an analysis of the existing. We conclude with a brief overview of our approach.

Keywords Entrepreneurial learning · Learning process · Serious games · Entrepreneurship education

R. Lamrani (✉) · E.H. Abdelwahed · S. Chraibi · S. Qassimi · M. Hafidi · A. El Amrani
Computer Systems Engineering Laboratory, Faculty of Sciences Semlalia,
Cadi Ayyad University, Boulevard of Prince My Abdellah B.P. 2390,
40000 Marrakech, Morocco
e-mail: rachid.lamrani@ced.uca.ac.ma

E.H. Abdelwahed
e-mail: abdelwahed@uca.ac.ma

S. Chraibi
e-mail: chraibi@uca.ac.ma

S. Qassimi
e-mail: qassimi.sara@gmail.com

M. Hafidi
e-mail: meriem00@gmail.com

A. El Amrani
e-mail: elamrani.abdelaziz@yahoo.fr

Á. Rocha et al. (eds.), *Europe and MENA Cooperation Advances in Information and Communication Technologies*, Advances in Intelligent Systems and Computing 520, DOI 10.1007/978-3-319-46568-5_8

1 Introduction

Entrepreneurship, constitutes a main lever of local development, economic growth and social stability for a country, allows to enlarge and increase new employment opportunities which provide great business potential. Therefore, the majority of countries gave it great importance. Morocco, one of them, aligned in the same direction, owed on one hand to the constraints of the employment market (public and private) who is unable to satisfy all the needs of recruitment, whose number is in constant increase, but also, due to the desire for reducing the unemployment rate [1].

To remedy this, we are interested in studying one of several solutions, could promote entrepreneurship. A solution, date since a long time, but, constitutes until instant, one of the most active research domains in EIAH. It's the use of games, where the integration of a so-called intention "serious", in order to train, to lead, recruit, to simulate … [2–5]. Serious games, provokes an importance and immense result [5, 6], with purpose, creating a context of entertainment and lucidity, then, the user could benefit from an air entertaining, however, purpose doesn't stop any more in this point, but also includes and inserts the capacity to advance, to take up challenges, to augment its personal strategically level, … [7, 8].

"Serious games" is part of a broad framework, we focus our study to their uses in a learning context, which we benefit from the acquisition of new knowledge's, a learning encourage and deep enough, and therefore succeed to improve short-comings and surpass problems of demotivation [9, 10].

The present contribution focuses on the use, the exploitation of "serious games", their proven interest, in particularly the entrepreneurship field. First of all, we present, the concept of "serious games", their efficiencies, and their areas of applications. On the other hand, we present a theoretical framework, where we highlight, the context of the entrepreneurship generally, the importance of spiritual development to undertake, precisely at national level, etc. By the suite, we specify across a study of the existent, the contribution of the "serious games" in entrepreneurship. At the end, we describe our approach.

2 Serious Games, Playfulness with Serious Purpose

Attainment of the objectives, requires a good method to use and follow, and through it, the motivation and envy increases, therefore we could succeed in what we wanted to.

In our article, we're talking about using games, particularly "serious games". Thus, if we only take the game concept solely, it proves a great importance for the development of our personality, in addition, it's clear that if we prevents a child from playing, we strongly risk to come out with an incomplete personality, given the great importance of play in the development of reflection expression, cooperation, trade and imagination … [11, 12]. Otherwise, the game is a natural way to learn. Starting, therefore, from this point, researchers began for a vision to change and improve learning and make it easier and fun, further, the use of gaming dynamics applied to corporate learning.

Serious games, a category combining the fun aspect, entertainment and serious intent. It combines the strength of the game themselves, the environment involving this game and professional settings, to achieve desired goals. Furthermore, it's constitute a blended of applications developed using advanced gaming technologies, having common, design approaches and expertise as the classical game (real-time 3D; simulation of objects, people, environments …).

Their first applications returned to military purposes where the training of militants to further develop their capacities, as confirmed by the great success of America's Army, multiplayer shooter developed by the US Army [13, 14], but afterwards, they were across various fields. For example, at the medical area, multitude of serious games have been designed, among them, Pulse!, serious game, allowing nurses and doctors training [15]. SIMUrgences, a real-time 3D module, enable doctors' training to care of patients in cardiac emergencies [16]. Moreover, many others serious games developed concerning pedagogical purposes, notably Mathematical sciences, Chemistry, Music education, … [4, 17–19].

Serious games, constitute also, a significant tool to enhance youth awareness, the example of 2025 ex machina, which aims to provide a critical view for the use of fixed and mobile Internet among adolescents [20].

In contrast to serious games, Business games take the form of games set play situations to discover the different facets of Business activity, thus, Oreal Business game, ranks among the best known examples [21].

A state of the art has been already presented, where we recall gamification as well, we reveal serious game concept, their potential and applications, and we highlight our proposed idea [22].

It's true that the use of serious games differs from user to another, but through a well-designed SG, the user doesn't waste time just for the game pleasure but with establishing desired goals.

3 Entrepreneurship

Entrepreneurship since its origins, were observed and analyzed by sociologists, historians, psychologists, economists … including Joseph Schumpeter, an Austrian economist, that through these works, was able to develop and to create an impressive richness, but also with other researchers, which situated it, as a factor in economic prosperity [23].

A question so simple, yet so difficult at the same, it's to propose a precise definition of entrepreneurship, seen that it includes sectors and different realities [24]. Therefore, literature has a large number and great wealth. By simplifying its meaning, we could say that's constitute the process to lead a project in all its dimensions, teamwork, cooperation, face challenges, solve problems, make profits and ensure survival in the long term. ….

Thus, the most developed countries in the world, is those promoting entrepreneurship among its population even more, through a set of values building entrepreneurship spirit. As example, the sense of initiative, of responsibility …

Also, education play a key role in the enhancement of entrepreneurial culture, and for this purpose, we mention some projects have substantially interested in this regard including, at primary level [25], in high school [26, 27], at University [28] as well as, an entrepreneurship doctoral education [29] …

The growth rate in Morocco remains insufficient to be in charge of the large numbers of job seekers who arrive every year, in fact, the official unemployment rate now stands at 10.1 % at end-September 2015, against a 9.6 % year earlier, based on the report issued by the High commission for Planning [30].

By this reason, entrepreneurship remains a major lever for the employability of young people in Morocco. In this respect, the role of some public and private initiatives launched reached, through reform a broad strategy in order to stimulate entrepreneurship. Maroc Entrepreneurs, a nonprofit organization that aims to contribute to Morocco's economic development through the main levers: make discover the world of entrepreneurship and socio-economic issues of Morocco, also set a synergy between companies based in Morocco and the Moroccan expertise abroad [31]. Beyond this association, other events were based on the entrepreneurship ecosystem in Morocco, giving as an example the conference organized by Attijariwafa Bank Foundation [32].

INJAZ Al-Maghrib, a recognized association of public utility, contributing to the emergence of a new generation of entrepreneurs. INJAZ Al-Maghrib adapts in Moroccan context, programs bringing learners to create a junior company, in order to prepare for the challenges of working life [33].

Cadi Ayyad University, in collaboration with the University of Portland (USA) organized two events in recent years, where the last was in March 2016, entitled the 2nd edition of the Marrakech International Entrepreneurship Event, an event on entrepreneurship in academia [34].

By way of information, these examples are cited in order to favor and perceive the situation of entrepreneurship in Morocco, having regard, it's has a remarkable importance, and thus implicitly indicate why we have chosen to continue at that sense.

4 Related Work

During this section, we mention some projects promoting entrepreneurial learning:

[35] Proposes to develop a comprehensive overview of relevant SGs available on the market, and identify through expert analysis, key benefits and issues related to their teaching adoption of entrepreneurship, particularly for scientific university learners.

[36] Describes a serious game, aimed at fostering an entrepreneurial spirit among youth. It's dedicated for a business tourism management in a complex market, then players have to compete with other companies. Regarding the technical tools upon they were based on, a Bayesian network, which provide a probabilistic explanation of the markets, starting from the strategic choices and structure of different groups. In addition, a data warehouse was created to analyze the simulation data, it can be accessed using Saiku, an open source tool.

[37] Emphasizes on those issues: market segmentation and market dynamics; financial and organizational management… The game titled PNP Village, divided into seven levels and in each level learners will acquire new business concepts.

Explorer "Serious Game for Immersive Entrepreneurs" an online serious game, which provides to learners, the opportunity to develop entrepreneurial skills, in order to manage their own companies in the future [38]. Similarly, SimVenture, serious game of corporate business, which allows learners to create and run a virtual company and therefore, develop an entrepreneurial spirit [39].

Practice plays a very interesting role for the development of learner's personality and creativity, however we note that practice without theory is blind and theory without practice is powerless, likewise, serious games remains as a solution among others, to practice, promote and improve entrepreneurial skills.

5 Proposed Approach

5.1 Overview

Our primary goal is to achieve a serious game, promoting an entrepreneurial spirit among youth. It constitute an environment which facilitates the acquisition of skills through play. In fact, a game that allows learners to leave the academic context and adopt realist practices.

Before we talking about neither a management, nor prescribed tasks, learners must first define innovative ideas, aimed to choose one to continue with.

What we mean by defining the project, reside in the choice of name, a meaningful slogan, staff recruitment, … and therefore, learner's must necessary taking into account, precautions in choices. Thus, starting from this point that the role of the teacher, the notes generated in parallel taking importance. In addition, the game will mention concepts clarifications, as like how to choose a good slogan, recruitment principles (that's important to clarify the enterprise needs, the overall operation of each position, conditions related each post …). Once the learners define the general context and understand the role of each position and function, the next phase, is to distribute the duties and tasks between them.

Subsequently, from the process of reflection and acquisitions developed throughout gaming, learners while collaborating together, have to achieve a Business Model.

Table 1 Key features of our approach

Target users	Covered areas	Synthesis
Youth wanting to promote the spirit of entrepreneurship	Entrepreneurship concepts (market study, business plan …). A context to develop good Communication skills, risk-taking, leadership, time management …	Game support learners, to develop capacities to confidently manage their business

Market research, an indispensable step for any business creation, it's serve as the foundation for all marketing decisions. Thereof, it's resides in several inseparable phases, first determine the nature of the market (geographical size, growth rate …), identify customers (determine their needs, their desires …), competition analysis (identify differentiation factors, their financial statistics …). However, learners will pass through all these steps, to ensure that the project can be commercially feasible.

Following the market research, learners should be able to write a Business Plan, includes costed and operational description of the business model, and describing the business vision, strategy and the expected profitability.

Any business must have capital to grow and develop, and therefore, during this stage learners understand, what role the investment is.

At this level, the game effectively begins, where learners exploit business. They are interested in creating products and their delivery to customers.

By the way, after each phase, the learner should prove what he learned, through an evaluation (QCM as example).

We present a table summarizes the main features (Table 1):

5.2 *Serious Game Model*

Above, we propose our serious game model, composed of several levels: *Educational goals*, a model of educational goals that the learner must be able to demonstrate during the game, moreover, it represents the capabilities which must possess all after the educational activities.

Learners manage their business, then each of them will take decisions, while keeping alive the cooperative spirit. Accordingly, we will attempt to enrich the *learner model* (Monitoring learner progress …).

Taking into account the learner model, and according to the educational goals (Educational goals), a *generation model*, whose function, generating scenarios (suites of activities) tailored to the needs of learners.

An *adaptive model*, consist of a set of rules leading to dynamically change the game behavior, this last, regarded in some points: the display (presentation), content (changing the game difficulty) …

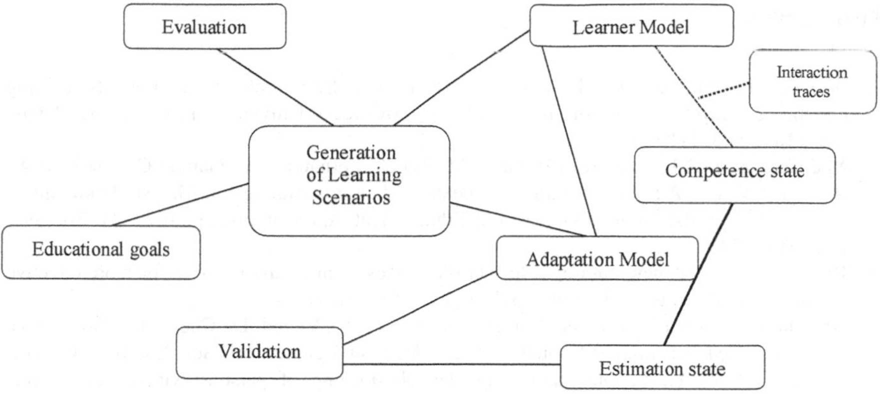

Fig. 1 Serious game model definition, illustrating our approach

During gameplay, we will try each time to assess the level of learning, starting from several types of questions and interactive exercises, could ensure the achievement of objectives.

In terms of (*model validation*), we will lead to interpret the results, in order to, generate an estimate of success chance or failure (*state estimation*) (Fig. 1).

6 Conclusion

In order to realize Morocco's entrepreneurial potential, we must ensure on the support and encouragement of young people to discover their creative talent, to develop their spirit of innovation and to inspire them to undertake. Initiatives, began in this sense, whose purpose, the identification, promotion, encouragement of competences, as well as the dissemination of entrepreneurial values among youth. Therefore, we decided to orientate our researches in that vein, by studying what impact can have the introduction of serious games on entrepreneurship. Thus, this article constitute the first step in our approach, where on the one hand, we have shown and clarified the interest of serious games generally, their contributions in various areas, notably regarding learning context. On the other hand, we tried to introduce the concept of entrepreneurship generally, highlight how important is, especially in Morocco. The study of the existing, a crucial step where we examine the impact of serious games to youth practical situations, including a purpose, having aware best practices of entrepreneurship, starting from the analyses of some projects. Lastly, we proposed and described our approach. As a perspective, we plan subsequently to implement it, then propose it as a tool, helping to grow tomorrow's entrepreneurs.

References

1. Ayegou, J., Mahrek, F., Rajraji, A., Talbi, M.: Self-employment: towards making entrepreneurship teaching more beneficial at the Moroccan University. Procedia Soc. Behav. Sci. **116**, 3410–3416 (2014)
2. Martínez-Durá, R.J., Arevalillo-Herráez, M., García-Fernández, I., Gamón-Giménez, M.A., Rodríguez-Cerro, A.: Serious games for health and safety training. In: Ma, M., Oikonomou, A., Jain, C.L. (eds.) Serious Games and Edutainment Applications, pp. 107–124. Springer, London (2011)
3. Raybourn, E.M.: A new paradigm for serious games: transmedia learning for more effective training and education. J. Comput. Sci. **5**, 471–481 (2014)
4. Annetta, L., Lamb, R., Minogue, J., Folta, E., Holmes, S., Vallett, D., Cheng, R.: Safe science classrooms: teacher training through serious educational games. Inf. Sci. **264**, 61–74 (2014)
5. Kwon, J., Lee, Y.: Serious games for the job training of persons with developmental disabilities. Comput. Educ. **95**, 328–339 (2016)
6. Bozanta, A., Kutlu, B., Nowlan, N., Shirmohammadi, S.: Effects of serious games on perceived team cohesiveness in a multi-user virtual environment. Comput. Hum. Behav. **59**, 380–388 (2016)
7. Ma, M.: Introduction to serious games development and applications. Entertain. Comput. **2**, 59–60 (2011)
8. Freitas, S., Liarokapis, F.: Serious games: a new paradigm for education? In: Ma, M., Oikonomou, A., Jain, C.L. (eds.) Serious Games and Edutainment Applications, pp. 9–23. Springer, London (2011)
9. Marfisi-Schottman, I.: Méthodologie, modèles et outils pour la conception de Learning Games. Thèse Dr. En Inform. Lyon Fr. INSA Lyon. **339** (2012)
10. Mitgutsch, K.: Serious learning in serious games. In: Ma, M., Oikonomou, A., Jain, C.L. (eds.) Serious Games and Edutainment Applications, pp. 45–58. Springer, London (2011)
11. Rossignol, F.A., Moal, C.L., Griffoul, K., Soler, V., Marzouk, V.: Jouer et apprendre en maternelle. Réseau Canopé. **189** (2015)
12. Métra, M.: Le jeu dans le developpement affectif, cognitif, corporel et social de l'enfant. UFAIS Lyon. **9** (2006)
13. Brougère, G.: Le jeu peut-il être sérieux? Revisiter Jouer/Apprendre en temps de serious game. Aust. J. Fr. Stud. **2**, 117–129 (2012)
14. Djaouti, D., Alvarez, J., Jessel, J.-P., Rampnoux, O.: Origins of serious games. In: Ma, M., Oikonomou, A., Jain, C.L. (eds.) Serious Games and Edutainment Applications, pp. 25–43. Springer, London (2011)
15. Pulse Serious Game. http://serious.gameclassification.com/FR/games/1017-/index.html
16. SIMUrgences. http://goo.gl/vzcugz
17. Hapgood, M.P.J.: Zombie Division: Intrinsic Integration in Digital Learning Games. 2005 Hum. Centred Technol. Workshop Brighton UK, pp. 45–48 (2005)
18. Shui, L.: A serious game designed for senior high school students chemistry study. In: IEEE International Games Innovations Conference in Vancouver, BC, pp. 236–240 (2013)
19. Baratè, A., Bergomi, M.G., Ludovico, L.A.: Development of serious games for music education. J. E-Learn. Knowl. Soc. **2**, 89–104 (2013)
20. 2025EXmachina. http://www.2025exmachina.net/
21. Fedbusiness. http://www.fedbusiness.fr/actualites-fed-business/business-et-serious-games-jouez-le-jeu
22. Lamrani, R., Abdelwahed, E.H.: Learning through play in pervasive context : a survey. In: 12th ACS/IEEE International Conference on Computer Systems and Applications AICCSA (2015)
23. Donnellon, A, et al.: Constructing entrepreneurial identity in entrepreneurship education. Int. J. Manag. Educ. (2014)
24. Jarniou, C.L.: Le Grand Livre de l'Entrepreneuriat (2013)

25. Huber, L.R., Sloof, R., Praag, M.V.: The effect of early entrepreneurship education: evidence from a field experiment. Eur. Econ. Rev. **72**, 76–97 (2014)
26. Elert, N., Andersson, F.W., Wennberg, K.: The impact of entrepreneurship education in high school on long-term entrepreneurial performance. J. Econ. Behav. Organ. **111**, 209–223 (2015)
27. Sánchez, J.C.: The impact of an entrepreneurship education program on entrepreneurial competencies and intention. J. Small Bus. Manag. **51**, 447–465 (2013)
28. Başçı, E.S., Alkan, R.M.: Entrepreneurship education at universities: suggestion for a model using financial support. Procedia Soc. Behav. Sci. **195**, 856–861 (2015)
29. Brush, C.G., Duhaime, I.M., Gartner, W.B., Stewart, A., Katz, J.A., Hitt, M.A., Alvarez, S.A., Meyer, G.D., Venkataraman, S.: Doctoral education in the field of entrepreneurship. J. Manag. **29**, 309–331 (2003)
30. Leconomistemaghrebin (2015). http://www.leconomistemaghrebin.com/2015/11/10/maroc-taux-chomage-officiel/
31. Maroc Entrepreneurs. http://www.marocentrepreneurs.com/
32. Attijariwafa Bank Foundation Conference. http://goo.gl/xzcmf4
33. Injaz Al-Maghrib. http://injaz-morocco.org/
34. Marrakech International Entrepreneurship Event. http://www.mieevent.com/
35. Bellotti, F., Berta, R., Gloria, A.D., Lavagnino, E., Antonaci, A., Dagnino, F., Ott, M., Romero, M., Usart, M., Mayer, I.S.: Serious games and the development of an entrepreneurial mindset in higher education engineering students. Entertain. Comput. **5**, 357–366 (2014)
36. Allegra, M., Guardia, D.L., Ottaviano, S., Grande, V.D., Gentile, M.: A serious game to promote and facilitate entrepreneurship education for young students. In: International Conference "Education and Educational Technologies" Rodhes Island, Greece, 16–19 July 2013. In: Long, C.A., Mastorakis, N.E., Mladenov, V. (eds.) Recent Advances in Education and Educational Technologies, pp. 16–19 (2013)
37. Gentile, M., La Guardia, D., Dal Grande, V., Ottaviano, S., Allegra, M.: An agent based approach to designing serious game: the PNPV case study. Int. J. Serious Games **2** (2014)
38. ENTRE EXPLORER. http://www.entrexplorer.com/
39. SimVenture. http://simventure.co.uk/

Educational Data Mining: A Literature Review

Carla Silva and José Fonseca

Abstract The adoption of learning management systems in education has been increasing in the last few years. Various data mining techniques like prediction, clustering and relationship mining can be applied on educational data to study the behavior and performance of the students. This paper explores the different data mining approaches and techniques which can be applied on Educational data to build up a new environment give new predictions on the data. This study also looks into the recent applications of Big Data technologies in education and presents a literature review on Educational Data Mining and Learning Analytics.

Keywords EDM · Prediction · Clustering · Relationship mining · Learning management systems

1 Introduction

A lot of research is going nowadays on the data-mining field. Educational Data Mining is a major research field also known as EDM. It aims at devising and using algorithms to improve educational results and explain educational strategies for further decision making. This paper discusses some of the data mining algorithms applied on education related areas. These algorithms are applied to extract knowledge from educational data and study the attributes that can contribute to maximize the performance. In fact, learning initially started in the class room and

Track2—ARTIFICIAL INTELLIGENCE IN EDUCATION Distributed Artificial Intelligence in education (DAIED) and Web-based AIED systems.

C. Silva (✉) · J. Fonseca
Centre of Technologies and Systems (CTS) of Uninova, Lisbon, Portugal
e-mail: silvacarla.uab@gmail.com

J. Fonseca
e-mail: jmrf@uninova.pt

Á. Rocha et al. (eds.), *Europe and MENA Cooperation Advances in Information and Communication Technologies*, Advances in Intelligent Systems and Computing 520, DOI 10.1007/978-3-319-46568-5_9

was based on behavioral, cognitive and constructivist models [1, 2]. Behavioral models rely on observable changes in the behavior of the student to assess the learning outcome. Cognitive models are based on the active involvement of teacher in the learning process. In the constructivist models, the students have to learn on their own from the available knowledge. A new termed "*Connectivism*" which is characterized as the "*amplification of learning, knowledge and understanding through the extension of personal network*" appeared in the recent years. According to Siemens, learning is no longer an internal, isolated activity [2, 3]. It is considered to be an act in a network of nodes which improves the learning experience of students and reduces the need for the direct involvement of a Professor. Actually, traditional learning environments have gradually mutated into community based learning environments [4].

1.1 Data Mining, a Concept and a Challenge

Educational data mining can be defined as "An emerging discipline concerned with developing methods for exploring the unique types of data that come from educational settings and using those methods to better understand students, and the settings which they learn in" [5]. EDM is the process of transforming raw data compiled by educational systems in useful information that can be used to take informed decisions and answer research questions. But the development of data mining and analytics in the field of Education was fairly late, compared to other fields. However, the challenge for educational data mining of online learning is due to its specific features on data. While many types of data have sequential aspects, the distribution of educational data over time has unique characteristics; for instance, a skill may be encountered many times during a school year, but separated over time and in the context of quite different activities [6]. Additionally, educational data mining methods have been successful at modeling a range of phenomena relevant to student learning in online intelligent systems and models are achieving better accuracy every year and are being validated to be more generalizable over time. There are important aspects that need to be discussed to justify the unique development for educational data, which is the growing realization that not all key information is stored in one data stream; the improvement in model quality, driven by continuing improvements in methodology and the importance in existence that there are more published examples of detectors than there are of detectors being used to drive intervention, like *Ellucian* [6, 7] which provided Professors with reports of whether students were at risk of dropping or failing a course, and scaffolded Professors in how to intervene, leading to better outcomes for learners. Research in education [8] has resulted in several new pedagogical improvements. Computer-based technologies have transformed the way we live and learn. Today, the use of data collected through [6] these technologies is supporting a second-round of transformation in all areas and learning with different achievements.

Data mining is a powerful new technology with great potential to help Schools and Universities focus on the most important information in the data they have collected about the behavior of their students and potential learners [9]. Data mining involves the use of data analysis tools to discover previously unknown, patterns and relationships in large data sets. These tools can include statistical models, mathematical algorithms and machine learning methods. These techniques are able to discover information within the data that queries and reports can't effectively reveal.

1.2 Literature Review

Many investigations have been carried out to demonstrate the importance of the "Data Mining" techniques in education, demonstrating that this is a new concept for the purpose of extracting valid and accurate information about the behavior and effectiveness in the learning process [10, 11].

In the field of education techniques "Data Mining" has also been used to analyze the curriculum and subject of the current research topics, as well as to analyze the students performance [12]. There have been several investigations made under this proposed study object. For example, Bhardwaj used the Naïve Bayes algorithm to predict student performance based on 13 variables [13]. The results were used to build a model that is used to predefine the students who are at risk of failure and thus activate a guidance and counseling program. Varghese et al. [14] in their research used the "K means" algorithm to cluster 8000 students based on five variables (input average in the University average scores of the tests/exams, average scores of papers, seminars notes and notes the work by frequency). The results showed a strong relationship between attendance and student performance. Gulati and Sharma [15] claim that knowledge through analysis by "Data Mining" can improve the education system in orientation, student performance and organizations management. Ayesha et al. [16] directed a study on evaluation, taking into account the evolution of learning and analysis of tests at the beginning and end of the courses. Bresfelean [17] conducted a study based on students' results and how ease of these can be provided. Cortez and Silva [18] conducted a research on the education system in Portugal and the results showed that a good and accurate prediction can be achieved. This is established by development tools that help improve the management of education in schools and the effectiveness of learning, which is a very important return. According to Sun [9], the result of the relationship between assessment and learning is an important tool to monitor and guide a quality education. Noaman and Al-Twijri [19] published a recent study applied to the entry requirements of the University of Saudi Arabia. They used algorithms and with techniques they have developed and a model that fits the public and the variables that describe it. They took into account input admission to the frequency of notes in previous education, admission notes and even the characteristics that describe the needs of the University. Some studies show the impact of the use of Moodle by applying Data Mining [20]. Sun [9] describes the different data mining techniques that can be applied to promote student learning on digital platforms. Aslam and Ashraf [21] used clustering algorithm to provide a model of student learning. Some investigations [22] discussed how data worked for Data improving the education system and enface knowledge in the classroom. Vince Kellen in his case study, described the implementation of a structured analysis tool for Data Mining—SAP's HANA at the University of Kentucky, which estimates a value "k-score" for each student. This value will determine the involvement and subsequent guidance for good student performance. Grafsgaard et al. [23] developed a system that recognizes facial expressions based on frustration or understanding of students in the classroom. They also used algorithms to detect unspoken behaviors and associate them to the knowledge acquired. Lee [24] describe also a record for the use of human behavior prediction models.

2 Approaches of Data Mining in Educational Data

Data mining is the field of computer science that aims to find out different potential factors and patterns to help decision making.

The model in Fig. 1 intends to design the Educational Data Mining. In this way, Data Mining can facilitate Institutional Memory. Data Mining [25], also popularly known as Knowledge Discovery in Databases, refers to extracting or "mining" knowledge from large amounts of data. An educational system typically has a large number of educational data. This data [26] may be students' data, teachers' data, alumni data, resource data, etc. EDM focuses on the development of methods for exploring the unique types of data that come from an educational context. These data come from several source, including data from traditional face-to-face class room environment, educational software, online courseware, etc.

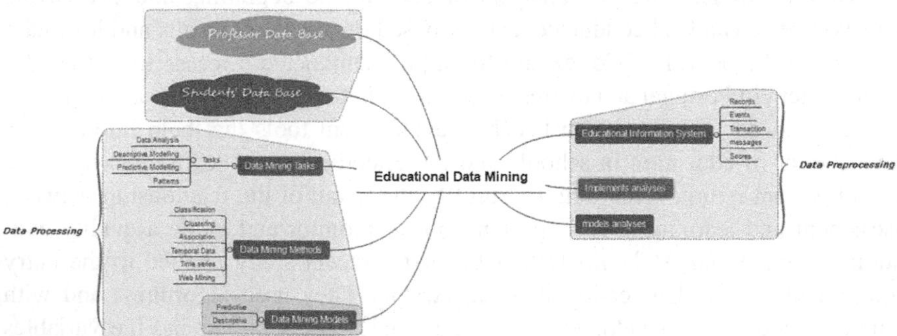

Fig. 1 Intelligent system model for educational data mining

Data mining techniques are used to operate on large volumes of data to discover hidden patterns and relationships helpful for decision-making. Various algorithms and techniques such as Classification, Clustering, Regression, Artificial Intelligence, Neural Networks, Association Rules, Decision Trees, Genetic Algorithm, Nearest Neighbor method etc., are used for knowledge discovery from databases.

2.1 *Clustering*

Clustering can be defined as the identification and classification of objects into different groups, or more precisely, the partitioning of a data set into subsets (clusters) so that the data in each subset (ideally) share some common trait of similar classes of objects (Fig. 2).

2.2 *Classification*

Classification models describe data relationships and predict values for future observations (Fig. 3). Classification is the task of learning a target function that maps each attribute set X to one of the predefined class labels Y. There are different classification techniques, namely Decision Tree based Methods, Rule-based Methods, Memory based reasoning, Neural Networks, Naïve Bayes and Bayesian Belief Networks, Support Vector Machines. In classification [26] test data is used to estimate the accuracy of the classification rules. If the accuracy is acceptable, the

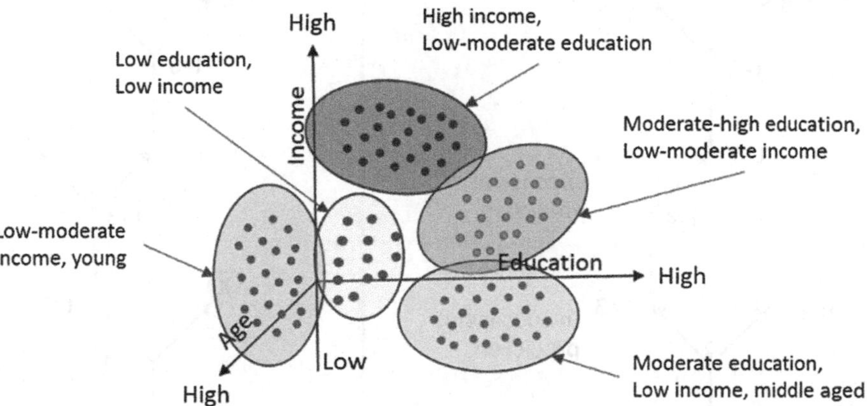

Fig. 2 Example of K means clustering using R ([27] Silva presented a cluster analysis of outcomes and incomes in education on Europe)

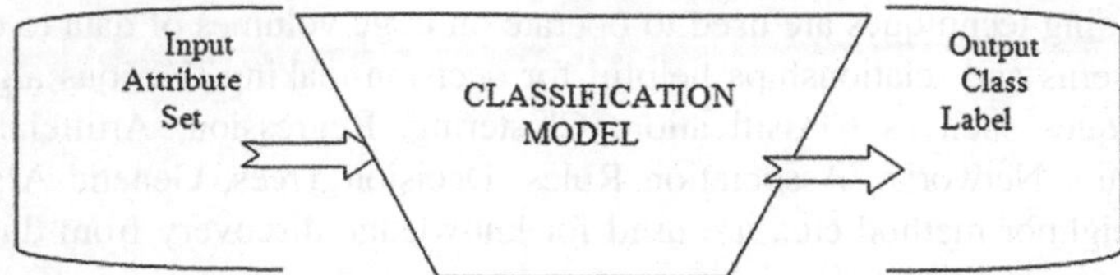

Fig. 3 Classification as a task

rules can be applied to the new data tuples. The classifier-training algorithm uses these pre-classified examples to determine the set of parameters required for proper discrimination.

2.3 Predication

Regression techniques (Fig. 4) can be adapted for predication [25]. Regression analysis can be used to model the relationship between one or more independent variables and dependent variables. In data mining, independent variables are attributes already known and response variables are what we want to predict.

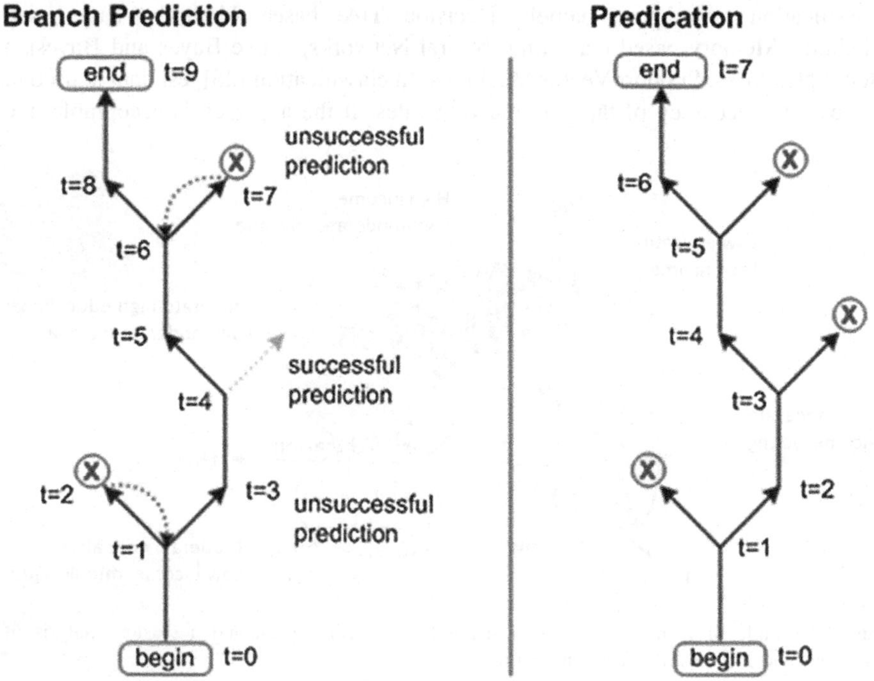

Branch Prediction
end t=9
unsuccessful prediction
t=8 t=7
t=6
t=5
successful prediction
t=4
t=2 t=3
unsuccessful prediction
t=1
begin t=0

Predication
end t=7
t=6
t=5
t=4
t=3
t=2
t=1
begin t=0

Fig. 4 Prediction as a task

Unfortunately, many real-world problems are not simply prediction. Therefore, more complex techniques (e.g., logistic regression, decision trees, or neural nets) may be necessary to forecast future values.

3 Future Scope

There are increasing research interests in using data mining in education. This new emerging domain, called Educational Data Mining, concerns with developing methods that discover knowledge from data originated from educational environments. Data mining is a tremendously vast area that includes employing different techniques and algorithms for pattern finding. This paper is just a simple review to this emerging field and aims to highlight the importance of its study.

References

1. Bell, F.: Connectivism: its place in theory-informed research and innovation in technology-enabled learning. Int. Rev. Res. Open Distance Learn. **12**(3), 98–118 (2011)
2. Siemens, G.: Connectivism: a learning theory for the digital age. Int. J. Instr. Technol. Distance Learn. **1**, 1–8 (2014)
3. Siemens, G.: Connectivism: a learning theory for the digital age **2000**, 1–15 (2004). http://www.elearnspace.org/Articles/connectivism. htm
4. Ertmer, P.A., Newby, T.J.: Behaviorism, cognitivism, and constructivism: connecting yesterday's theories to today's contexts. Perform. Improv. Q. **26**(2), 43–71 (2013)
5. Kumar, S.A., Vijayalakshmi, M.: A novel approach in data mining techniques for educational data. In: 3rd International Conference on Machine Learning Computing (ICMLC 2011) A, no. Icmlc, pp. 152–154 (2011)
6. Baker, R.S.: Educational data mining: an advance for intelligent systems in education. AI Educ. 78–82 (2014)
7. Baker, R.S., Inventado, P.S.: Educational data mining and learning analytics. In: Learning Analytics: From Research to Practice, pp. 61–75. Springer, New York (2014)
8. Sin, K., Muthu, L.: Application of big data in education data mining and learning analytics—a literature review. ICTACT J. Soft Comput.: Special Issue Soft Comput. Models Big Data **5**(4), 1035–1049 (2015)
9. Sun, H.: Research on student learning result system based on data mining. Int. J. Comput. Sci. Netw. Secur. **10**(4), 203–205 (2010)
10. Ramaswami, M., Bhaskaran, R.: A CHAID based performance prediction model in educational data mining. Int. J. Comput. Sci. Issues **7**(1), 10–18 (2010)
11. Ramaswami, M., Bhaskaran, R.: A study on feature selection techniques in educational data mining. J. Comput. **1**(1), 7–11 (2009)
12. Permata Alfiani, A., Ayu Wulandari, F.: Mapping student's performance based on data mining approach (a case study). Ital. Oral Surg. **3**, 173–177 (2015)
13. Kumar, V.: An empirical study of the applications of data mining techniques in higher education. Int. J. Adv. Comput. Sci. Appl. **2**(3), 80–84 (2011)
14. Varghese, J., Bindiya, M., Tomy, J., Poulose, U.A.: Clustering student data to characterize performance patterns. Int. J. Adv. Comput. Sci. Appl. 138–140 (2010)

15. Gulati, P., Sharma, A.: Educational data mining for improving educational quality. Int. J. Comput. Sci. Inf. Technol. Secur. **2**(3), 648–650 (2012)
16. Ayesha, S., Mustafa, T., Raza, A., Sattar, Khan, M.I.: Data mining model for higher education system. Eur. J. Sci. Res. **43**(1), 24–29 (2010)
17. Breşfelean, V.P.: Analysis and predictions on students' behavior using decision trees in weka environment. In: Proceedings of the International Conference on Information Technology Interfaces, ITI, pp. 51–56 (2007)
18. Cortez, P., Silva, A.: Using data mining to predict secondary school student performance. In: 5th Annual Future Business Technology Conference, vol. 2003, no. 2000, pp. pp. 5–12 (2008)
19. Al-Twijri, M.I., Noaman, A.Y.: A new data mining model adopted for higher institutions. Procedia Comput. Sci. **65**, 836–844 (2015)
20. Romero, C., Ventura, S., García, E.: Data mining in course management systems: moodle case study and tutorial. Comput. Educ. **51**(1), 368–384 (2008)
21. Aslam, S., Ashraf, I.: Data mining algorithms and their applications in education data mining. Int. J. **7782**, 50–56 (2014)
22. Rashan, A.P.K.H.: Data Mining Applications in the Education Sector (2011)
23. Grafsgaard, J.F., Wiggins, J.B., Boyer, K.E., Wiebe, E.N., Lester, J.C.: Predicting learning and affect from multimodal data streams in task-oriented tutorial dialogue. In: Proceedings of 7th International Conference on Educational Data Mining, pp. 122–129 (2014)
24. Lee, S.J., Liu, Y., Popović, Z.: Learning individual behavior in an educational game: a data-driven approach. In: Proceedings of 7th International Conference on Educational Data Mining, pp. 114–121 (2014)
25. Ranadive, F., Surti, A.Z.: Hybrid agent based educational data mining model for student performance improvement **4**, 45–47 (2014)
26. Swamy, M., Hanumanthappa, M.: Predicting academic success from student enrolment data using decision tree technique. Int. J. Appl. Inf. Syst. **4**(3), 1–6 (2012)
27. Silva, C.: Does education matter? Vocational education and social mobility strategies in young people of Barcelona and Lisbon. A comparative study. ULHT (2014)

Individualized Learning Path Through a Services-Oriented Approach

Mohamed Bendahmane, Brahim El Falaki and Mohammed Benattou

Abstract In the most learning systems, pedagogical activities are presented in a static way without considering the specifics or learner goals. However, customize the learning environment and adapt learning path to learners' profile improve learning quality. In our proposed system, we believe that learning path's individualization depends on collected traces' activities in learning environment. We propose in this paper a model-oriented services to offer to each learner an individualized learning path to acquire the targeted skills. The system will be implemented as composed services norms and standards.

Keywords Service-oriented architecture · Learning path · Web service · Individualization · Trace analysis · IMS LIP

1 Introduction and Context

In an E-learning environment, learners tend to have the same goal but they have different characteristics and predispositions to achieve it. Thus, an optimal learning path for one learner is not necessarily the same for the other [1]. Consequently, individualized learning path is inevitable. Consequently, individualized learning paths is crucial to manage learner differences. To achieve this individualization we propose a system based learners traces' analysis to regulate learning path.

M. Bendahmane (✉) · M. Benattou
Telecommunications Systems and Decision Engineering Laboratory,
Faculty of Sciences, Ibn Tofail University, Kenitra, Morocco
e-mail: med_bendahmane@yahoo.fr

M. Benattou
e-mail: mbenatou@yahoo.fr

B.E. Falaki
Computer Science Department, Mohammadia Engineering School,
Mohammed Vth University, Rabat, Morocco
e-mail: elfalaki.brahim@gmail.com

Á. Rocha et al. (eds.), *Europe and MENA Cooperation Advances in Information and Communication Technologies*, Advances in Intelligent Systems and Computing 520, DOI 10.1007/978-3-319-46568-5_10

Our proposal [2] consists to individualize learning path to a learners' capacities by implementing an orchestrated web component in a service-oriented architecture (SOA).

To enable reuse and operability, the environment will be designed according to standards, such as IMS-LD [3], IMS-QTI [4] and IMS-LIP [5].

The next section deals with individualization in e-learning. The two ensuing Sects. 3 and 4 concerns traces' analysis and modeling learner. The proposed system will be presented in Sect. 5, and we terminate with a conclusion and perspectives.

2 The Learning Individualization

The concept of individualization in E-learning has become an important field of research in recent years. Generally, individualization aims to adapt contents and services offered to the user to promote the quality of his interactions with the system [6]. Individualize the learning path in education field is providing each learner the feeling that the training is designed to meet their expectations, taking into account his limits and capabilities.

In our proposal, we believe that adaptation is based on the identification of the learner, his/her ability, prior knowledge and current performance for the acquisition of competencies. Thus, we stipulate that two elements are essential, namely a learner modeling and a relevant analysis of traces' activities.

In this perspective we propose a system based on learners' traces analysis. This analysis will conclude with a regulation of the learning path based on targeted skills and learner profile.

3 Trace Notion

In the literature, there are several names to describe the monitoring of online learners, among them we find the term "tracking" and "trace". The trace definition differs according to its context, the research area, its role and its use.

We present a table with different definitions of trace concept (Table 1).

In this work, we define a trace as a result of data exchange observations and interaction between the actor and the system [2].

Traces can be collected from several sources: videos, databases, XML or Log files.

Traces can be classified into three categories [14]: those relating to collaborative learning "social traces", "activities' traces" from activities proposed to learners and "cognitive traces" to assess knowledge and skills.

In our work, we will collect activities' traces from e-learning platform database for classification and analysis.

Table 1 Definitions of trace concept

Authors	Definition
Champin	Objects temporal sequence of and operations mobilized by the user when using the system is called trace of use [7]
Choquet and Iksal	A pedagogical object as well as the resources or educational scenarios [8]
Courtin and Mille	A feedback base for learning actors [9]
Cram	Any computer object in which accumulate data about interactions between a computer system and its user [10]
Jermann	An observation or a recording learner interaction with a system for analysis [11]
Pernin Settouti	Actors activities index in a learning situation whether or not instrumented [12] The trace is defined as an observed temporal sequence [13]

4 Modeling the Learner in Competence Based Approach (CBA)

In CBA, the learner is central and the learning environment must take into account his needs and expectations for the acquisition of a competency. To adapt pedagogical path to the needs of learners and follow their progress in e-learning platforms, we require collecting data on these learners [15].

The data relevant recognized, modes of production and operation have been the subject of numerous publications [16] that led to the learner model creation.

4.1 Definition

The learner model is a data structure (as defined in the computer) that characterizes the knowledge acquired by the learner [17]. The five main features shown in the user model are [18] (Fig. 1):

This model provides information on the environment to be adapted to each user and be refreshed explicitly.

In our contribution, the learner model is applied when analyzing learners' traces. It will contribute to the learning path regulation based on skills and learner profile. The learner model can be implemented using standard models. In our proposal, we adopt IMS Learner Information Package (LIP-IMS) specifications, which toe our vision that takes into account the CBA as a reference.

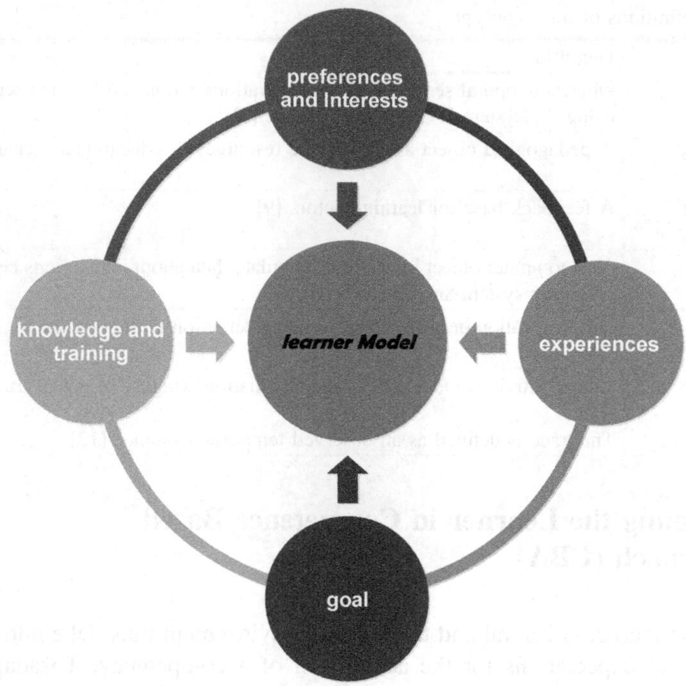

Fig. 1 The five main features shown in the user model

4.2 *Modeling Learner Using IMS-LIP*

It is necessary to record learner individual competencies in a persistent and standard way to ensure the provision of competence-based learning services. It facilitates also the interaction with the learner. Thus, the learner can find learning activities that meets his needs to achieve the desired competencies.

In our proposed system, we adopt the IMS-LIP specifications. IMS LIP (Learner Information Package) is defined in a XML structure (Fig. 2) for the exchange of data between systems learners including learning management systems [19].

5 The Proposed System

The model proposed is based on reusability and autonomy of web services, all orchestrated in a services-oriented architecture (SOA). The idea is to offer services that exploit the learners activities traces to generate indicators. The interpretation of these indicators will enable teachers engineers to adapt the educational path scenarios to learners needs.

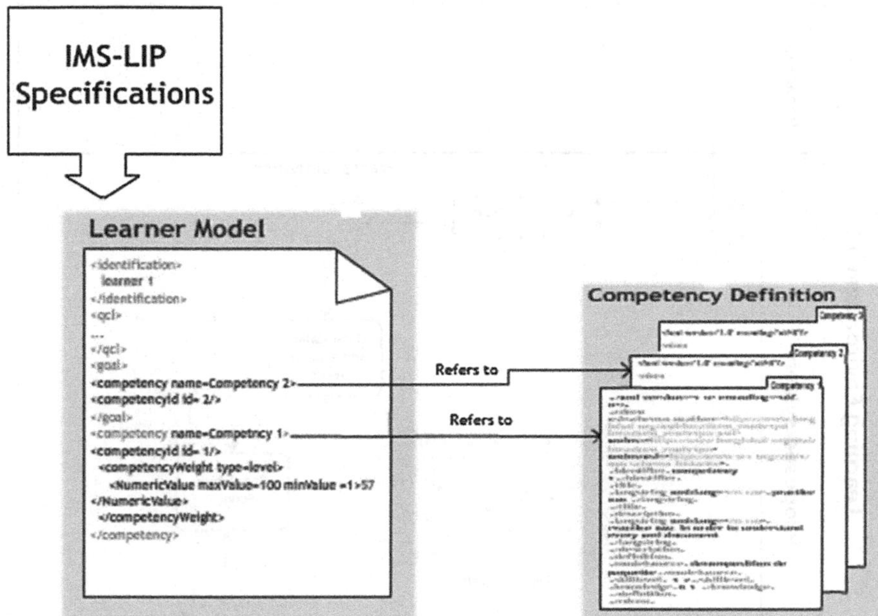

Fig. 2 Learner model using competency definition [20]

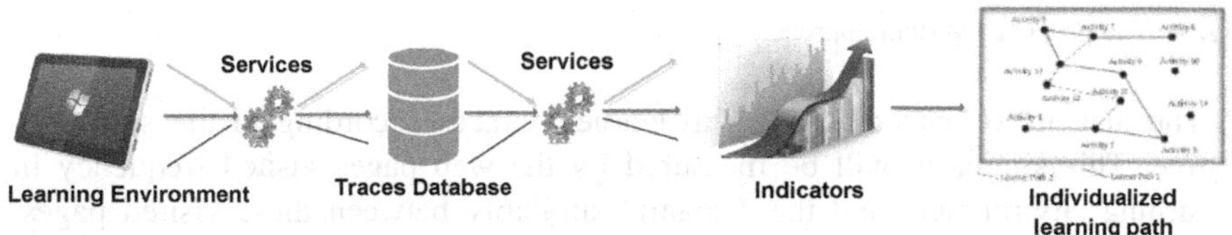

Fig. 3 Overview of suggested model

Figure 3 shows an Overview of suggested model.

According to the competency-based approach, the goal of each pedagogical scenario is to enable learners to acquire some skills. Therefore, the teacher prepares its course by offering a progression of activities according to targeted skills. These skills will be decomposed into criteria by the pedagogue.

Our model (Fig. 4) provides a "collection" service that is based on this criteria to collect learners activities' traces. These traces are usually stored in log files, XML files or databases according to the learning environment. Traces' extraction from log files are based on data format. For XML files or database, SQL language will be helpful.

The service "classification and analysis" aims to classify learners according to their preferences.

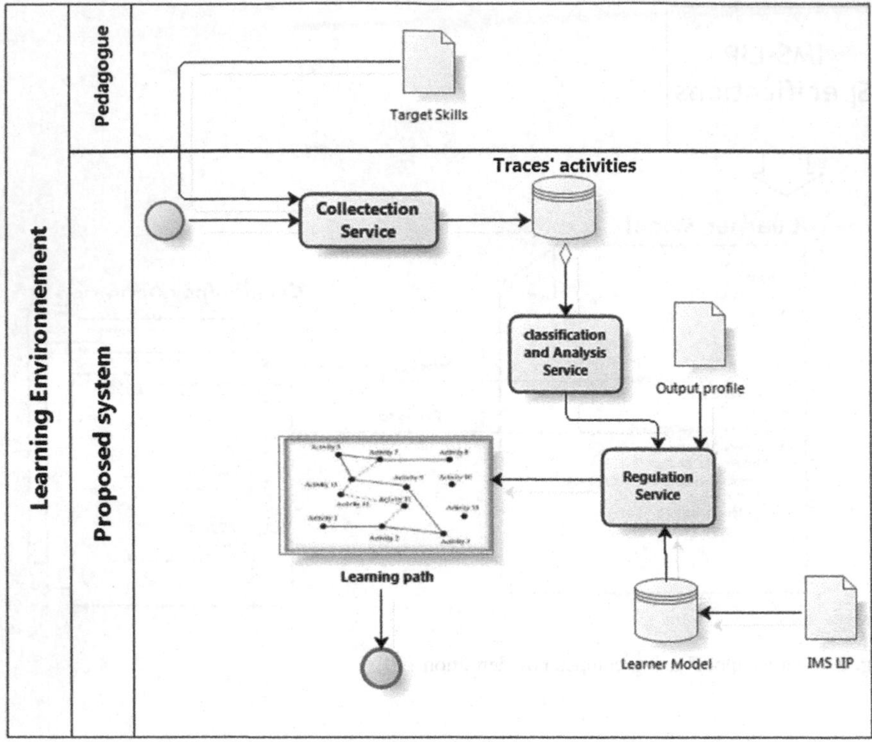

Fig. 4 The proposed system approach

The idea is to analyze and group learners' traces according to the similarity degree. This similarity will be measured by the web pages visited frequency in e-learning environment and the semantic similarity between these visited pages. These traces' analysis will consider the learner model based on IMS LIP and produce a set of performance indicators according to indicators' types desired by educator.

These indicators analysis will allow teachers to individualize learning path for each learner or learners group by changing the proposed learning activities' order. The "regulation" service will be in charge of this task.

6 Implementation

In Service oriented architecture the goal is to provide "an open Platform" for the development, deployment, interaction and management of distributed e-services. [21]. The model of web services [22], is defined as an architecture calling upon a set of standardized protocols (Fig. 5). The orchestration of services is carried out by IMS LD specifications.

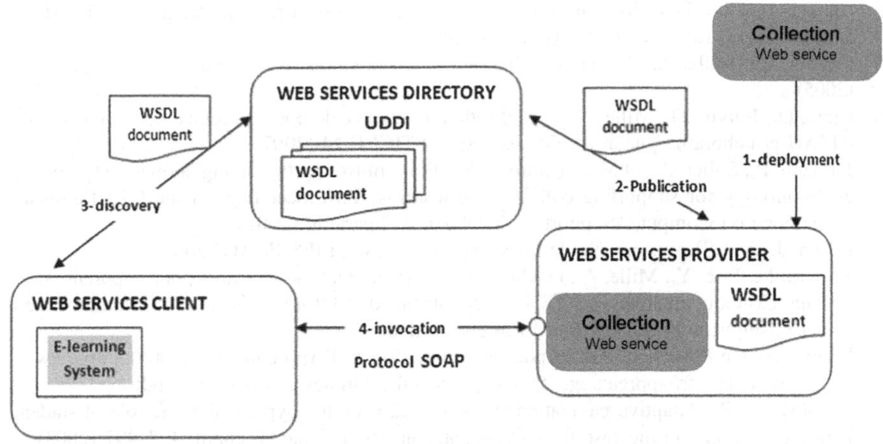

Fig. 5 Collection web service

7 Conclusion

One of the most important objectives in e-learning is to offer learners an adapted learning path to their needs. Several studies have focused on pedagogical individualization according to several ways. Our proposal is different, it is based on web services' independence and reusability to implement four components that are responsible for traces' collection, classification, analysis and exploitation. These traces are collected from learners activities.

In perspective, we expect experimentation and results' analysis to judge the suitability of our proposal.

References

1. Perrenoud, Ph: Construire des compétences dès l'école. ESF, Paris (2000)
2. Bendahmane, M., Elfalaki, B., Benattou, M.: Services-oriented model for the regulation of learning. Int. J. Soc. Behav. Educ. Econ. Bus. Ind. Eng. **10**(7) (2016)
3. IMS LD IMS. IMS Learning Design Information Model, IMS Global Learning Consortium, from http://imsglobal.org/learningdesign/
4. IMS Question and Test Interoperability: Version 2.1 public draft (revision 2) specification (2006). http://www.imsglobal.org/question/
5. IMS Learning Design Information Model, Version 1.0 Final Specification. http://www.imsglobal.org/learningdesign/ldv1p0/imsld_infov1p0.html#1529256
6. Stewart, A., Niederee, C., Mehta, B.: State of the art in user modeling for personalization in content, service and interaction. DELOS Report on Personalization, NSF (2004)
7. Champin, P.-A., Prié, Y., Mille, A.: MUSETTE: a framework for knowledge from experience. EGC'04, RNTI-E-2, Cepadues Edition, Clermont-Ferrand, France, pp. 129–134 (2004)

8. Choquet, C., Iksal, S.: Modeling tracks for the model driven reengineering of a tel system. J. Inter. Learn. Res. (JILR) **18**, 161Ŕ184 (2007)
9. Courtin, C., Mille, A.: Tracer pour interpréter les situations d'apprentissage avec les TICE (2005)
10. Cram, D., Jouvin, D., Mille, A.: Visualisation interactive de traces et réflexivité: application à l"EIAH collaboratif synchrone eMédiathèque. STICEF **14** (2007)
11. Jermann, P., Soller, A., et Muehlenbrock, M.: From mirroring to guiding: a review state of the art technology for supporting collaborative learning. In: Proceedings of the First European Conference on Computer-Supported Collaborative Learning (2001)
12. Pernin, J.-P.: CSE, un modèle de traitement de traces. CLIPS-IMAG (2005)
13. Settouti, L., Prié, Y., Mille, A., et Marty, J-C.: Système à base de traces pour l'apprentissage humain. Colloque international TICE, Technologies de l'Information et de la Communication dans l'Enseignement Supérieur et l'Entreprise (2006)
14. Diagne, F.: Un Modèle de Traces pour la Supervision de l'Apprentissage (2006). http://www.grappa.univ-ille3.fr/~ppreux/egc2006/actes/modelisation-des-connaissances.pdf
15. Brusilovsky, P.: Adaptive navigation support in educational hypermedia: the role of student knowledge level and the case for meta-adaptation. Br. J. Educ. Technol. 487–497 (2003)
16. Brusilovsky, P.: Adaptive hypermedia. User Model. User-Adap. Inter. **11**, 87–110 (2001). Kluwer Academic Publishers
17. Bruillard, E.: Les machines à enseigner. Hermès (1997)
18. Brusilovsky, P.: Methods and technique of adaptive hypermedia. User Model. User-Adap. Inter. **6**, 87–126 (1996)
19. CEN, Learning Technologies Standards Observatory (2009). http://www.cen-ltso.net
20. CEN/ISSS cwa 15455: A European model for learner competencies. ICS 03180; 35.240.99 (2005)
21. Rich Powers CSC 9010, Service Oriented Architecture, Spring (2008)
22. Zhuge, H., Liu, J.: Flexible retrieval of web services. J. Syst. Softw. (2004). ISSN 107-116

An Ontology to Assess the Performances of Learners in an e-Learning Platform Based on Semantic Web Technology: Moodle Case Study

Badr Hssina, Belaid Bouikhalene and Abdelkrim Merbouha

Abstract In this paper, we created a system to assess the skills of learners on an e-learning platform using semantic web technologies. Indeed, to supervise learners' activities in an e-learning platform is a major challenge for teachers and tutors alike. Our approach is based on standards that fall within the area of the semantic web as ontologies, the JENA API, and the SPARQL query language that aim to help a tutor in the monitoring of the activities of learners. The main idea behind our approach is that an ontology can be useful not only as a learning tool, but it can also be used to assess learners' skills.

Keywords E-learning · JENA · Ontology · OWL · Semantic web · SPARQL

1 Introduction

In this work, we will focus on how the semantic web [1], in particular ontologies [2], can improve the effectiveness of e-learning platforms [3] to satisfy their users. The approach of the Semantic Web adds metadata to Web resources that describe their contents and features, such metadata should be based on ontologies to be shared and provided with operational interpretations. Ontologies are one of the most important foundations of the Semantic Web approach for e-learning. Indeed, E-Learning, like all other Web services, can benefit from the new vision of the Semantic Web but mostly based on the potential of ontologies. The contributions of ontologies for e-learning systems are numerous, citing for example: need for

B. Hssina (✉) · A. Merbouha
TIAD Laboratory, Computer Sciences Department, Sultan Moulay Slimane University, FST, Beni-Mellal, Morocco
e-mail: hssina.badr@hotmail.fr

B. Bouikhalene · A. Merbouha
LMACS Laboratory, Mathematics Department, Sultan Moulay Slimane University, FST, Beni-Mellal, Morocco

Á. Rocha et al. (eds.), *Europe and MENA Cooperation Advances in Information and Communication Technologies*, Advances in Intelligent Systems and Computing 520, DOI 10.1007/978-3-319-46568-5_11

archiving and research [4], need for sharing [5], need for reuse of learning objects [6], indexing of education materials based on ontology [7], semantic enrichment of users' profile [8]. Using a semantic resource such as the ontology could be a way to enrich the learning data to respond more precisely to complex questions.

The aim of this work is to use the methods and semantic web tools for the exploitation of learner data to predict their performance in training. Ontologies now occupy an important place in the field of e-learning. As a matter of fact, formal and consensual nature of an ontology allows it to be spread in a community and helps make interoperable applications by explicitly representing the semantics of the data.

Our contribution is to create an ontology to annotate learning activities on an e-learning platform. Subsequently, we have populated our ontology manually. The language used for web development of ontology is OWL (Ontology Web Language). We have exploited the Jena API to execute SPARQL queries tutor.

2 Related Work

The use of ontology in e-learning particularly in IR (Information Retrieval) is one of the current issues of the Semantic Web, which aims to overcome the limitations of the search based on keywords. The first works date from the 1990 [9–11]; the use of ontologies for IR has really taken off with the advent of the Semantic Web, which advocates a semantic characterization of Web resources [12].

In IR systems using a model based on ontology, the index is built on the basis of the concepts present in the ontology and not based on words found in the documents. We are talking about conceptual indexing based on an ontology. Compared to an index based on the keywords, an index based on ontologies has two advantages [13]:

- The presentation of research results can be carried out according to the categories in the ontology;
- The formulation and refinement of queries can be based on a structured vocabulary provided by the ontology.

The use of ontologies in e-learning is about different works, such QBLS (Question Based Learning System) [14] that is an aid to solving issues of tutorials. It uses an educational ontology where the lesson is seen as a network card (definition, example, formalization, precision) and abstract resources (theme, idea, notion of courses) where each abstract resource refers to one or more files.

3 Semantic Web in e-Learning

The European Commission defines e-learning as: "the use of new multimedia technologies and the Internet to improve the quality of learning by facilitating access to resources and services, as well as exchange and remote collaboration".

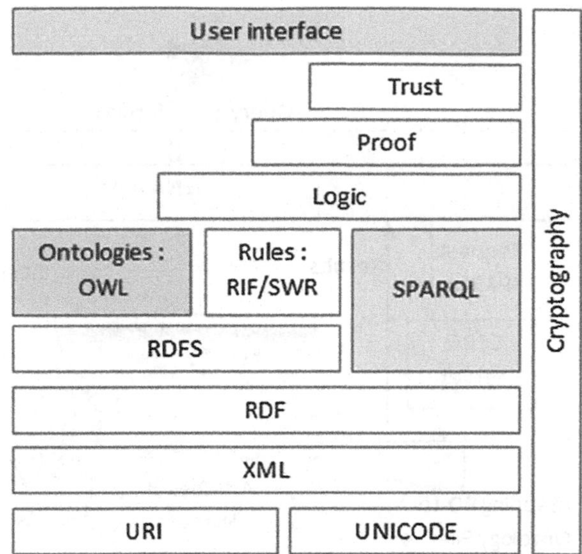

Fig. 1 The layered model of the semantic web

E-learning can benefit from the advantages provided by the semantic web like ontologies.

The architecture of the Semantic Web is based on a layered model proposed by W3C (Web standards body) [15], this model is shown in Fig. 1.

Thus, the Semantic Web can be treated as a suitable platform for implementing an e-Learning system, as long as it provides all means for the development of an ontology (learning), annotation based ontology learning materials, their composition courses and active delivery of courses through the learning portals.

4 Architecture of Our System

Our system is developed using two approaches as shown in Fig. 2. In the first we created ontology to annotate the activities of learners on an e-learning platform Moodle [16]. To run SPARQL queries on our ontology, we have populated our ontology manually. The web language used for developing ontology is OWL (Web Ontology Language) in the ontologies editor Protégé [17]. The tutor uses a graphical interface based on the Jena API to evaluate the work of learners.

The second approach is based on the mapping of the database to ontology. We converted our MySQL database of Moodle in Turtle format [18] using the D2RQ server [19].

An ontology is an explicit formal description of concepts in a domain [20]. For the semantic web, ontology is a structured body of knowledge that provides a common understanding, both structured and shared of a field or a task that can be

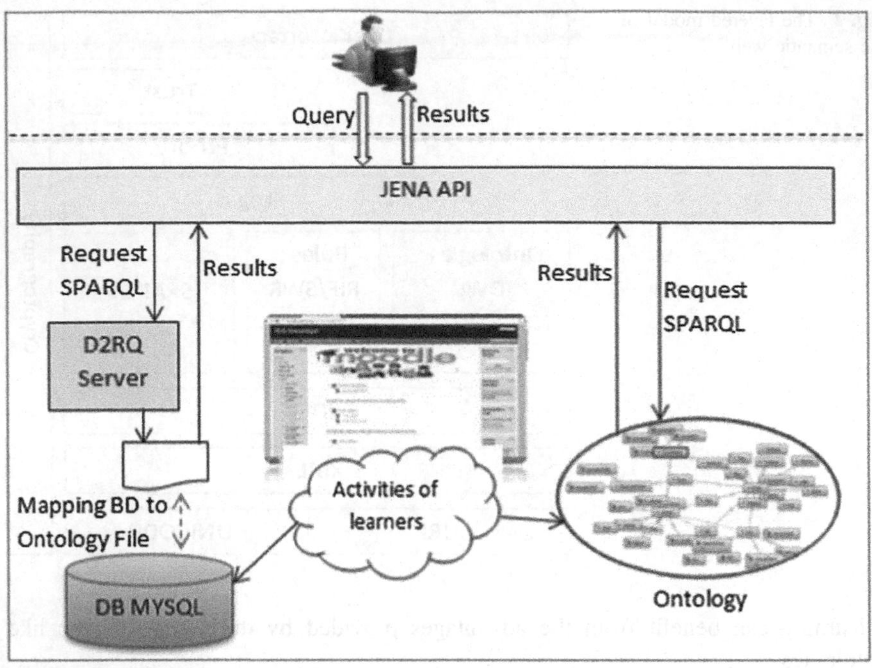

Fig. 2 Overall system architecture

used for communication between agents or to annotate semantic web resources. Compared to the thesaurus [21], ontologies are complex structures. In addition to the hierarchy of concepts based on relationships of hypernymy/hyponymy (Is-A), or meronymy (Part-Of) [22], they can contain any type of relationship found useful as well as constraints on the field concerned.

4.1 Creation of Our Ontology

The ontology we designed aims to make an explicit formal description of some modules of an e-learning platform (Moodle as a case study). Our ontology specifically addresses the elements that help the assessment of the learners' activities. It is therefore not intended to describe exhaustively the field of online learning environments and remains at a fairly high level description to be the most general possible, in order to represent all cases of operations by a user. Modeled classes are: Tutor Learners, Exam, Note, Wiki, etc.

The Fig. 3 shows the structure of our ontology developed under Protégé.

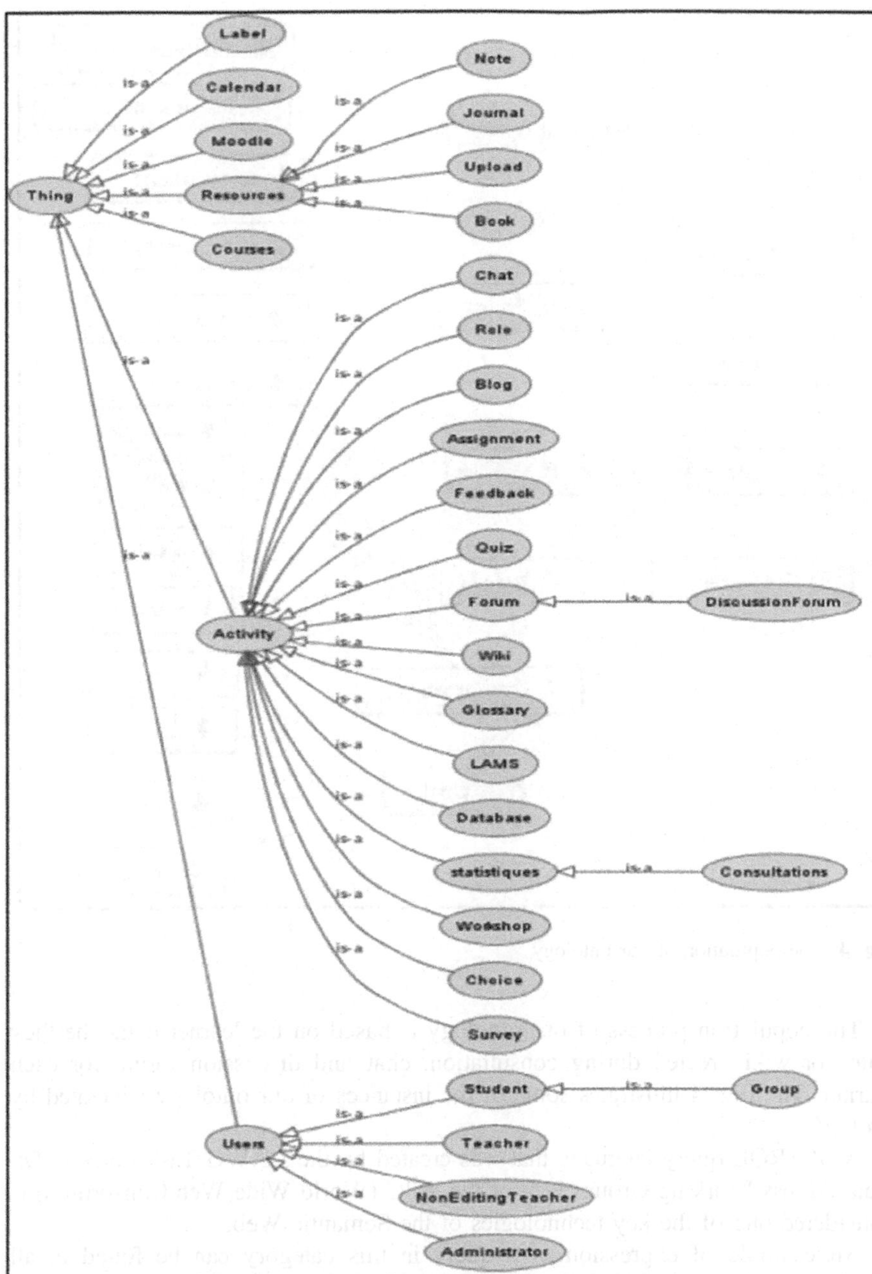

Fig. 3 The structure of our ontology

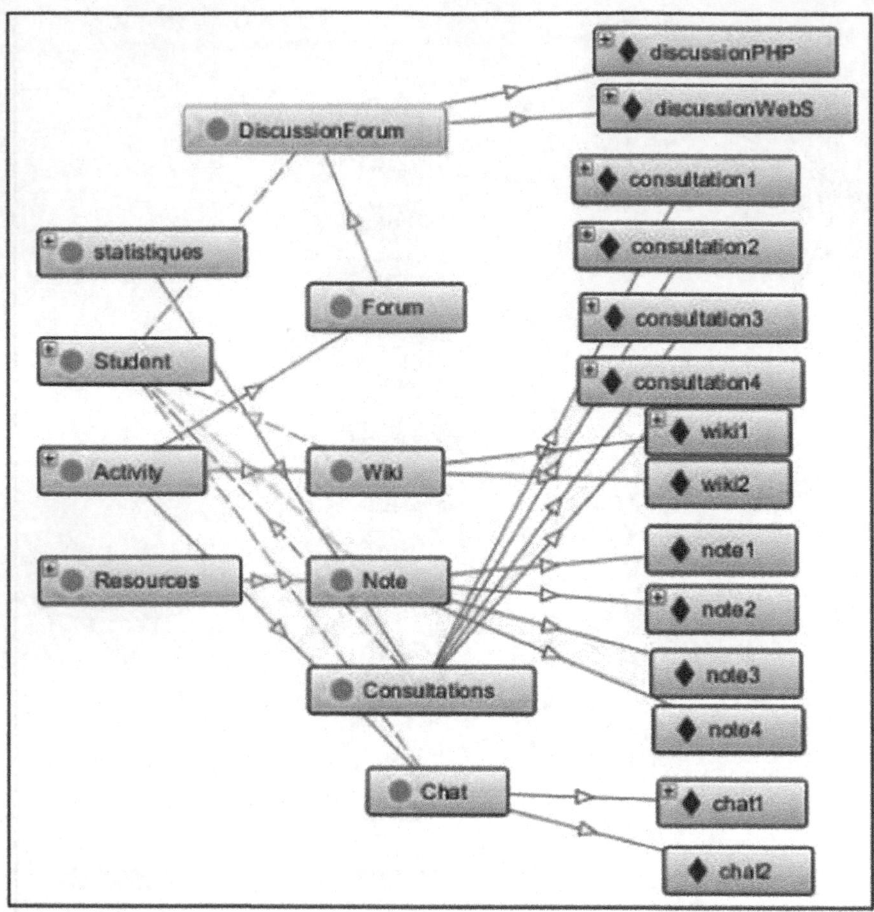

Fig. 4 The population of our ontology

The population process of our ontology is based on the learner data—be they notes or wiki created during consultation, chat and discussion forum for each learner. The Fig. 4 illustrates some of the instances of our ontology generated by Protégé.

A SPARQL query language that was created by the DAWG Task Force (RDF Data Access Working Group) [23] of the W3C (World Wide Web Consortium) is considered one of the key technologies of the Semantic Web.

An example of expression of a query in this category can be found in all documents related to the concept C and in all documents related to concept C′ linked to C by relation R.

Examples of SPARQL queries:

```
SELECT ?s ?max
   WHERE { ?s rd:nombreConsultation ?max}
   ORDER BY DESC(?max) LIMIT 1
   SELECT ?nombreConsultation
   WHERE { ?subject rd:has_consultation ?object.
   ?subject rd:nom ?nom.
   ?object rd:nombreConsultation ?nombreConsultation.
   filter(?nom = "sami")}
```

4.2 Mapping Data Base to Ontology

The D2RQ language of association is a declarative language for the association diagrams of a relational database to RDF Schema vocabularies and OWL ontologies. The association defines a virtual RDF graph that contains information from the database used (in this case MYSQL). This association is described in a mapping file to be generated by the user.

The D2RQ platform is a system for accessing relational databases. It offers access based on RDF for the content of relational databases.

4.3 Interface User JENA

Jena is an open source framework which is initiated initially by HP Labs [24]. Jena consists of a set of tools implemented in Java that provides:

- A programming API for managing data (RDF, RDF Schema, DAML + OIL and OWL) Semantic Web applications
- An AQL query language which is an implementation of the SPARQL language
- A relational structure for storing persistent data in RDF, RDF Schema, DAML + OIL and OWL
- A parser RDF/XML,
- An inference engine.

The database case can be explored using SPARQL (Fig. 5).

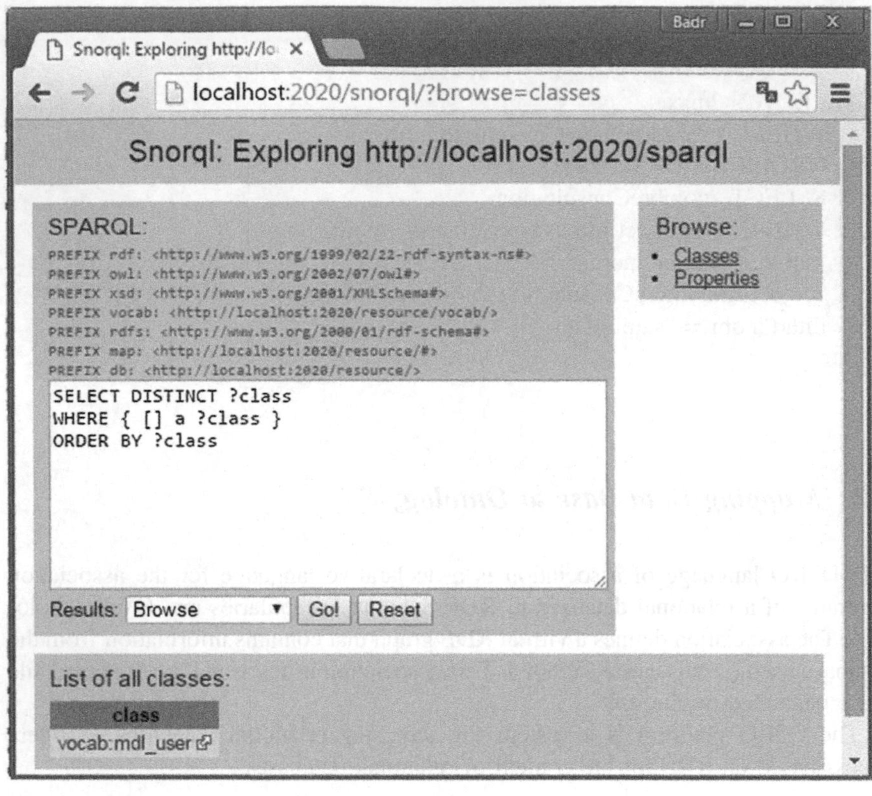

Fig. 5 D2R server running at http://localhost:2020

5 Conclusion

In this work, we propose this ontology to assess the performances of learners in an E-Learning Platform based on Semantic Web Technology. Ontology, which is most often described using OWL, may need to be represented in multiple ways depending on the use we want to do. The possibilities of expression of these different paradigms are different, the possibilities of expression of these different paradigms are different, to bring them closer we need to import new concepts in the world ontologies.

In this article we have become familiar with the different semantic web technologies namely OWL, RDF and SPARQL to annotate modules of e-learning platform with the view of assessing the learner's performance.

References

1. Berners-Lee, T., Hendler, J., Lassila, O.: The semantic web. Sci. Am. **284**(5), 28–37 (2001)
2. Maedche, A.: Ontology Learning for the Semantic Web, vol. 665. Springer Science & Business Media (2012)
3. Henze, N., Dolog, P., Nejdl, W.: Reasoning and Ontologies for Personalized E-Learning in the Semantic Web. Educational Technology & Society **7**(4), 82–97 (2004)
4. Garlatti, S., Prié, Y.: Adaptation et personnalisation dans le Web sémantique. Interaction, Intell. Inf. (2004)
5. Shadbolt, N., Hall, W., Berners-Lee, T.: The semantic web revisited. Intell. Syst. IEEE **21**(3), 96–101 (2006)
6. Pernin, J.P., Lejeune, A.: Dispositifs d'apprentissage instrumentés par les technologies: vers une ingénierie centrée sur les scénarios. In: Technologies de l'Information et de la Connaissance dans l'Enseignement Supérieur et de l'Industrie, pp. 407–414. Université de Technologie de Compiègne (2004)
7. Pernin, J.P., Lejeune, A.: Modèles pour la réutilisation de scénarios d'apprentissage. TICE Méditerranée, Nice (2004)
8. Bouzeghoub, M., Kostadinov, D.: Personnalisation de l'information: aperçu de l'état de l'art et définition d'un modèle flexible de profils. CORIA **5**, 201–218 (2005)
9. Devcndra, L.: Ontology driven e-learning Doctoral dissertation, Department of Mechanical Engineering, University of Moratuwa Sri Lanka (2007)
10. Snae, C., Brueckner, M.: Ontology-driven e-learning system based on roles and activities for Thai learning environment. Interdiscip. J. Knowl. Learn. Objects **3**(1), 1–17 (2007)
11. Gaeta, M., Orciuoli, F., Ritrovato, P.: Advanced ontology management system for personalised e-Learning. Knowl.-Based Syst. **22**(4), 292–301 (2009)
12. Sintek, M., Decker, S.: TRIPLE—a query, inference, and transformation language for the semantic web. In: The Semantic Web—ISWC 2002, pp. 364–378. Springer, Berlin (2002)
13. Uschold, M., King, M.: Towards a methodology for building ontologies, pp. 15–30. Artificial Intelligence Applications Institute, University of Edinburgh, Edinburgh (1995)
14. Dehors, S., Faron-Zucker, C., Stromboni, J.P., Giboin, A.: Des annotations Sémantiques pour apprendre: l'Expérimentation QBLS. Actes de la Journée thématique WebLearn, plate-forme AFIA, 31 (2005)
15. Fensel, D., Van Harmelen, F., Horrocks, I., McGuinness, D.L., Patel-Schneider, P.F.: OIL: an ontology infrastructure for the semantic web. IEEE Intell. Syst. **2**, 38–45 (2001)
16. Dougiamas, M., Taylor, P.: Moodle: using learning communities to create an open source course management system (2003)
17. Knublauch, H., Musen, M.A., Rector, A.L.: Editing description logic ontologies with the Protégé OWL plugin. In: International Workshop on Description Logics, vol. 49 (2004)
18. World Wide Web Consortium: OWL 2 web ontology language document overview (2012)
19. Bizer, C., Cyganiak, R.: D2r server-publishing relational databases on the semantic web. In: Poster at the 5th International Semantic Web Conference, pp. 294–309 (2006)
20. Lacher, M.S., Groh, G.: Facilitating the exchange of explicit knowledge through ontology mappings. In: FLAIRS Conference, pp. 305–309 (2001)
21. Moreira, A., Alvarenga, L., de Paiva Oliveira, A.: thesaurus and ontology: a study of the definitions found in the computer and information science literature, by means of an analytical-synthetic method. Knowl. Organ. **31**(4), 231–244 (2004)
22. Alfonseca, E., Manandhar, S.: Extending a lexical ontology by a combination of distributional semantics signatures. In: Knowledge Engineering and Knowledge Management: Ontologies and the Semantic Web, pp. 1–7. Springer, Berlin (2002)

23. Pérez, J., Arenas, M., Gutierrez, C.: Semantics and Complexity of SPARQL. In: International Semantic Web Conference, vol. 4273, pp. 30–43 (2006)
24. Grobe, M.: Rdf, jena, sparql and the 'semantic web'. In: Proceedings of the 37th Annual ACM SIGUCCS Fall Conference: Communication and Collaboration, pp. 131–138. ACM (2009)

Toward a Framework for Designing Adaptive Educational Hypermedia System Based on Agile Learning Design Approach

Amal Battou, Omar Baz and Driss Mammass

Abstract Adaptive Educational Hypermedia systems (AEHS) have provided new perspectives for access to information and enhance learning. However AEHS have advantages and positive impact on learning, there are still defies to their design and production. In this paper, we propose an Agile Learning Design method to support the design and production of AEHS. It is based on agile practices from software engineering and on practices of learning design. We illustrate our approach with an experiment that validates the proposed method through its application in the design of an AEHS called (ALD-AEHS) and creation of one of the most important component of AEHS: The User Model.

Keywords AEHS · Learning design · Instructional design · Agile learning design · Individual training · ALD-AEHS · User model

1 Introduction

Adaptive educational hypermedia was one of the first application areas for Adaptive Hypermedia [1]. The use of such environments, offers to learners personalized content, presentation, and navigation support adapted based upon various features such learner's data, usage data and environment data of individual learners.

The design and authoring of AEHS is more complicated than the design of regular educational hypermedia [1]. Furthermore, one of the challenges faced by developers of AEHS has been how to design and create quality and pertinent

---

A. Battou (✉) · O. Baz · D. Mammass
IRF-SIC Laboratory, IBN ZOHR University, B.P.28/S, Agadir, Maroc
e-mail: ambattou@gmail.com

O. Baz
e-mail: o.baz@uiz.ac.ma

D. Mammass
e-mail: mammas@univ-ibnzohr.ac.ma

Á. Rocha et al. (eds.), *Europe and MENA Cooperation Advances in Information and Communication Technologies*, Advances in Intelligent Systems and Computing 520, DOI 10.1007/978-3-319-46568-5_12

component of AEHS, able to build courses based on goals, preferences and knowledge of an individual user and use this throughout the interaction for adaptation of the content to the needs of that user.

This is due to the fact that AEHS deals with diverse backgrounds, such as software developers, web application experts, content developers, domain experts, instructional designers, user modeling experts, pedagogues, etc. Moreover, these systems have presentational, behavioral, pedagogical and architectural aspects that need to be taken into account. To make matters worse, most AEHA are designed and developed from scratch, without taking advantage of the experience from previously developed applications, because the latter's design is not codified or documented. As a result, development teams are forced to 're-invent the wheel' [2]. Keppell et al. [3] recommend that "Academic teachers should be encouraged to model and share learning designs within their own university, partner institutions and symposiums and conferences in higher education" to enhance learning and teaching through technology-enhanced learning.

Various works have been presented in the literature in order to support the design of AEHS [2, 4, 5]. Thus, there are several Design learning methods presented in the literature, such as ADDIE, OULDI, Design thinking, Xproblem, etc. In this work, we focus on one of the recent works proposed to design AEHS, which is called Agile Learning Design.

The purpose of this paper aims to present a framework for designing AEHS based on Agile Learning Design. The remainder of this paper is organized as follows. The Sect. 2 provides the background and a summative review of the most common design learning methods in the literature. The Sect. 3 describes a case study based on Agile Learning Design approach. The Sect. 4 provides discuss of some results of this work. Finally, a conclusion and future work are presented in the Sect. 5.

2 Background and Related Work

Rawsthorn [6] claimed that "computer technologies and related practices and methods have had a significant influence over Instructional Design methods. One of the major trends is the influence of Software Development Life Cycle methodologies over Instructional Design methodologies. This influence is evident in the ADDIE, Dick and Carey, and other Instructional Design methodologies".

In this section, we first present an overview of the two concepts: Instructional Design and Learning Design. Then, we outline the most used approaches. Then, we give a description about methods using in the field of Learning Design. Finally, we give a comparison between those methods.

2.1 Learning Design Versus Instructional Design

The terms Learning Design and Instructional Design are used interchangeably depending on the nature of the work and the environment in which it is carried out [7]. Below, we present an overview of the two concepts.

Instructional Design. The concept of Instructional Design arrived in the literature of technology for learning in the late of 1950. Instructional Design is the process by which instruction is improved through the analysis of learning needs and systematic development of learning materials. Merril et al. [8] define the Instructional Design as the practice of creating instructional experiences which make the acquisition of knowledge and skill more efficient, effective, and appealing.

In addition, Instructional Design may be supposed of as a framework for developing modules or lessons that increase and enhance the possibility of learning and encourages the engagement of learners so that they learn faster and gain deeper levels of understanding [9].

Learning Design. Learning Design as a research field has emerged in the last recent years, as a methodology for both articulating and representing the design process and providing tools and methods to help designers in their design process [10].

Beetham [11] defines the Learning Design as: a set of practices carried out by learning professionals… defined as designing, planning and orchestrating learning activities which involve the use of technology, as part of a learning session or program.

Learning Design aims to enable reflection, refinement, change and communication by focusing on forms of representation, notation and documentation [12]. It can take place at a number of levels: from the creation of a specific learning activity, through the sequencing and linking of activities and resources, to the broad curriculum and program levels.

Learning Design can be represented in several ways; each representation will articulate particular aspects of the learning that the designer anticipates will take place. Four main types of representations are identified: verbal, textual, visual, or data-based. Many tools can be used for implementation such LAMS (learning Activity Management System), MOT+ (Modeling using Object type), Reload, etc.

Discuss. As presented above, we conclude that Learning Design and Instructional Design are meticulously aligned but have distinct concentrations. Conole [13] claimed that Instructional designers design instruction to meet learning needs for a particular audiences and setting. Learning Design, in contrast, takes a much broader perspective and sees design as a dynamic process, which is ongoing and inclusive; taking account of all stakeholders involved the teaching-learning process.

As the Learning Design is boarder then the instructional Design, we will use only the term "Learning Design" in the sections below even if some methods use the term "Instructional Design".

2.2 Learning Design Methods

As we highlight above, the field of Learning Design has gained importance in the literature. According to our reading, we can classify the Learning Design approaches in two large categories. The first one intended at developing a Learning Design Specification for machine interpretation and execution. This was the direction adopted specially by IMS Learning Design. It seeks a formal educational mark-up language that can document a single or multiple learner experience in a computer readable and sharable (XML) format [13]. Instructors reproach to this category, that implementations of the full specification conducted to date are limited. Furthermore, this orientation does not make pedagogic design and learner activity explicit in a human-readable form.

The second category, that matches our vision, adopts a more general interpretation of Learning Design. It focuses on pedagogy and the activity of the student rather than the content. This approach advocates a process of 'design for learning' by which one arrives at a plan, structure or design for a learning situation, where support is realized through tools that support the process (e.g. software applications, websites) and resources that represent the design (e.g. designs of specific cases, templates) [12].

Various toolkit and model for mapping pedagogy and tools for effective learning design were proposed. In the section below, we will present the most cited methods in the literature.

ADDIE Model. ADDIE model [14] is the most common model used for creating instructional materials is the ADDIE Model. This acronym stands for the five phases contained in the model (Analyze, Design, Develop, Implement, and Evaluate). Each phase has an outcome that feeds into the subsequent phase.

1. Analyze: identify instructional goals and tasks, analyzing learner characteristics;
2. Design: develop learning objectives, choose an instructional approach, define performance objectives, develop assessment instruments, and develop instructional strategy;
3. Development: designers and developers start the production and the testing of the methodology being used in the project.
4. Implementation: deliver instructional materials; apply instructional activities; formative evaluation.
5. Evaluation: consists of two parts: (1) *Formative evaluation* is present in each stage of the ADDIE process. (2) *Summative evaluation* consists of tests designed for domain specific criterion-related referenced items and providing opportunities for feedback from the users.

Dick and Carey. The Dick and Carey method [15] is constituted by a series of steps, all of which will receive input from the preceding steps and will provide output for the next steps. All of the components work together in order for the user to produce effective instruction.

1. Assess Needs to Identify Goal(s): Determine the instructional goals.
2. Conduct Instructional Analysis: Determine the required skills, knowledge, and attitudes.
3. Analyze Learners and Contexts: Analyze the context in which the learners will learn the skills and they will use them.
4. Write Performance Objectives: Determine the conditions under which the skills must be performed, and the validation criteria.
5. Develop Assessment Instruments: Develop assessments to measure the learners' ability to perform the skills.
6. Develop Instructional Strategy.
7. Develop and Select Instructional Materials.
8. Develop and Construct Formative Evaluation of Instruction.
9. Design and Conduct Summative Evaluation.
10. Revise instruction.

OULDI approach. Open University Learning Design Initiative was led by the Institute of Educational Technology at The Open University. The initiative aims to provide support for the entire design process; from gathering initial ideas, through consolidating, producing and using designs, to sharing, reuse and community engagement [16].

The OULDI approach specifies three aspects of design [17]:

1. Collaboration and dialogue—mechanisms to encourage the sharing and discussing of learning and teaching ideas.
2. Representation—identification of different types of design representation and use of a range of tools to help visualize and represent designs.
3. Theoretical perspectives—the development of a body of empirical research and conceptual tools to help guide the design decision-making process and to provide a shared language to enable comparisons to be made between different designs.

Agile Learning Design approach. The Agile Learning Design is an iterative model of learning design that focuses on collaboration and rapid prototyping. Agile Learning Design can be adapted to fit the needs of the learning and training community by providing an ethos for the design of learning [18].

The flow of agile Learning Design may contain several cycle. Each cycle consists of problem analysis in the first phase, followed by the development of a single feature of the final product. Once this single small part of your course is finished you can start testing and evaluating the efficiency and the return on investment of this part. If the results are satisfying a new iteration begins, until the course or the project are fully finished, otherwise the designer has to take one step back, understand what went wrong, and correct.

The agile practices are combined with Learning Design, assisting and guiding the design and creation of AEHS [19]:

1. Active users participation: users are involved in the development process, helping to identify and solve problems and mistakes and providing rapid feedback to the team
2. Collaborative development: All team members constantly interact and communicate throughout the development process, promoting a collaborative and productive environment
3. Architecture/Design envisioning: Initial software architecture and requirements are designed at the beginning of a project to identify and think through critical issues
4. Iterative modeling/design: Software functionalities are designed at the beginning of an iteration to identify team's strategy for that iteration
5. Model/Design storming: Software functionalities are designed on a just-in-time (JIT) basis to reflect on specific aspects of team's solution
6. Early and continuous Evaluation: Testing and validation activities are conducted at the beginning of the project and extend throughout the development process

2.3 *Discuss*

We highlight that all of the frameworks presented above are development methodologies that are leveraged to guide Learning Design teams through a project of eLearning. The philosophies of those frameworks methodologies share many of the same practices. All of them include analysis, design, development, implementation, and evaluation as part of their process.

The study of these four Approaches allowed us to make the following comparative in Table 1.

Table 1 Summary of the approaches presented above

	ADDIE approach	Dick and Carey	OULDI	Agile LD
Process of development	Analysis, design development implementation evaluation	Identify, conduct, analyze, write performance, develop, design, revise, design	Vision, gather, assemble, run, evaluate, adapt	Align, get set iterate and implement leverage, evaluate
Type of process	Linear development process	Iteratively and parallel	Iteratively and linear	Short iteration
Implication of users	Users specify all requirement at start	Users specify all requirement at start	Users specify all requirement at start	Users embedded throughout the process
Delivery	All at once delivery	All at once delivery	All at once delivery	Constant delivery

We notice that the ADDIE, the carey and Dick and the OULDI approaches, although, they aim to make the design more explicit, they don't specify the steps and guidelines for a Learning Design process.

However, the Agile Learning Design approach has distinct characteristics that set it apart from the rest. The use of Agile Learning Design permits an incremental organization, flexible schedule, collaborative and transparent process. Moreover, the Agile Learning Design method allows designs to be modified, repurposed and evolved according to the needs of users emerging during development. Furthermore, it focuses on the final client which is in our case the learners and their interactivity with the system.

3 ALD-AEHS: A Case of Study

In this section, we illustrate our approach with an experiment that validates the proposed framework through its application in the design of an AEHS and creation of one of the most important component of AEHS: User Model, which we call Agile Learning Design for User Model (ALD-UM).

The research of the grounds for the other module of this AEHS model will be further described in detail after validation.

3.1 The Design of User Model

The agile Learning Design method used to implement the User Model is organized in four phases:

Establish the initial content of the User Model. In this stage, we use as a starting point, the User Model giving in generic AEHS that allows changing several aspects of the system, in reply to certain characteristics (given or inferred) of the user [20].

The User Model in AEHS includes two type of information grouped in two domains:

1. Domain Independent Data (DID): are composed of two elements: the Psychological Model and the Generic Model of the Student Profile, with an explicit representation [21]. These data are more permanent which allows the system to know beforehand which the characteristics that it must adapt to [22]. The DID include several aspects such initial user knowledge, objective and plans, cognitive capacities, learning styles, preferences, academic profile (technological studies versus economical studies and management, knowledge of literature, artistic capacities, etc.), etc.
2. Domain Dependent Data (DDD): information referring to the specific knowledge that the system judges that the user possesses on the domain. Martins [23]

say that the components of the DDD correspond to the Domain Model with three-level functionality: (a) Task level, with the objectives/competences of the domain that the user will have to master. In this case, the objectives or intermediate objectives can be altered according to the evolution of the learning process; (b) Logical Level, which describes the user knowledge of the domain and is updated during the student's learning process; (c) Physical Level, that registers and infers the profile of the user knowledge.

Those two elements and theirs contents were discussed with prospective learners, and the member of our team to approve the initial architecture of the UM, presented (Fig. 1).

Plan and create the structure. In this stage, we agree the content of the UM in adequacy with our learning context. We highlight that we can refine this model (add or delete some content) since we can do iterative design.

Implement the component. In this stage, we start the implementation, we agree the technologies that we will use to implement our UM and the design of the user interface.

Two different types of techniques are used to implement the Student Model: Knowledge and Behavioral based. The Knowledge-Based adaptation typically results for data collected through questionnaires and studies of the user, with the purpose to produce a set of initial heuristics. The Behavioral adaptation results from the monitorization of the user during his activity [23].

For the DID, we developed a form from which we will collect all the information about DID (Fig. 2)

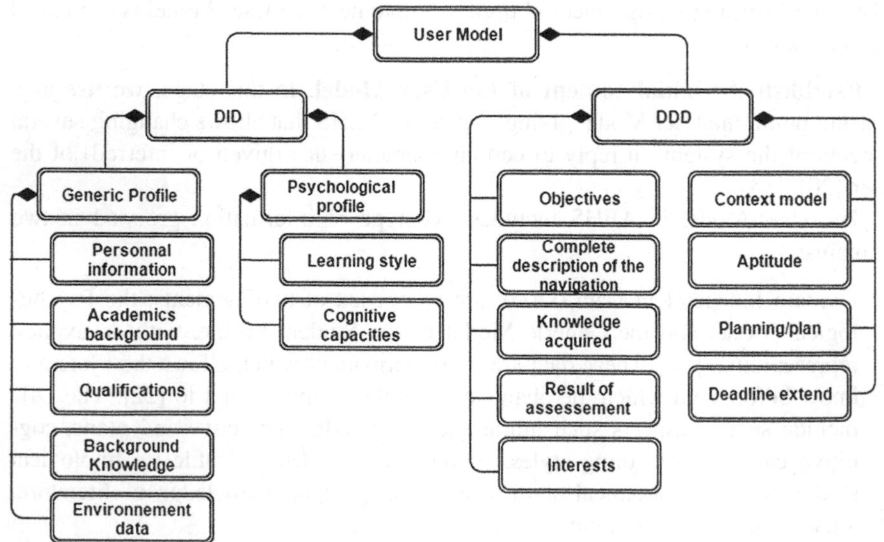

Fig. 1 Characteristic used in ALD-UM

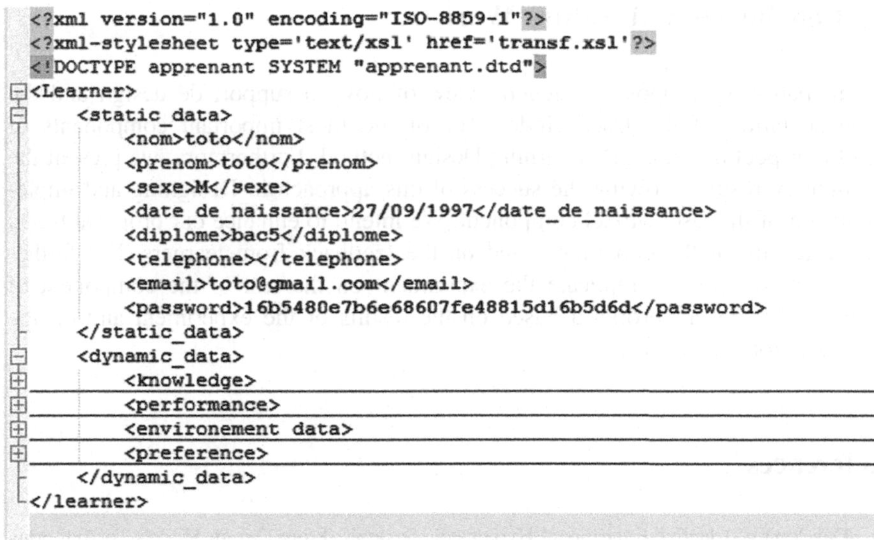

```
<?xml version="1.0" encoding="ISO-8859-1"?>
<?xml-stylesheet type='text/xsl' href='transf.xsl'?>
<!DOCTYPE apprenant SYSTEM "apprenant.dtd">
<Learner>
    <static_data>
        <nom>toto</nom>
        <prenom>toto</prenom>
        <sexe>M</sexe>
        <date_de_naissance>27/09/1997</date_de_naissance>
        <diplome>bac5</diplome>
        <telephone></telephone>
        <email>toto@gmail.com</email>
        <password>16b5480e7b6e68607fe48815d16b5d6d</password>
    </static_data>
    <dynamic_data>
        <knowledge>
        <performance>
        <environement data>
        <preference>
    </dynamic_data>
</learner>
```

Fig. 2 Example of an XML file the output of DID form

Evaluate. In this step evaluates and approves the work. Some learners create their account in the component of UM, fill in the form and evaluate the initial version of the UM. In this stage, we focus on remarks and feedback of learners. We collect all information that ca be and used to improve the succeeding iteration and to contribute to the constant enhancement process.

4 Results and Discuss

The first version of the framework presented in previous section, has already been implemented and tested to validate our approach. We notice that we have tested and validated only the implementation of the DID information. For the DDD, we have to implement our User Model in a complete AEHS to validate it. The other components of the ALD-AEHS will be subject of other publication.

For the part tested, we highlight the Agile Learning Design method allows designs to be modified, repurposed and evolved according to the needs of users emerging during development. In terms of the applicability of the method, the preliminary results indicate that the method is useful, easy to use. Furthermore, it focuses on the final client which is in our case the learners and their interactivity with the system.

However, we notice that in this preliminary system, the number of the students with whom we work is limited, due to lack of availability.

5 Conclusion and Future Work

In this paper we proposed a general view of how to support de design and the implementation of the User Model, one of the most important components of AEHS, respecting the Agile Learning Design method. Furthermore, we present the preliminary results showing the success of this approach in Designing and implementation of the User Model component. We intend to enhance our proposal based on the results of the experiment and on the feedback from learners. For further validation, we plan to replicate the experiment on the rest of the component of AEHS, enhance our proposal based on the results of the experiment and on the feedback from learners.

References

1. Developing Adaptive Educational Hypermedia Systems: From Design Models To Authoring Tools. http://www.pitt.edu/~peterb/papers/KluwerAuthBook.pdf
2. Retalis, R., Papasalouros, A.: Designing and generating educational adaptive hypermedia applications. Educ. Tech. Soc. **8**(3), 26–35 (2005)
3. Keppell, M., Suddaby, G., Hard, N.: Good Practice Report: Technology-Enhanced Learning and Teaching. Australian Learning and Teaching Council, Sydney (2011)
4. Grigoriadou, M., Papanikolaou, K., Kornilakis, H., Magoulas, G.: INSPIRE: an intelligent system for personalized instruction in a remote environment. In: Bra, P.D., Brusilovsky, P., Kobsa, A. (eds.) Proceedings of Third workshop on Adaptive Hypertext and Hypermedia, July 14, 2001, pp. 13–24. Technical University Eindhoven, Sonthofen, Germany (2001)
5. Stern, M.K., Woolf, B.P.: Adaptive content in an online lecture system. In Brusilovsky, P., Stock, O., Strapparava, C. (eds.) Adaptive Hypermedia and Adaptive Webbased systems, pp. 225–238. Springer, Berlin (2000)
6. Rawsthorn, P.: Agile Methods of Software Engineering Should Continue to Have an Influence Over Instructional Design Methodologies (2005)
7. Devilee, A.: What is instructional design? http://instructionaldesign.com.au/content/differences-between-educational-design-learning-design-and-instructional-design
8. Merrill, M.D., Drake, L., Lacy, M.J., Pratt, J.: Reclaiming instructional design. Educ. Technol. **36**(5), 5–7 (1996)
9. Clark, D.: Instructional design. http://www.nwlink.com/~Donclark/hrd/learning/development.html
10. Conole, G.: An overview of design representations. In: Dirckinck-Holmfeld, L., Hodgson, V., Jones, C., de Laat, M., McConnell, D., Ryberg, T. (eds.) Proceedings of the 7th International Conference on Networked Learning (2010)
11. Beetham, H.: Review of the Design for Learning programme phase 2. JISC Design for Learning programme report, (2008). http://www.jisc.ac.uk/whatwedo/programmes/elearningpedagogy/designlearn.aspx
12. Cross, S. Conole, G.: Learn About Learning Design *OU Learn About Series*, Milton Keynes (2009). http://www.open.ac.uk/blogs/OULDI/wp-content/uploads/2010/11/Learn-about-learning-design_v7.doc
13. Conole, G.: Learning Design: making practice explicit. In: Connected Conference, Sydney (2010)
14. ADDIE model. http://educationaltechnology.net/the-addie-model-instructional-design/
15. Dick and Carey. http://www.instructionaldesign.org/models/dick_carey_model.html

16. A Review of Curriculum Design at the Open University 2008–09: OULDI-JISC Project Baseline Report. http://www.open.ac.uk/blogs/OULDI/wp-content/uploads/2010/11/OULDI_baseline_Report_Final_v1.doc
17. OULDI project. http://jiscdesignstudio.pbworks.com/w/page/29228368/Open%20University%20Learning%20Design%20Initiative%20%28JISC-OULDI%29%20project
18. Agile Learning Design. http://www.nwlink.com/~donclark/agile/agile_learning_design.html
19. Arimoto, M.M., Barroca, L., Barbosa, E.F.: An agile learning design method for open educational resources. In: 2015 IEEE Frontiers in Education Conference Proceedings, pp. 1897–1905. IEEE (2015)
20. Brusilovsky, P.: Adaptive hypermedia. User Model. User Adap. Inter. **11**(1/2), 87–110 (2001)
21. Kobsa, A.: Generic user modeling systems. User Model. User-Adap. Inter. **11**(1–2), 49–63 (2001)
22. Vassileva, J.: A task-centred approach for user modeling in a hypermedia office documentation system. In: Brusilovsky, P., Kobsa, A., Vassileva, J. (eds.) Adaptive Hypertext and Hypermedia, pp. 209–247. Kluwer Academic, Dordrecht (1998)
23. Martins, A.C., Faria, L., Vaz de Carvalho, C., Carrapatoso, E.: User Modeling in Adaptive Hypermedia Educational Systems. Educational Technology and Society, vol. 11(1), pp. 194–207 (2008)

An Approach for the Identification and Tracking of Learning Styles in MOOCs

Brahim Hmedna, Ali El Mezouary and Omar Baz

Abstract This paper is devoted to describe a preliminary draft of our approach that aims to identify and track learners' learning styles based on their behavior and actions they perform in a MOOC environment. Adaptation arises with intensity in MOOCs. Indeed, it has been proved that MOOCs can benefit from the advantages of learning styles as a way to provide an adaptive navigational guidance to learners. In this approach, we use neural networks for the identification and tracking of learner's learning styles in MOOCs so as to increase learners' engagement and satisfaction. The purpose of this paper is to examine the point of view of literature and solution to integrate an adaptive system in MOOC.

Keywords MOOC · Learning styles · Neural network · Adaptation

1 Introduction

Over the past couple of years, MOOCs (Massive Online Open Courses) have emerged as a powerful contender for the next new education technology [1]. MOOC provides a new way of learning, which is open, participatory, distributed and lifelong [2]. It is an emerging format of online courses designed for a large number of participants that uses an open access via the web, in addition to traditional course material such as videos, reading and problem sets. MOOCs provide interactive user forms that help build a community for the students and professors. MOOCs are recent development in distance education.

B. Hmedna (✉) · A.E. Mezouary · O. Baz
IRF-SIC Laboratory, FSA, Ibn Zohr University, Agadir, Morocco
e-mail: br.hmedna@gmail.com

A.E. Mezouary
e-mail: elmezouaryali@gmail.com

O. Baz
e-mail: o.baz@uiz.ac.ma

Á. Rocha et al. (eds.), *Europe and MENA Cooperation Advances in Information and Communication Technologies*, Advances in Intelligent Systems and Computing 520, DOI 10.1007/978-3-319-46568-5_13

MOOCs take multiple forms, in the literature review there are two principal kinds of MOOCS, namely: cMOOC, xMOOC [3]. The first one, cMOOC, was led in 2008 by Downes and Siemens, is based on the connectivism learning theory and focuses on knowledge construction and creation and puts much emphasis on creation, autonomy and social network learning [4]. The second one, xMOOC started in 2012, is closer to traditional teaching process and concept, and focusing on knowledge dissemination and duplication, as well as such learning methods of video, homework and test [5].

Another taxonomy of MOOC has been proposed by Clark who has distinguished eight types of MOOCs based on their functionalities [6]: transferMOOCs are xMOOCs, madeMOOCs implement more crafted and challenging assignments, synchMOOCs are synchronous MOOCs with fixed start days, end days, and deadlines for assignments, asynchMOOCs are asynchronous MOOCs with no (or frequent) start days, and tend to have no or looser deadlines for assignments, adaptiveMOOCs are MOOCs that use adaptive algorithms to present personalized learning experiences, groupMOOCs are MOOCs starting with small and collaborative groups of learners, connectivistMOOCS are cMOOCs, and miniMOOCSs are shorter MOOCs for contents and skills that do not require a semester structure.

Learners have several ways of learning, Massive open online courses (MOOC) can take advantage of automatic identification and tracking of learners learning styles so as to use this learning system information to build an automatic recommendation system for MOOC.

It has been shown that providing learners with learning resources and activities that suit their preferences and learning styles increases learner's satisfaction [7], improve learning performances (effectiveness) and save time (efficiency).

To identify learner's learning styles, many systems ask learners to complete questionnaires, which is not appropriate because learners tend to choose answers arbitrarily when questions are too long. Therefore, we introduce an approach, which combines collaborative approach (questionnaire), and automatic (learners' behavior) ones to identify and track learners learning styles.

In this regard, we believe that MOOCs environments can take advantage of these different forms of learning by recognizing the style of each individual learner and adapting the content of courses to match this style. In light of these factors, we want to explore how neural network can be used in the context of a MOOC as a practical method of identification and tracking of learners learning style based on their behaviors.

The rest of the paper is organized as follows. In Sect. 2 theoretical background on learning styles is presented. In Sect. 3 description of how neural network can be used to identify the learning styles of learners. Finally, future works and conclusions are given in Sect. 4.

2 Related Works

Making learners aware of their learning styles and providing them with learning resources that match their individual learning styles has positive impact on their learning progress [8].

Defining the concept of learning styles is not simple task. A literature review highlight the multiple definitions that exist:

The term learning style refers to the way in which an individual concentrates on processes, internalizes, and retains new and difficult information [9]. Smith and Dalton [10] defined learning style as a unique and Habitual behavior of acquiring knowledge and skills through every day study or experience. While Felder and Silverman [11] described it as the way in which persons receive and process information. Moreover, Kolb [12] had his own opinion as to what a learning style is. He defined it as the process of creating knowledge through the transformation of experience.

In the last decade, many different kinds of learning style models have been proposed, some of these learning style models have been found more applicable for online learning than others [13]: Kolb's learning style model [14]. The Honey and Mumford's learning style model [15] and Felder and Silverman's learning style model [16].

In the following, we present two popular learning style models and also studies of several research works that combines Learning Styles and MOOCs and finally explain in light of these studies the importance of our work.

2.1 *VARK Learning Style Model*

Fleming's VARK model [17] is one of the most popular models, which divides learner's preferences for learning into four categories: Visual (V), Auditory (A), Reading/writing (R), and Kinesthetic (K), each of these categories is defined in Table 1.

Table 1 Description of VARK learning preferences

Preference	Description
Visual (V)	Visual learners are those who learn best with visual artifacts like diagrams and pictures
Auditory (A)	Auditory learners are those that learn with oral stimulations, such as talking and listening
Reading/writing (R)	Reading and writing learners prefer printed words to gain knowledge
Kinesthetic (K)	Kinesthetic learners are those who learn by experience

2.2 Kolb's Learning Style Inventory

The Kolb experiential learning model (Fig. 1) was introduced by Kolb [12] this model is presented as a transformation process beginning from reflection and ending by experimentation. The Kolb learning cycle is based on four-stage:

- Concrete experience (CE)—feeling
- Reflective observation (RO)—watching
- Conceptualization (AC)—thinking
- Active Experimentation (AE)—doing

Essentially, the four learning styles are combinations of these four stages. For example, Converger (AC/AE) is a combination of abstract Conceptualization and Active Experimentation [12].

To identify learning styles, Kolb developed a learning style inventory based on the experiential model; the inventory measures an individual's relative emphasis on four learning abilities:

- Accommodator (CE/AE): Prefers practical hands-on approach to problems.
- Converger (AC/AE): Attracted to technical tasks and problems.
- Diverger (CE/RO): Prefer to watch rather than do, tending to gather information and use imagination to solve problems.
- Assimilator (AC/RO): Interested in ideas and abstract concepts.

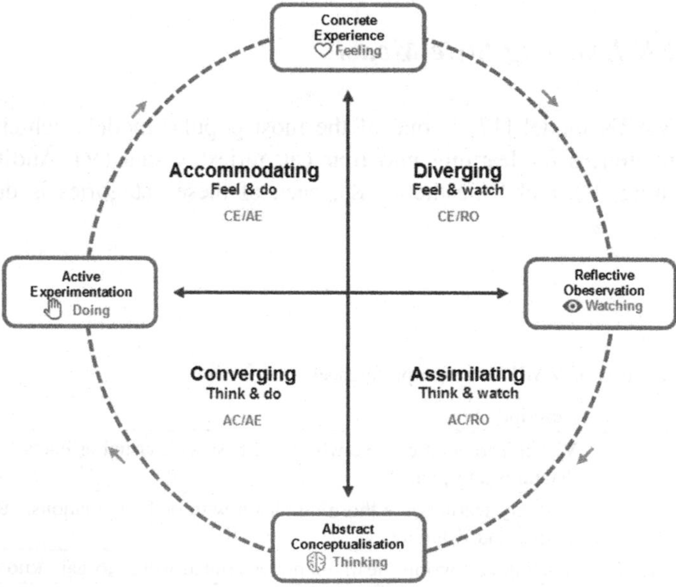

Fig. 1 Kolb experiential learning model

2.3 Felder-Silverman Learning Styles Model

The Felder-Silverman learning style model (FSLSM) was created by Richard Felder and Linda Silverman in 1988. It focuses on aspects of learning styles on engineering students.

FSLSM describes learning styles in more detail by characterizing each learner according to four dimensions; each of these dimensions is defined as below:

- **Sensory Versus Intuitive**: Sensory learners like learning facts and solving problems with known methods while intuitive prefer discovering possibilities [16].
- **Visual Versus Verbal:** visual learners learn best from what they can see such as graphics, images, and flow charts, verbal learners prefer to learn from words, regardless whether they are spoken or written.
- **Active Versus Reflective:** Active learners like to try things out or do something active. Reflective learners prefer thinking about things on their own [16].
- **Sequential Versus Global**: Sequential learners learn in small steps when global learners understand things in large steps [16].

Based on this model a corresponding psychometric assessment instrument was created. It was called the Felder-Solomon's Index of Learning Styles (ILS). It is a 44-item questionnaire where learners' personal preferences for each dimension are expressed with values between +11 to −11 per dimension, with steps ±2 (Fig. 2). This range comes from the eleven questions that are posed for each dimension [18].

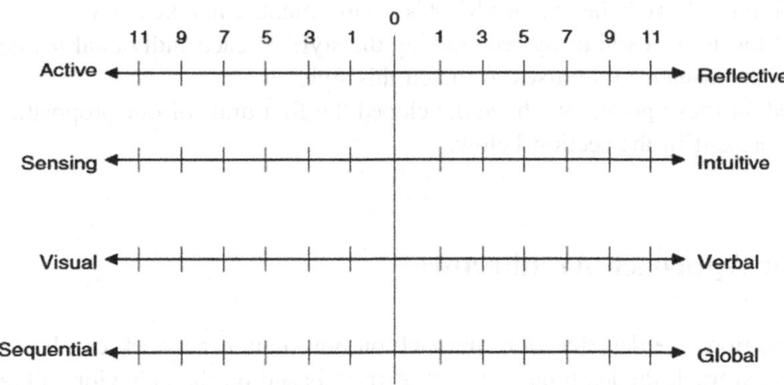

Fig. 2 Index of learning styles (ILS)

2.4 Learning Styles' Application in MOOCs

In this paper we are particularly interested by MOOC environments. In the literature review different works (studies) combine Learning Styles and MOOCs. In this regard,

Fasihuddin et al. [19] proposes a framework to personalize open learning environments based on the theory of learning styles and particularly the Felder and Silverman learning style model (FSLSM). The framework provides adaptive navigational support through sorting and hiding the learning materials based on learners' learning styles and the involved preferences.

Another work by Sonwalker [20] proposes an adaptive MOOC (aMOOC) that offers adapted learning content based on learning styles. The author proposes the learning cube that illustrates organization of learning materials as text, graphics, audio, video, animations, and simulations according to different learners' learning styles.

2.5 Learning Style Benefits

The section above makes explicit the particular interest of research community considering the benefits of learning styles in online environment especially In MOOC.

Indeed, it has been shown that providing learners with learning resources and activities that suit their preferences and learning styles increases learner's satisfaction [7].

In this regard, we believe that MOOCs environments can take advantage of these different forms of learning by recognizing the style of each individual learner and adapting the content of courses to match this style.

Based on these points, we have developed the first draft of our proposition that we will present in the section below.

3 Our Approach Architecture

In this section, we describe our approach on how neural network can be used to identify and track the learning styles of learners based on the behavior and actions they perform in an MOOC environment.

The architecture of our approach and its components can be seen in Fig. 3 and are described in the following

Our approach is consisting of five stages: Data Collection—Pre-processing—Feature Extraction—Classification—Adaptation (recommendation).

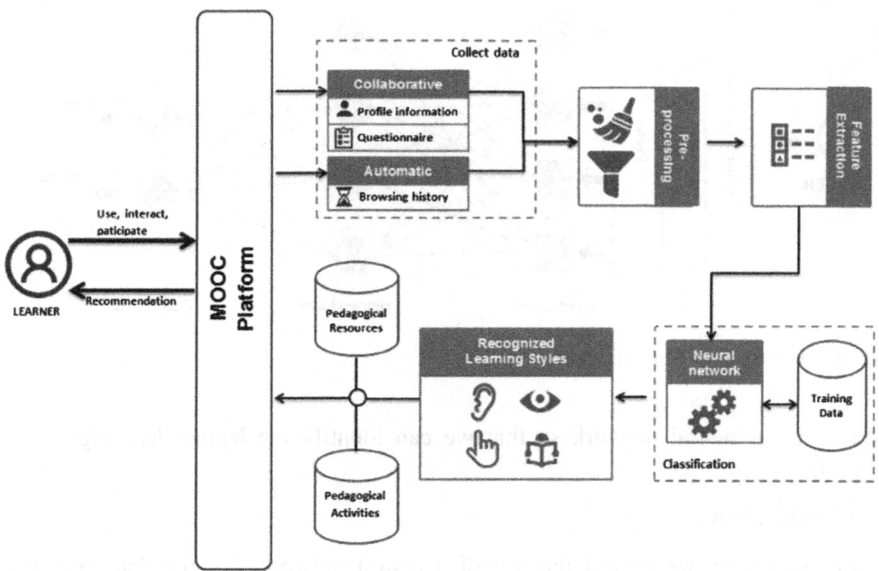

Fig. 3 A process of identification of learning styles using neural networks

- Data collection

During the first stage, our goal is to collect data; we will gather data by two different ways: collaborative and automatic [21].

In the collaborative approach, learners are asked to provide their preferences explicitly by filling in a questionnaire, such as the ILS questionnaire [22].

In the automatic approach, we use the behavior of the learners and their actions with the systems while they are learning [22].

- Pre-processing

Pre-processing operation is the first step performed on raw data collection. This step aims to:

- Clean data collected of low-quality information.
- Transform the data into a clean format which can be used by our system.
- Prepare data for analysis.

- Feature extraction

After Pre-processing, a feature extraction method will be applied to extract the most appropriate characteristics that can be used to identify learning styles of learners.

This stage aims at creating vectors from the characteristics of each learner. These characteristics are gathered from the data collection stage. These vectors serve as an

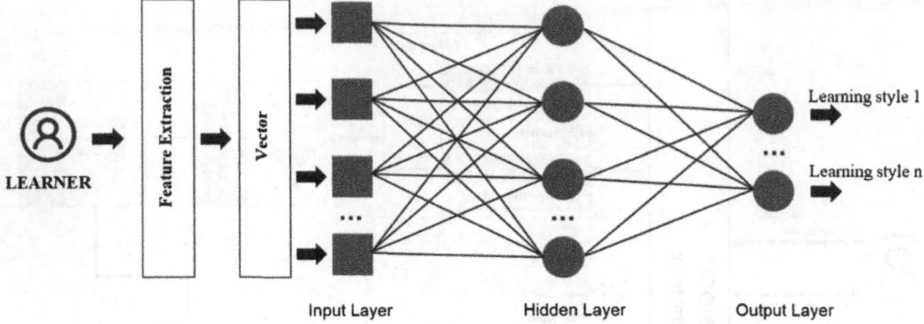

Fig. 4 The neural network architecture

input for our neural network so that we can identify the learner learning styles (Fig. 4).

- Classification

In this phase, we expect the use of a neural networks for the detection and recognition of learning styles. Two classification types can be defined: supervised and unsupervised [23] classifications. In this research, we rely on a supervised classification which consists of two processes: training and testing process.

Step 1: Training

In this step we aim to create training dataset from fixed dimension vectors, these are obtained from characteristics of each learner. After this, learning Dataset will be presented to the neural network for learning about the properties for each learner styles.

Step 2: Testing

This step consists of creating a model that can measure the performance and accuracy of test dataset.

- The process of this step begins when new learners enroll to the platform.
- The system will then extract vectors from the characteristics of these learners
- These vectors will constitute the input for our trained neural network (step 1).
- Finally, our neural network seeks to predict the learner styles closer that belongs to each learner.

- Adaptation

After identifying learning style for learners, we aim to provide relevant content to the learner according to his learning style through navigational support.

Briefly, the adaptation can be done as follows:

- Identification of learners who have the same learning style.
- Create learners' clusters based on their profiles.
- Recommendation of appropriate resources for each cluster (not individuals) via navigational support [21].

4 Conclusion and Future Works

This paper shed light on the relation between learning styles, MOOC environments, and machine learning.

Based on the finding, our research proposed to design a suggestion to adapt MOOC environments through traces analysis. The adaptation of these environments aim to provide relevant content to the learner via navigational support. Concrete implementation of these general suggestions and validations of their effectiveness are however left as future works. These experimentations will be done on courses prepared and hosted on Ibn Zohr University servers in Morocco.

In this work we first introduce learning styles theories. In addition, we described our methodology on how neural network can be used to identify the learning styles of learners based on the actions they perform in MOOC.

References

1. Russell, M., Klemmer, S., Fox, A., Latulipe, C., Duneier, M., Losh, E.: Will massive online open courses (moocs) change education?. In: CHI '13 Extended Abstracts on Human Factors in Computing Systems (2013)
2. Bansal, N.: Adaptive recommendation system for MOOC. Doctoral dissertation, Indian Institute of Technology, Bombay (2013)
3. Daniel, J.: Making sense of MOOCs: musings in a maze of myth, paradox and possibility. J. Interact. Media Educ. **3** (2012)
4. Downes, S.: A true history of the MOOC. Stephen's Web site (2012)
5. Downes, S.: Places to go: connectivism and connective knowledge. Innovate: J. Online Educ. (2008)
6. Clark, D.: MOOCs: taxonomy of 8 types of MOOC. Donald Clark Paln B (2013)
7. Graf, S., Tzu-Chien, L.: Supporting teachers in identifying students' learning styles in learning management systems: an automatic student modelling approach. J. Educ. Technol. Soc. **12**, 3–14 (2009)
8. Graf, S., Shuk, K., Liu, T.C.: Identifying learning styles in learning management systems by using indications from students' behaviour. Adv. Learn. Technol. **2008**, 482–486 (2008)
9. Dunn, R.: Rita Dunn answers questions on learning styles. In: Educational Leadership. pp. 15–19. (1990)
10. Smith, P.J., Dalton, J.: Getting to grips with learning styles (2005). http://johnwatsonsite.com/Share-Eng/SIC/Getting_to_grips_with_learnng_styles.pdf
11. Felder, R.M., Silverman, L.K.: Learning and teaching styles in engineering education. In: Engineering Education, pp. 674–681 (1988). http://www4.ncsu.edu/unity/lockers/users/f/felder/public/Papers/LS-1988.pdf

12. Kolb, D.A.: Experiential learning: Experience as the source of learning and development (2014)
13. Kuljis, J., Liu, F.: A comparison of learning style theories on the suitability for elearning. Web Technol. Appl. Serv. **2005**, 191–197 (2005)
14. Kolb, A., Kolb, D.A.: Kolb's learning styles. In: Encyclopedia of the Sciences of Learning. Springer US, pp. 1698–1703 (2012)
15. Honey, P., Mumford, A.: The Manual of Learning Styles (1992)
16. Felder, R.M., Silverman, L.K.: Learning and teaching styles in engineering education. Eng. Educ. **78**, 674–681 (1988)
17. Fleming, N.D.: Teaching and learning styles: VARK strategies. (2001)
18. Graf, S., Viola, S.R., Leo, T., Kinshuk.: In-depth analysis of the Felder-Silverman learning style dimensions. J. Res. Technol. Educ. 79–93 (2007)
19. Fasihuddin, H., Skinner, G., Athauda, R.: A framework to personalise open learning environments by adapting to learning Styles. In: 7th International Conference on Computer Supported Education, vol. 1, pp. 296–305 (2015)
20. Sonwalkar, N.: The first adaptive MOOC: a case study on pedagogy framework and scalable cloud architecture part I. In: MOOCs Forum, pp. 22–29 (2013)
21. Brusilovsky, P.: Methods and techniques of adaptive hypermedia. In: Adaptive hypertext and hypermedia, pp. 1–43. Springer, Netherlands (1998)
22. Fasihuddin, H., Skinner, G., Athauda, R.: Towards an adaptive model to personalise open learning environments using learning styles. In: Information, Communication Technology and System (ICTS), 2014 International Conference, pp. 183–188 (2014)
23. Villaverde, J.E., Godoy, D., Amandi, A.: Learning styles' recognition in e-learning environments with feed-forward neural networks. J. Comput. Assist. Learn. **22**(3), 197–206 (2006)

Toward Incorporating Bio-signals in Online Education Case of Assessing Student Attention with BCI

Mohammed Serrhini and Abdelamjid Dargham

Abstract Bio-signals acquired with sensors technologies are increasingly gaining attention beyond the classical medical domain, into a new paradigms such education. Attention is bio signal that can be measured and checked by Brain Computer Interface technology (BCI) through alpha wave (8–13 Hz) and beta wave (14–30 Hz) frequency measurement. Attention and learning are very dependent on one another. Student with attention deficits often have learning disabilities. According to many teachers' and professional researchers, it has been found that the student's attention is reducing. This paper talks about how to assess student attention in online education, during learning process students' attention is controlled by attention assessment system EEG based on (BCI). Attention Data are stored in database, and used for Signal processing algorithms to understand student knowledge advancement.

Keywords Online education · BCI · Attention in education · Bio signals · Human machine interaction · Mind brain · Education

1 Introduction

Researchers have now acquired so much information about how the brain learns that a new academic discipline has been born, called "educational neuroscience" or mind, brain, and education science, this emerging discipline of Mind, Brain, and Education (MBE) explore the benefits as well as the difficulties involved in integrating neuroscience into educational policy and practice.

M. Serrhini (✉) · A. Dargham
Faculty of Sciences, University Mohammed First Oujda,
BV Mohammed VI, Oujda, Morocco
e-mail: Serrhini@gmail.com

A. Dargham
e-mail: abdelmajid.dargham@gmail.com

Á. Rocha et al. (eds.), *Europe and MENA Cooperation Advances in Information and Communication Technologies*, Advances in Intelligent Systems and Computing 520, DOI 10.1007/978-3-319-46568-5_14

This field explores how research findings from neuroscience, education, and psychology can inform our understandings about teaching and learning, and whether they have implications for educational practice. Neuroscience research shows that the changes in the brain that underlie learning occur when experiences are active [1, 2]. With student-centered learning approaches, students are empowered to engage in active learning experiences that are relevant to their lives and goals. When a student is passively sitting in a classroom where the teacher is presenting decontextualized information that he/she is not paying attention to, the brain is not learning. The brain can focus on only one task at a time. Each shift of the brains attention requires increased mental effort and incurs a loss of information in working memory of the first task. In effect, the individual ends up doing two tasks poorly rather than one task well. Attention is an important part in the learning process both in real life circumstances and in computer-based instruction and provides the basis that informs motivation modeling [3, 4]. The brain not only juggles tasks, it also juggles focus and attention. When people attempt to perform two cognitively complex tasks such as driving and talking on a phone, the brain shifts its focus (people develop "inattention blindness").

Today, bio-signals acquired with sensors technologies are increasingly gaining attention beyond the classical medical domain, into a paradigm, which using the physical computing analogy [5], can be described as physiological computing. Physical computing, that deals with the study and development of systems that sense and react to the human body. The modern uses of bio-signals have become an increasingly important topic of study within the global engineering community and consequently, many evidences show that bio-signals are clearly a growing field of interest, where recent applications include: Human-Computer Interaction (HCI), which involve the interface between the user and the computer [6]; Quantified-self, giving people new ways to deal with medical problems or improve their quality of life; and many other disciplines.

The continuous or relaxing rhythms of the brain produce bio-signals called Brain Waves, they, are classified by frequency bands. Different brain waves frequencies correspond to behavioral and attentional states of the brain, and a traditional classification system has long been used to explain these different Electroencephalogram (EEG) rhythms. EEG is a measure of the brain's voltage fluctuations as detected from scalp electrodes sensors of the cumulative electrical activity of neurons.

Online Education (OE) used in modern engineering learning to help students studying remotely through the Internet [7]. Institutions may also show interest to apply MBE approach in OE. Nowadays, digital technology evolution allows facilitating integration of neuroscience bio-signal, it open new ways to include remotely some student mind parameters into online education, this make learning process as real as possible and more instructive. Attention and concentration plays a big role in learning process. In conventional classroom education guided by human teacher, student is usually alerted about risk that he/she incurs because of his/her inattentiveness. In distant education, intelligent tutoring system can plays teacher role to provide personalized teaching sessions and feedbacks for the specific needs of each student, like alerting him about his/her low attention level. This becomes crucial for the success of future OE projects.

In this chapter, we introduce use of MBE approach to enhance student attention in online education. This paper is organized as following, after introduction, Sect. 2 will introduce readers with background materials to understand, the role of attention in learning, what is EEG bio-signal, brain computer interface (BCI) technology to capture and process this attention waves in order to check student concentration during interaction with learning material in online education, third section discuss proposed system overview (Architectural, Practical implementation, Use and gain), Sect. 6 we present some results with discussion.

2 Background

2.1 Attention and Concentration in Education

Attention is an important aspect of the learning situation [3, 4, 8]. Keller's strategies for attention underline "getting and sustaining attention" [9] before attracting on other strategies to motivate the student. Attention is one of the most intensely studied topics, and remains a major area of investigation within education, psychology and neuroscience. Keller's ARCS [8, 10] model for example, considers attention as the most fundamental element towards achieving motivation in the classroom (ARCS stands for Attention, Relevance, Confidence and Satisfaction). Areas of active investigation involve determining the source of the signals that generate attention, the effects of these signals on the tuning properties of sensory neurons, and the relationship between attention and other cognitive processes like working memory, learning and vigilance.

In computer mediated learning, attention has generated a growing body of research with the aim of identifying students' attention and responding appropriately given low states. In order to identify and reply researchers have employed Artificial Intelligence methods that allow personalizing the interaction. Artificial Intelligence in Education (AIED) has dealt with recognition of attention. For recognition side, researchers have employed two main methodologies: modeling using physiological clues [11, 12] and employing user-generated data [13, 14]. The results provide an indication of attention states during the interaction between a learner and an educational system. On the reaction side, researchers have investigated different ways of proposing corrective feedback if the detection shows low levels of attention.

The various approaches taken in AIED research have carried about benefits that translated into learning advances. Giving the relevance of attention in the learning gains, our approach considers reading user's attention (recognition) using physiological inputs. The physiological inputs, however, will be based on a Brain Computer Interface capable of measuring the learners' attention levels based on neural activity. We have chosen to combine these reading with intelligent notification System represented by notification alerts for each levels of attention of the students interacting with OE to determine when the user is paying attention to or not by only using the BCI.

2.2 EEG Attention Wave's Bio-signal

An electroencephalogram (EEG) is a measure of the brain's voltage fluctuations as detected from scalp by electrodes that can be used to measure an electrical signal of the human body, such as a brain wave. It is an approximation of the cumulative electrical activity of neurons. EEG signal changes according to the brain activity states. Depending on these states, we can distinguish several rhythms (waves). EEG activity has been used mainly for clinical diagnosis and for exploring brain function (attention, meditation, stimulus, etc.).

Attention is the cognitive process of selectively concentrating on one aspect of the environment while ignoring other things. Attention has also been referred to as the allocation of processing resources, it is a brain activity. A brain activity produces electrical signals that can be measured from the human scalp by Hans Berger from 1929 [15]. When subject must keep attention, brain wave signal always appear in frontal lobe and parietal lobe (Fig. 1 shows brain attention lobe) of the human brain when he/she is in an alert situation as mentioned by [16, 17].

The continuous or resting rhythms of the brain, "brain waves", are categorized by frequency bands. Different brain wave frequencies correspond to behavioral and attentional states of the brain, and a traditional classification system has long been used to characterize these different EEG rhythms are (Alpha waves, Beta, Theta, Delta, Lambda, and Vertex waves).

Generally, Alpha and Beta Waves Studied since 1920s found in Parietal and Frontal Cortex, Relaxed mean Alpha has high amplitude, Excited mean Beta has high amplitude, alpha waves (8–12 Hz) correlate with relaxation or rest state (Fig. 2

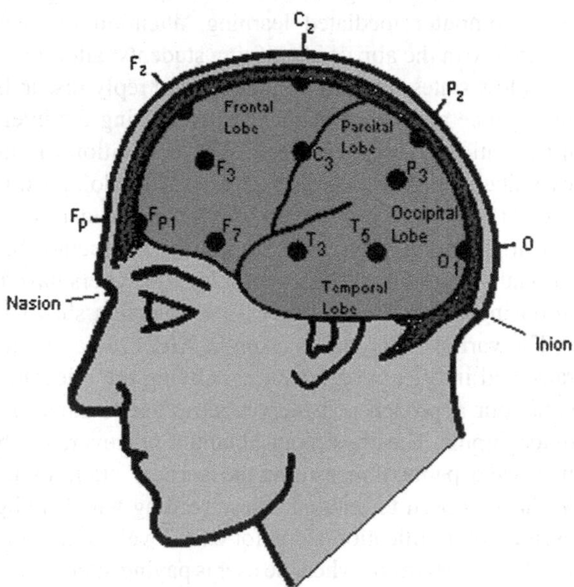

Fig. 1 Brain lobes

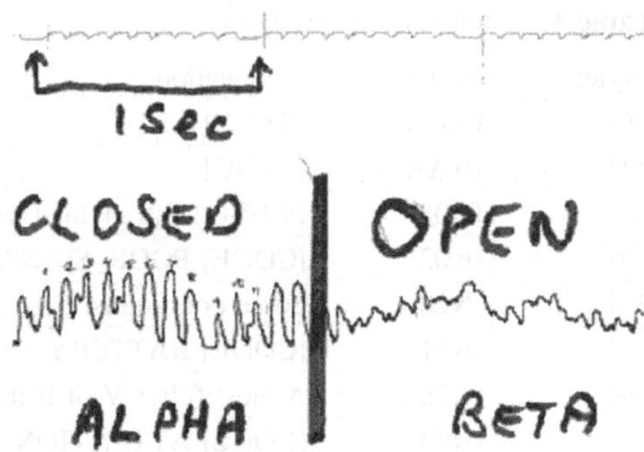

Fig. 2 Eyes alpha and beta signal

shows an example)., while beta waves (13–30 Hz) correlate with mental attention, concentration and active thinking. However, human brain waves generally include all of these waves and vary dynamically.

2.3 *Brain Computer Interfaces (BCI)*

In recent years, we can observe a growing interest in BCI. The main advantage of the communication between brain and computer is its "directness". BCI, are input devices that use the brains' electrical activities to allow communication between users and computers. Normally, BCI's are used to activate commands based on specific reading or to measure neural activity of interest such as attention, anxiety or relaxation, and stimulus etc.

User wears headset with EEG sensors on it to record neural activity. The 26 sensor electrodes were organized according to the 10–20 standards for EEG location. The sensors were recorded as interleaved channels of signed 32 bit integers at a rate of 500 samples per second.

The channels were separated into individually named files and converted to (American Standard Code for Information Interchange) ASCII format for simplicity of loading on different systems for furthers processing Table 1.

There are two type of BCI:

Invasive: It is the brain signal reading process which is applied to the inside of grey matter of brain.

Non Invasive: It is the most useful neuron signal imaging method which is applied to the outside of the skull, just applied on the scalp.

There are three main consumers-devices commercial competitors in this area which selling a non invasive BCI.

Neural Impulse Actuator from April 2008,

Emotiv Systems 2009,

NeuroSky 2009 developed MINDSET easy to use BCI headset for less than 100 dollars.

Table 1 Example of Packet Data

Byte:	Value	Explanation
[0]	0xAA	[SYNC]
[1]	0xAA	[SYNC]
[2]	0x08	PLENGTH] (payload length) of 8 bytes
[3]	0x02	[CODE] POOR_SIGNAL Quality
[4]	0x20	Some poor signal detected (32/255)
[5]	0x01	[CODE] BATTERY Level
[6]	0x7E	Almost full 3 V of battery (126/127)
[7]	0x04	[CODE] ATTENTION eSense
[8]	0x30	eSense Attention level of 48 %
[9]	0x05	[CODE] MEDITATION eSense
[10]	0x12	eSense Meditation level of 16 %
[11]	0xE3	[CHKSUM] (1's comp inverse of 8-bit Payload sum of 0x1C)

2.3.1 NEUROSKY Headset

NeuroSky startup has developed technologies based on a non-invasive, dry, bio sensor to read electrical neuron-triggered activity in the brain to determine states of attention and relaxation. NeuroSky headset is used as wearing device, this low-cost, easy to use headset developed for leisure, non-clinical human-computer interaction. Neurosky is a neural activity produces a faint electrical signal that constitutes the basis for EEG-based NeuroSky readings. To do so, it detects these signals using three dried electrodes and decrypts them by applying algorithms to disambiguate multiple signals and give coherence to the interpretations. TinkGear is the technology inside every NeuroSky product that includes onboard chip that processes all data and provides these data to software's and applications in digital form for further data processing and commands, an example of provided data are given in Table 1.

To us, the originality of using NeuroSky in our research is its movability and easiness of use and the potential to apply it as an input device to for physiological, brain-generated relevant information.

3 System Overview

In online education proposed system that will assess student attentiveness with BCI during his interaction learning materials as depicted in Fig. 3, all the interaction is accessed by students through internet via a common Web Browser (Internet Explorer, Mozilla Firefox, Google chrome etc.) and grants control of the simulation materials or laboratory equipment's using a user-friendly Graphical User Interface (GUI). Through this GUI students learn remotely and they receive notification alert of the lack of attention during learning process. To begin session student launches Web Browser and wears his/her Neurosky BCI headset. A high-level Web

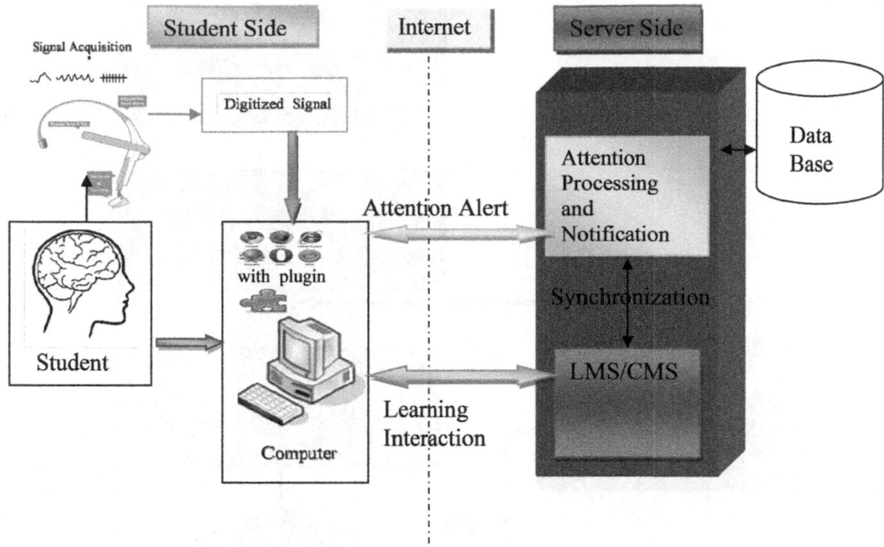

Fig. 3 System architecture

Application programming interface (API) was developed to capture attention and concentration wave's values, send them to attention processing algorithm, and return feedbacks to student.

4 System Use and Gain for Student

4.1 How Student Use System

As shown in experiment process flowchart (Fig. 4) student access system after successfully login via a common login PHP web page, firstly Java-script program will automatically check the presence of the needed Plug-ins in his/her browser, if not installed, a web-link is provided to download and to install them, the update of these plug-in is assured automatically in all majors browsers, then system will invite student to connect his headset, Neurosky headset is plug and play technology so it can be easy pared with student system, thereafter student choose the experiment or simulation that he/she want to practice, the system will remember him to do Pre-check test of the headset (physical state, battery power level, and other abnormalities) battery level is checked also during interaction via byte 5 Data stream (Table 1). Subsequently system will check if the head-set is correctly primed, if not, student is prompted to re-check his headset, an additional helps materials can be provided to help him/her to perform this task. If no problem the system will synchronizes the start of the learning material and ITS, during

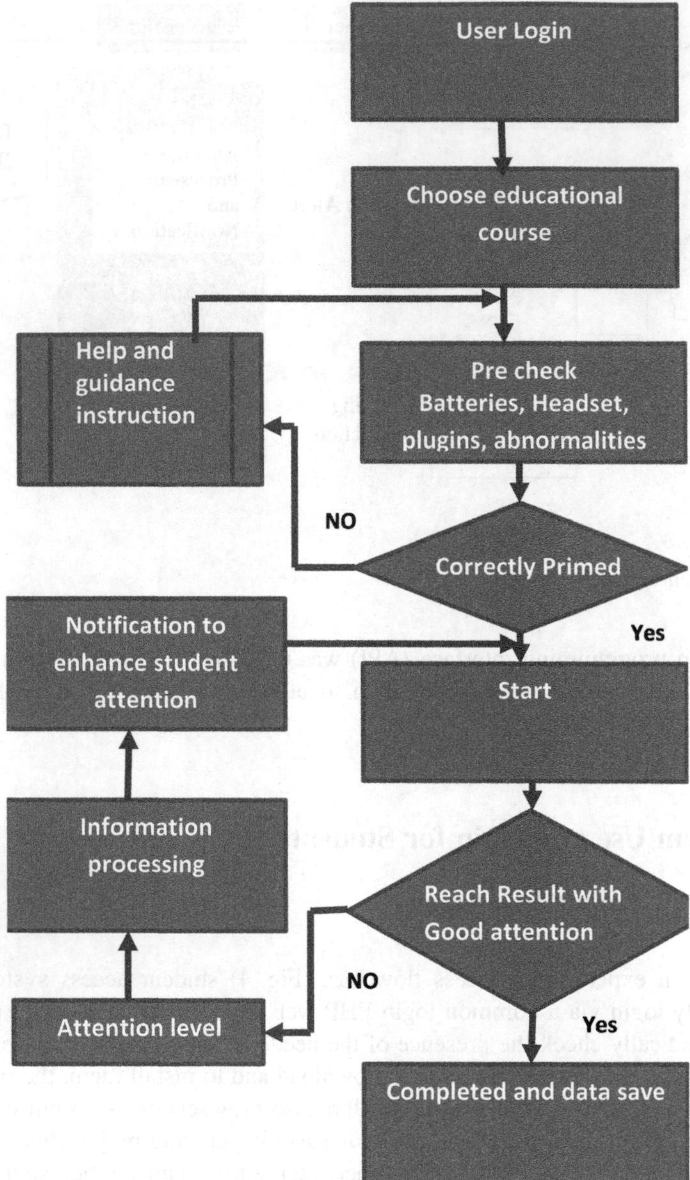

Fig. 4 Flow chart of system use

interaction with Web-Lab experiment/simulation materials, ITS modules read EEG Data acquisition through TCP/IP protocol, and parses student signal to detect his/her attention level dynamically.

For any attention detected abnormalities, Attention processing modules will find appropriate reaction according to predefined strategies until good results in learning process are reached with good attention level. Teachers or institution may be interested also in further data analysis to understand which parts of e-learning system provoke lack of students attention, because of bad system design like (loud various alarms, wrong positioning of webcam, indistinguishable number in counter, etc.), this will help for future system materials enhancement; data can be stored in database, and used for Signal processing, data analysis, and data presentation, this can be programmed in (Math-Lab, Ms-Excel, OpenVibes, etc.). There are some common digital signal processing algorithms often used for EEG study by the researchers, for example, Fast Fourier Transform (FFT), Bispectral Power Spectrum and Power Spectral Density (PSD), Wavelet Transform (WT), filter theorems etc.

4.2 Strategies and Reaction for Attention Student Gain

Main gain of this approach is that student learns online with good attention. By controlling his/her attention and to make students post-training attention level better than pre-training, in his research, Posner and Rothbart [18] thought that the attention training can not only enhance attention but also extend the influence to the other cognitive functions needed in our life, for example, intelligence. Posner and Raichle [19], also thought that attention training is a special repetitive practice concept with respect to some kind of specific cognitive function and make the efficiency of the subject's attention better. In OE attention and concentration plays a big role in learning process.

Student gain after use of such proposed system is not limited only to determine (detect) low or high levels of attention, but to improve or sustain learner's attention, this system should also provide feedback for the learner attention (react). Our model of attention considers Keller's approaches to sensitively deliver feedback aimed at educating or supporting the learner's attention. Table 2 presents Keller's strategies.

Table 2 Outputs taken from Keller's ARCS (REF) model

Attention value	Reaction type
0	1: Alerting student about the problems to receive attention signals
1–20	2: Change status quo like asking the learner to return at a later time or take a pause
21–40	3: Ask learner about lack of his/her attention, is the student tired? Does he/she found any troubles? Would he/she like to come back at a later time?
41–66.66	4: No notification attention is good
66.67–83.33	5: Learner whether he/she wants to explore more material or examples
83.34–100	6: Student receives felicitation message from teacher

5 Methodology, Experiences, and Results Discussion

Validation and usability of our methodology. Experience was conducted among students in the university enrolled in C programming course, an online course was developed, to get a better understanding of the way students use the learning materials and the way attention notifications enhances student attentiveness and concentration state during learning, data was collected by observations of a student's focus group. There were 40 subjects including 19 females and 21 males, all these subjects are taking C language programming course for first time, they are healthy and in good mental condition, most were familiar with computers, portable media players and smart phones. Their ages are between 18 and 26 years old.

Students interact with learning material around 9 min. Their attention and concentration level was measured and recorded for each taken chapter. Students get immediate alert notification when their attention level is low.

The results obtained as shown in results graphics (Figs. 5 and 6), it is clearly visible how well the student's attention level is enhanced noticeably from first test to third one. Notification plays consistently big role to improve students' attention with value of 12 % in each test, and around 24 % in global.

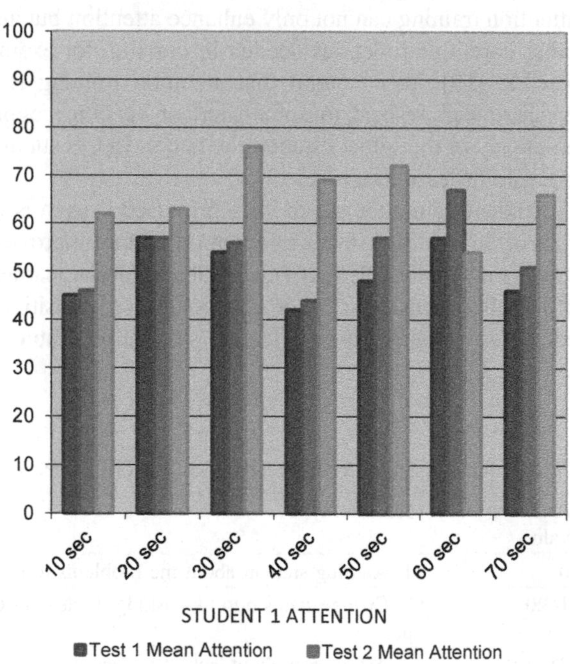

Fig. 5 Global attention results

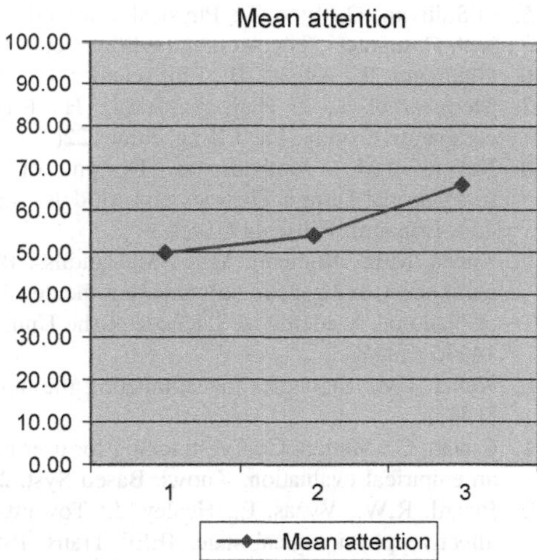

Fig. 6 Mean attention enhancement

6 Conclusion

Todays, all designed numerical devices and systems (Computers, smartphones, TV, Cars etc.) are integrating bio-signals to enhance human interaction. This article shows incorporating Bio-signals in online education, by combining neural input with interaction traits assisted with attention alerting system. The prototype of attention proposed on this paper is based on input from a Brain Computer Interface (BCI) named Neurosky. This headset offers the possibility of detecting brain waves associated to levels of attention. Given the importance of attention in education, we propose a model targeting attention in order to give personalized feedback to student and training him during learning. The innovation of our approach includes of combining a novel form of computer input with online education. Work for the future consists on integrating more brain waves such (Movement Imagination, P300, SSVEP and ERP-Analysis). Allow us to conclude the suitability of using it in educational with computers and whether our model approach can be used to enhance learning level.

References

1. Weinberger, N.M.: Cortical plasticity in associative learning and memory. In: Byrne, J.H. (ed.) Learning and Memory: A Comprehensive Reference. Oxford, UK, Elsevier Ltd (2008)
2. Winer, J.A., Schreiner, C.E. (eds.): The Auditory Cortex. Springer, New York, NY (2011)
3. Malone, T.: What makes Things Fun to Learn? A Study of Intrinsically Motivating Computer Games. Xerox Palo Alto Research Center, Palo Alto (1980)
4. Pintrich, P.R., De Groot, E.V.: Motivation and self—regulated learning components of classroom academic performance. J. Educ. Psychol. **82**(1), 33–40 (1990)

5. O'Sullivan, D., Igoe, T.: Physical Computing Sensing and Controlling the Physical World with Computers. Thomson, 1st edition (2004)
6. Graimann, B., Allison, B., Pfurtscheller, G.: Brain-Computer Interfaces. Springer (2011)
7. Mougharbel, I., El Hajj, A., Artail, H., Riman, C.: Remote lab experiments models: a comparative study. Int. J. Eng. Educ. **22**(4) 2006
8. Keller, J.M.: Motivational Design of Instruction. In: Reigeluth, C.M. (ed.) Instructional-Design Theories and Models: An Overview of Their Current Status, pp. 383–434. Erlbaum, Hillsdale (1983)
9. Rueda, M.R., Rothbart, M.K., McCandliss, B.D., Saccomanno, L., Posner, M.I.: Training, maturation, and genetic influences on the development of executive attention. In: Proceedings of National Academy of Sciences of the United States of America, vol. 102(41), pp.14931–14936 (2005)
10. Keller, J.M.: Strategies for stimulating the motivation to learn. Perform. Instr. J. **26**(8), 1–7 (1987)
11. Conati, C., Merten, C.: Eye-tracking for user modeling in exploratory learning environments: an empirical evaluation. Knowl. Based Syst. **20**(6), 557–574 (2007)
12. Picard, R.W., Vyzas, E., Healey, J.: Towards machine emotional intelligence: analysis of affective physiological state. IEEE Trans. Pattern Anal. Mach. Intell. **23**(10), 1175–1191 (2001)
13. Qu, L., Johnson, W.L.: Detecting the learner's motivational states in an interactive learning environment. In: AIED2005. IOS Press (2005)
14. del Soldato, T., du Boulay, B.: Implementation of motivational tactics in tutoring systems. Int. J. Artif. Intell. Educ. **6**, 337–378 (1995)
15. Berger, H.: On the electroencephalogram of man. In: Gloor, P. (ed.) The Fourteen Original Reports on the Human Electroencephalogram. AMBterdam (1969)
16. Robertson, I.H., Garavan, H.: Vigilant attention. In: Gazzaniga, M.S. (ed.) The Cognitive Neurosciences III, pp. 631–640. MIT, New York (2004)
17. Fan, J., McCandliss, B.D., Fossella, J., Flombaum, J.I., Posner, M.I.: The activation of attentional networks. Neuroimage **26**, 471–479 (2005)
18. Posner, M.I., Rothbart, M.K.: Educating the Human Brain. Washington, DC, American Psychological Association (2007)
19. Posner, M.I., Raichle, M.E.: Images of Mind. Scientific American Library, NY (1996)

Part II
Artificial Intelligence in Education

Units' Categorization Model: The Adapted Genetic Algorithm for a Personalized E-Content

Naoual Chaouni Benabdellah, Mourad Gharbi and Mostafa Bellafkih

Abstract This paper presents the model of units' categorization, which aims to improve course materials' difficulties and to maintain learners' motivation. Moreover, it presents the application of the adapted genetic algorithm to compute the average and the maximum fitness function that reveal the pertinence of a new parameter. However, the purpose is to propose the most pertinent and adapted units to a learner. The relevance of a unit is quantified by the success of several learners and units' belonging to the right pedagogical sequence. The results obtained by the implementation of the adapted genetic algorithm are based on the proposed model. Therefore, there is a thorough analysis of the convergence of the average fitness function with the maximum fitness function.

Keywords Genetic algorithm • E-Content adaptation • Learners' motivation

1 Introduction

Most of the Learning Management Systems (LMS) that are currently used by educational institutions do not fit the learner's needs [1]. Many personalized systems consider learners' preferences, interests, browsing behaviors, and abilities [2]

N.C. Benabdellah (✉)
LIMIARF and STRS Laboratories, Faculty of Science, Mohammed V University, and National Institute of Post and Telecommunication (INPT), Rabat, Morocco
e-mail: naoual.chaouni@gmail.com

M. Gharbi
LIMIARF Laboratory, Faculty of Science, Mohammed V University, Rabat, Morocco
e-mail: mourad.gharbi@fsr.ac.ma

M. Bellafkih
STRS Laboratory, National Institute of Post and Telecommunication (INPT), Rabat, Morocco
e-mail: Bellafkih@inpt.ac.ma

Á. Rocha et al. (eds.), *Europe and MENA Cooperation Advances in Information and Communication Technologies*, Advances in Intelligent Systems and Computing 520, DOI 10.1007/978-3-319-46568-5_15

in providing personalized services. Nonetheless, learners' motivation is usually neglected as an important factor. Students enrolled in foundational computer science courses and adopted maladaptive profiles learned less than those who adopted adaptive learning profiles [3].

Motivation involves two aspects: intrinsic and extrinsic [4]. Intrinsic is the result of an expectation and challenges goals. Although, extrinsic motivation [5] is the individual attitude toward a clear direction, reward and recognition, it is usually based on a social pressure and competition. The proposed presentation of units in this paper follows the requirements of ARCS model [6, 7]. The attention is satisfied by the numerous resources used to design a unit including self-assessments. In addition, the relevance is the objective of the proposed distribution of units in five categories. Furthermore, the satisfaction is the fourth and the last element that belongs to ARCS model; it refers to the positive feeling that a learner has towards their learning experience. The course designed within this sequence has the objective to maintain the height of learners' motivation [8].

Genetic algorithms are inspired by natural behavior of selection individuals. In 1960, his colleagues and students from Michigan University, worked together and are the origin of the research on genetic algorithm [9]. Besides, many projects have the subject to apply genetic algorithm to personalize e-course [10–12] by adapting the algorithm like using the forcing legality concept before fitness value calculation step. Authors proposed an improvement of genetic algorithms for solving multi-objective optimization problems. In each generation, the best population is updated [13].

This paper is divided into four sections. After the introduction, an overview on the proposed model and the adapted genetic algorithm are given. In the third section, two tests cases and findings in each case are exposed. In the fourth section the summary of contributions and some future works are cited.

2 Proposed Model and Genetic Algorithm

In this section, the proposed model composed of five categories is explained, its validation by using ASK (Attitudes, Skills and Knowledge) is verified. Besides, based on the returns defined and the advancement of learning in his learning and regarding to the model the problem is formulated and the genetic algorithm is explained with the major adaptations. To confirm the validity of the proposed model is necessary.

2.1 Proposed Model

The course we want to develop and personalize is language development in computer science. It is designed by chapters. The units are represented as Sharable

Content Object SCO. Each SCO contains a mini test as well as a self assessment, objective of the unit, and the core which is the message that any learner should memorize because they will be tested on it. Units in chapters are classified in a pedagogical sequence. There are five categories or chapters: awareness, exposure, appropriation, assimilation and production.

Awareness is the first category that includes units that introduce the course to learners and its context as an objective. The beginning as an implicit learning is for to give a generalization of the course materiel. It answers the following questions: 'Why is it important to study this course?' and 'If a learner masters the content, what profession he can have in future?'

The exposition is the second category that contains facts and concepts in order to develop student knowledge and skills for the next level [14]. The proposed category considers the main knowledge that students must remember. At this level, the comprehension is not yet sought.

The second category is the appropriation. It covers all basic concepts that the learner needs at that stage.

In addition, direct access to the knowledge does not guarantee assimilation. Neither the speed of access to knowledge nor the services reduced the difficulty linked to the knowledge [14]. The fourth category is where the important ideas composed must be assimilated. Production in this study concerns the ability of learners to produce complex tasks.

As it was explained before, the awareness chapter contains generalizations. If a learner fails in a unit, he is asked to review this last. He must validate it for pursuing his study. Moreover, It considered that for a learner who cannot make associations between concepts and their uses, assimilate a coding problem, and produce a complex code must go back to revise or study the chapter or the category of exposition. In addition, when a learner succeeds in all the units starting from awareness chapter, exposure, appropriation, assimilation and production he validates the e-course content. Possible returns are presented from a unit in a specific category to another unit from different category (Fig. 1).

2.2 Problem Formulation

The problem formulation is as the following:

- A— → Awareness, B— → Exposure, C— → Appropriation, D — → Assimilation, E— → Production
- (x, y)— → denotes units in {A, B, C, D, E}
- Ap— → is a learner that is represented by Ap [GS, Cp, S, F] // one learner having a specific profile GS and making many passages on units in sequence and/or not Cp, made several Successes S and failures F.
- Cp— → (Cpr, Cpn), Cpr denotes the number of time that unit is in the right sequence. That match with the model proposed in Fig. 1. Cpn denotes the

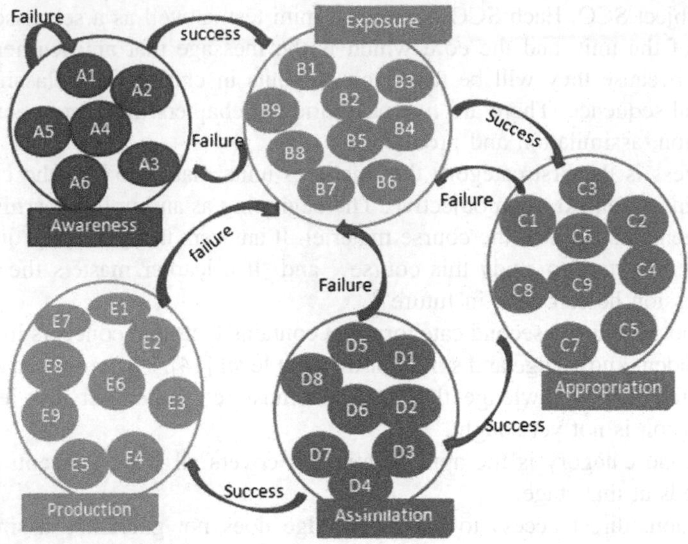

Fig. 1 Categories in sequence and possible returns

number of the time that the unit is in wrong sequence that not matches to any passages defined in the model. Cp is the sum of both Cpr and Cpn.

- S— → denotes the sum of success in the unit.
- F— → denotes the sum of failure in the unit.
- Apk = {Ap1, Ap2……..Apk}— → denotes K learners (the number of learners).
- GSl = {GS1, GS2…., GSl}—- → denotes learners' profile.

Each profile is represented by a global score which is obtained using the Fuzzy Logic Method [15]. To compute the global score (GS) four criteria are considered: positioning test score, elapsed time in that positioning test, unit test score, and elapsed time for unit's learning. Therefore, in this paper the learner's profile value is considered as a random value taken from a list of values. GS parameter is computed as a variable in fitness function as well as Cp of units' sequence, success, and failure.

2.3 Adapted Genetic Algorithm

The genetic algorithm begins by creating a random initial population. The initial population is in our case the units' course. These are represented in a graph. The purpose is to look for the best path for each learner's profile.

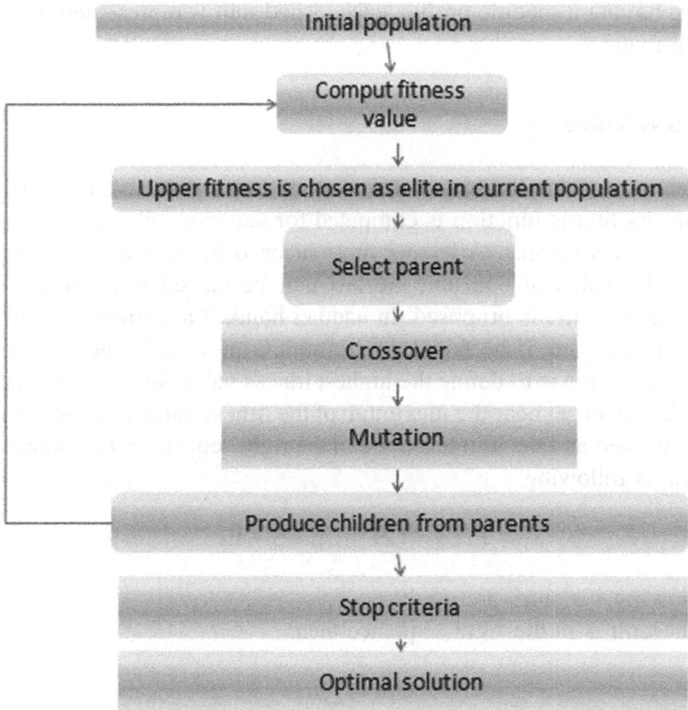

Fig. 2 Genetic algorithm process

The algorithm creates a sequence of some selected population which represents a new population. At each step, the algorithm is based on the current selection of units to create the next new selection or population. To create the new population, the algorithm executes the following steps shown in Fig. 2.

Accordingly, this paper proposes an adaptive genetic algorithm. The concept of pedagogical sequence operator is included in fitness value calculation with a value of success and a value of failure.

2.3.1 Initial Population

The algorithm starts from a population of initial solution randomly generated. The number of chromosomes as the initial population represents the number of units' course [16]. Units are encoded as [U1k, U2k, …, Unk], n number of units visited by a learner Ap having a specific profile, each visited unit is represented by 0 value or 1 value in a matrix.

2.3.2 Fitness Value

Fitness value is named individual evaluation function or objective function whereby for each unit its fitness function is computed for success, failure, and for a visited unit in pedagogic sequence. A learner must succeed in all units in a category. On one hand, if he failed in one unit the last unit he tacked is checked first; if he succeeded the unit after is proposed. In another hand, if a learner failed in one unit the last unit is checked; if he failed the last unit is proposed. The objective of the global fitness function is to define the highest fitness value for a unit in a population F(U). The need is to get near the maximum of the fitness value of a unit, where most of learners succeed and the unit has to be in the right sequence. The fitness function is computed as following:

$$F(U) = GS * (Cp + S - F) \quad (1)$$

The following variables are used:
When the unit is in the right sequence then:

$$Cpr = \sum \alpha 1 \quad (2)$$

When the unit is not in the right sequence then:

$$Cpn = - \sum \alpha 1 \quad (3)$$

Global calculation of Cp is

$$(Cpr + Cpn) \quad (4)$$

If a learner succeeds in a unit then:

$$S = \sum \alpha 2 \quad (5)$$

If a learner failed in a unit then

$$F = \sum \alpha 3 \quad (6)$$

U = [U1k, U2k, …, Unk], k represents number of units.

0 <= GS <= 1, we remind that GS is the global score obtained thanks to fuzzy method.

2.3.3 Selection

Starting from an initial population, the selection step creates parents to crossover. The best parents are chosen from their fitness. The way the selection process is made by one of several methods. The selection method, roulette wheel is applied. In fact, better the unit is, more it has the chance to be selected. The roulette wheel method consists of calculating the sum of all units' fitness, and then it generates random number from 0 to the sum computed. In addition, when the sum fitness is greater than the random number the roulette wheel algorithm returns the current unit. However, selection happens every generation.

2.3.4 Crossover

Crossover operator is a basic concept of genetic algorithm. There are many possibilities to make crossover to the units in a population. Uniform crossover method is applied. Thus, the principal of this last is that parts are selected randomly from the first parent and the second parent to create a new parent. The objective of crossover is to ensure the convergence to the local minimum/maximum.

2.3.5 Mutation

Mutation operator is used to explore more units. The rate of the probability is set to 0.003. A big value is not appropriate to not lose the population coverage. The objective of the mutation is to satisfy the diversity. The frequency of using mutation is considerably less than using crossover and selection.

3 Findings

Matlab is used to analyze the efficiency of the model to propose the pertinent and the relevant unit that correspond to learners. The program of genetic algorithm is detailed [16]. Fitness function is coded to fit the design model of units' categories. An application of genetic algorithm was proposed that may be adapted using multi-variables [17]. The adapted fitness function in this work is tested based on the Mitchell's proposed algorithm. The convergence between average function value and the maximum is obtained.

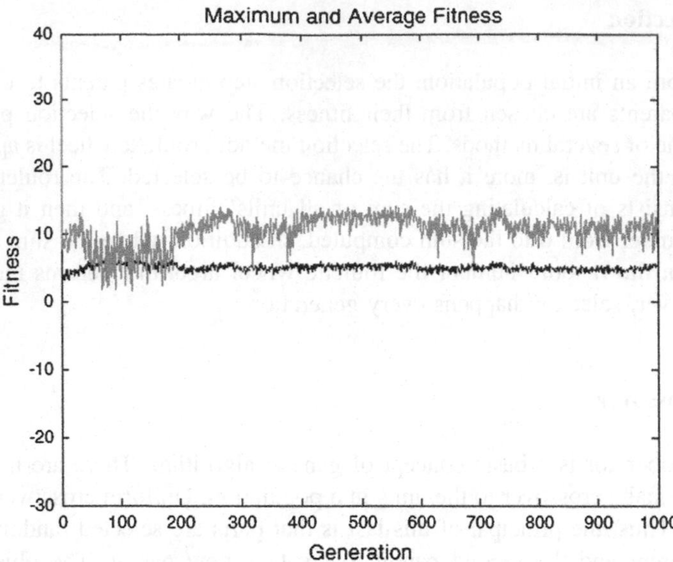

Fig. 3 Considering 150 units, the convergence is observed from 100 to 200 generations

3.1 *Case 1*

The number of iterations is set to 1000. And the number of units is 150. $\alpha 1 = 0.8$ is the constant that will be added each time the unit is in the right sequence (the same constant will be subtracts if the unit is in the wrong sequence) as it shown in formula 2 and 3 previously defined in fitness value section; $\alpha 2 = 0.6$ is the constant that will be added in case of learner's success (see formula 5); $\alpha 3 = 0.2$ is the constant that will be subtract in case of learner's failure (see formula 6). In the Fig. 3 the convergence is obtained from 100 generation to 200 generation. With these parameters to vary their values is recommended to observe the impact on the results.

3.2 *Case2*

In this case, the impact of the constants α1, α2, α3 is discussed in term of the functions convergences. The number of units is set to 100; the number of iteration is kept the same which is 1000; and the constants' values are $\alpha 1 = 1$; $\alpha 2 = 0.8$; $\alpha 3 = 0.2$.

However, in Fig. 4, the convergence is started from 300 generation to 500 generation. By increasing α1 the convergence is kept longer but it starts from 300 generation. It stands during 200 generations.

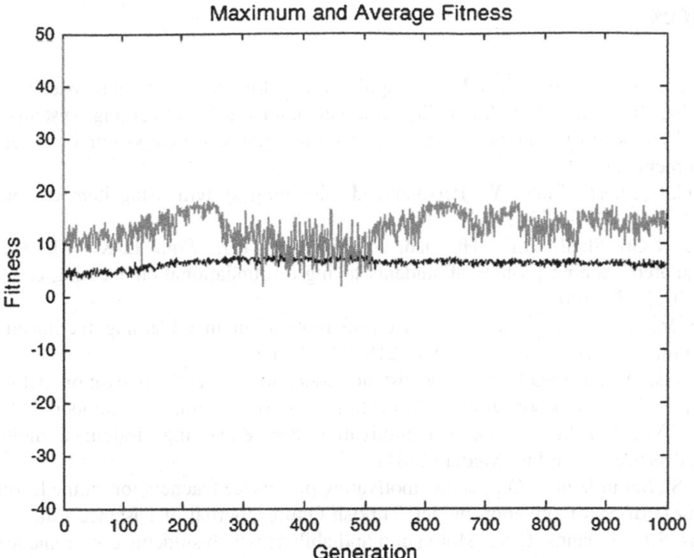

Fig. 4 For 100 units, number of generations 1000 and α1 = 1> α2 = 0.8 > α3 = 0.2, the convergence of average fitness function with the maximum function is obtained from 300 to 500 generations

4 Conclusion

In this study, a pedagogical model is proposed of the units' categorization. The objective is to maintain learner's motivation during the learning path. To analyze the pertinence of the proposed units based on the model (awareness, exposure, appropriation, assimilation, production) the genetic algorithm is applied. In fact, the fitness function was adapted to apply the genetic algorithm by taking into consideration the categorization of units and the sequence of their successions, successes in each unit, failures, and the global score which differentiates between learner's profiles.

Moreover, we need to trace the evolution of the learning algorithm to analyze the pertinent units and from which category they are proposed to the learner. Before this present experience in this paper, ant colony algorithm was used but it doesn't allow the traceability of the belonging of the units to the categories of the model. With genetic algorithm the belonging is shown during the learning process of the algorithm.

Accordingly, the convergence of the average fitness function with the maximum shows that the proposed units are the most pertinent in term of the successes and their sequence. When we increase the constant α1 of Cp criteria of pedagogical sequence of the unit, we obtain the convergence.

References

1. Graf, S., Kinshuk, K.B., Khan, F.A., Maguire, P., Mahmoud, A., Rambharose, T., Shtern, V., Tortorella, R., Zhng, Q.: Adaptivity and personalization in learning systems based on students' characteristics and context. In: The 1st International Symposium on Smart Learning Environment (2012)
2. Chen, C., Lee, H., Chen, Y.: Personalized e-learning system using item response theory. Comput. Educ. 237–255 (2005)
3. Nelson, K.G., Shell, D.F., Husman, J., Fishman, E.J., Soh, L.-K.: Motivational and self-regulated learning profiles of students taking a foundational engineering course. J. Eng. Educ. (2015) 74–100
4. Law, K.M., Lee, V.C., Yu, Y.T.: Learning motivation in e-learning facilitated computer programming courses. Comput. Educ. 218–228 (2010)
5. Goulimaris, D.: The relation between distance education students' motivation and satisfaction. Turkish Online Journal of Distance Education, (2015) (Volume: 16 Number: 2 Article 2)
6. Keller, J.M., Suzuki, K.: Learner motivation and e-learning design: a multinationally validated process. J. Educ. Media (2004)
7. Izmirli, S., Sahin Izmirli, O.: Factors motivating preservice teachers for online learning within the context of arcs motivation model. Turkish Online Journal of Distance Education (2015)
8. Hoskins, S.L., Johanna, C.V.: Motivation and ability: which students use online learning and what influence does it have on their achievement? Br. J. Educ. Technol. 177–192 (2015)
9. Holland, J.H.: Adaptation in Natural and Artificial Systems. University of Michigan Press (1975)
10. Li, J.-W., Chang, Y.C., Chu, C.-P., Tsaic, C.C.: A self-adjusting e-course generation process for personalized learning. Expert Syst. Appl. 3223–3232 (2012)
11. Azough, S., Bellafkih, M., Bouyakhf, H.: Adaptive e-learning using genetic algorithms. Int. J. Comput. Sci. Netw. Secur. (2010)
12. Truong, H.M.: Integrating learning styles and adaptive e-learning system: current developments, problems and opportunities. Comput. Hum. Behav. **55**, 1185–1193 (2016)
13. Long, Q., Wu, C., Huang, T., Wang, X.: A genetic algorithm for unconstrained multi-objective optimization. Swarm and Evolutionary Computation (2015) Paper in press
14. Webb, E., Jones, A., Barker, P., van Schaik, P.: Using e-learning dialogues in higher education. Innovations in Education and Teaching International (2004)
15. Benabdellah, N.C., Gharbi, M., Bellafkih, M.: Learner's profile definition: fuzzy logic application. Int. J. Comput. Sci. Electron. Eng. (2013)
16. Guo, C., Yang, X.: A programming of genetic algorithm in Matlab7.0. Mod. Appl. Sci. **5**(1) (2011)
17. Mitchell, M.: An Introduction to Genetic Algorithms, pp. 62–75. Cambridge, Massachusetts London, England, Fifth printing 3 (1999)

Researching of the Problem of Solution Automation of Software Systems Compatibility

Denis Polikashin, Arslan Enikeev and Victor Georgiev

Abstract During software development is often a lack of backward compatibility between different versions of software components. It will lead to: violation of the ecosystem that is built around this software component, uncoupling community of programmers who use this software in their projects, which entails the general slowdown in the implementation of projects related to this ecosystem and the development of the ecosystem itself; the use in the development of software components that are absolute, which in turn entails problems with safety; the economic costs, during the implementation of a software system migration to a new version of the software component.

Terms, symbols and abbreviations

Term/concept	Symbol
Problem(s) leading to the loss of backward compatibility between program versions	PBC
Target software	Software whose code must be modified to pass to the new version of the used program component
Abstract syntax tree	AST
Programming language	PL

D. Polikashin · A. Enikeev (✉) · V. Georgiev
Kazan Federal University, Kazan, Russia
e-mail: a_eniki@inbox.ru

Á. Rocha et al. (eds.), *Europe and MENA Cooperation Advances in Information and Communication Technologies*, Advances in Intelligent Systems and Computing 520, DOI 10.1007/978-3-319-46568-5_16

1 Introduction

Modern software development methodology is based on several key principles:

1. modularity and code reuse;
2. short development cycles;
3. continuous implementation.

These practices are widespread and have a huge positive effect, increasing the dynamics of software development and its qualitative characteristics.

Open source ideology and such projects as github, sourceforge, stackoverflow [Appendix 12, 21, 22] and other analogues thereof, directly contribute to this beneficial effect. They allow to involve into solving the problem not only thousands of individual developers from around the world, but also the experience of key IT companies and research laboratories, such as Google, Yandex, IBM, Intel through software components provided by them [Appendix 13–15, 16, 23]. Since the beginning of their appearance, the latest ideas and developments get directly into the hands of programming community and developers start to use them in projects all over the world. All this leads to positive effects such as the increase of the dynamics of the appearance of final software and components for its creation and their adaptation to changing conditions.

However, careful application of the best practices of the present and the past does not guarantee the absence of problems in the production of software and its further support and refinement [1–5, 8–10]. History of developing software products shows common appearance of architectural and ideological problems, that can be solved only by a change in the architecture of these programs. In some cases, these changes are impossible to be made without losing backward compatibility with its previous version. In the case of software components, at the time these problems are revealed they are generally present in many other programs. That consequently leads to numerous negative effects. Obvious examples are the releases of popular software components such as OpenCV, Node.js, AngularJS, Ruby, Qt, DirectX [Appendix 11, 17, 18–22].

Depending on whether a library is distributed in source or binary form, two notions of interface compatibility are relevant.

1. Source compatibility ensures that every program that compiles the old library implementation is also compiled with the new library implementation.
2. Binary compatibility ensures that every program that links to the binary form of the old library also links to the binary form of the new implementation.

In the following, only source compatible library implementations are considered.

The some researches give some methods, techniques and softwares to verify backward compatibility or equivalence of programs [1–10].

In this regard, it is necessary to create a relevant tool for the implementation of the automatic transfer of the application source code to the actual version of the library used.

Therefore, the aims of this work are:

1. To identify the concepts and components involved in the migration of software systems between different versions of libraries used.
2. To identify key architecture requirements to be met by the software implementing the automation of the solution to the problem of transition between the versions of the libraries without backwards compatibility.
3. To identify architectural solutions allowing to solve the problem based on identified needs.

2 Research Domain

The migration of software systems between the different versions of libraries used in them affects a number of different elements.

The first are programming languages. Considering programming language, namely its syntax, is one of the main elements to pay close attention to, because it requires direct interaction not only in searching for differences leading to the PBC between software component versions, but also in migration of the source target program between different versions of the library used. Along with the basic syntax of PL, frameworks expanding their syntax are no less important (examples: Qt, Groovy, Kotlin, etc.) and other specialized libraries (examples: OpenCV, OpenCL, IntelTBB, Boost) building their own ecosystem around them.

Secondly, it is the set of possible changes to the source code of a software component as the specific subset of these changes leads to a lack of inter-version backward compatibility [4].

Third, these are the algorithms of engine PBC search. It is necessary to take into account the peculiarities of each of the PBC to make the best of their automatic detection [1–10].

Fourth, these are the algorithms of PBC elimination in the source code of the target software. These algorithms are designed to simplify the target code migration from one version of the involved library to another, both by giving detailed instructions and tips on carrying out this process to the developer, and automatic target application code transformation to eliminate PBC where it is possible.

Fifth, these are ways of obtaining baseline data to compare the versions of the libraries. The data source may be found in the library source code if there is any, and in alternative sources of documentation files in different formats or in the course of a direct dialogue with the developer.

3 Basic Requirements to the Architecture

Taking into account than in the process of software development aimed at solving the PBC a large number of different components should be considered, such as:

4. syntax:
   (a) programming languages;
   (b) frameworks and specialized software libraries;
   (c) formats of documentation submission;
5. classes of problems leading to the loss of compatibility between the versions of the libraries;
6. algorithms of search and elimination of PBC.

That's why the architecture should support the following conditions.

- Scalable—that is to have the ability of adding new entities and functions into the system without disturbing its basic structure. At that making the most probable changes should require the least effort that is to provide an easy way to add the functionality of the following types:
  - The support of not only new programming languages, frameworks, but of other syntactic constructs including their new versions and PL standards.
  - New ways of PBC search and elimination.
- Flexible—that is to allow parallel operation of the components responsible for different functions. Also the aspects of reliability, productivity, scalability are important.

4 System Architecture

According to Mark Richards "Software architecture patterns" report, to meet the above-stated requirements, microkernel architecture of the application was chosen [6] (Fig. 1).

This architecture consists of two components: the main system (kernel) and plug-ins (Fig. 2). The kernel contains minimum of the business logic, but directs loading, offloading and launching of the necessary plug-ins. Thus, the plug-ins are unrelated to each other.

As the plug-ins can be developed independently of each other, the developed system will provide maximum flexibility and scalability. Independent development of plug-in provides an easy testing of each of them, that increases the overall reliability of the whole system.

Productivity of applications built on the basis of this architecture, is directly dependent on the number of connected and active modules. Also, in case of

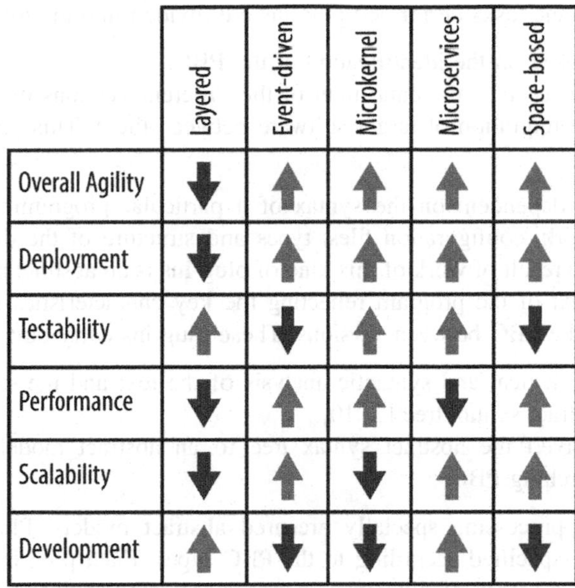

Layered
Event-driven
Microkernel
Microservices
Space-based
Overall Agility
Deployment
Testability
Performance
Scalability
Development

Fig. 1 Architecture of the application

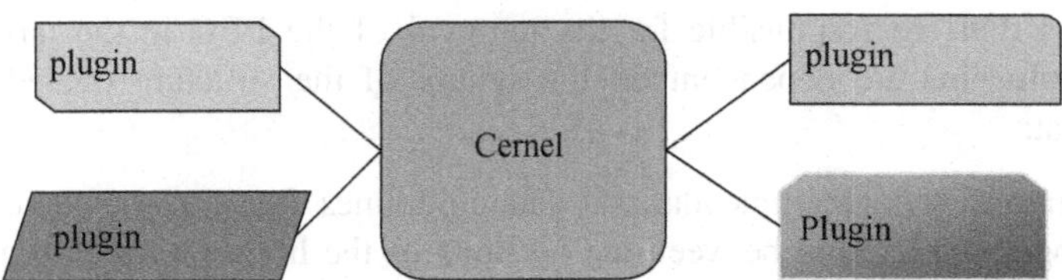

Fig. 2 Components

necessity of greater performance, this architecture can be easily modified in microservicing by taking the main logic of the plug-in out to a separate service. In this case, a plug-in is used only for communication of the microservice with the program kernel.

Examples of successful microkernel architecture are such software development tools like Eclipse IDE, SublimeText.

The kernel of the system provides a command line interface of the kernel adjustment and the actions of search and download of the plug-ins from the external storage. Also, the main purpose of the kernel is work with the loaded plug-ins:

1. Install, activation, deactivating and deinstallation.
2. Configuration.
3. Coordination of plug-ins.
4. Interaction between plug-ins and user.

According to the tasks performed plug-ins are divided into the following types:

1. Plug-ins focused on the identification of the PBC.
   The program receives the data input of the different versions of the library to carry out the migration of target software between them. This plug-in class is divided into:
   (a) Plug-ins dependent on the syntax of a particular programming language (framework, configuration files, types and structure of the documentation file). The result of work of this kind of plug-ins is an abstract representation of the text of the program reflecting the key characteristics necessary for finding the PBC between versions. These plug-ins carry out:
      (i) The lexical and syntactic analysis of the text and the structure of an abstract syntax tree [2, 10].
      (ii) Convert the abstract syntax tree to an abstract model suitable for searching PBC.
   (b) Plug-ins processing specially prepared abstract models. Plug-ins of this class are specified according to the PBC types. Each plug-in of this class conducts a search of differences between the texts of different versions of the test program and semantic analysis to find a strictly defined PBC. The result of work of this type of plug-ins is a list of changes leading to the loss of backward compatibility [1, 7, 10].
2. Plug-ins that are responsible for the removal of the PBC in the target code. These plug-ins are dependent on the syntax of the structure treated and are aimed at:
   (a) Bringing the recommendations and guidelines for implementation of the program migration between the versions of the library by the programmer. Recommendations can be presented in a separate document, as well as integrated into the source code of the target program.
   (b) Implementation of the automated transformation of the source code of the program and target configuration files of software build system.

The plug-ins working with the source code or documentation text admit syntactically correct constructions. Plug-ins are not intended to perform the functions of static and dynamic code analysis to detect errors.

5 Conclusion

This article provides the results of a study aimed at identifying the basic concepts and components involved in the migration of the source code applications from earlier to more recent versions of the libraries used in them.

It also defines basic requirements for the tool architecture allowing programmers to transfer automatically the application source code to the actual version of the library used.

Basic architectural solutions are stated. The article also contains a classification of the components of the developed software and distinguishes basic directions of further work.

Further work in this direction is to test the functionality of the considered architectural solutions by creating working models from the components of a software system presented in this work.

References

1. Fluri, B., Wursch, M., Pinzger, M., Gall, H.C.: Change distilling: tree differencing for fine grained source code change extraction. IEEE Transactions on Software Engineering №11, pp. 725–743 (2007)
2. Neamtiu, I., Jeffrey, S. Foster, Hicks M.: Understanding source code evolution using abstract syntax tree matching. In: MSR '05 Proceedings of the 2005 International Workshop on Mining Software Repositories, New York, USA (2005)
3. Ponomarenko, A., Rubanov, V.: Automated verification of shared libraries for backward binary compatibility. In: VALID '10: Proceedings of the 2010 Second International Conference on Advances in System Testing and Validation Lifecycle. IEEE Computer Society (2010)
4. Ponomarenko, A., Rubanov, V.: A combined technique for automatic detection of backward binary compatibility problems. In: PSI'11: Proceedings of the 8th International Conference on Perspectives of System Informatics, Springer. (2011)
5. Ponomarenko, A., Rubanov, V.: Automated analysis of feedback Linux binary compatibility libraries. In: Seventh-Conference of Developers of Free Software. Pereslavl-Zalessk, 2010 (2012)
6. Richards, M.: Software architecture patterns, O'Reilly (2015)
7. Saukh, A.M.: Semantic analysis of some aspects of the source code of programs on the basis of formal syntax and semantics of the specifications. Appl. Discrete Math. № 5 (adj.), 110–111 (2012) Irkutsk
8. Turbin, A.: Complementary Hashing Subsets. Seventh-conference of developers of free software, Pereslavl-Zalessk (2010)
9. Welsch, Y., Poetzsch-Heffter, A.: Verifying backwards compatibility of object-oriented libraries using boogie. In: Proceedings of the 14th Workshop on Formal Techniques for Java-like Programs, pp. 35–41. NY:ACM New York, USA (2012)
10. Wursch, M.: Improving Abstract Syntax Tree based Source Code Change Detection, Software Evolution and Architecture Lab (Harald Gall), Zurich: University of Zurich (2006). http://www.merlin.uzh.ch/publication/show/2913

Appendix Article

11. AngularJS Developer Guide.Migrating from 1.3 to 1.4. https://docs.angularjs.org/guide/migration
12. GitHub. https://github.com/
13. Google Code Project. https://developers.google.com/

14. IBM DeveloperWorks. http://www.ibm.com/developerworks/
15. Intel Developer Zone. https://software.intel.com
16. Microsoft Developer Technologies. https://dev.windows.com
17. Migrating to Direct3D 11. https://msdn.microsoft.com/en-us/library/windows/desktop/ff476190(v=vs.85).aspx#direct3d_9_to_direct3d_11
18. OpenCV 3.0 API Changes. http://docs.opencv.org/master/db/dfa/tutorial_transition_guide.html#gsc.tab=0
19. Qt API Changes. http://doc.qt.io/qt-5/sourcebreaks.html
20. Ruby version policy changes starting with Ruby 2.1.0. https://www.ruby-lang.org/en/news/2013/12/21/ruby-version-policy-changes-with-2-1-0/
21. Sourceforge. http://sourceforge.net/
22. StackOverflow. http://stackoverflow.com/
23. Yandex Technologies Catalog. https://tech.yandex.ru/

Digit Recognition Using Different Features Extraction Methods

Khalid Zine Dine, M'barek Nasri, Mimoun Moussaoui, Soukaina Benchaou and Fouad Aouinti

Abstract The Numeral features extraction consists of transforming the image into an attribute vector, which contains a set of discriminated characteristics for recognition, and also reducing the amount of information supplied to the system. Several characteristic extractions methods have been proposed in the literature. These characteristics can be digital or symbolic. Mainly, we distinguish two approaches, statistical and structural. In this paper, we are interested in a comparative study of these four methods: profile projection, zoning, cavities and freeman chain code. Digit recognition is carried out in this work through k nearest neighbors. We evaluated our scheme on handwritten samples of the MNIST database and we have achieved very promising results.

Keywords Pattern recognition • Feature extraction • Profile projection • Cavities • Zoning • Freeman chain code • K nearest neighbors

K.Z. Dine (✉) · M. Nasri · M. Moussaoui · S. Benchaou · F. Aouinti
MATSI Laboratory, Superior School of Technology,
Mohammed I University, Oujda, Morocco
e-mail: khalid.zinedine.eia@gmail.com

M. Nasri
e-mail: nasrihome@gmail.com

M. Moussaoui
e-mail: moussaoui.mim@gmail.com

S. Benchaou
e-mail: soukaina.benchaou@gmail.com

F. Aouinti
e-mail: aouinti.fouad@gmail.com

Á. Rocha et al. (eds.), *Europe and MENA Cooperation Advances in Information and Communication Technologies*, Advances in Intelligent Systems and Computing 520, DOI 10.1007/978-3-319-46568-5_17

1 Introduction

The recognition of handwritten script is a difficult task due to the different handwriting qualities and styles that are subject to inter-writer and intra-writer variations. Many recognition systems in many applications have been proposed in recent years where higher recognition accuracy is always desired [1]. Typically, the recognition systems are adapted to specific realization of numerous industrial applications, in particular the numeral recognition which is used in several sectors such as postal sorting, bank check reading, order form processing, etc. [2]. Such system can be divided into three main steps: preprocessing step, feature extraction and selection step, and classification step.

The steps involved in preprocessing are binarization, cropping, normalization, and thinning. In binarization the given image is converted into binary image by computing the average threshold value of the image. Normally the image data is not aligned to the center, so the images are cropped. In normalization all the sample images of the dataset are set to a predefined size since all the images may not have the same size. In thinning the normalized image is reduced to a single pixel thickness.

The Numeral features extraction is a delicate process and it is crucial for a good numeral recognition [3]. It consists of transforming the image into an attribute vector, which contains a set of discriminated characteristics for recognition, and also reducing the amount of information supplied to the system. These characteristics can be digital or symbolic. Mainly, we distinguish two approaches, statistical and structural.

In the literature, several works have been proposed for features extractions such as freeman coding [3], invariant moments [4], Zernike moments [5], Fourier descriptors [6], Loci characteristics [7], and contour direction histogram. A feature set made to feed a classifier can be a mixture of such features.

Classification is the step of decision, which realizes the recognition. It consists of partitioning a set of data entity into separate classes according to a similarity criterion. Different methods are proposed in this context includes neural networks, support vector machines [1], k nearest neighbors, kmeans, hidden Markov models (HMM), etc.

This paper focuses on feature extraction and classification. The performance of a classifier can rely as much on the quality of the features as on the classifier itself. In order to validate our contributions, we have used in this work a database of 6500 handwritten digits, provided by MNIST database, 650 images of each digit. The used database is split into two sets, 5700 digit images for learning and 800 digit images for test.

This work is organized as follows: Sect. 2 describes the feature extraction procedure for the four feature sets. The k nearest neighbors technique of classification is discussed in Sect. 3. The proposed system for digits recognition is presented in Sect. 4. The result of simulations and comparisons are introduced in Sect. 5. Finally, we give a conclusion.

2 Features Extraction

The feature extraction parameter is a step of great importance. If poorly designed, it will be difficult or impossible to carry out effective recognition. After the preprocessing step, the preprocessed image is represented by a matrix of pixels which can be of very large size. So, it will be useful to represent objects by characteristics containing the necessary information. This operation is called features extraction.

A total of four feature extraction algorithms were used which are: The Zoning, the Profile Projection, the Concavities and freeman chain code. The methods are described below.

2.1 Profile Projection

Profile projection is a statistical method successfully used in recognition of handwritten characters [3]. The preprocessing stage consists of binarizing the numeral input image which is presented in grey level. The next stage is to only preserve the numeral position in image by cropping it. Final stage is to fix the size of cropped numeral image.

This method calculates the number of pixels (distance) between the left, bottom, right, top edge of the image and the first black pixel met on this row or column. The dimension of the obtained attribute vector is twice the sum of the number of rows and columns associated to the image of the numeral (Fig. 1).

2.2 Zoning

Zoning is a popular method used in character recognition tasks. In this method, the character images are divided into x zones of predefined sizes and then features are

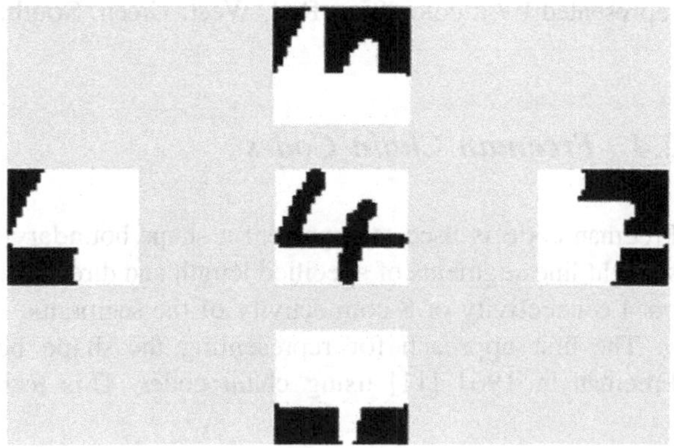

Fig. 1 The four profile projection of digit '4'

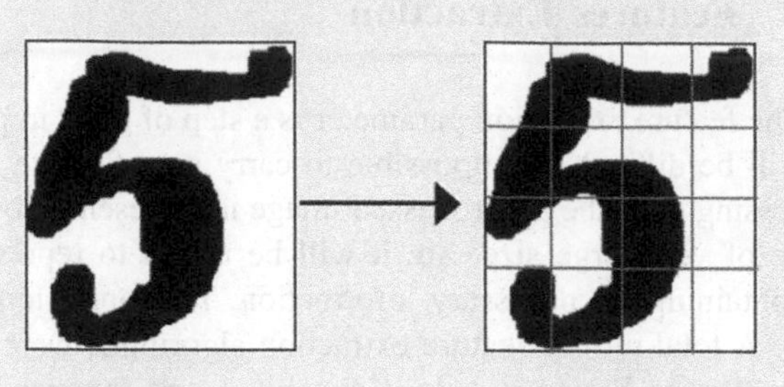

Fig. 2 4 × 4 zoning of digit '5'

computed for each of these zones. Zoning obtains local characteristics of an image instead of global characteristics. Here, we have divided the preprocessed character images into 16 zones (4 × 4) and then pixel density features were computed for each of the zones Fig. 2. The average pixel density was calculated by dividing the number of foreground pixels by the total number of pixels in each zone i [8, 9].

$$D(i) = \frac{\textit{Number of forground pixels in zone i}}{\text{Total number of pixels in zone i}} \tag{1}$$

2.3 Cavities

Almost all the digits include cavities among which the number and the position vary according to the digit and cast iron. We define 5 types of cavities: the West, the North, the South, and Central. It supposes the definition of 4 cardinal directions in the image: the West towards the left, the North upward, etc. [10]. In our work, we use as characteristic parameters, the relative surfaces of the various types of cavities (cavities are normalized by dividing them by the total surface of cavities, and number of cavity).

Figure 3 shows the cavities detected on the numbers. Each type of cavity is represented by a color (Est: Red, West: Green, South: blue, Central: purple).

2.4 Freeman Chain Codes

Freeman code is used to represent a shape boundary by a connected sequence of straight-line segments of specified length and direction. This representation is based on 4-connectivity or 8-connectivity of the segments.

The first approach for representing the shape boundary was introduced by Freeman in 1961 [11] using chain codes. This technique is based on contour

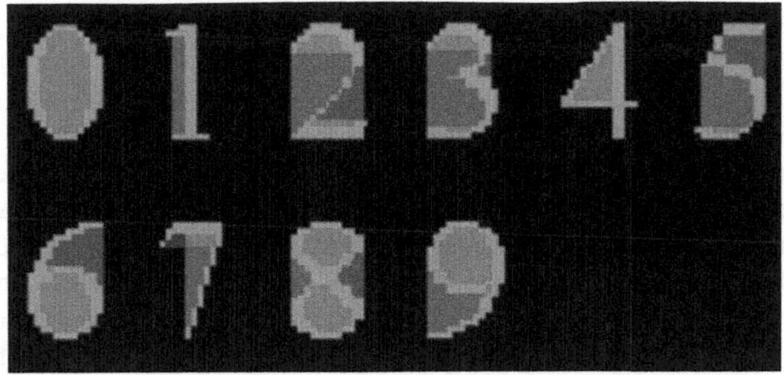

Fig. 3 The different cavities

detection, which is an important Image processing technique. A chain code or Freeman code can be generated by following a boundary of an object in a clockwise direction and coding the direction of each segment connecting every pair of pixels by using a numbering scheme. A code formed from a boundary shape is referred to as a Freeman code [3].

The Freeman code of a boundary depends on the starting pixel P(x0, y0); it was selected as being the first pixel encountered in the boundary of the object. Our aim is to find the next pixel in the boundary or the first coding. There must be an adjoining boundary pixel at one of the eight locations surrounding the current boundary pixel. By looking at each of the eight adjoining pixels, we will find at least one that is also a boundary pixel. Depending on which one it is, we assign a numeric code between 0 and 7 as already shown in Fig. 4. For example, if the pixel found is located at the right of the current location of pixel, a code "0" is assigned. If the pixel found is directly to the upper left, a code "3" is assigned. The process of locating the next boundary pixel and assigning a code is repeated until we return to the starting pixel. Since the size of the chain code varies for different digits, we normalize it to get a feature vector of size 120.

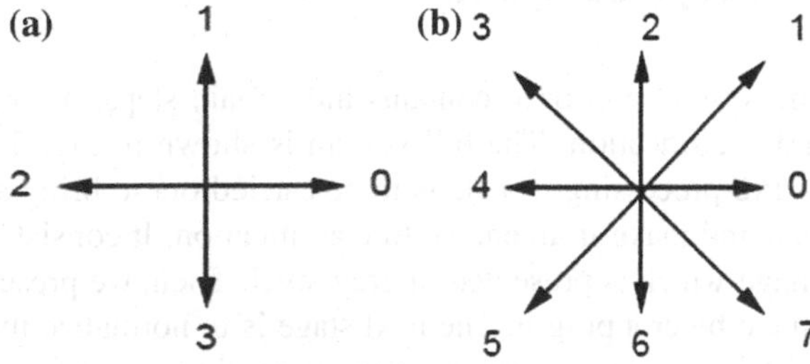

Fig. 4 4-connectivity (**a**) and 8-connectivity (**b**) of Freeman code

3 Classification: K Nearest Neighbors

Classification is the final stage of digit recognition task in which digit images are assigned unique labels based on the extracted features.

There are many techniques of classification. In this study, we have used the k nearest neighbors (KNN) classifier. It has been widely used in handwritten numerals recognition tasks [12] for its conceptual simplicity, theoretical elegance and its robustness. KNN is a method, which was inspired from the closest neighbor rule. It is based on computing the distance between the test sample and the different learning data samples and then attributes the sample to the k nearest neighbors.

This method is apparently simple and easy to program when it is compared to neural network and support vector machine methods. It requires however a large amount of memory to store and examine the data.

Algorithm : K nearest neighbors

-Let L = {(x′, c)/x′ ∈, ℝN, c = 1, 2,…, $\mathcal{C}$} the learning set.
-Let x the example to determine its class.
1: **For each** object x′ of the set L
- Compute the distance between the object x and the object x′, d(x, x′).
- Classify the different distances in increasing way.

End
2: **For every** {x' ∈ KNN(x)} **do:**
- Identify the most frequent class.

End
3: Assign to x the identified class.

---

4 Proposed System

The system proposed contains three main steps, preprocessing, features extraction and classification. The full system is shown in Fig. 5.

Pre-processing of data is to be carried out to bring about some uniformity in the data and make it amenable to classification. It consists of binarizing the digit input image which is presented in grey level. Then, we preserve only the digit position in image by cropping it. The next stage is to normalize the image in a predefined size.

After preprocessing, features extraction is carried out by one of the four techniques described below. For classification task we are used the KNN method.

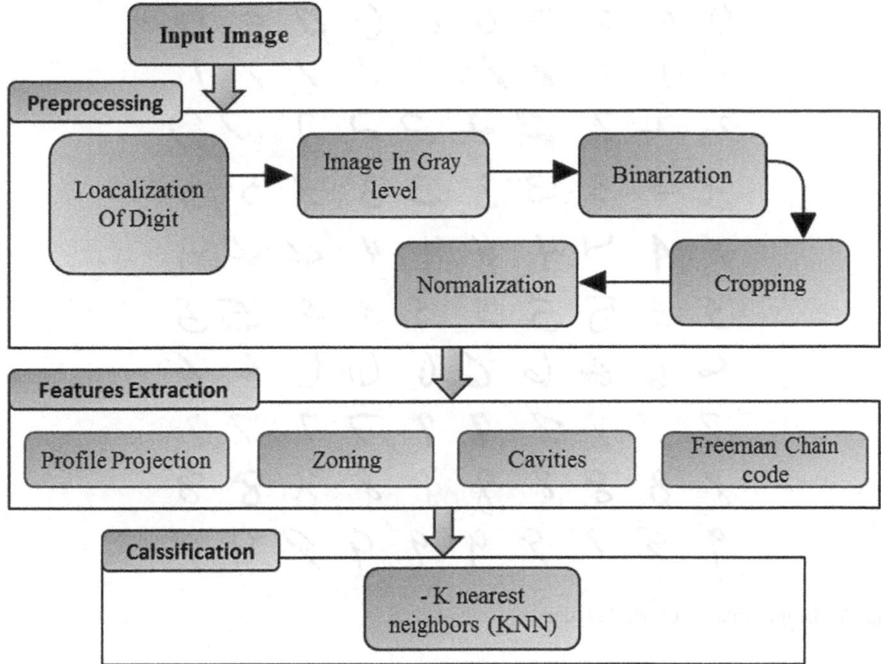

Fig. 5 Scheme for digits recognition

5 Experimental Results and Comparative Study

This section presents the experiments obtained by each feature extraction method separately. We have used a database of 6500 digits, provided by different writers. A sample of the database is shown in Fig. 6. The database is divided into two sets, one set of 5700 digits is used for learning and the remaining 800 digits are used for the test stage. Classes are equiprobables.

In this test, several experiments were carried out to determine the recognition rate according to the size of normalization using the different features extraction methods.

For classification task we used the KNN method. For the parameter K, we have varied K from 1 to 9 and we have retained the value of K that gives the best recognition rate.

Table 1 summarizes the overall accuracy of the system with each of the four feature sets.

From the experiment, the best recognition rate of 95 % was obtained for the freeman chain code features. The next highest accuracy was obtained for zoning. Compared to other feature sets, they need feature vectors of larger size. The zoning based features provide good recognition accuracy with a relatively small feature vector size of 25.

Fig. 6 Sample images of MNIST data

Table 1 Recognition results of different feature sets

Features extraction methods	Size of normalization	Feature size	Recognition rate %
Zoning	40 × 40	16	90.5
	45 × 45	25	**94**
	50 × 50	100	88
Profile projection	40 × 40	160	88.75
	45 × 45	180	88
	50 × 50	200	**91.5**
Freeman chain code	40 × 40	120	89
	45 × 45	120	93
	50 × 50	120	**95**
Cavities	32 × 32	5	**86**

The cavities features gave the lowest recognition results among all the features used in this study. The recognition rate of 86 % was obtained using this cavities feature.

6 Conclusion

In this work we have presented a comparative study of four methods of extracting features for handwritten digits recognition that are profile projection, zoning, cavities and freeman chain code features. The image representing the digit is initially preprocessed; it is binarized, cropped, brought to a fixed size by normalization. One of these features set is applied to a feature vector. These features were classified using a K nearest neighbors classifier. The simulations have obtained promising results. We have obtained the best recognition accuracy of 95 % using freeman chain code features. This accuracy can be further improved by using a larger dataset for training. Also reduction of attributes vector can further improve the recognition rate. In our next work we will try to minimize the size of vector attributes by using a genetic approach to increase the rate and time of recognition.

References

1. Boukharouba, A., Bennia, A.: Novel feature extraction technique for the recognition of handwritten digits. Appl. Comput. Inf. (2015)
2. Cruz, R.M.O., Cavalcanti, G.D.C., Tsang, I.R.: Handwritten digit recognition using multiple feature extraction techniques and classifier ensemble. In: 17th International Conference on Systems, Signals and Image Processing, pp. 215–218 (2010)
3. Benchaou, S., Nasri, M., El Melhaoui, O., Bouali, B.: New structural approach for numeral recognition based on mathematical morphology and Freeman code. Mediterr. Telecommun. J. **5**(2), 159–163 (2015)
4. Nagare, A.P.: License plate character recognition system using neural network. Int. J. Comput. Appl. **25**(10), 36–39 (2011)
5. Bakar, N.A., Shamsuddin, S.M., Ali, A.: An integrated formulation of Zernike representation in character images. In: Trends in Applied Intelligent Systems, pp. 359–368. Springer, Berlin, Heidelberg (2010)
6. Rajput, G.G., Mali, S.M.: Marathi handwritten numeral recognition using Fourier descriptors and normalized chain code. Int. J. Comput. Appl. Recent Trends Image Process. Pattern Recogn. 141–145 (2010)
7. El Melhaoui, O., El Hitmy, M., Lekhal, F.: Arabic numerals recognition based on an improved version of the loci characteristic. Int. J. Comput. Appl. **24**(1) (2011)
8. Impedovo, D., Pirlo, G.: Zoning methods for handwritten character recognition: a survey. Pattern Recogn. **47**(3), 969–981 (2014)
9. Chacko, A.M.M.O., Dhanya, P.M.: A comparative study of different feature extraction techniques for offline Malayalam character recognition. In: Computational Intelligence in Data Mining, vol. 2, pp. 9–18. Springer, India (2015)
10. Burel, G.: Introduction au traitement d'images: Simulation sous Matlab. Hermès Science Publ. (2001)
11. Freeman, H.: Boundary encoding and processing. In: Picture Processing and Psychopictorics, pp. 241–266. (1970)
12. Bernard, M., Fromont, E., Habrard, A., Sebban, M.: Handwritten digit recognition using edit distance-based KNN. In: Teaching Machine Learning Workshop (2012)

A Multiple Ontologies Based System for Answering Natural Language Questions

Anas El-Ansari, Abderrahim Beni-Hssane and Mostafa Saadi

Abstract Due to the massive growth of information on the Web, information retrieval systems come to play a more critical role. Most of these systems are based on content matching rather than the meaning, therefore the returned results are not always relevant to the user. To solve this problem, the next generation of information retrieval systems focus on the meaning of the user query and search data using ontologies that provide the vocabulary and structure associated with metadata. In this work we present a Question Answering system which combines multiple knowledge bases, with a Natural Language parser to transform questions into SPARQL queries or other query language. We demonstrate the feasibility to build such a semantic QA system and the accuracy and relevance of the returned results.

Keywords Semantic Web · Ontology · Question answering system · Natural language processing

1 Introduction

The Web is a global information space. With A rapidly increasing rate of information available to users through the Web, there is a pressing need for efficient information retrieval systems such as search engines, question answering systems, etc. Nowadays,

A. El-Ansari (✉) · A. Beni-Hssane
LAROSERI Laboratory, Computer Science Department Sciences Faculty,
Chouab Doukkali University, El-Jadida, Morocco
e-mail: anas.elansari@gmail.com

A. Beni-Hssane
e-mail: abenihssane@yahoo.fr

M. Saadi
Departement Informatique & Telecoms, Ecole Nationale des Sciences
Appliquées (ENSA), Université Hassan 1er - Settat, Khouribga, Morocco
e-mail: saadi_mo@yahoo.fr

Á. Rocha et al. (eds.), *Europe and MENA Cooperation Advances in Information and Communication Technologies*, Advances in Intelligent Systems and Computing 520, DOI 10.1007/978-3-319-46568-5_18

those systems develop very fast and successfully. However they still suffer from a lack of accuracy and the relevance of the provided results is just not up to the mark. To solve this problem, Ontologies and semantic web are becoming a pivotal methodology to represent any domain-specific conceptual knowledge in order to promote the semantic capability of an information retrieval system [1].

The Semantic Web aims to extend the current web standards and technology so that all the Web contents and information can be processed by machines. The use of ontology in the search process provides an interaction between machine and human.

The traditional search is based on term matching techniques, which helps retrieving all resources containing the user's query terms. While in a semantic search, queries that can be expressed in several ways, will be mapped on the semantic level to define topics related to the user informational need that must be retrieved from the web [1].

The question answering (QA) systems [2] aims at providing precise answers to the user's questions. For example, for a question such as (What is the capital of Morocco?), traditional term matching search systems might return a large number of web pages about Morocco and the user would have to dig into these web pages to find the answer. While an efficient QA system would directly answer the question with the name of the capital "Rabat". For that, a QA system needs an efficient natural language question processing mechanism to understand the users question and a semantic data source to get the exact answer for the question.

The QA system we propose in this paper, transforms the user's questions in natural language to a query language (SPARQL or MQL). The last is then used to interrogate different online Knowledge Bases to return an exact answer to the user.

The paper is organized as follows: Section 2 describes the NL question processing. Section 3 presents the semantic data sources we used (the knowledge bases). Section 4 explains the proposed QA system. And finally, the conclusion and directions for future work.

2 Related Work

Researches in the field of Question Answering has been advanced in the past couple of years [2]. With the semantic web technologies a domain-specific QA system working on a specific technical domain can make use of the specific domain-dependent ontology to recognize the true meaning included in a natural language text. So we realize that the ontology plays a pivotal role in a technical domain.

One of the typical examples of a QA system is Jeeves [3] which allows users to ask natural language questions and returns a list of matching questions to which it knows the answer. Another example [4], is a research in information processing that has focused on health care consumers. These users often have a frustrated experience while seeking online information.

Other works such as [5, 6], have proved the feasibility of implementing an ontology based question answering system. However, the degree of complicity of these systems is considerable. And also the relevance of the answers is not optimal. GINSENG [7], a guided-input natural language search engine, and Cuebee [8] progressively guide the scientists by suggesting concepts and relationships that decompose the question into an RDF triple, which is then internally translated into a SPARQL query. This process demands more effort from the user and is a time consuming task.

Most of the studies focusing on ontology based QA systems use a domain local ontology which cannot answer a wide range of questions. In our research we made use of some global knowledge bases offering a huge amount of data to be interrogated and that are available for online access. Those data sources offers a wider range of relevant answers. We also focus on the simplicity of the system, using a simple graphical user interface assisted with autocomplete feature and an error handling component. Our system also offers the user the possibility to interrogate other knowledge base by transforming his question to a SPARQL query that he can copy to the other endpoints.

3 Natural Language Question Processing

The Natural Language Processing (NLP) is a research field that explores how computers can be used to understand and process natural language text or even speech to do useful things [9]. Once the user enter his question in a natural language, the system must process it and transform it to the query language. The question can be classified as follows:

- What—objects specification or an activity definition
- Who—object or person specification
- When—date
- Where—geographical location; ...

There are some frameworks that can process natural language question and transform it to a query language such as NLTK [10] and Quepy [11].

In our system we made use of the last one, Quepy, which is a python framework because it can be easily adapted to different question types and query languages. Quepy uses NLTK tagger which is a linguistic tool to analyze natural language questions. It's composed of: a tokenizer, a part-of-speech tagger and a lemmatizer.

So, once the user enters his question, for example: "Who is Bill Gates?" the framework runs NLTK tagger on the string and returns a list of quepy.tagger.Word objects. The transformation from natural language text to the SPARQL query is done by first using a special form of regular expressions:

```
person_name = Group(Plus(Pos("NNP")), "person_name")
regex = Lemma("who") + Lemma("be") + person_name + Question(Pos("."))
```

And then using a convenient way to express semantic relations:

```
person = IsPerson() + HasKeyword(person_name)
definition = DefinitionOf(person)
```

The rest is handled automatically by the framework to finally produce this SPARQL:

```
SELECT DISTINCT ?x2 WHERE {
   ?x0 rdf:type foaf:Person.
   ?x0 rdfs:label "Bill Gates"@en.
   ?x0 rdfs:comment ?x1.
}
```

The SPARQL query is then sent to the knowledge base server which will return the answer. The system offer the user the possibility to choose and search in multiple knowledge bases. The next section will describe those knowledge bases.

4 Semantic Data Sources

Semantic knowledge base is a machine-readable resource for the dissemination of information, generally online or with the capacity to be online. A knowledge base is not a static collection of information, but a dynamic resource that may have a learning capacity, as part of an artificial intelligence [12].

Knowledge bases are playing a major role in optimizing the intelligence of Web and search systems and in supporting information integration [12]. Today, most knowledge bases cover only specific domains, created by relatively small groups of knowledge engineers and specialists, and are very cost intensive to keep updated as domains change.

In this system we used multiple knowledge bases in order to increase the chance of getting answers to every user's question. For that, the system interrogates tree large scale publicly available knowledge bases that cover a wide range of domains.

The first one is DBPedia [13]; The English version of this knowledge base describes 4.58 million things, out of which 4.22 million are classified in a consistent ontology, including 251,000 species, 1,445,000 persons, 735,000 places, 411,000 creative works, 241,000 organizations and 6,000 diseases.

Another important knowledge base used in our system is the Freebase [14]. A large collaborative knowledge base composed mainly by its own community members. This online collection of structured data was collected from many sources, including individual contributions, wikis, etc. Freebase aimed to create a global resource that allow people and machines to access and process information more effectively [15]. We also made use of the LOD knowledge base [16]. The used knowledge bases are connected with other data sets to cover even a wider range of domains. Other knowledge bases can be easily added to the system to cover more domains if needed.

5 Proposed Semantic QA System

Our Question Answering System includes three components: question processing based on the Quepy framework, knowledge base Interrogator and answer processing.

The Question Processing component's job is to analyze and transform a natural language question to a SPARQL query. SPARQL (Protocol and RDF Query Language) [17] is an RDF query language, that is, a semantic query language for knowledge bases, able to retrieve and manipulate data stored in Resource Description Framework (RDF) format.

The knowledge base Interrogator is responsible for sending the query to the SPARQL endpoint of the knowledge base selected by the user. This model returns a set of RDF results. The Answer Processing component handles the returned results to filtrate and transform into a natural language result.

5.1 System Architecture

In this section, we present a high level overview of how the whole system works as pictured in Fig. 2. The aim of QA systems is to find exact and correct answers for user's questions. In addition to the graphical user interface (GUI), our QA system contain three main components:

1. Question processing
2. Knowledge base Interrogator
3. Answer processing

Using the GUI the user enters a question in natural language and hits the search button. The Question Processing component analyses the user's question to extract the keywords to be used in the query and prepares an adapted SPARQL and MQL query. The user then is asked to choose the knowledge base to interrogate. According to the user's choice, the Question Processing component will send either the SPARQL query (For DBpedia and Linked Open Data) or the MQL query (For Freebase

The knowledge base Interrogator uses the SPARQL/MQL query to interrogate the selected knowledge base via the endpoint and then returns an RDF answer.

The Answer processing component gets the RDF answer and process it to extract the exact answer and any related information possible and transform it into a natural language answer. The answer is finally returned to the user via the GUI (Fig. 1).

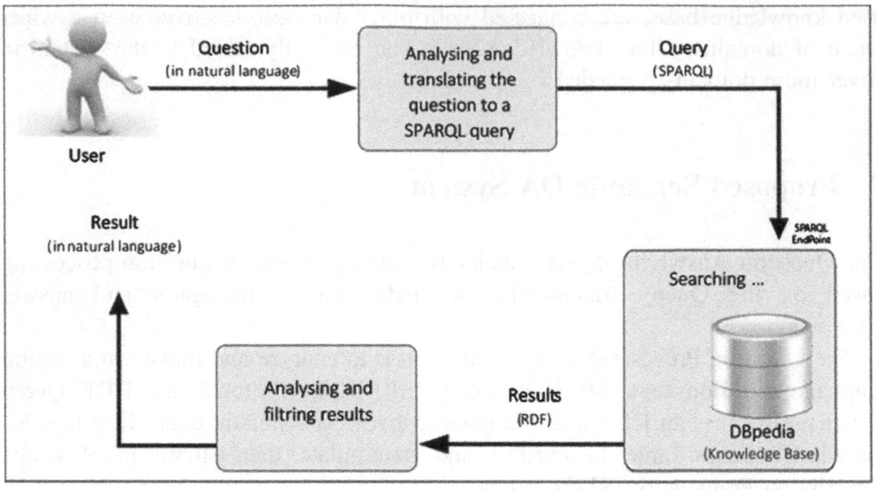

Fig. 1 Design Architecture of the QA system

5.2 Simulation

By accessing the interface of the application, you will find, like all search engines, a text field where you type your question (in English) and you search with the little orange button. You can also enjoy the "autocomplete" feature that helps you type the question (Fig. 2).

Once you click the search button, three other buttons appear asking you to choose which knowledge base you want to interrogate: Freebase—DBPedia—Linked Open Data (LOD) (Fig. 3).

The answer to a question is not necessarily the same in all three knowledge bases it is also possible that you find the right answer in a single database or two. This is the reason why we used the three in this system instead of single one.

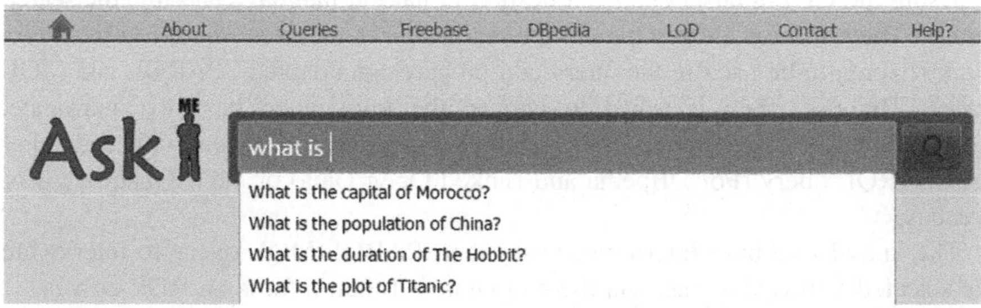

Fig. 2 User interface of the QA system (1) while typing the question

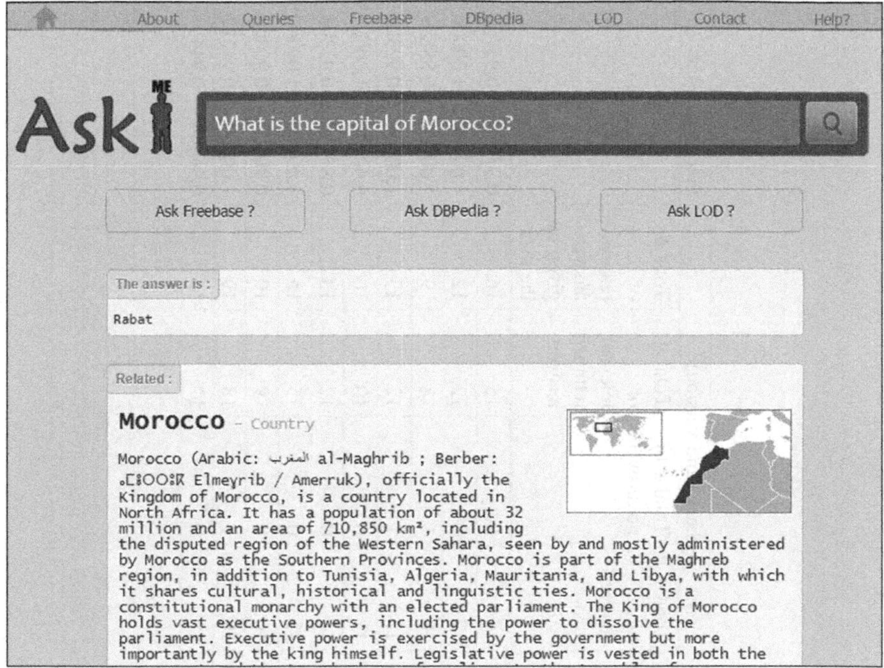

Fig. 3 User interface of the QA system (2) after hitting the search button and choosing the knowledge base to interrogate

Sometimes the search engine cannot transform a question in natural language to a SPARQL or MQL query. For one reason or another, either because the question contains errors, a problem of capitalized names, or it was not well formulated.

5.3 Evaluation

After the implementation phase, we conducted some initial experiments on two versions of the system with the help of ten volunteers. The first version was based on a single ontology (DBpedia), while the second was based on the tree knowledge bases combined (DBpedia, Freebase and LOD). To evaluate the precision and relevance of the returned results we use the precision and recall method [1]. For that, we asked ten subjects (S_i) to use the system and ask different questions. Once the experience was done we saved the users feedback in the following table (Table 1).

Table 1 Experimentation results

Subject	User's feedback (1st version)				Precision	Recall	User's feedback (2nd version)				Precision	Recall
	Total # of returned	Total # of no-answers	Total # of relevant returned answers	Total # of relevant given by the user			Total # of returned answers	Total # of no-answers	Total # of relevant returned answers	Total # of relevant answers given by the user		
S1	12	0	10	12	0.83333	0.83333	16	0	15	16	0.93750	0.93750
S2	10	1	10	11	1.00000	0.90909	12	0	12	12	1.00000	1.00000
S3	5	2	4	7	0.80000	0.57143	8	1	8	9	1.00000	0.88889
S4	13	0	11	13	0.84615	0.84615	12	1	12	13	1.00000	0.92308
S5	8	1	8	9	1.00000	0.88889	11	0	10	10	0.90909	1.00000
S6	12	0	12	13	1.00000	0.92308	14	0	12	13	0.85714	0.92308
S7	11	0	9	11	0.81818	0.81818	15	0	15	15	1.00000	1.00000
S8	6	0	5	6	0.83333	0.83333	10	0	9	10	0.90000	0.90000
S9	15	2	14	17	0.93333	0.82353	18	1	18	19	1.00000	0.94737
S10	15	1	13	16	0.86667	0.81250	18	0	17	18	0.94444	0.94444

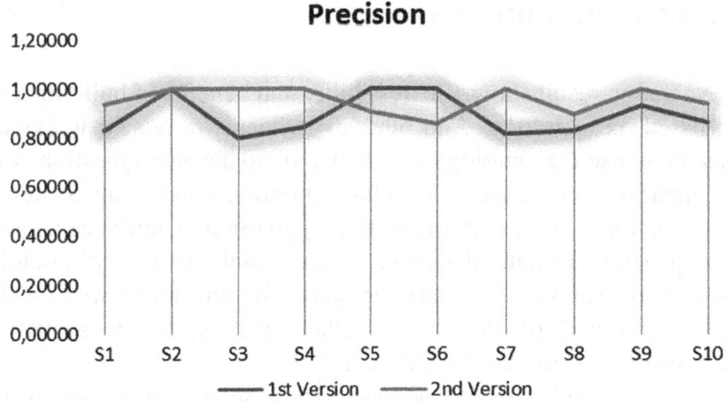

Fig. 4 Precision—calculated for both versions based on each user's feedback

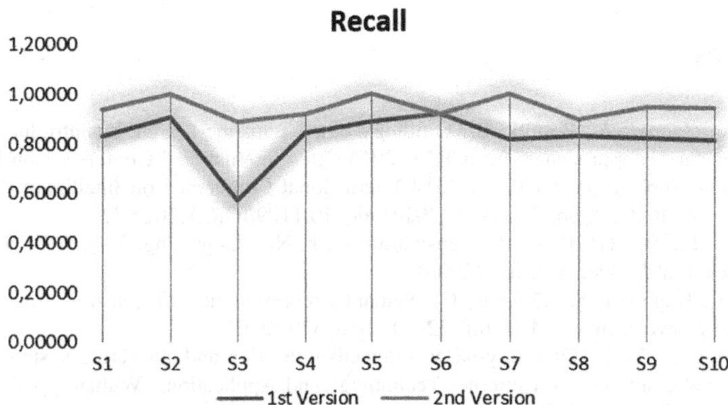

Fig. 5 Recall—calculated for both versions based on each user's feedback

The goal of our ontology based semantic search system is to maximize precision and accuracy of the results by combining different knowledge bases. And we can see from the experimentation results in the following figures that the precision and recall values increases in the second version where we combined different knowledge bases (Figs. 4 and 5).

The precision graph shows how useful the search results are, while the recall graph shows how complete the results are. The experimentation results shows that when combining multiple ontologies, the system relevance rate increases, however an important number of questions didn't have answers in the used knowledge bases so there is still room for improvements.

6 Conclusion and Future Work

The initial evaluation result shows the feasibility and benefits of building a semantic QA system based on Ontology. And other experiments in related works do prove that it is feasible to use the Ontology-based method to develop Question Answering Systems. Comparing our system with other Question Answering Systems in the Related Work section. Our system offers an easy to use user interface and offers the user to enter questions in natural language. Also combines multiple ontologies to increase the relevance and range of answers so our system can answer a wider range of user's question in multiple domains. Another advantage of this system is that it returns exact answers to the user's questions.

We have implemented a natural language question answering system based on multiple ontologies however there are still many features to implement, on the realized system, such as error handling, or a result relevance rating component.

References

1. Sudha Ramkumar, A., Poorna, B.: Ontology based semantic search: an introduction and a survey of current approaches. In: ICICA, 2014, 2014 International Conference on Intelligent Computing Applications (ICICA), 2014 International Conference on Intelligent Computing Applications (ICICA), pp. 372–376 (2014). doi:10.1109/ICICA.2014.82
2. Voorhees, E.: The TREC question answering track. Nat. Lang. Eng. **7**(4), 361–378 (2006)
3. Askjeeves. http://askjeeves.com/ (2000)
4. Laura, A., Dagobert, S., Thomas, C.: Semantic representation of consumer questions and physician answers. Int. J. Med. Inf. **22**(10), 513–529 (2006)
5. Liu, L., Qi, Q., Li, F.: Ontology-based interactive question and answering system. In: 2010 International Conference on Internet Technology and Applications, Wuhan, pp. 1–4 (2010). doi:10.1109/ITAPP.2010.5566132
6. Guo, Q., Zhang, M.: Question answering based on pervasive agent ontology and Semantic Web. J. Knowl.-Based Syst. **22**(6), 443 (2009). doi:10.1016/j.knosys.2009.06.003
7. Bernstein, A., Kaufmann, E., Kaiser, C.: Querying the semantic web with Ginseng: a guided input natural language search engine. In: 15th Workshop on Information Technologies and Systems. Las Vegas, NV: SSRN (2005)
8. Mendes, P.N., McKnight, B., Sheth, A.P., Kissinger, J.C.: TcruziKB: Enabling complex queries for genomic data exploration. In: 2008 IEEE International Conference on Semantic Computing, pp. 432–439. IEEE (2008)
9. Chowdhury, G.G.: Natural language processing. Ann. Rev. Inf. Sci. Technol. **37**, 51–89 (2003). doi:10.1002/aris.1440370103
10. NLTK Project. http://www.nltk.org/ (2015)
11. Quepy Framework. https://quepy.readthedocs.org/en/latest/ (2012)
12. Swartout, B., et al.: Toward distributed use of large-scale ontologies. In: Proceedings of the Tenth Workshop on Knowledge Acquisition for Knowledge-Based Systems (1996)
13. DBpedia Project. http://dbpedia.org/about
14. MetaWeb, FreeBase. https://www.freebase.com/
15. WikiPedia: Freebase. https://en.wikipedia.org/wiki/Freebase
16. LOD diagram. http://richard.cyganiak.de/2007/10/lod
17. WikiPedia: SPARQL. https://en.wikipedia.org/wiki/SPARQL

Toward a Model Driven Approach for Generating Multi Platform Applications with ZeroCouplage Framework

Sarra Roubi, El Hassane Ettifouri, Mohammed Erramdani and Toumi Bouchentouf

Abstract It is crucial for companies to have the latest technology. But with their exponential growth, migration and adoption of these new technologies can be very expensive and requires more effort and learning which can lead to big losses. ZeroCouplage is a multi platform Framework that is introduced to solve this problem. Indeed, we code once and the application is built for desktop, web and mobile as well. However, using a new framework can often be time consuming and make the task more difficult. That's why we used the model driven engineering to automate the whole process. We propose a meta model for this framework and automate the entire process up to code generation.

Keywords Model driven engineering · Meta model · Code generation · Multi platform

1 Introduction

Nowadays, development of applications for multiple platforms is far more articulated issue. Companies must keep up with this rapid evolution and offer their solutions and services on the various existing supports in order to reach the maximum possible users and customers. However, for each platform, the whole process should be repeated and re adapted and corrections must be operated for the various

S. Roubi (✉) · M. Erramdani
MATSI Laboratory, Mohammed First University, Oujda, Morocco
e-mail: roubi.sarra@gmail.com

M. Erramdani
e-mail: m.errmadani@gmail.com

E.H. Ettifouri · T. Bouchentouf
SIQL Team, LSEII Laboratory, Mohammed First University, Oujda, Morocco
e-mail: h.ettifouri@gmail.com

T. Bouchentouf
e-mail: tbouchentouf@gmail.com

Á. Rocha et al. (eds.), *Europe and MENA Cooperation Advances in Information and Communication Technologies*, Advances in Intelligent Systems and Computing 520, DOI 10.1007/978-3-319-46568-5_19

supports. Moreover, the capitalization is difficult to reach with one dedicated team for the project. All these facts make the task time consuming and very tedious.

On the one hand, Responsive web design has been introduced among others as solution to these limitations. The concept gathers a set of principles of design and technologies, which allow a Web site to auto-adapt according to the screen size of the support used to consult it from a computer, a tablet or a Smartphone. This responsive web design ensures the ergonomic by optimizing the interface for each terminal [1] but it is dedicated to web application. Also, it has some drawbacks; among, the users are obliged to download the HTML/CSS code in a useless way [1, 2], the images resizing requires the CPU calculations, which can also slow down the support and the total loading rate of the page.

As an alternative, a new framework; ZeroCouplage, was proposed by [3]. The main aim of Framework ZeroCouplage is to code once and choose the target platform without neither injecting a new code in the existing application nor having knowledge in the programming languages for Web, mobile or Desktop since it is based only on the language java.

On the other hand, MDE principles are being used for development of application in order to successfully address the development and evolution of these applications.

In [4], it is shown how MDD/MDA principles are applied in the Web Domain to define models. Moreover, the method OOH4RIA [5] proposes a model driven development process that extends OOH methodology employing the OOHDM conceptual and navigational scheme. Also, a Rich Internet Application for web based product development was presented in [6]. Besides, [4] proposes an approach for modeling a desktop application and generating the whole MVC swing code.

Besides, a Rich Internet Application for web based product development was presented in [6]. Also, an extension of the IFML OMG standard was proposed for generating mobile application front end [7].

These methods use models to separate the platform independent model (PIM) design of web systems from the platform-dependent (PSM) implementations as much as possible.

However, these proposed approaches are related to a specific platform. In this paper, we propose a model driven approach for a multi platform framework Zerocouplage. We define a model independently from any execution platform and we generate the code for desktop, web and Mobile applications.

The paper is organized as follows. Sections 2 and 3 present respectively the concept of ZeroCouplage framework and the Model Driven Engineering approach. In Sect. 4 we present the Model Driven Engineering Approach applied to ZeroCouplage and the resulting model. Finally we conclude and present future work.

2 Multi Platform Framework: ZeroCouplage

ZeroCouplage is a new framework that aims to create the same application for different platforms. We first present the principle and design pattern adopted before presenting the MDA approach that we propose for automating the whole process.

2.1 ZeroCouplage, the Concept

To face the problem of multi platform targeted applications, ZeroCouplage is introduced as a solution for Java developers. Indeed, they develop the application once with a well known language Java, then the application is generated for the Mobile, Desktop or Web.

ZeroCouplage meets the major needs of the development of IT projects which are: reusability, maintainability and cost reduction. The developer does not need any specific skills on JAVA/JEE or Android to develop applications on both web and mobile platforms.

ZeroCouplage is based on the design pattern M2VC; a derivative of the MVC Pattern. It provides zero coupling between the presentation layer and the business layer; hence the name ZeroCouplage. It also allows rapid switching between different platforms just by changing the settings in the configuration file.

2.2 M2VC Design Pattern Structure

Model-Virtual-View-Controller (M2VC) design pattern is a derivative of the Model-View-Controller (MVC) design pattern; it is intended for the development of applications multi support, such as web, mobile and desktop support. M2VC and MVC share the same concept of the model, but the M2VC controller uses a virtual view to manipulate the concretes views, it loads the concrete view, according to the selected support [3].

Each concrete view of the platform target implements the virtual view that manages the navigation between the pages of the same support. The suitable graphic component is loaded, Fig. 1.

3 Model Driven Engineering

Model Driven Engineering enables the automation of the development process and based on models throughout this process. In fact, it offers three levels of abstraction and allows defining models:

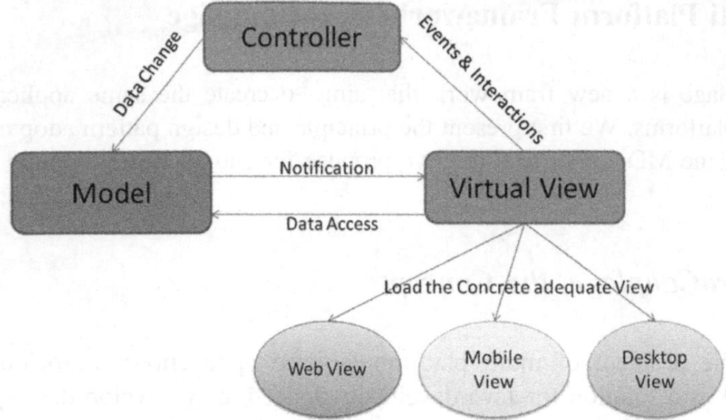

Fig. 1 M2VC design pattern principle

- **Computation Independent Model (CIM)**: describes the functional requirements of the application regardless of implementation details and gives a vision of what is expected from the application.
- **Platform Independent Model (PIM)**: allows the extraction of the common concept of applying, independently of the target platform allowing its mapping on one or more platforms.
- **Platform Specific Model (PSM)**: combines the specifications in the PIM with the details needed to determine how a system uses the chosen platform.

Then, the transition between these levels is automated by defining transformation engines. These transformations may affect the transition from one model to another or code generation. However, for each platform target, we need to specify a different meta model with the appropriate transformation engine, then make the code generation, Fig. 2.

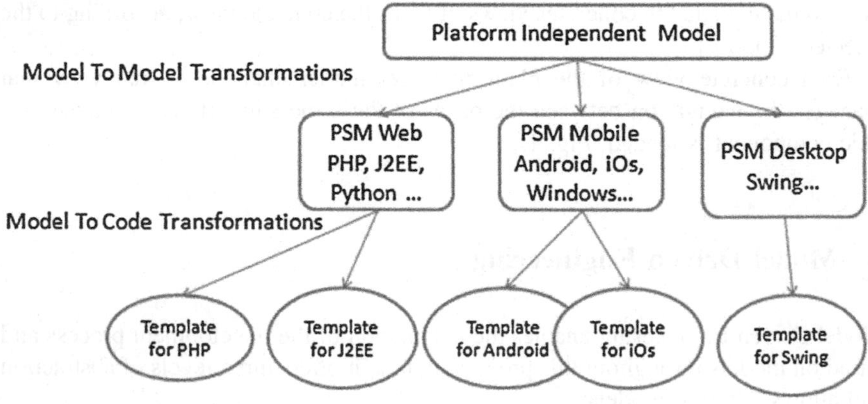

Fig. 2 Model driven architecture approach: levels of abstraction

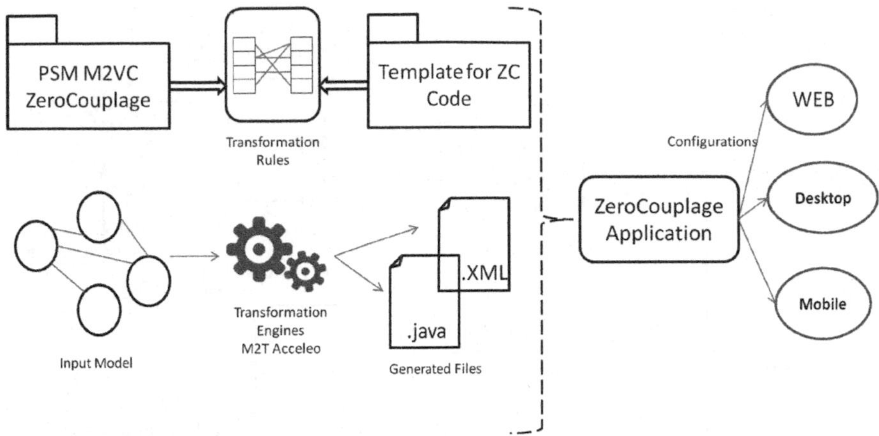

Fig. 3 Model driven approach using ZeroCouplage framework

4 Model Driven Approach for ZeroCouplage

As shown in Fig. 2, the use of MDE principles can be quite hard and time consuming to be able to cover a majority of existing platform. We combined the MDA with the ZercoCouplage and the proposed process is described in Fig. 3.

4.1 Proposed Meta Model for ZeroCouplage

We defined a meta model for the ZC framework respecting the M2VC pattern as described in previous section. We focused on the graphical aspect of the application. Indeed, each concrete view implements the virtual view that manages the navigation between the pages of the same support, and allows for a bidirectional exchange of data between the concrete view and Model.

The main element is the ZC_Component derived to several graphical components and layout managers. All these elements form the Graphical User Interface. Figure 4 shows the graphical part of the proposed meta model.

In the meta model, the view part is composed of one or several pages that holds a layout to help organize the components. All the components inherit form the parent class ZC_Component. We focused on the graphical part in the modeling phase and described all the graphical components and their relationship to improve the quality of the GUI and reduce the manual modifications at the end.

Then, we modeled the configuration part and the other layers of the M2VC Design Pattern. Indeed, the business package holds all the beans, actions and validation elements. All these meta classes are gathered under the ZeroCouplage

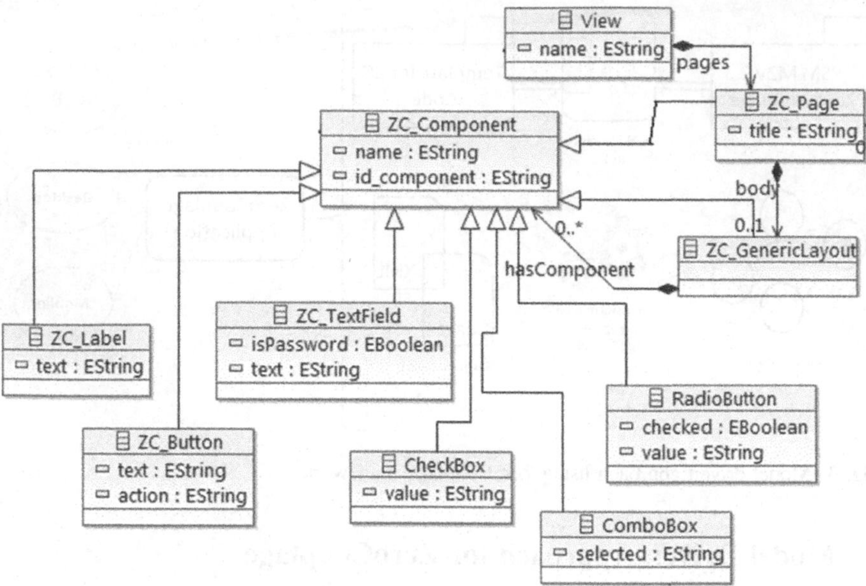

Fig. 4 Model driven approach using ZeroCouplage framework

root element that gives the mappings and the configuration to generate the code for the appropriate platform; desktop, mobile or web.

4.2 Model Result Describing ZeroCouplage

After completing the meta modeling phase, we used the Eclipse Modeling Framework to generate the code of meta model of the framework, which contains the editor also. This allows us to instantiate models describing the application with all elements of the layers of the view, model and controller as well as validation.

The input file, as shown in Fig. 5 describes an application for registration form. This file represents the model that describes the future applications in terms of ZeroCouplage elements. This file represents the input to the transformation engine to generate the code files. Indeed, it contains all information for the application especially for the view part. Also, it contains the elements to generate the mapping between the views and the business layer.

Finally, to automate the whole process, the transformation engine describing a Model To Text transformation using one of the known OMG standards, acceleo for instance, will be integrated to the process and the whole code application will be generated from one input model file only.

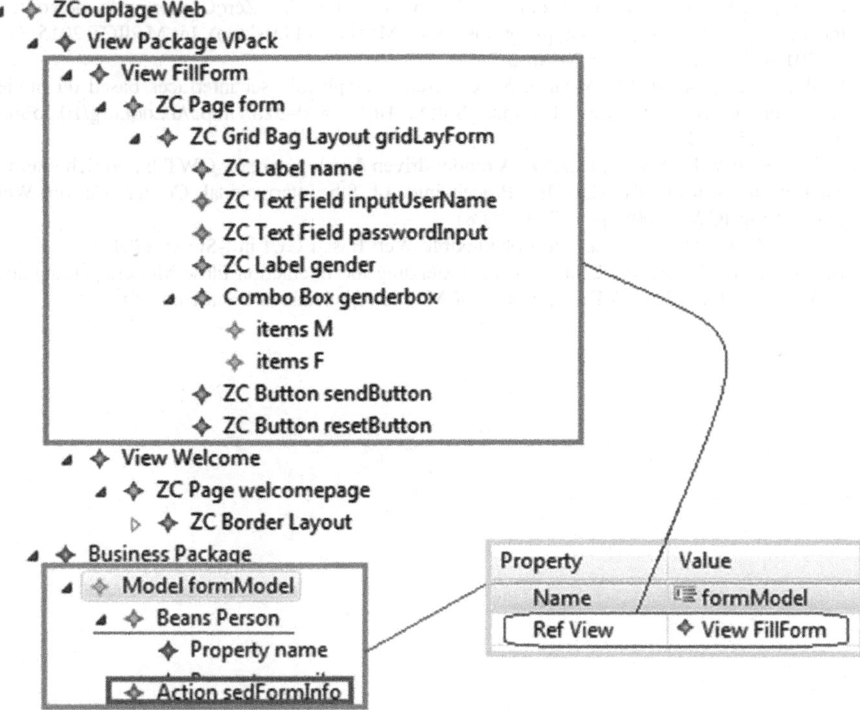

Fig. 5 Result mode for a ZeroCouplage application

5 Conclusion

In this paper we presented a new model driven approach for generating a multi platform application. Indeed, we proposed a meta model describing ZeroCouplage framework which can be deployed and visualized on several supports (Web, mobile-web, mobile-native and desktop) without any limits of the Web-Responsive.

For future works, the resulting model file will be used as input for transformation engine to be able to generate the source code files for the application. Also, we can realize a graphical tool to simplify more the task for the user.

References

1. Groves, R.M., Heeringa, S.G.: Responsive design for household surveys: tools for actively controlling survey errors and costs. J. R. Stat. Soc.: Ser. A (Stat. Soc.) **169**, 439–457 (2006). doi:10.1111/j.1467-985X.2006.00423.x
2. Bryant, J., Jones M.: Pro HTML5 Performance, Responsive Web Design (2012)

3. Ettifouri, E.H., Rhouati, A., Dahhane, W., Bouchentouf, T.: ZeroCouplage framework: a framework for multi-supports applications (Web, Mobile and Desktop). In: MedICT 2015, vol. 2. (2015). ISBN: 978-3-319-30296-6
4. Roubi, S., Erramdani, M., Mbarki, S.: Generating graphical user interfaces based on model driven engineering. Int. Rev. Comput. Softw. **10**(5), 520–528. http://dx.doi.org/10.15866/irecos.v10i5.6303
5. Meli, S., Gmez, J., Prez, S., Daz, O.: A model-driven development for GWT-based rich internet applications with OOH4RIA. In: Proceedings of 8th International Conference on Web Engineering ICWE 2008, pp. 1323 (2008)
6. Ahmed, Z., Popov, V.: Integration of Flexible Web Based GUI in I-SOAS (2010)
7. Brambilla, M., Mauri, A., Umuhoza, E.: Extending the Interaction Flow Modeling Language (IFML) for Model Driven Development of Mobile Applications Front End (2014)

QVT Transformation Rules to Get PIM Model from CIM Model

Imane Essebaa and Salima Chantit

Abstract The Model Driven Architecture (MDA) approach introduces a clear separation of system requirements and their implementation. This approach uses models at the center of the development of software systems. The MDA approach is based on models transformations; in the literature, several works have summarized the MDA approach to the passage from PIM to PSM then to code. But very little works have contributed on CIM to PIM transformations. Thus, our proposal aims to provide a method of transforming CIM to PIM using QVT transformation rules and respecting the two levels aspects as specified by the Object Management Group (OMG). Our approach proposes to represent a CIM level by Business Use Case Diagram and Activity Diagram while after applying transformation rules, the PIM level is represented by Class Diagram and Sequence Diagram.

Keywords Model Driven Architecture · Models transformation · QVT · CIM · PIM · PSM

1 Introduction

The Model Driven Architecture approach aims to separate the views and concerns; MDA approach have three views represented by models: Computation Independent Model (CIM) contains system requirements independently of any computation information, Platform Independent Model (PIM) focuses on only the operation of the system without details of a particular platform, and finally a Platform Specific Model (PSM) that depends on technical platform.

I. Essebaa (✉) · S. Chantit
Laboratoire Informatique Mohammedia, Computer Science Departement, Faculté des Sciences et Techninques de Mohammedia, BP 146, 20650 Mohammedia, Morocco
e-mail: imane.essebaa@gmail.com

S. Chantit
e-mail: salima.chantit@gmail.com

Á. Rocha et al. (eds.), *Europe and MENA Cooperation Advances in Information and Communication Technologies*, Advances in Intelligent Systems and Computing 520, DOI 10.1007/978-3-319-46568-5_20

Models transformations from one level to the other one constitute the key of the MDA; several works have focused on the transformation between PIM and PSM then code generation, whereas few has addressed the CIM to PIM transformations, and the existing works don't propose a complete solution to ensure this transformation, existing works can be classified to works which cover all the aspects of the CIM and PIM level specified by the OMG but they don't define complete transformation rules or don't ensure the traceability and automation, and works that don't cover all the aspects of CIM and PIM.

In order to automate transformations between CIM and PIM, we investigate this paper to present an approach that covers the different viewpoints of CIM and PIM levels, a set of transformation rules are defined using QVT to build PIM from CIM.

This paper is organized as follows: Sect. 2 presents an MDA approach and their levels, it contains also a brief presentation of QVT transformation language. Section 3 presents our evaluation of some of related works, in Sect. 4 we present our proposal approach followed by Sect. 5 that illustrate our approach in a case study, we finally conclude with future works and conclusion of this work.

2 Model Driven Architecture

2.1 Modeling Levels in MDA

The MDA (Model Driven Architecture) is an initiative of the OMG (Object Management Group) released in 2000 [1] The basic idea of the MDA approach is the separation of the functional system specifications and its implementation on a particular platform.

The MDA approach lies in the context of the Model Driven Engineering which involve the use of model and metamodels in the different phases of development lifecycle of an application [2], MDA defines three viewpoints

- CIM (Computation Independent Model): the objective of this model is to represent the application in their environment independently of any computation information.
- PIM (Platform Independent Model): the role of the PIM is to give a static and dynamic vision of the application regardless of the technical conception of it.
- PSM (Platform Specific Model): This model depends on technical platforms; it represents a template of code that facilitates code generation.

2.2 Models Transformation

CIM, PIM and PSM models are the main levels of the MDA approach, each of them contain information necessary for the generation of the source code for the application. The code is obtained by automatic generation from the PSM, the PSM is obtained by successive transformations of models CIM to PIM and PIM to PSM.

Execution of transformations ensure a link traceability between different models of the MDA approach. This link is a guarantee of quality in the software development process in MDA.

The MOF (Meta Object Facility) is normalized by OMG [3], it allows the definition of transformation rules and modeling languages, it also specifies the structure and syntax of metamodels. In this context the OMG proposes a standard transformation language QVT (Query/view/transformation) to define transformation rules from models to models.

3 Related Works

In the context of MDA approach, several methods of model transformations were proposed, specially between PIM level and PSM level, however limited works were done between CIM level and PSM one.

After the analyze of the previous works we deduced that the transformation rules proposed are not complete, indeed those methods may be classified into two categories; researches who cover different aspects of CIM and PIM but not defining complete transformation rules and researches that not cover the various aspects of the CIM and PIM levels.

Kherraf et al. in his proposition [4] uses to model the CIM by UML activity Diagrams which are detailed to get the system requirements which are transformed to be modeled as system components in PIM level. This method is based on modeling CIM using only the Activity Diagram, and after the transformation it doesn't cover the behavioral aspect.

In their method [5] Kardoš et al. modeled the CIM level by Data Flow Diagram (DFD), and PIM level was modeled by fours UML diagrams: Use Case Diagram, Activity Diagram, Sequence Diagram, and the domain models.

Wu et al. Method [6] describes how to get PIM level from CIM one, and the transformation PIM into PSM, in this method CIM is represented by Use Case Diagram, Activity Diagram and robustness diagram, while the PIM one was presented by Sequence Diagram and Class diagram, this method we note that the authors covered all the aspects of CIM and PIM but they didn't propose a complete transformation rules, neither traceability and automation of transformation.

In their proposition Kriouile et al. [7] modeled the CIM level by two diagrams: Business Process Model and Use Case Diagram, while the transformation of the CIM generates Domain Class Diagram and Sequence Diagram, this method covers the CIM and PIM aspects, but it doesn't ensure a completeness of transformation rules neither the automation of transformation.

It should be noted that in the literature that we have found, it exists many methods that are limited only in modeling CIM level without proposing transformation rules to get PIM level [8], it also exists works [9] that used a method called

TFM4MDA (Topological Functioning Modeling for Model Driven Architecture), it uses formal mathematical foundations of topological functioning model.

The results of our analyze are presented in Table 1, the columns represents an evaluation criterion while the rows represent studied methods.

4 Proposed Approach

This section presents the approach of the CIM Modeling and its transformation to PIM Model, our approach consists of:

- Modeling CIM through UML Business Use Case Diagram to cover the structural and functional aspect, and UML Activity Diagram to cover the dynamic aspect.
- Obtain the static viewpoint of PIM from CIM level through a vertical transformation C2P1 detailed in Table 2, the result of this transformation is represented with Class Diagram.
- Obtain the dynamic viewpoint of PIM from CIM level through a vertical transformation C2P2 detailed in Table 3, the result of this transformation is represented with Detailed Sequence Diagram.
- Transformation rules that map elements of one metamodel to the elements of another metamodel, are defined by using the QVT transformation language.

The Fig. 1 below describes this approach.

4.1 CIM Architecture in Our Approach

In his book "MDA in action" Xavier Blanc [10] consider the CIM level as the most important and complex entity, any changes of requirements in CIM level will reflect the PIM and the PSM levels.

After the analysis of the previous related works we didn't find any consensus and rules on how CIM should be presented and even how many models could represent CIM.

In practice, the first question to address in order to represent CIM level is "what should be represented in this model?".

According to the OMG specifications and criteria, we deduce that there are two topics of CIM level: System requirements and the process between system stakeholders, moreover those topics determine the three aspects of CIM level: Functional, Static and Behavioral.

In our approach we have opted to present this level by two models: Business Use Case Diagram to describe the functional requirements as well as a static view of the

Table 1 Results of analyze and evaluation

Methods studied	CIM level			PIM level		CIM2PIM transformation		
	Behavioral	Functional	Static	Dynamic	Static	Automation rules	Rules traceability	Completeness of rules
Kherraf et al. [4]	Y	N	N	N	Y	N	P	N
Kardoš et al. [5]	P	N	N	Y	Y	N	N	N
Wu et al. [6]	Y	Y	Y	Y	Y	N	N	P
Kriouile et al. [7]	Y	Y	Y	Y	Y	N	P	N
Sharifi et al. [8]	N	Y	Y					
Erika et al. [9]	N	Y	Y					

Legend: *Y* Yes; *N* No; *P* Partial

Table 2 UCD2CD transformation QVT rules

Rule	Transformation rule	Source model	Target model
1	Actor2Class	Actor	Class
2	DataObject2Class	DataObject	Class
3	Associations2Associations	Associations	Associations
4	UseCase2Operation	UseCase	Operation

Table 3 UCD&AD2SD transformation QVT rules

Rule	Transformation rule	Source model	Target model
1	Actor2Actor	Actor	Actor
2	UsecaseSystem2System	UsecaseSystem	System
3	Include2Ref	Include	Ref
4	Extend2Alt	Extend	Alt
5	DataObject2LifelineObject	DataObject	LifelineObject
6	Action2Message	Action	Message
7	NodeObject2LifelineObject	NodeObject	LifelineObject
8	FlowFinal2break	FlowFinal	Break

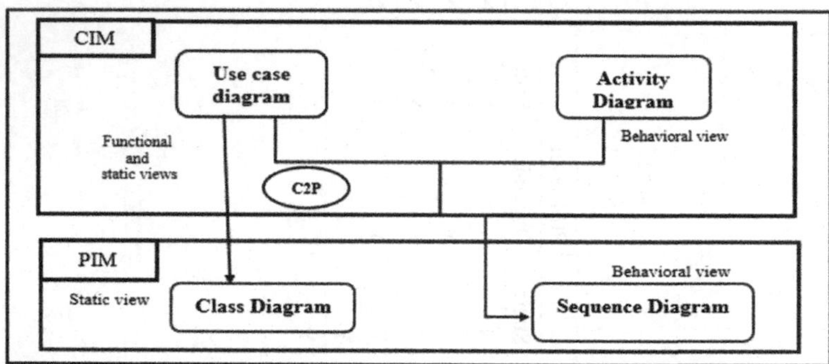

Fig. 1 Overview of the approach

system, and the Activity Diagram to represent different sequences and activities of the system.

UML Business Use Case Diagram

UML (Unified Modeling Language) defines a Use Case as: "the specification of a set of actions performed by a system, which yields an observable result that is, typically, of value for one or more actors or other stakeholders of the system" [11].

We choose Business Use Case diagram for several reasons; it identifies functionality of system and how the system works making links between actors and functionality which cover the functional aspect, and in order to cover the static aspect, we have extended the Use Cases diagram by adding "DataObject" element

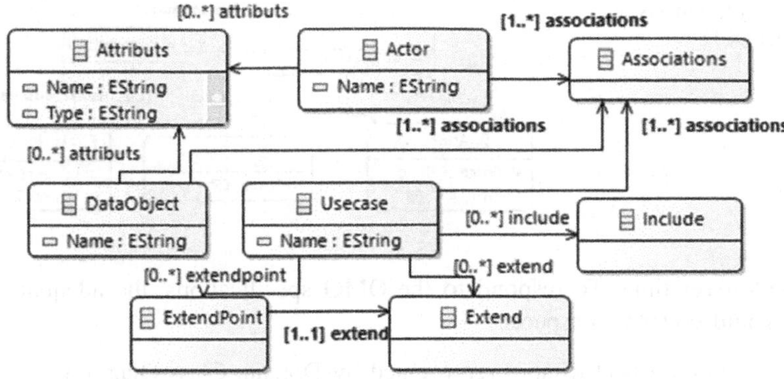

Fig. 2 Main fragments of metamodel UCD

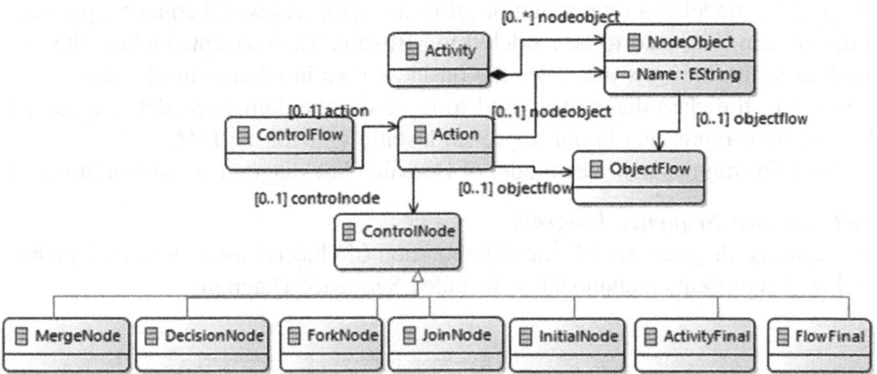

Fig. 3 Main fragments of metamodel Activity Diagram

A simplified version of the metamodel of the Business Use Case Diagram is represented in the Fig. 2.

UML Activity Diagram
Activity diagram is UML behavior diagram which shows flow of control or object flow with emphasis on the sequence and conditions of the flow.

We choose the activity diagram to represent a dynamic view of system in the CIM level, it allows to understand the process of different features of the system.

Figure 3 shows the metamodel of the activity diagram:

4.2 PIM Architecture in Our Approach

The PIM, called Model of analysis and design, represents the business logic specific for a system. It depicts the functioning of entities and services. It must be

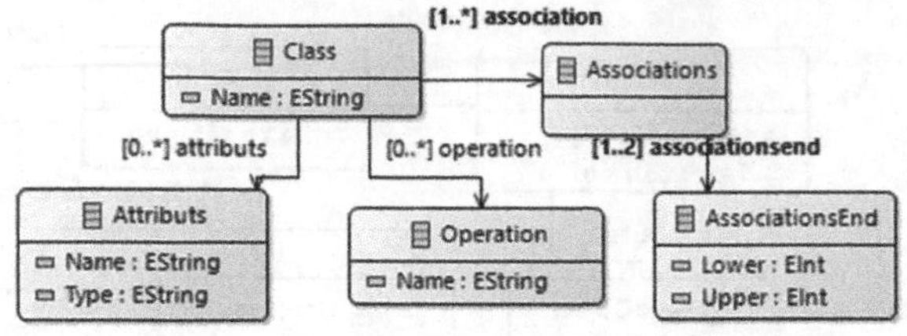

Fig. 4 Main fragments of metamodel DCD

sustainable over time. To respond to the OMG specifications, the adequate PIM model should cover two aspects:

- The structural aspect (Static) represented by Domain Class Diagram.
- The dynamic represented by Sequence Diagram.

UML Domain Class Diagram

The domain model is a representation of meaningful real-world concepts pertinent to the domain that need to be modeled in software. The concepts include the data involved in the business and rules the business uses in relation to that data.

In UML, the class diagram is used to represent the domain model, the domain class diagram represents in our approach a static view of the PIM.

The main fragments of metamodel of Domain class diagram are shown in Fig. 4.

UML Detailed Sequence Diagram

The sequence diagram models the collaboration of objects based on time sequence, the Fig. 5 represents metamodel of Detailed Sequence Diagram.

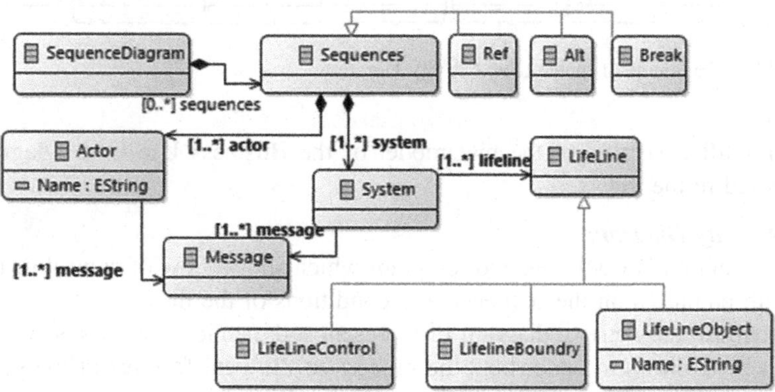

Fig. 5 Main fragments metamodel Detailed Sequence Diagram

```
mapping UseCaseDiagram::Usecase::UseCase2Operation() : Class::Operation{
   result.Name := self.Name;
   }
mapping UseCaseDiagram::Actor::Actor2Class() : Class::Class {
   result.Name :=self.Name;
   result.association +=self.associations.map associations2association();
   result.attributs +=self.attributs.map attributs2attributs();
   }
mapping UseCaseDiagram::DataObject::DataObject2Class() : Class::Class{
   result.Name := self.Name;
   result.attributs +=self.attributs.map attributs2attributs();
   result.association +=self.associations.map associations2association();
   }
mapping UseCaseDiagram::Attributs::attributs2attributs() : Class::Attributs {
   result.Name := self.Name;
   result.Type := self.Type;
   }
```

Fig. 6 Part of transformation QVT rules of UCD2CD

```
mapping UseCaseDiagram::Actor::Actor2Actor() : sequence::Actor{
    result.Name := self.Name;
}
mapping UseCaseDiagram::DataObject::DataObject2LifelineObject() : sequence::LifeLineObject{
    result.Name := self.Name;
}
```

Fig. 7 Part of transformation QVT rules of UCD&AD2SD

Transformation approach from CIM to PIM

CIM Modeling

In this step, the requirements of system are represented firstly through UML Business Use Case Diagram, and secondly their process is represented with UML Activity Diagram.

Obtaining the PIM static view from the CIM

$C2P_1$ is a vertical transformation which consists of transforming the UML-UCD conformed to its metamodel to a Class Diagram conformed to its metamodel. Those transformation rules are presented in Table 2.

A part of those rules are described with QVT Language in the Fig. 6.

Obtaining the PIM dynamic view from the CIM

The CIM2PIM transformation, noted $C2P_2$, is a vertical transformation aims to transform UML-UCD and UML Activity diagram to Detailed Sequence Diagram that represent the dynamic aspect of the PIM. Those transformation rules are presented in Table 3.

A part of those rules are described with QVT Language in the Fig. 7.

5 Case Study

In this section we are going to illustrate our approach throw an example, we choose as an example a system of E-Library that models an interaction between system and customers. Any customer can access to the web site and search one book, they can read it online or download it, they can also request a new book by filling a form, customers must connect with their account or subscribe if it is their first visit of the web site. To implement our approach, we must start by modeling the requirements with UML-UCD (Fig. 8).

After modeling all system requirements, we are going to focus on "Choose book" use case and implement an activity diagram to describe its process (Fig. 9).

UML-UCD transformation to UML Class diagram
From the Business Use Case model, by applying the mapping rules presented in Table 2 we get the Class Diagram bellow (Fig. 10)

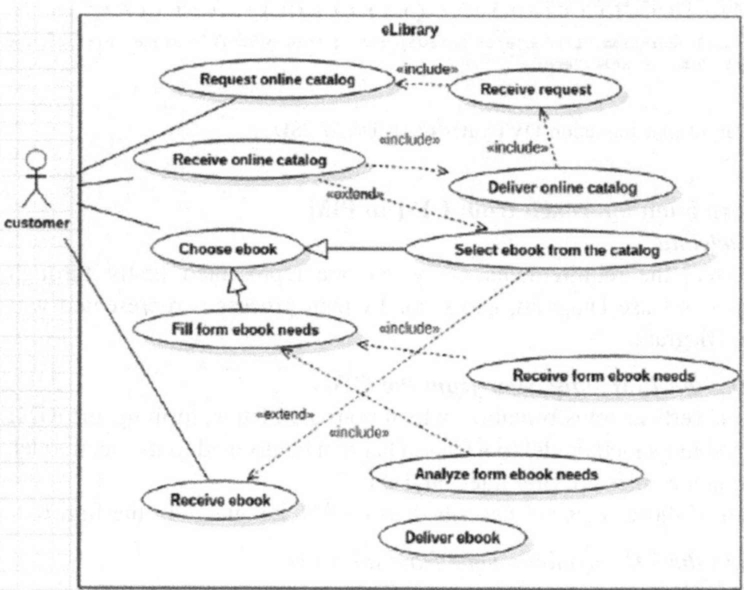

Fig. 8 Business Use case diagram of E-library

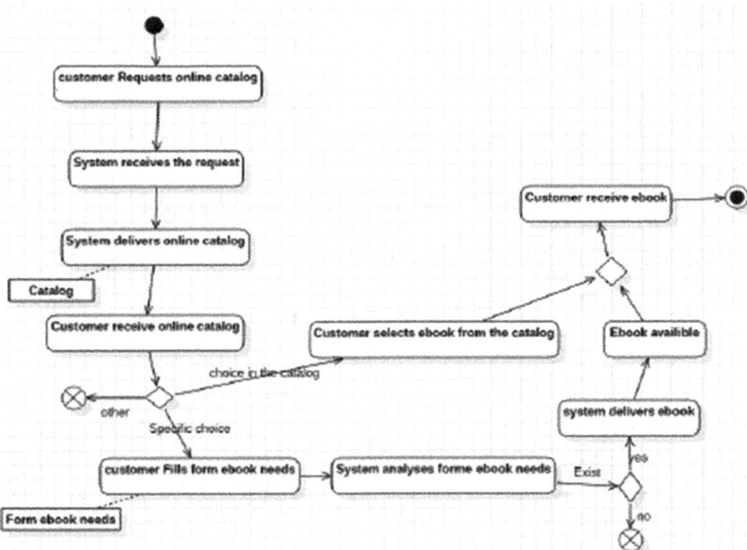

Fig. 9 Activity diagram of "choose book"

Fig. 10 Class Diagram of the system "E-Library"

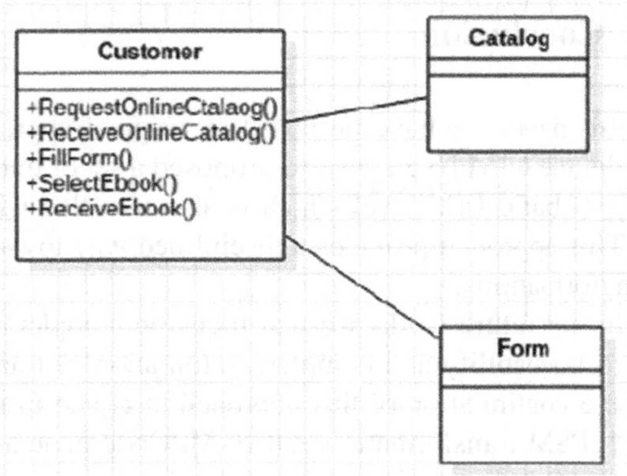

UML-UCD and Activity Diagram transformation to Detailed Sequence Diagram The dynamic aspect of PIM is represented through a Sequence Diagram; this diagram it is a result of transformation $C2P_2$ presented in Table 3 which allow to transform UML Business Use Case Diagram and Activity Diagram to a Sequence Diagram (Fig. 11).

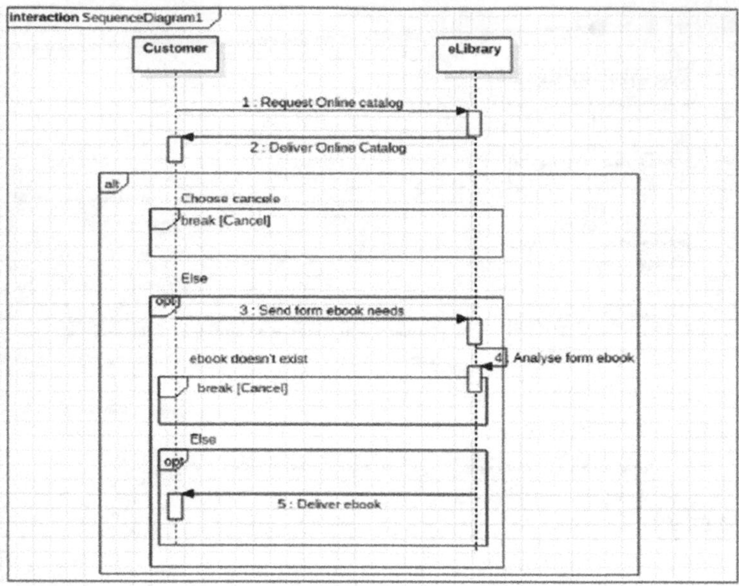

Fig. 11 Sequence diagram of "choose book"

6 Conclusion

In this paper we have proposed an original approach based on UML diagrams to model the CIM level, we also proposed a set of transformation rules using QVT that help to build PIM level which is also modeled with UML diagrams.

This approach provides a disciplined way towards automation of the CIM2PIM transformations.

In our future works we will focus on completing transformation rules to ensure more traceability and complete automation of transformation, we aim also to propose a continuation of this approach to cover all the MDA transformation namely PIM2PSM transformation and PSM2Code generation.

References

1. Miller, J., Mukerji, J.: MDA Guide Version 1.0.1. OMG (2003)
2. Soley, R.: Model driven architecture (MDA), draft 3.2. Rapport technique, disponible sur: http://www.omg.org/cgibin/doc?omg/00-11-05 (2000)
3. OMG: Meta Object Facility (MOF)2.0Query/View/Transformation Specification. http://www.omg.org/spec/QVT/1.0/PDF (2009)
4. Kherraf, S., Lefebvre, E., Suryn, W.: Transformation from CIM to PIM using patterns and archetypes. In: 19th Australian Conference on Software Engineering (2008)

5. Kardoš, M., Drozdová, M.: Analytical method of CIM to PIM transformation in Model Driven Architecture (MDA). J. Inf. Organ. Sci. **34**, 89–99 (2010)
6. Wu, J.H., Shin, S.S., Chien, J.L., Chao, W.S., Hsieh, M.C.: An extended MDA method for user interface modeling and transformation. In: The 15th European Conference on Information Systems, pp. 1632–1641, June 2007
7. Kriouile, A., Addamssiri, N., Gadi, T.: An MDA method for automatic transformation of models from CIM to PIM. Am. J. Softw. Eng. Appl. **4**(1), 1–14 (2015). doi:10.11648/j.ajsea.20150401.11
8. Sharifi, H.R., Mohsenzadeh, M.: A new method for generating CIM using business and requirement models. World Comput. Sci. Inf. Technol. J. (WCSIT) **2**(1), 8–12 (2012)
9. Osis, J., Asnina, E., Grave, A.: Computation independent modeling within the MDA. In: IEEE International Conference on Software-Science, Technology and Engineering, 2007. SwSTE 2007, pp. 22–34. IEEE (2007)
10. Xavier, B.: MDA en Action: Ingénirie Logicielles Dirigée par les Modèles. Eyrolles, Paris (2005)
11. OMG: OMG Unified Modeling Language TM (OMG UML), Superstructure, http://www.omg.org/spec/UML/2.4.1/Superstructure. Accessed Aug 2011

Features Extraction for Offline Handwritten Character Recognition

Soukaina Benchaou, M'barek Nasri and Ouafae El Melhaoui

Abstract Offline handwritten character recognition has been one of the most challenging research areas in the field of image processing and pattern recognition in the recent years. Handwritten character recognition is a very problematic research area because writing styles may vary from one user to another. This paper throws light on four different feature techniques, Zoning, Profile projection, Freeman chain code and Histograms of oriented gradients for handwritten vowels recognition. The recognition is carried out in this work through K nearest neighbors and fuzzy min max classification methods. The best recognition rate of 96 % was obtained using Histogram of oriented gradients features.

Keywords Offline handwritten character · Feature extraction · K nearest neighbors · Fuzzy min max

1 Introduction

Machine simulation of human reading has been a very challenging research field since the advent of digital computers. The main reason for such an effort was not only the challenges in simulating human reading but also the possibility of efficient applications in which the data present on paper documents has to be transferred into machine-readable format. Recognition of printed and handwritten information present on documents like cheques, envelopes, forms, and other manuscripts has a

S. Benchaou (✉) · M. Nasri · O. El Melhaoui
Labo MATSI, EST, University Mohammed 1, Oujda, Morocco
e-mail: soukaina.benchaou@gmail.com

M. Nasri
e-mail: nasrihome@gmail.com

O. El Melhaoui
e-mail: wafa19819@gmail.com

Á. Rocha et al. (eds.), *Europe and MENA Cooperation Advances in Information and Communication Technologies*, Advances in Intelligent Systems and Computing 520, DOI 10.1007/978-3-319-46568-5_21

variety of practical and commercial applications in banks, post offices, libraries, and publishing houses.

The character recognition system goes through three main steps that are: preprocessing, features extraction and classification.

The preprocessing phase is a current step in a recognition system. It consists of discarding the imperfections, reducing the analyzed area and correcting the size of the particular image.

The features extraction is a delicate process and is crucial [1] in achieving high recognition performance. It consists of transforming the image into an attribute vector, which contains a set of discriminated characteristics for recognition, and also reducing the amount of information supplied to the system. Mainly, we distinguish two approaches, statistical and structural.

Several methods of feature extraction for character recognition have been reported in the literature such as invariant moments [2], Zernike moments [3], freeman coding [4], Fourier descriptors [5], Loci characteristics [6], Projection histograms [7], etc.

Classification is the step of decision, which realizes the recognition. It consists of partitioning a set of data entity into separate classes according to a similarity criterion. Different methods are proposed in this context includes neural networks [8], support vector machines [9], k nearest neighbors, k-means, fuzzy min max, etc.

In this paper, we present a comparative study of handwritten vowels recognition using four different feature techniques—Zoning, Profile projection, Freeman chain code and Histograms of oriented gradients.

The recognition is carried out in this work through K nearest neighbors and fuzzy min max classification methods.

The performance of these features is evaluated using a database of handwritten vowels, provided by various categories of writers. The used database is split into two sets, one set of 300 characters is used for learning and the remaining 200 numerals are used for the test stage.

The paper is organized as follows: Sect. 2 presents the different features extraction methods used in this work. The k nearest neighbors and fuzzy min max classifiers are discussed in Sect. 3. The result of simulations and comparisons are introduced in Sect. 4. Finally, we give a conclusion.

2 Features Extraction Methods

2.1 Zoning

Zoning is a statistical method of features extraction. It's widely used for characters recognition because of its simplicity and its good performance. This method

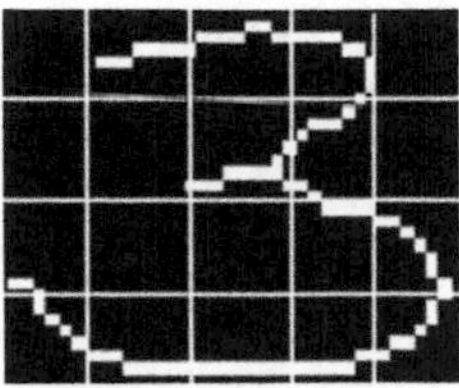

Fig. 1 Division of numeral "3" into 20 zones

consists of subdividing horizontally and vertically the pattern image into ($m * n$) zones of equal sizes (Fig. 1). For every zone, the density is calculated by dividing the number of pixels which represent the character on the total number of pixels in this zone i. We obtain an attribute vector of $m * n$ components for every pattern.

$$d(i) = \frac{\textit{Number of foreground pixels in zone i}}{\textit{Total number of pixels in zone i}} \tag{1}$$

Zoning method requires a phase of preprocessing. It consists of binarizing the numeral input image which is presented in grey level, and then we preserve the numeral position in image by cropping it. The next stage is to normalize the image in a predefined size and finally to skeletonize the resultant image.

2.2 *Profile Projection*

Profile projection is a statistical method. It consists of calculating the number of pixels between the left, bottom, right, top edge of the image and the first black pixel met on this row or column. The dimension of the obtained attribute vector is twice the sum of the number of rows and columns associated to the image of the numeral (Fig. 2).

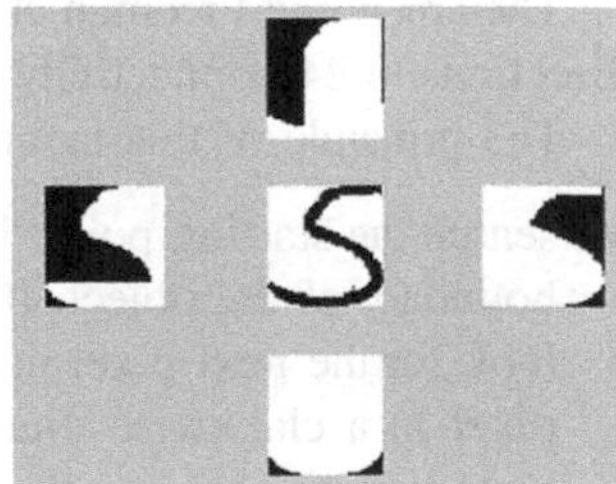

Fig. 2 The four profile projections of numeral '5'

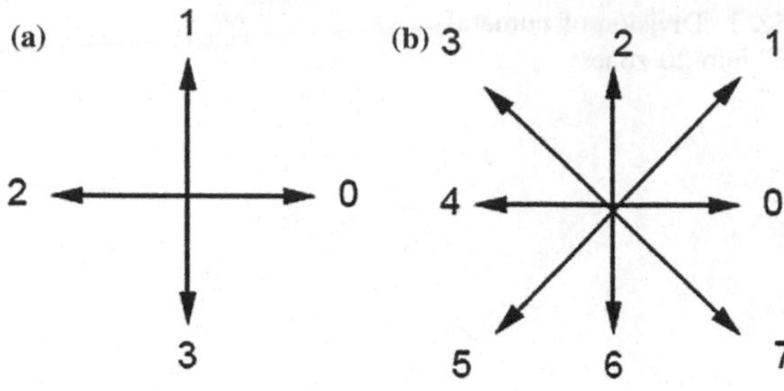

Fig. 3 4-connectivity (**a**) and 8-connectivity (**b**) of Freeman code

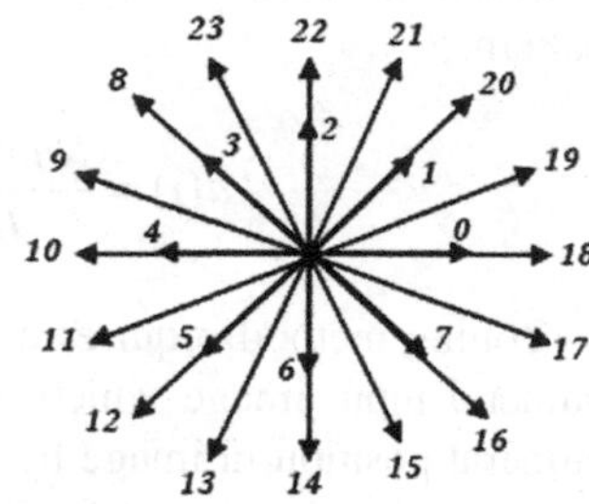

Fig. 4 Direction numbers for 24-directional chain codes

This method requires the same preprocessing stage than the zoning method except the skeletonization.

2.3 *Ameliored Freeman Chain Code*

Freeman code, introduced by Freeman in 1961 [10], is used to represent a shape boundary by a connected sequence of straight-line segments of specified length and direction. This representation is based on 4-connectivity or 8-connectivity of the segments (Fig. 3).

The ameliored Freeman chain code [4] concerns the extension of the chain codes directions to 24-connectivity (Fig. 4) instead of 8-connectivity.

The principle of this technique has to go through the following steps:

- search the starting point; it's selected as being the first pixel encountered in the boundary of the object. It's called the current boundary pixel.
- look for the next pixel in the eight locations surrounding the current boundary pixel in a clockwise direction. At least one pixel will be find, that is also a boundary pixel.
- assign a numeric code between 0 and 7 as already shown in Fig. 3b.

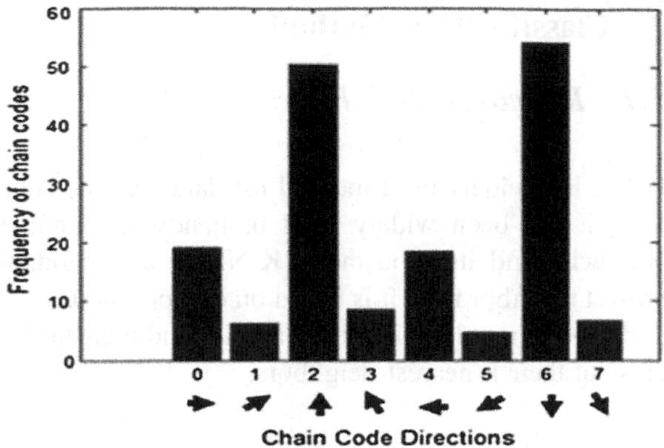

Fig. 5 Histogram of gradient orientation

- If there is no pixel, look for the first pixel in the 24 connectivity encountered in the boundary.
- assign a numeric code between 8 and 23 (Fig. 4).

The process of locating the next boundary pixel and assigning a code is repeated until we return to the starting point.

2.4 Histogram of Oriented Gradients

Histogram of Oriented Gradient (HOG) descriptors, introduced by Dalal and Triggs [11], are feature descriptors used in computer vision and image processing for the purpose of object detection. The principle of the Histogram of Oriented Gradient descriptors is that local object appearance and shape within an image can be described by counting the occurrences of gradient orientations or edge directions in localized parts of an image. For computing these features, the image is divided into cells and histograms of gradient directions are formed for each of these cells. The combination of these histograms then represents the descriptor (Fig. 5).

This technique requires a phase of preprocessing similar to the profile projection method. It is to convert color images to grayscale, binarise these images, crop them and finally normalize them in a predefined size.

3 Classification Methods

3.1 K-Nearest Neighbors (K-NN)

K-NN is a widely used method for data classification. Proposed in 1967 by Cover [12], it has been widely used in handwritten numerals recognition [13] for its simplicity and its robustness. K-NN is a method, which was inspired from the closest neighbor rule. It is based on computing the distance between the test sample and the different learning data samples and then attributes the sample to the frequent class of their k-nearest neighbors.

Algorithm 1: K-nearest neighbors
- Let $L = \{(x', c)/x' \in \mathbb{R}^N, c = 1,2,\ldots,\mathscr{C}\}$ the learning set. $\mathscr{C}$ is the number of classes. - Let x the example to determine its class. 1: **For each** object x' of the set L - Compute the distance between the object x and the object x', $d(x, x')$. - Classify the different distances in increasing way. **end** 2: **For every** $\{x' \in KNN(x)\}$ **do** : - Identify the most frequent class. **end** 3: Assign to x the identified class.

3.2 Fuzzy Min Max Classification (FMMC)

Fuzzy Min Max Classification, FMMC, is a classification method introduced by Simpson [14] in 1993, based on neural network architecture. FMMC contains three layers, input, output and hidden layers. The number of neurons in the input layer is equal to the dimension of the data representation space. The number of neurons in the hidden layer increases in time, with respect to the creation of prototypes. The number of neurons in the output layer is equal to the number of classes known initially.

The learning process is made in three steps, expansion overlapping and contraction repeated for each training input pattern [15]. These phases are controlled by two parameters; the sensitivity and the vigilance factor that control the maximum size of created hyperboxes. This method is well detailed in [16].

Algorithm 2: FMMC

1: Initial values for γ and θ.
2: Assume that the first input pattern form a hyperbox B_l, defined by its min point (V_1, W_1) where $V_1 = W_1 = O_1$.

3: **Repeat**
- Select a new input pattern (O_i, d_k) of the set A, identify a hyperbox for the same class, and provide the highest degree of membership. If a hyperbox cannot be found, a new hyperbox is formed and added to the neural network.
- Expand the new created hyperbox B_j defined by a couple of points (v_{jn}^-, w_{jn}^-), where $w_{jn}^- = \max(w_{jn}, a_{in})$ and $v_{jn}^- = \min(v_{jn}, a_{in})$, $1 \le n \le N$.
-Compute the size T where:
$$T = \frac{1}{N}\left(\sum_{n=1}^{N}(\max(w_{jn}, a_{in}) - \min(v_{jn}, a_{in}))\right)$$
* If $T > \theta$, a new hyperbox is created. Otherwise, we check if there is an overlap between the hyperbox B_j expanded in the last expansion step, and the hyperboxes which represent other classes than that of B_j.
- If an overlap between two hyperboxes of two different classes has been detected, a contraction of the two hyperboxes is carried out.

4. **Until** stabilization of hyperboxes.

4 Experimental Results and Discussions

In this section, the results of different feature sets for offline handwritten character recognition are presented. The performance of these features is evaluated using a database of handwritten vowels, provided by various categories of writers. The used database is split into two sets, one set of 300 characters is used for learning and the remaining 200 numerals are used for the test stage (Fig. 6).

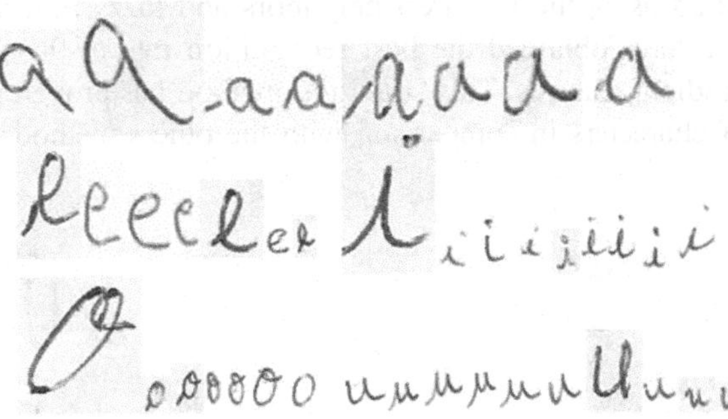

Fig. 6 A sample of the database of handwritten vowels

Table 1 Recognition results by k-NN and FMMC classifiers using different features methods

Features extraction methods	Recognition rate by K-NN (%)	Recognition rate by FMMC (%)
Zoning	90	88
Profile projection	93	92
Ameliored Freeman chain code	89	89
Histogram of oriented gradients	**96**	**93**

In test phase, several experiments were carried out to determine the recognition rate according to the normalization size using Zoning, Profile projection, Ameliored Freeman chain code and Histogram of oriented gradients as features techniques.

Table 1 summarizes the overall recognition rate of the system with each of the four feature techniques. The experiment results show that the best recognition rate was obtained using the histogram of oriented gradients features for both classification methods. It is equal to 96 % by k-NN and 93 % by FMMC. The next highest rate is obtained by profile projection. The zoning and the ameliored Freeman chain code features give a lower recognition results among all the features used in this study.

This study shows that the structural approach is well suited for the handwritten recognition, it gives a discriminative characteristics to the character that allow a good classification. This is the case for the Histogram of oriented gradients method, but not for the ameliored Freeman chain code in our experiments. This is due to the discontinuities that can persist at the edge detection that makes confusion between the letters.

5 Conclusion

The work presented in this paper analyses the performance of offline character recognition using four feature sets—zoning, profile projection, ameliored Freeman chain code and histogram of oriented gradient features. These features were classified using the k-nearest neighbors and fuzzy min max as classification methods. We have obtained the best recognition rate of 96 % using histogram of oriented gradient features. This structural method has proven its performance in recognition of characters in comparison with the others methods.

References

1. Benne, R.G., Dhandra, B.V., Hangarge, M.: Tri-scripts handwritten numeral recognition: a novel approach. Adv. Comput. Res. **1**(2), 47–51 (2009). ISSN: 0975–3273
2. Papakostas, G.A., Koulouriotis, D.E., Karakasis, E.G., Tourassis, V.D.: Moment-based local binary patterns: a novel descriptor for invariant pattern recognition applications. Neurocomputing 358–371 (2013)
3. Abu Bakar, N., Shamsuddin, S., Ali, A.: An integrated formulation of zernike representation in character images. In: García-Pedrajas, N., Herrera, F., Fyfe, C., Benítez, J., Ali, M. (eds.) Trends in Applied Intelligent Systems. Lecture Notes in Computer Science, pages, vol. 6098, pp. 359–368. Springer, Berlin/Heidelberg (2010)
4. Benchaou, S., Nasri, M., El Melhaoui, O.: New structural approach for numeral recognition based on mathematical morphology and Freeman code. Mediterr. Telecommun. J. **5**(2) (2015)
5. Rajput, G.G., Mali, S.M.: Marathi handwritten numeral recognition using fourier descriptors and normalized chain code. In: IJCA Special Issue on Recent Trends in Image Processing and Pattern Recognition RTIPPR (2010)
6. El Melhaoui, O.: Arabic numerals recognition based on an improved version of the loci characteristic. Int. J. Comput. Appl. **24**(1) (2011)
7. Chacko, A.M.M.O., Dhanya, P.M.: A comparative study of different feature extraction techniques for offline Malayalam character recognition. In: Proceedings of the International Conference on CIDM Computational Intelligence in Data Mining (2014)
8. Benchaou, S., Nasri, M., El Melhaoui, O.: Neural network for numeral recognition. Int. J. Comput. Appl. **118**, 23–26 (2015)
9. Gao, X., Guan, B., Yu, L.: Handwritten digit recognition based on support vector machine. In: International Conference on Information Sciences, Machinery, Materials and Energy (ICISMME), pp. 941–944 (2015)
10. Freeman, H.: Boundary encoding and processing. In: Picture Processing and Psychopictorics, pp. 241–266 (1970)
11. Dalal, N., Triggs, B.: Histograms of oriented gradients for human detection. In: Proceedings of the 2005 IEEE Computer Society Conference on Computer Vision and Pattern Recognition (CVPR'05), vol. 1, pp. 886–893 (2005)
12. Gil-Pita, R., Yao, X.: Evolving edited k nearest neighbor classifiers. Int. J. Neural Syst. **18**(6), 459–467 (2008)
13. Bernard, M., Fromont, E., Habardand, A., Sebban, M.: Handwritten Digit Recognition using Edit Distance based KNN (2012)
14. Simpson, P.K.: Fuzzy min-max neural networks. Part1: Classification. IEEE Trans. Neural Netw. **3**, 776–786 (1992)
15. Quteishat, A., Lim, C.P.: A modified fuzzy minmax neural network with rule extraction and its application to fault detection and classification. Applied Soft Computing, vol. 8, pp. 985–995. Elsevier, (2008)
16. Benchaou, S., Nasri, M., El Melhaoui, O.: New approach of features extraction for numeral recognition. Accepted for publication in Int. J. Pattern Recogn. Artif. Intell. (2015)

Construction of an Evaluation Tool for the Use of the Smart City by Music Professionals and Music Teaching

Soraya Mohamed Mohamed, Laila Mohamed Mohand and María del Carmen Olmos Gomez

Abstract This research has been carried out through a questionnaire with the necessary criteria of reliability and validity, with established technical requirements, based on the metric and theoretical principles that allow to maximize the validity of the inferences made from the instrument used, and whose application and corresponding data allow us in decision making [11], relating to the knowledge of the educational and professional population linked to the music, through the smartcity tools. We claim to develop an instrument that allows us to know how they access to different information and what new technologies are commonly used and known in their field in order to detect possible needs, solve them and improve them in a positive way. The elaboration of this instrument collects such information and has been designed with adaptability to ICT, validity and reliability; psychometric features, which allow a correct use of the measures that are obtained.

Keywords Smart city · TIC · Music education

1 Introduction

From the middle of the last century, society has undergone significant growth regarding the use of new technologies. This increase can be seen in the various activities of daily life, from a simple purchase of a bus ticket, listening to the radio, turn on the television, go to the bank to make a transfer [1].

S.M. Mohamed (✉) · L.M. Mohand · M. del Carmen Olmos Gomez
Faculty of Education and Humanities, University of Granada, Granada, Spain
e-mail: sorayasori@correo.ugr.es

L.M. Mohand
e-mail: lafu@ugr.es

M. del Carmen Olmos Gomez
e-mail: mcolmos@ugr.es

Á. Rocha et al. (eds.), *Europe and MENA Cooperation Advances in Information and Communication Technologies*, Advances in Intelligent Systems and Computing 520, DOI 10.1007/978-3-319-46568-5_22

The inclusion of Information and Communication Technology (ICT) in educational processes is an improvement in all facets of education. Therefore, their audiovisual resources, as well as the technological ones, should be considered, and those who facilitate communication [3]. The victory in the teaching-learning process is not only limited to the availability of the existing technology but also the pedagogical design [13].

Gonzalez [7], exhibits that currently we are experiencing a big change in our society, without realizing, and without fully knowing all this broad technology that surrounds us, these new changes force us to be always informed and updated about the new information technologies. The media and technology are currently part of our daily life. The content that is created and consumed in these media represents a meaningful impact on the social, emotional, cognitive and physical development of the human being.

ICT have led to a relevant change in the organizational, methodological and educational aspects in teaching environments. In addition to the well-known resources such as smartphones and applications tailored to needs, we can mention others like: e-mail, various computer programs, the rise of the social software (Facebook, Instagram, Twitter, Tuenti, Flicker, My Space, personal blogs, websites…). There is a large number of fields where these tools are used for educational purposes, taking an important role in teaching and promoting self-learning for its innovative and motivating character [2]. Additionally, they offer to the educational community a set of useful tools as much for individual development as well as they foster the group work.

There are several field studies on the management of ICT and its use through them for getting information about activities and musical events that employ different methodologies and tools for collecting data such as those developed by Moya et al. [10], which focused on determining the level of knowledge, use and attitudes towards ICT by future teachers, adolescents interactions that occur in the information society as well as investigations and gender analysis in the digital native field in the use of technology and their effectiveness in teaching and learning of Music Education [6, 15].

2 Purpose and Objectives

To approximate to the work done by music teachers of ICT in their music classes, an analysis has been carried out bearing in mind different documentary sources as well as official documents. Moreover, a survey technique has been used, specifically, the questionnaire. As a result, the primary objective was the design and establishment of the psychometric properties that support their reliability and validity to transfer the results to its reference population.

Table 1 Distribution of the sample

Educational stages	Education	%
Primary	36	22.36
High School	43	26.71
University	22	13.66
University Teacher	8	4.97
Conservatory	26	16.15
Music	26	16.15
Total	161	100

2.1 *Population and Sample*

The study population is the faculty of Music Education in the different educational stages both in formal and non-formal education. The choice of the sample was made not by using a non-probabilistic sample but accidental or causal. The sampling units have been classrooms and institutions that are specified as training centers for students in the music field.

Below are the data with which we have worked (Table 1).

2.2 *Construction of the Questionnaire*

During the completion of the questionnaire for the Faculty of Music Education a number of questionnaires have been taken as a reference. Questionnaires already developed in the works of Martínez et al. [9] that were adapted to the level of knowledge, use and attitudes towards ICT by future teachers, and it was also used the questionnaire carried out by Ubovich [16] to learn about the use of technology and its effectiveness in teaching and learning of Music Education used.

The validation of the test has been done before the empirical development of the same. The reliability and validity are indispensable in the development of measuring instruments for proper implementation of the same.

The development of the most appropriate items formed the final questionnaire as it would be validated by a group of experts.

During this process the number of cases lost at the different responses of the experts were studied and, finally, a descriptive study of the items was conducted. Then the means and medians were analyzed to find out the index of difficulty in polytomous items [8]. In addition, it was also analyzed the variance of each item in order to see the discrimination index [14]. Items that have reduced variance will benefit from reduced discriminatory value and cooperative stance to the reliability and validity of the test. In the study, the maximum variance is when half of the subjects have selected the favorable external option and the other half the unfavorable of the polytomous items.

Through the qualitative analysis of the items, the items that were unclear, the ones which were not written correctly and those that didn't answer to the proposed objectives to evaluate were excluded.

Information about their completion was examined in addition to the supplementary information section. The rest of the items were prepared according to a Likert scale of values between 1 and 4, according to the following interpretation: 1 "nothing"; 2 "something"; 3 "quite" and 4 "very much". On this scale developed for this research they were chosen values of 1 to 4 in order to avoid affirmative or negative positions on the items. The questionnaire consists of 29 items in four blocks.

2.3 Factor Analysis

The purpose of factor analysis is to determine the validity of builder questionnaire and the orthogonal rotation method used is the Equamax, that simultaneously synthesized factors and variables [4, 12].

Previously to the factorization process, it has been delimited the degree of adaptation of the values obtained. For this we used four indicators: correlation matrix, Bartlett's sphericity test and the test of adequacy of sampling of Kaiser-Meyer-Olkin (KMO), and the residuals.

- The correlation matrix between each variable, "determinant" is 1.42E-009. This value is reduced so that is considered the presence of variables with correlations among themselves very exalted. This allows the realization of the factor analysis.
- Through Bartlett's sphericity test it has been obtained that the value of X^2 is 3045.648, which for 406 g.l. is significant to 0.000. This test indicates that you are able to perform factor analysis.
- The test of adequacy of sampling Kasier-Meyer-Olkin (KMO) must be a value between zero and one, and the value must be above 0.5 [15]. It was obtained from measurement of KMO = 0.900. This value indicates that the results can be considered significant.

The establishment of the commonalities is the initial step, since it establishes the proportion of variance explained by the components. In this study, twenty-nine items are explained by components, there are no values close to zero, but they oscillate between 0.516 and 0.828.

By the method of extraction of components six factors have been obtained, which explains the 68.051 % of the total variance. Cattel's sedimentation test shows significantly that the optimal number of factors that are above the value 1 are six (Fig. 1).

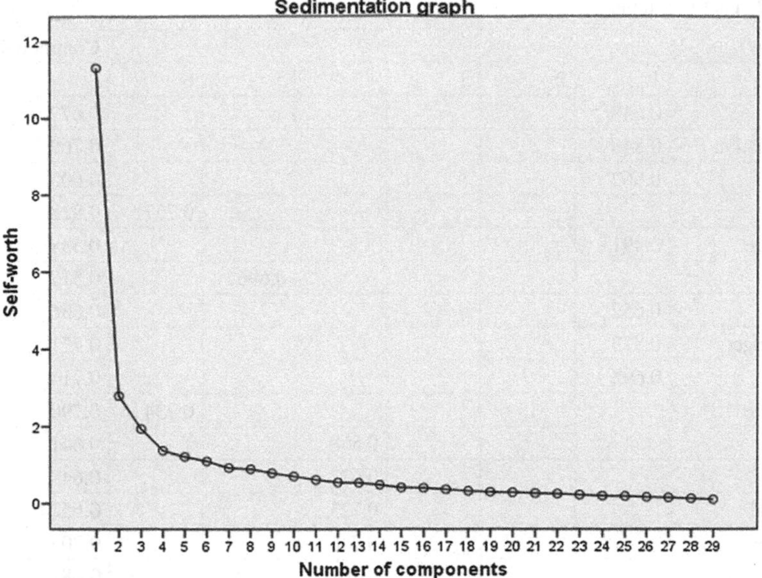

Fig. 1 Sedimentation graph

The matrix component identifies variables associated to factors. The criterion used to define the required limit on the degree of correlation between variable and factor is the one proposed by Comrey [5] of 0.3. This limit sets those variables that are below that value, less than 10 % of the variance in common with the factor while variables that are above this value are considered to have weight factor in the component.

The following table shows that the commonalities of the items that form the scale are located above 0.30. Variables that represent the highest percentages correspond to video editing programs that the teacher knows 0.828 and video editing programs that employed in teaching 0.790 (Table 2).

Questionnaire items of use of the Information and Communications Technology by teachers of Music Education (ICT) and its link with the smart city are grouped into six factors:

The first factor relates to the knowledge of computer programs and employment in teaching practice of teachers with 39.022 % of total variance and the items that are set out below:

1. Knowledge of basic programs (Word, Excel and PowerPoint).
2. Interpersonal relation programs.
3. Blog, chat and forum.
5. Knowledge of video-sharing websites (YouTube).
7. Use of basic programs (Word, Excel and PowerPoint).
8. Use of Interpersonal relation programs.
9. Use of blogs, chats and forums.

Table 2 Factorial table

Factors							Commonality
	1	2	3	4	5	6	
Basics	0.658						0.677
Interrelation	0.848						0.769
Blog	0.662						0.603
Video						0.757	0.828
Youtube	0.491						0.538
Smart					0.696		0.517
Word	0.652						0.656
Messenger	0.823						0.771
Chat	0.693						0.714
Pinnacle						0.734	0.790
Online				0.568			0.641
ICT				0.677			0.649
Musical				0.723			0.652
Culture				0.746			0.703
Events				0.701			0.681
Process		0.578					0.755
Results		0.713					0.759
Resources		0.687					0.610
Learning		0.754					0.738
Music		0.784					0.745
City					0.804		0.771
Emerging			0.507				0.669
Work			0.552				0.721
Effectiveness			0.595				0.686
Overwhelms					0.410		0.516
PROS			0.695				0.547
Platforms			0.748				0.625
Conclusions			0.734				0.656
Project					0.792		0.749

The second factor describes the impact that ICT has in the teaching process during his professional activity with 9.623 % of total variance and the following items:

16. ICT helps my learning process.
17. They improve the academic performance of students.
18. They replace traditional resources.
19. They increase the Music Education learning.
20. Dynamic Music Education teaching and involves saving time.

The third factor refers to the result obtained from the ICT according to learning style with 6.682 % of total variance and the following items:

22. I'm aware of ICT that are emerging.
23. I prepare my work using ICTs.
24. Use of ICT to enhance the effectiveness of teaching.
26. Before working with ICT, I analyze pros and cons.
27. I know what other teachers think through media platforms.
28. I get conclusions of my work with ICT.

The fourth factor makes mention to the use made of ICT in their personal field in order to realize if they are used to know the musical events in their city and as increase of their musical genre with 4.751 % of total variance and the following items:

11. Use of video-sharing websites (YouTube).
12. I use ICT to share musical tastes.
13. Extension of the music genre through ICT.
14. Search for information about musical culture.
15. Get to know the musical events in the city.

The fifth factor is focused on knowing if Music teachers know the Smart city and also to check if they are familiar with their different uses in music, with 4.183 % of total variance and the following items:

6. Smart city as a tool of the musical event information.
21. Smart city as a valuable tool to acquire musical knowledge.
25. The use of ICT overwhelms me.
29. The Smart city will positively affect the musical knowledge of students.

The sixth factor refers to the knowledge and use of programs for audio and video editing with 3.790 % of total variance and the following items:

4. I know audio and video editing programs.
10. Use audio and video editing programs.

3 Discussion of Results

The Questionnaire of Evaluation of the Use of the Smart City by Music Professionals and Music Education has shown a good structure factor, developed on five factors with a high internal consistency and precise stability. The exploratory factor analysis of main components identified a structure in six factors that explained the 68.051 % of the variance, but for reasons of adjustment, it has been decided to establish five factors for greater convergence. The name of the factors have been done looking for the links between the variables according to the highest saturations, to establish within which factor these variables are encompassed.

Based on the results obtained and as conclusion we establish that a useful and reliable questionnaire has been created to assess the use of the Smart City by music professionals and Music Education. In conclusion, it is recommended for use in similar conditions to those established for validation.

References

1. Aragón, R.: Estilos de aprendizaje: Uso de los blogs de educación (2009)
2. Area, M.: Algunos principios para el desarrollo de buenas prácticas pedagógicas con las TICs en el aula. Comunicación y Pedagogía: Nuevas Tecnologías y Recursos Didácticos **222**, 42–47 (2007)
3. Bautista, A., Alba, C.: ¿Qué es Tecnología Educativa?: Autores y significados. Revista Píxel-bit **9**, 4 (1997)
4. Berrocal, E., Olmedo, E., Olmos, Mª.C.: Validation of an Evaluation Tool for Shared Experience in Intercultural Secondary Classrooms through a Structural Equation Model. Procedia 94 **114**, 244–256 (2013)
5. Comrey, A.L.: Manual de análisis factorial. Cátedra (1985)
6. Flores, J.M.: Nuevos modelos de comunicación, perfiles y tendencias en las redes Sociales. Revista científica iberoamericana de comunicación y educación **33**, 73–81 (2009). doi:10.3916/c33-2009-02-007
7. González, A.: La tecnología actual en nuestra sociedad (2013). http://www.tribunasalamanca.com/noticias/la-tecnologia-actual-en-nuestra-sociedad/1369849795
8. Lukas, J.F.: Análisis de ítems y de tests con ITEMAN. Servicio de publicaciones de la Universidad del País Vasco, Bilbao (1998)
9. Martínez, M.D.V.M., Hérnandez, J.R.B., Hérnandez, J.A.H., Gutiérrez, R.C.: Análisis de los estilos de aprendizaje y las TIC en la formación personal del alumnado universitario a través del cuestionario REATIC. Revista de Investigación Educativa **29**(1), 137–156 (2011)
10. Moya, Mª., Hernández, J.R., Hernández, J.A., Cózar, R.: Análisis de los estilos de aprendizaje y las TIC en la formación personal del alumnado universitario a través del cuestionario REATIC. Revista de Investigación Educativa **29**(1), 137–156 (2011)
11. Muñiz, J.: Las teorías de los tests. Teoría Clásica y Teoría de Respuesta a los Ítems. Papeles del Psicólogo **31**(1), 57–66 (2012)
12. Olmedo, E., Aguaded, E., Berrocal, E., Buendía, L., Expósito, J., Sánchez, Ch., Carmona, M.: Constructing an instrument for evaluating group relations in intercultural secondary school classes. Int. J. Assess. Eval. **21**, 11–21 (2014)
13. Paily, M.U.: Creating constructivist learning environment: role of "web 2.0" technology. Int. Forum Teach. Stud. **9**(1), 39–50 (2013). http://search.proquest.com/docview/1346942900?accountid=14542
14. Renom, J.: Diseño de tests. IDEA, Barcelona (1992)
15. Rodríguez, S., Gallardo M.A., Ruiz, F., Olmos MªC.: Investigación educativa: análisis de datos cuantitativos y cualitativos en la metodología de encuesta. Granada: Ed. Grupo Editorial Universitario (2008)
16. Ubovich, B.A.: Utilization and effectiveness of technology in Music Education (Doctoral dissertation, American Conservatory of Music) (2015)

Model Transformations in the MOF Meta-Modeling Architecture: From UML to CodeIgniter PHP Framework

Oualid Betari, Mohammed Erramdani, Sarra Roubi, Karim Arrhioui and Samir Mbarki

Abstract Over the last few years, with the increased importance of the internet in many domains, web development industry has seen ground breaking changes. To solve the challenge of business and technology change, models have become increasingly important in constructing application systems. For example, OMG's Model Driven Architecture (MDA) uses models as building blocks to support application development. In this paper, we present the application of the MDA approach to model the CodeIgniter PHP framework. We developed the models used for transforming Platform Independent Model (PIM) to Platform Specific Model (PSM), using a UML class diagram as a source model to generate an XML file containing the core components of a CodeIgniter PHP framework.

Keywords Model driven architecture · Unified modeling language · Platform independent model · Platform specific model · Codeigniter · PHP · Framework

O. Betari (✉) · M. Erramdani · S. Roubi
MATSI Laboratory, Superior School of Technology,
Mohammed First University, Oujda, Morocco
e-mail: beta.oualid@gmail.com

M. Erramdani
e-mail: m.erramdani@gmail.com

S. Roubi
e-mail: roubi.sarra@gmail.com

K. Arrhioui · S. Mbarki
MISC Laboratory, Faculty of Sciences, Ibn Tofail University, Kenitra, Morocco
e-mail: arr.karim@gmail.com

S. Mbarki
e-mail: mbarkisamir@gmail.com

Á. Rocha et al. (eds.), *Europe and MENA Cooperation Advances in Information and Communication Technologies*, Advances in Intelligent Systems and Computing 520, DOI 10.1007/978-3-319-46568-5_23

1 Introduction

The coding methodology, affected by the rapid development of internet for web based application, indicates a higher need of sustainability and maintainability. PHP, a scripting tool for web that enable dynamic interactive web development, provides an intuitive, compiled fast, cross platform, open source, flexibility as well as required minimal setup [1]. Furthermore, PHP has become one of the most powerful programing languages for developing web applications. In order to solve the problems created by the growing complexity of the projects, Several methods for writing PHP codes such as Object Oriented Programming (OOP), Procedural PHP coding and Model View Controller (MVC) pattern have been proposed. This rapid development gave birth to several frameworks such as CodeIgniter PHP framework. As the time has gone popular PHP frameworks have just got bigger and better and become handy tool for developers to build giant application effortlessly.

Computing infrastructures are expanding their reach in every dimension. New platforms and applications must interoperate with legacy systems while virtual enterprises span multiple companies. New implementation platforms are continually coming down the road, each claiming to be "the next big thing".

To handle these changes, the Object Management Group (OMG) proposes the Model Driven Architecture as a solution. This approach supports evolving standards in application domains. MDA provides an open, vendor-neutral approach to the challenge of interoperability, building upon and leveraging the value of OMG's established modeling standards: Unified Modeling Language (UML); Meta-Object Facility (MOF) [2].

In this paper, we consider the unison of the solutions presented by the introduction of the PHP frameworks and the use of models in developing application, as we will apply the MDA approach to model the CodeIgniter PHP framework.

This paper is organized as follows: in the second section, we present a preview to the MDA approach and the CodeIgniter PHP framework. UML and CodeIgniter meta-models and the transformation rules of our approach are introduced in the third section. In the fourth section, we discuss related work. Last section concludes this paper and presents the future directions of our work.

2 MDA and CodeIgniter

2.1 Model Driven Architecture

In late 2000, OMG members first reviewed the document entitled "Model Driven Architecture" and decided to form an architecture team to produce a more formal statement of the MDA [2].

MDA addresses the challenges of today's highly networked, constantly changing systems, providing an architecture that assures portability, cross-platform

Interoperability, platform independence, domain specificity and productivity [2]. Its architecture [2, 3] is divided into three layers:

- In the first layer, we find the standard UML (Unified Modelling Language), MOF (Meta-Object Facility) and CWM (Common Warehouse Meta-model).
- In the second layer, we find a standard XMI (XML Metadata Interchange), which enables the dialogue between middleware.
- The third layer contains the services that manage events, security, directories and transactions. The last layer provides frameworks which are adaptable to different types of applications namely Finance, Telecommunications, Transport, medicine, E-commerce and Manufacture, etc.).

Note that the term "architecture" in Model-driven architecture does not refer to the architecture of the system being modeled, but rather to the architecture of the various standards and model forms that serve as the technology basis for MDA.

MDA consists of three general types of models, structured into three basic layers, as shown in Fig. 1 [4].

These model types are briefly described as follows [5].

- Computation Independent Model (CIM): It represents a high-level specification of the system's functionalities. It shows exactly what the system is supposed to do, but hides all the technology specifications.
- Platform Independent Model (PIM): It allows the extraction of the common concept of the application independently from the platform target.

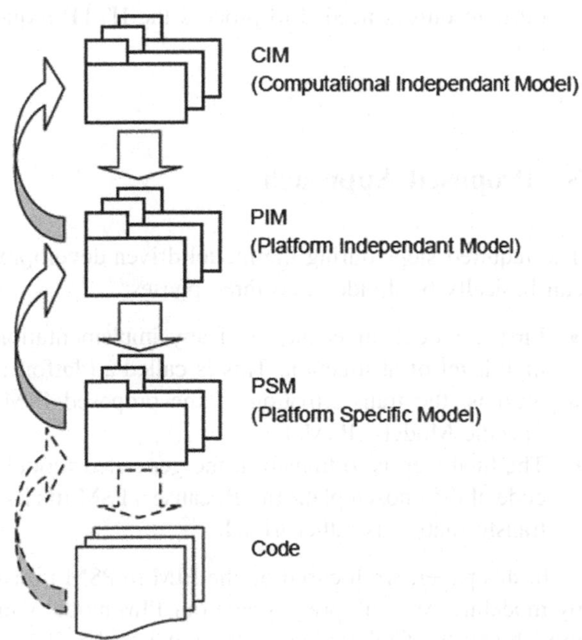

Fig. 1 Model driven architecture layers

- Platform Specific Model (PSM): It combines the specifications in the PIM with the details required of the platform to stipulate how a system uses a particular type of platform which leads to include platform specific details.

2.2 *CodeIgniter MVC (Model View Controller)*

CodeIgniter is an Application Development Framework—a toolkit—for people who build web applications using PHP. Its goal is to enable to develop projects much faster than writing code from scratch, by providing a rich set of libraries for commonly needed tasks, as well as a simple interface and logical structure to access these libraries.

CodeIgniter is based on the Model-View-Controller development pattern. MVC (Model View Controller) is a software approach that separates application logic from presentation. In practice, it permits your web pages to contain minimal scripting since the presentation is separate from the PHP scripting [6].

- The Model represents the data structures. Typically, the model classes will contain functions that help retrieve, insert, and update information.
- The View is the information that is being presented to a user. A View will normally be a web page, but in CodeIgniter, a view can also be a page fragment like a header or footer. It can also be an RSS page, or any other type of "page".
- The Controller serves as an intermediary between the Model, the View, and any other resources needed to process the HTTP request and generate a web page [7].

3 Proposed Approach

The required steps during the model-driven development with the UML approach can basically be divided into three phases:

- First, a model, independent of any implementation technology, is built with a high level of abstraction. This is called a Platform Independent Model (PIM).
- Next is, the transformation of the proposed PIM into one or more Platform Specific Models (PSMs).
- The final step is to transform the generated model respecting the PSM into the code of the chosen platform. Because a PSM fits its technology very closely, this transformation is rather trivial.

In this paper, we focused on the PIM to PSM transformation using an approach by modeling. We will present next our PIM model which consists of a UML Meta-model, and the CodeIgniter Meta-model as the PSM model.

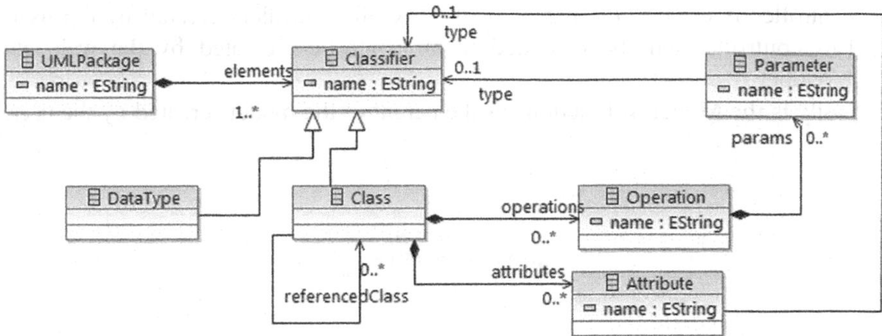

Fig. 2 Simplified UML meta-model

3.1 UML Meta-Model

The UML specification has the Kernel package which provides the core modeling concepts of the UML, our source Meta-model structures a simplified UML model based on a package containing the data types and classes. The classes contain structural features represented by attributes, and behavioral features represented by operations. The Fig. 2 presents the source meta-model.

- UmlPackage: is an abstract element of UML used to group together elements that are semantically related. This meta-class is connected to the meta-class Classifier.
- Classifier: This is an abstract meta-class which describes set of instances having common features. This meta-class represents both the concept of class and the concept of data type.
- Parameter: describes the order, type and direction of arguments that can be given when the operation is invoked. It explains the link between Parameter meta-class and Classifier meta-class.

3.2 CodeIgniter Meta-Model

The developed Meta-model target structure contains the core of the CodeIgniter framework, represented by the model, view and controller packages. This last package is composed of helpers, libraries and the loader which connects it to the database. The modeling of CodeIgniter PHP framework model is shown in the Fig. 3.

- Router recovers the URL and decomposes it into actions, its role is to find the controller to call.

- Controller is used as the parent of the overall controllers created by the user. This controller will be extended to all controllers created by the web site developer.
- Lastly is the Model. It functions as the parent of the models created by the user.

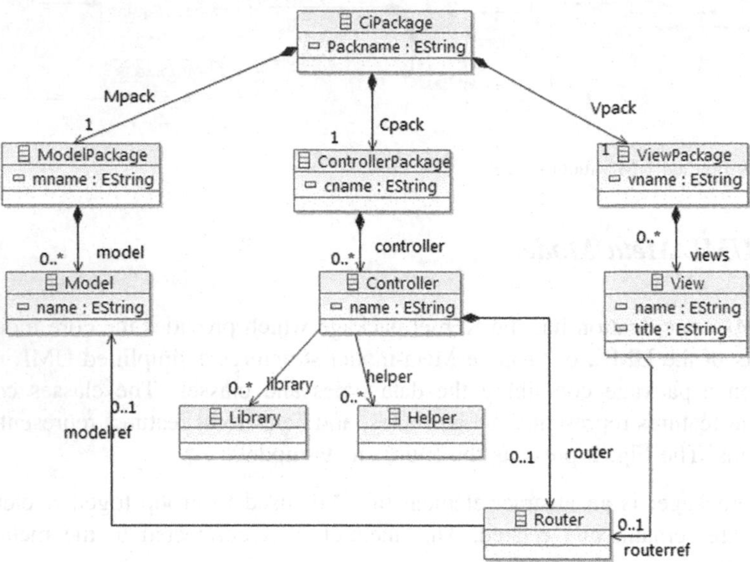

Fig. 3 CodeIgniter meta-model

Fig. 4 UML meta-model instance

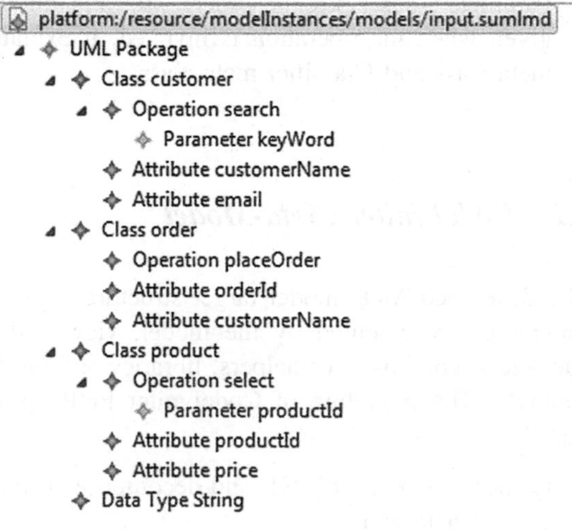

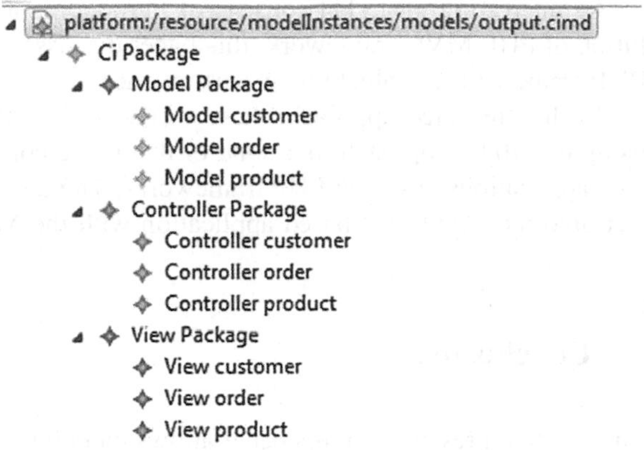

Fig. 5 CodeIgniter meta-model instance

3.3 *Models Instances*

In our case study, as an input, we will use an UML model that represents part of the information system of a commercial company. This file, shown in the Fig. 4, describes the elements needed for the transformation.

The output of the model to model transformation should be generated as shown in the Fig. 5, this file contains the major elements of the CodeIgniter Framework.

4 Related Work

There are several works related to the presented approach. Meta-modeling architectures based on the MDA approach can be found in other domains of computer science.

A lot of ideas for this work have been inspired by the paper [8] which applies of the MDA approach to generate, from the UML model, the Code following the MVC2 pattern using the standard MOF 2.0 QVT (Meta-Object Facility 2.0 Query-View-Transformation) as a transformation language.

In our approach, that uses as well the UML model as a source for transforming PIM to PSM, we went for a PHP Framework based on MVC design pattern as our target model.

In a different work, an MDA process have been used to automatically generate the multidimensional schema of data warehouse in [9], This process uses model transformation using several standards such as Unified Modeling Language, Meta-Object Facility, Query View Transformation and Object Constraint Language, from the UML model.

An overview of the MDA has been treated in the European MDA Workshops [10]. The goal of the workshops was to understand the foundations of MDA, to share experience in applying MDA techniques and tools, and to outline future research directions.

An interesting idea [11] consists of introducing an empirical study of the evolution of PHP MVC framework, this paper discusses the MVC based most famous PHP frameworks, evaluate their performance.

As the presented approach, these works dealt with modeling different systems using the MDA approach, but none of the above considered the use of MDA with web applications based on PHP frameworks. The developed models will be the first step into modeling web based application with the MDA approach.

5 Conclusions

The solution presented in this paper allows modeling web applications based on the CodeIgniter PHP. For this, we used the MDA approach, which provides the added advantage of improving application quality and portability, while significantly reducing costs and time-to-market. This approach has demonstrated its efficiency through enabling to build a complete model describing sufficiently a MVC CodeIgniter application.

The transformation from the UML model to the CodeIgniter model will be the subject of a study thereafter.

Afterwards, we will consider applying the solution of the MDA approach to other PHP frameworks in order to propose a comparative study.

References

1. Bergmann, S., Kniesel, G.: GAP: generic aspects for PHP. In: EWAS'06 (2006)
2. Object Management Group, Model Driven Architecture (MDA), MDA Guide rev. 2.0. http://www.omg.org/cgi-bin/doc?ormsc/14-06-01/
3. Mbarki, S., Erramdani, M.: Towards automatic generation of MVC2 web applications. INFOCOMP **7**, 88–91 (2008)
4. Blanc, X.: MDA en action: Ingénierie logicielle guidée par les modèles. Eyrolles, Paris (2005)
5. Roubi, S., Erramdani, M., Mbarki, S.: Model-Driven Transformation for GWT with Approach by Modeling: from UML Model to MVP Web Applications. In: MEDICT (2015)
6. CodeIgniter Documentation Website. https://www.codeigniter.com/docs/
7. Pitt, C.: Pro PHP MVC. Apress (2012)
8. Esbai, R., Erramdani, M., Mbarki, S., Arrassen, I., Meziane, A., Moussaoui, M.: Transformation by Modeling MOF 2.0 QVT: from UML to MVC2 Web model. INFOCOMP **10**, 01–11 (2011)
9. Arrassen, I., Meziane, A., Esbai, R., Erramdani, M.: QVT transformation by modeling. IJACSA **2**, 07–14 (2011)
10. Assmann, U., Aksit, M., Rensink, A.: Model Driven Architecture: European MDA Workshops: Foundations and Applications, Linkoping. Springer, Sweden (2005)
11. Olanrewaju, R., Islam, T., Ali, N.: An Empirical Study of the Evolution of PHP MVC Framework. In: Advanced Computer and Communication Engineering Technology (2015)

A Comparative Study of Three Population-Based Metaheuristics for Solving the JSSP

Abdelhamid Bouzidi and Mohammed Essaid Riffi

Abstract The Job shop-scheduling problem (JSSP) is known as the hardest combinatorial optimization problem. To solve it, means to find the schedule that have the optimal total execution time (i.e. makespan) by respecting some constraints; that is why JSSP had capture the interest of some researchers. They have proposed a significant number of metaheuristics by using the computational intelligence, thus to solve a real application based on JSSP, the best method should be acknowledged. The present paper aims to do a comparative study between three population-based metaheuristics for solving the JSSP, which are; the cat swarm optimization algorithm, the Cuckoo search algorithm, and the ant colony optimization, in order to evaluate the effectiveness of the three approaches; they were tested on a set of benchmark instances included in OR-Library. The Collected computation results proved that the ACO is more efficient to solve the JSSP.

Keywords Computational intelligent · Job shop scheduling problem, makespan · Ant colony · Cat swarm · Cuckoo search · Metaheuristic

1 Introduction

The Job Shop scheduling problem (JSSP) is one well-known optimization problem in operation research. Fischer and Thompson [2] had formulated this problem in 1963 as follow; the JSSP is a set of **m** machines, and **n** Jobs, each job J_i consist a

A. Bouzidi (✉) · M.E. Riffi
LAROSERI Laboratory, Faculty of Sciences, Chouaib Doukkali University,
EL Jadida, Morocco
e-mail: mr.abdelhamid.bouzidi@gmail.com

M.E. Riffi
e-mail: said@riffi.fr

Á. Rocha et al. (eds.), *Europe and MENA Cooperation Advances in Information and Communication Technologies*, Advances in Intelligent Systems and Computing 520, DOI 10.1007/978-3-319-46568-5_24

chain of **m** operations. Each operation have a precise processing time, and the machine where it will be executed is determined. Solve the JSSP problem, means find the schedule that have the minimal total execution time of all process called makespan, by respecting the following constraints:

- All jobs are independent, and available for processing at time zero.
- The machines are continuously available from time zero onwards
- Each machine can process one operation at a time.
- Each job can be manufactured at a specific moment on a single machine
- Operations of the same job cannot be processed concurrently.
- Each operation should respect its sequence in corresponding job.

The JSSP have numerous applications such as industry, manufacturing applications, and it belongs to class NP-hard [1] optimization problem, that is why researchers have proposed some methods to solve it. It was first resolved in 1989 by Carlier and Pinson [3]. After that, researchers suggested some methods, such as: simulated annealing [2], Tabu Search [3], and metaheuristic like genetic algorithm [4], Ant colony optimization [5], Particle swarm optimization [6], and Bee colony optimization [7], Cuckoo search algorithm [8], Cat swarm optimization algorithm [9]. The question is which method is more efficient to solve JSSP, the answer to that is to find the best optimal schedule that achieve the minimal execution time. That is why, this paper aims to study and compare three population-based metaheuristics, which are, the Cat Swarm Algorithm CSO [9], the Ant Colony Optimization ACO [5], and Cuckoo search algorithm CS [10], to know which method is most effective to solve the JSSP. The selected methods in this study were applied to solve some benchmark instances from OR-library [11], the results are collected, and the percentage error is calculated, to conclude each metaheuristic is more efficiency to solve real applications based the JSSP.

The rest of the paper is organized as follow; section two presents the formulation description of the JSSP. Section three, present the metaheuristics in study. Section four; shows the result and discussion. Finally, the conclusion.

2 Formulation of the Problem

Let set J = $\{J_1, J_2 \ldots J_n\}$, as the set of **n** Jobs, M = $\{M_1, M_2 \ldots M_m\}$ the set of m machine, and each job is composed of a set of **m** operations, $O_{Ji} = \{O_{Ji1}, O_{Ji2} \ldots O_{Jim}\}$. Each operation $O_{Jik} = \{m_{Jik}, t_{Jik}\}$ $(k \in [1, m])$ have a determined processing time $\mathbf{t_{Jik}}$, and the machine $\mathbf{m_{Jik}}$ where it will be executed. The solution of the problem is represented by a sequence of $n \times m$ operations that should be corrected to be realizable, by respecting all constraints of JSSP. The matrix INFO in Fig. 1 has $m \times n$ columns and four lines; this matrix represent information about each operation:

Fig. 1 Information matrix

$$\begin{pmatrix} o_1 & o_2 & o_3 & o_4 & o_5 & o_6 & o_7 & o_8 & o_9 \\ J_{o_1} & J_{o_2} & J_{o_3} & J_{o_4} & J_{o_5} & J_{o_6} & J_{o_7} & J_{o_8} & J_{o_9} \\ S_{o_1} & S_{o_2} & S_{o_3} & S_{o_4} & S_{o_5} & S_{o_6} & S_{o_7} & S_{o_8} & S_{o_9} \\ M_{o_1} & M_{o_2} & M_{o_3} & M_{o_4} & M_{o_5} & M_{o_6} & M_{o_7} & M_{o_8} & M_{o_9} \\ T_{o_1} & T_{o_2} & T_{o_3} & T_{o_4} & T_{o_5} & T_{o_6} & T_{o_7} & T_{o_8} & T_{o_9} \end{pmatrix}$$

O_i: The number of operations in schedule $(i \in [1, (n^*m)])$.
J_{o_i}: The job belonging to the operation o_i
S_{o_i}: The sequence of operation o_i in corresponding job.
M_{o_i}: The machine name where the operation o_i is processed.
T_{o_i}: The processing time of operation o_i.

For example, let's consider the following: 3 × 3 JSSP, where $n = 3$, $m = 3$, $J = \{J_1, J_2, J_3\}$, $M = \{1, 2, 3\}$, and for every J_i in J, $Ji = \{(m_{ik}, t_{ik})\}$ for $k \in [1, 3]$,

$$J_1 = \{(1, 1), (2, 2), (3, 1)\}$$
$$J_2 = \{(1, 8), (3, 3), (2, 6)\}$$
$$J_3 = \{(3, 4), (1, 8), (2, 3)\}$$

The representation of matrix of information is:

$$\begin{pmatrix} 1 & 2 & 3 & 4 & 5 & 6 & 7 & 8 & 9 \\ 1 & 1 & 1 & 2 & 2 & 2 & 3 & 3 & 3 \\ 1 & 2 & 3 & 1 & 2 & 3 & 1 & 2 & 3 \\ 1 & 2 & 3 & 1 & 3 & 2 & 3 & 1 & 2 \\ 1 & 2 & 1 & 8 & 3 & 6 & 4 & 8 & 3 \end{pmatrix}$$

Let propose a random solution:

1	5	6	9	2	3	8	4	7

To correct the random solution in Fig. 2, the matrix in Fig. 3 is used. Let consider "*new*" as the arranged vector/solution, and "*old*" as the current solution. The description of correction process are as follow:

1. Scan the vector solution "*old*"
2. The first available operation is added to task queue of "*new*" and deleted from "*old*"
3. Repeat (1) and (2), until the "old" vector size is zero. Here all operations in "*new*" are arranged, thus the valid vector solution by this process, is:

1	2	3	4	5	6	7	8	9

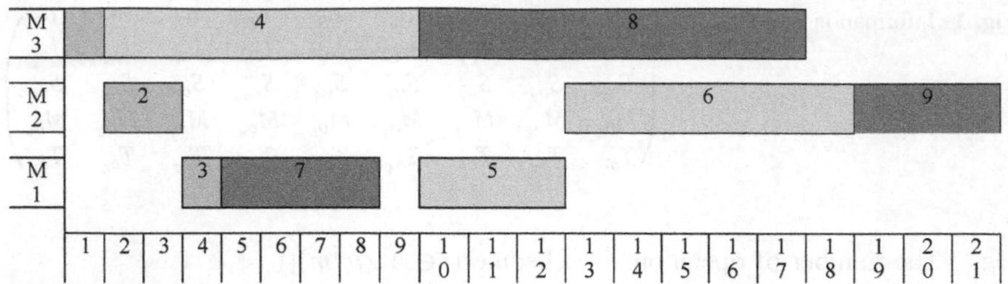

Fig. 2 Gantt chart

Fig. 3 CSO process

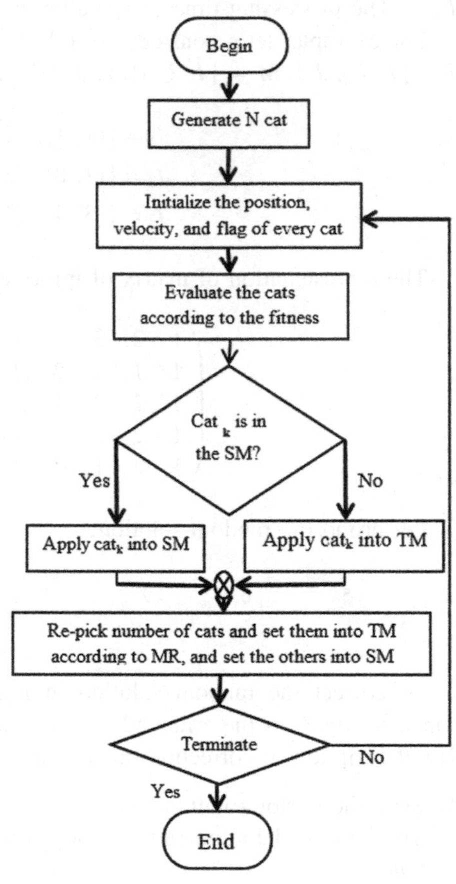

Finlay, to calculate the makespan of the correct solution. The Gantt chart is used:
The makespan of the solution presented in the Gantt chart is 21.

3 Metaheuristics Description

This part, describe the three metaheuristic used in this study, already applied to solve the JSSP, in each method, the solution is presented by a schedule, however, the process is different.

3.1 Ant Colony Optimization

The Ant Colony Optimization (ACO) is an evolutionary algorithm nature-inspired by the behavior of ants seeking a path between their colony and a source of food. Dorigo introduced the ACO in 1992 [12], to solve the Traveling salesman problem (TSP). The general process of ACO is:

```
Begin
      Initialization of the population
      While (!exit  condition) do
            Fori=1 until population_size do
                  Contructtraject(i)
            End for
            Update_pheromone()
      End_while
      Show the best optimal solution
End
```

Researchers have suggested some improvement to this method, to solve a variety of real applications, such as the ACO algorithm to solve the JSSP [5] proposed H. Nazif in 2015.

3.2 Cuckoo Search Algorithm

The cuckoo search (CS) algorithm is a nature-inspired population-based meta-heuristics. Proposed first by Yang and Deb [10] to solve multimodal functions. The CS bio-inspired by the obligate brood parasitism of some cuckoo species by laying their eggs in the nest of other host birds. The total process of the CS algorithm is

```
Begin
   Objective function f(x), x = (x_1,...,x_d)^T
   Generate initial population of n host nests xi (i =
1,2,..., n)
   while (t <MaxGeneration) or (stop criterion)
      Get a cuckoo randomly by Lévy flights
        evaluate its quality/fitness Fi
      Choose a nest among n (say, j) randomly
      if (Fi > Fj)
        replace j by the new solution;
      end
      A fraction (pa) of worse nests are abandoned and
  new ones are built;
      Keep the best solutions
        (or nests with quality solutions)
      Rank the solutions and find the current best
   end while
   Postprocess results and visualization
End
```

In 2014, Ouaarab et al. [13] have proposed an improved CS to solve combinatorial optimization problem that is TSP. Moreover, in 2015, the same authors had propose an improved CS [10], by including the Levy flights [14] methods to solve TSP. In the same year, they have used the last improved CS to solve the JSSP [8]. In the CS method, each egg in a nest represents a solution, and a cuckoo egg represents a new solution.

3.3 *Cat Swarm Optimization Algorithm*

Cat swarm optimization (CSO) is an evolutionary algorithm nature-inspired behavior of cats. This population-based metaheuristics was introduced in 2006 by Chu and Tsai [15]. In CSO algorithm, each cat has two modes; the seeking mode (SM) presents the cat when it is at rest, and the tracing mode (TM) presents the cat while tracing the path, according to its own velocity to chase a prey or any moving object. To combine these two modes, the mixture ratio (MR) was defined. The characteristic of each cat are; the position that presents the solution, the velocity, and the flag that defines the mode of the selected cat. The description of the total algorithm process of CSO is:

In 2013, the authors A Bouzidi and ME Riffi have proposed an improved the CSO to solve a combinatorial optimization problem, that is the TSP. and in 2014, the same authors had apply the CSO to solve the JSSP [9]. The solution is represented by the position of the cat, and the velocity, is a set of permutation, to change the position.

4 Results and Discussion

To test the performance of each metaheuristic, the three method in study (The ACO, CS, and CSO algorithms) have been applied to solve 15 benchmark instances problem from OR-library [11], the obtained results by CS, ACO, and CSO are presented in Table 1.

Table 1 shows the **instances** name, the number job **n** and the number machine **m**, best known solution (**BKS**) found by others algorithm, the best solution obtained by applying each method (**BS**) to the selected instance, and the percentage error (**%err**) value is obtained by:

$$\%\,\mathrm{err} = \frac{BS - \mathrm{BKS}}{\mathrm{BKS}} \times 100$$

To assess the collected results, the content of Table 1 is translated into the following graph (Fig. 4):

The Following graph present the calculated percentage error in Table 1, that shows clearly that ACO is efficient than the other methods. In addition, the CSO is more efficient than the CS.

By analyzing the variation of **%err**, after the application to solve some benchmark instances of JSSP, it can be shown clearly that the ACO algorithm is the best one to solve the JSSP problem.

Table 1 The results by apply ACO, CS, and CSO to solve JSSP

Instance	n × m	BKS	ACO		CSO		CS	
			BS	%err	BS	%err	BS	%err
abz5	10 × 10	1234	–	–	1234	0.00	1239	0.41
abz6	10 × 10	943	–	–	943	0.00	943	0.00
ft06	6 × 6	55	55	0.00	55	0.00	55	0.00
ft10	10 × 10	930	930	0.00	930	0.00	945	1.61
ft20	20 × 5	1165	1165	0.00	1175	0.86	1173	0.69
la01	15 × 5	666	666	0.00	666	0.00	666	0.00
la02	10 × 5	655	655	0.00	655	0.00	655	0.00
la03	10 × 5	597	597	0.00	597	0.00	604	1.17
la04	10 × 5	590	590	0.00	590	0.00	590	0.00
la06	10 × 5	926	926	0.00	926	0.00	926	0.00
la11	20 × 5	1222	1222	0.00	1222	0.00	1222	0.00
la16	10 × 10	946	946	0.00	946	0.00	946	0.00
la21	15 × 10	1040	1046	0.58	1053	1.25	1055	1.44
la26	20 × 10	1218	1218	0.00	1218	0.00	1218	0.00
la31	30 × 10	1784	1784	0.00	1784	0.00	1784	0.00

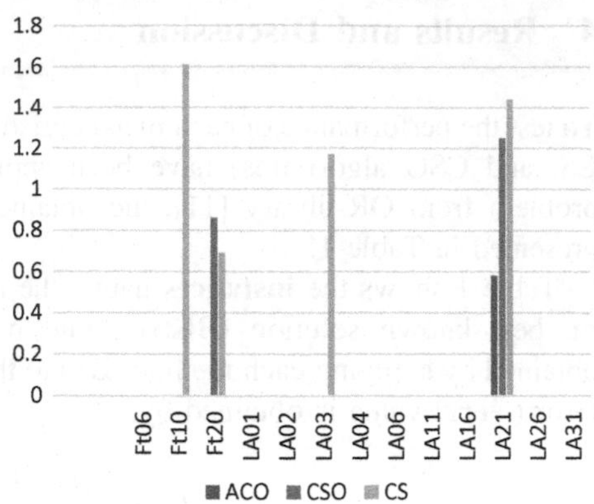

Fig. 4 %err by the application of CSO, ACO, and CS to JSSP

5 Conclusion

This paper presents the application of some metaheuristics to solve the Job shop-scheduling problem, to choose an efficient metaheuristic to solve real applications based on JSSP problem. The studied methods are Cuckoo search, Ant Colony Optimization, and the cat swarm optimization. This paper aims to compare the relative percentage error by the application of these metaheuristics to some benchmark instances. The computational results show that the ant colony optimization is more efficient than other metaheuristics in this study.

The future research, the hope is improve CSO to be more efficient to solve the JSSP, and apply this metaheuristic to solve real applications based on JSSP.

References

1. Garey, M.R., Johnson, D.S., Sethi, R.: The complexity of flowshop and jobshop scheduling. Math. Oper. Res. **2**(1), 117–129 (1976)
2. Van Laarhoven, P.J., Aarts, E.H., Lenstra, J.K.: Job shop scheduling by simulated annealing. Oper. Res. **40**(1), 113–125 (1992)
3. Dell'Amico, M., Trubian, M.: Applying tabu search to the job-shop scheduling problem. Ann. Oper. Res. **41**(3), 231–252 (1993)
4. Della Croce, F., Tadei, R., Volta, G.: A genetic algorithm for the job shop problem. Comput. Oper. Res. **22**(1), 15–24 (1995)
5. Nazif, H.: Solving job shop scheduling problem using an ant colony algorithm. J. Asian Sci. Res. **5**(5), 261–268 (2015)
6. Lian, Z., Jiao, B., Gu, X.: A similar particle swarm optimization algorithm for job-shop scheduling to minimize makespan. Appl. Math. Comput. **183**(2), 1008–1017 (2006)
7. Chong, C.S., Sivakumar, A.I., Low, M.Y.H., Gay, K.L.: A bee colony optimization algorithm to job shop scheduling. In: Proceedings of the 38th conference on Winter simulation, Winter Simulation Conference, 2006, pp. 1954–1961

8. Ouaarab, A., Ahiod, B., Yang, X.-S.: Discrete cuckoo search applied to job shop scheduling problem. In: Recent Advances in Swarm Intelligence and Evolutionary Computation, pp. 121–137. Springer, 2015
9. Bouzidi, A., Riffi, M.E.: Cat swarm optimization to solve job shop scheduling problem. In: 2014 Third IEEE International Colloquium Information Science and Technology (CIST), IEEE, 2014, pp. 202–205
10. Yang, X.-S., Deb, S.: Cuckoo search via Lévy flights. IN: World Congress on Nature and Biologically Inspired Computing, 2009. NaBIC 2009. IEEE, 2009, pp. 210–214
11. Beasley, J.E.: OR-Library: distributing test problems by electronic mail. J. Oper. Res. Soc. 1069–1072 (1990)
12. Dorigo, M.: Ph. D. Thesis, Politecnico di Milano, Italy, 1992
13. Ouaarab, A., Ahiod, B., Yang, X.-S.: Discrete cuckoo search algorithm for the travelling salesman problem. Neural Comput. Appl. **24**(7–8), 1659–1669 (2014)
14. Shlesinger, M.F., Zaslavsky, G.M., Frisch, U.: Lévy flights and related topics in physics. In: Levy Flights and Related Topics in Physics, vol. 450 (1995)
15. Chu, S.-C., Tsai, P.-W., Pan, J.-S.: Cat swarm optimization. In: PRICAI 2006: Trends in Artificial Intelligence, pp. 854–858. Springer, (2006)

Optimization of RDF Data Preprocessing for METIS Partitioning

Siham Benhamed and Safia Nait-Bahloul

Abstract The Resource Description Framework (RDF) developed by W3C is increasingly adopted to model data in a variety of scenarios, especially the published or exchanged data on Web. Managing a large volume of data is difficult because of size, heterogeneity, and additional complexity brought by the RDF reasoning. To take up the challenge of the massive size of RDF data, storage architectures and distributed processing are required. For this, the partitioning of data set is a widely adopted technique in many systems to deploy easily the distributed and parallel architectures. We study in this paper, the reduction of the dataset size to be partitioned into a step called preprocessing by improving data partitioning.

Keywords RDF graph · Preprocessing · Partitioning · MapReduce

1 Introduction

Semantic Web allows machines to understand the meaning of data to exploit better via standard frameworks such as RDF (Resource Description Framework) [1], which describes the resources published on the web formally to express knowledge. To effectively manage the increasing amount of these RDF data available on the

S. Benhamed (✉) · S. Nait-Bahloul
Department of Computer Science, Laboratory of LITIO, University of Oran, 1 Ahmed Ben Bella, PO Box 1524, El-M'Naouer, 31000 Oran, Algeria
e-mail: benhamed2007@yahoo.fr

S. Nait-Bahloul
e-mail: nait1@yahoo.fr

Á. Rocha et al. (eds.), *Europe and MENA Cooperation Advances in Information and Communication Technologies*, Advances in Intelligent Systems and Computing 520, DOI 10.1007/978-3-319-46568-5_25

web through very large sizes knowledge bases such as Cyc,[1] YAGO,[2] Bio2RDF,[3] Lexvo,[4] and DBpedia,[5] RDF data processing systems scalable and flexible are essential. However, the processing of such bases is very difficult and complex in terms of execution time and memory size. Although several approaches have been proposed for distributed RDF data processing, but their techniques are limited by the compromise between the complexity of massive data size and the efficiency of query.

The knowledge bases of semantic web are generally represented and expressed by triples RDF, which shape a graph for describing formally all the Web resources [1]. Each RDF triple consists of subject, predicate, and object. The set of RDF data having properties and classes according to a semantic is a standard which describes RDFS [2]. This is the first language definition of ontology which a key element of the Semantic Web. Ontologies can be considered as a data representative model of set concepts within a domain, by defining the relationships between concepts and specifying inference rules for reasoning on objects of area concerned [3]. Ontologies are used also to modeling the knowledge in a field presets. Query of databases knowledge that contains RDF data describing ontologies is a central requirement of the Semantic Web because semantic interpretation of queries is related firstly to knowledge representation and also to the power of language used. Several query languages of semantic data exist, with whom SPARQL [4] is the most used, since it supports different formats of complex queries for RDF data. SPARQL consists of three basic components: a query language, a transmission protocol, and an output format (a remote protocol that publishes queries and receives the results).

In order to improve these systems, we propose a new model that allows for processing and handling of RDF data at large scale. For this we use METIS [5] to partition data by reducing them upstream. By reducing the size of the file that contains the graph partitioning and process by METIS maximum in an additional step, called preprocessing phase.

The remainder of this paper is structured as follows: Sect. 2 presents our contribution with an overview of RDF data graph and an explanation of our pre-processing phase. Section 3 presents the partitioning phase. Finally, we present related work in Sect. 4 and conclude in Sect. 5.

---

[1]Cyc is an artificial intelligence project "AI". http://sw.opencyc.org/.

[2]YAGO (YetAnother Great Ontology) is a knowledge base derived from Wikipedia and other sources. http://www.mpi-inf.mpg.de/departments/databases-and-information-systems/research/yago-naga/yago/.

[3]Bio2RDF is a biological database. http://download.openbiocloud.org/release/3/release.html.

[4]Lexvo provides information on languages, words, characters, and other entities linked to human languages. http://www.lexvo.org/linkeddata/resources.html.

[5]DBpedia is a university project of exploration and automatic extraction of data derived from Wikipedia. http://wiki.dbpedia.org/Downloads2015-04#dbpedia-ontology.

2 Contribution

Our model (Fig. 1) articulates around four axes:

1. Represent the set of RDF data into graph to get a good representation of web knowledge.
2. Reduce the size of the graph during the preprocessing phase.
3. Apply a partitioning tool such as METIS [5] that uses multi-level algorithm in order to parallelize the processing to reduce the processing time of queries.
4. Optimize query of knowledge base in order to reduce the performed joins number in each partition because a SPARQL query involves a lot of join operation that requires a slow execution time.

In this paper, we present the pre-processing step which is based on the processing of data upstream, i.e. before partitioning of large mass data and distribution of treatment. This is achieved with the use of a parallel programming model as MapReduce [6]. This is considered as being one of popular parallel computing paradigm because of its scalability, reliability and integration of fault tolerance. This paradigm is realized for a wide variety of applications in the large data processing. To take better advantage of MapReduce for parallel processing of the graph, we will use lists to store our graph by adapting the adjacency lists structure that occupies less amount of memory compared with matrix and allowing us to store graphs in a more compact form. This will help us as well, to get the list of adjacent vertices with a linear complexity, which is a big advantage for the proposed algorithms in the preprocessing step that we propose.

2.1 RDF Data Graph

All of these triples establish a graph that expresses the relationship between resources and data to deploy semantics in the knowledge base. In our model, we rely on the RDF triple and RDF graph which are defined as follows (Fig. 2):

Definition of RDF triple. An RDF triple (s, p, o) $\in (U \cup B) \times U \times (U \cup B \cup L)$, with s represents the subject, p represents the predicate, and o represents the object. Thus,

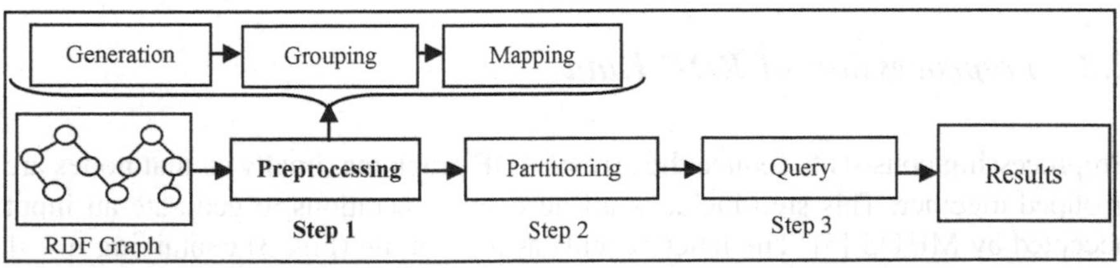

Fig. 1 General scheme of RDF data processing containing three main steps: preprocessing step, partitioning step, and the final phase of querying

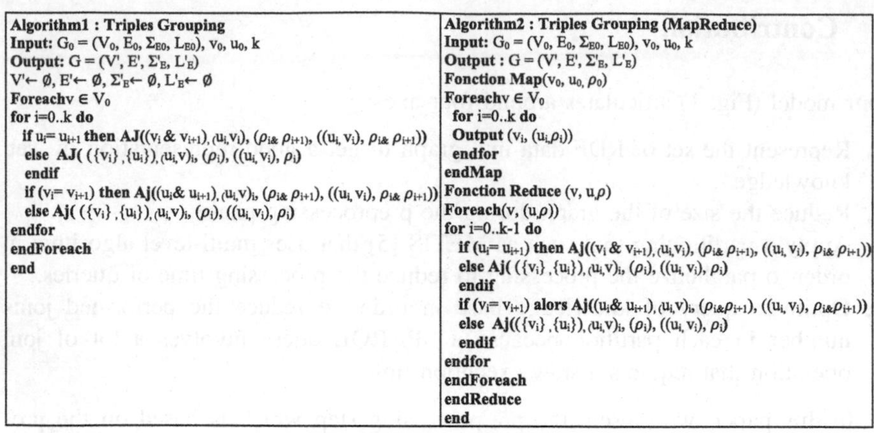

```
Algorithm1 : Triples Grouping
Input: G_0 = (V_0, E_0, Σ_E0, L_E0), v_0, u_0, k
Output: G = (V', E', Σ'_E, L'_E)
V'← ∅, E'← ∅, Σ'_E← ∅, L'_E← ∅
Foreachv ∈ V_0
for i=0..k do
  if u_i= u_i+1 then AJ((v_i & v_i+1), (u_i,v_i), (ρ_i& ρ_i+1), ((u_i, v_i), ρ_i& ρ_i+1))
   else  AJ( ({v_i} ,{u_i}), (u_i,v)_i, (ρ_i), ((u_i, v_i), ρ_i)
   endif
   if (v_i= v_i+1) then Aj((u_i& u_i+1), (u_i,v)_i, (ρ_i& ρ_i+1), ((u_i, v_i), ρ_i& ρ_i+1))
   else Aj( ({v_i} ,{u_i}), (u_i,v)_i, (ρ_i), ((u_i, v_i), ρ_i)
endfor
endForeach
end
```

```
Algorithm2 : Triples Grouping (MapReduce)
Input: G_0 = (V_0, E_0, Σ_E0, L_E0), v_0, u_0, k
Output : G = (V', E', Σ'_E, L'_E)
Fonction Map(v_0, u_0, ρ_0)
Foreachv ∈ V_0
 for i=0..k do
 Output (v_i, (u_i,ρ_i))
 endfor
endMap
Fonction Reduce (v, u,ρ)
Foreach(v, (u,ρ))
for i=0..k-1 do
  if (u_i= u_i+1) then Aj((v_i & v_i+1), (u_i,v_i), (ρ_i& ρ_i+1), ((u_i, v_i), ρ_i& ρ_i+1))
  else Aj( ({v_i} ,{u_i}), (u_i,v)_i, (ρ_i), ((u_i, v_i), ρ_i)
  endif
  if (v_i= v_i+1) alors Aj((u_i& u_i+1), (u_i,v)_i, (ρ_i&ρ_i+1), ((u_i, v_i), ρ_i&ρ_i+1))
  else  Aj((({v_i} ,{u_i}), (u_i,v)_i, (ρ_i), ((u_i, v_i), ρ_i)
  endif
endfor
endForeach
endReduce
end
```

Fig. 2 Algorithm of RDF triples grouping

U is the set of URI references, B is a set of blank nodes (bnode), and L is the set of literals. A triple is represented by a directed graph with labeled edges such as: $S \rightarrow^{P} O$.

Definition of graph RDF. An RDF graph is a directed and labeled multigraph with multiple edges, denoted G = (V, E, Σ_E, L_E) where V $\in (U \cup B \cup L)$ is the set of vertices of the graph that represents the subject or object of RDF triple. The set E $\in U$ is a multi-set of directed edges. (u, v) $\in$ E is a directed edge from u to v. $\Sigma_E \in U$ is set of edge labels (predicates) and L_E is the correspondence between a label $\rho \in \Sigma_E$ and its edge (u, v), (E àΣ_E ((u, v), ρ) $\in L_E$.

In this model, we must bring together all the data in RDF triples Group, so that intermediate results can be managed as sets of triples groups. This reduces the data set processed by grouping them into sets which reduces thereafter the number of intermediate results of each sub queries and reduces the number of join carried to obtain the final result.

Definition (Triples Group). Given an RDF graph RDF G = (V, E, Σ_E, L_E), S-TG of vertex v $\in$ V is a set of triples in which the subject is u, denoted s-TG (u) = {(v, ρ) | u $\in$ E, $\rho \in \Sigma_E$, u = v}. We call u the vertex of s-TG (u) grouping. Similarly, l'o-TG of v and p-TG of $\rho \in \Sigma_E$ are defined respectively as o-TG (v) = {(u, ρ) | v $\in$ E, $\rho \in \Sigma_E$, u = v} and p-TG (ρ) = {(u, v) | (u; v) $\in$ E}.

2.2 *Preprocessing of RDF Data*

Preprocessing consists to reduce the size of RDF graph maximally so that nodes are grouped together. This step includes all necessary operations to generate an input accepted by METIS [5]. The latter accepts as input a file (Fig. 3) containing n + 1 lines when the first line has general information on the graph header and the

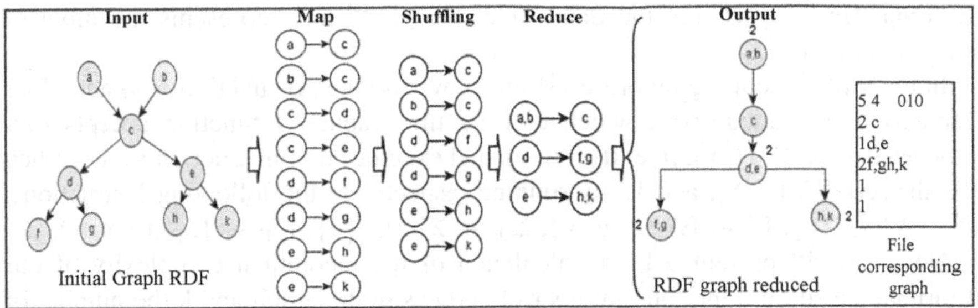

Fig. 3 Executing of various steps of Algorithm 2

remaining rows contain information about each graph vertex. Preprocessing contains three basic operations: generation, grouping, and mapping.

Generation of Initial Graph. The graph generator is a process producing the initial graph. For this, initially, it removes the value of blank nodes o ∈ V in the triple <s, p, o> and replaces it with the appropriate predicate. Then, in order to facilitate the partitioning of RDF graph by vertex, it removes the triples whose predicate value is rdf: type or with a sense of "type" as in [7], because this triples can generate unwanted connections in RDF graph and can make it more complex; which reduces the quality of graph partitioning significantly, since more the graph is connected, the harder it is to partition. This will prevent us to partition the related resources that are very important for data semantics, knowing that the purpose of METIS is to remove the links between resources to produce partitions containing predicates related as possible.

The last step, which executes the graph generator, is to ignore all the literal connection edges with string type, because they lead respectively to reductions in the running time of METIS and the memory.

Note. All data deleted, will be recover after partitioning for query step.

Grouping of Triples. In this stage it is necessary to group the sets of triples patterns to reduce their number and reduce respectively the size of the graph to treat. This step consists to build the set of triples groups with the same subject from the graph RDF.

This will allow us not to separate the linked data. The principle is to construct G' from G_0, such as the size of G' G is smaller than the initial graph G_0, for this, we must take all pairs of vertices (u_i, v_i) having the same value of the initial vertex u_i. For grouping the graph data using a MapReduce model in which we filter all the graph triples in Map step who return pairs <vi, (ui, ρi)> that represent respectively the object and the subject with the predicate matching, and the reduce phase must consolidate and combine pairs to reduce the number of records to be processed later, and therefore it must increase the effectiveness of treatment. The objective of Reduce phase is to maintain all the pairs which represent every group for which the aggregations are desired.

Grouping of related data sets is separated in Map function and the aggregation on these combined data is performed in the function Reduce. The result of this algorithm (Fig. 3) provide the data set of the graph that represents the input of mapping () phase.

In the triples grouping algorithm (Fig. 2), we use the AJ function who adds four elements of the quadruplet, which consists the graph. AJ function accepts four parameters inputs v, (uv), ρ, et ((u, v), ρ), and executes first instruction $v_= \{v\}$, then fills the sets V′, E′, Σ′, and L′ by running respectively the following instructions: $V' \leftarrow V' \cup \{v\}$, $E' \leftarrow E' \cup \{u, v\}$, $\Sigma'_E \leftarrow \Sigma'_E \cup \{\rho\}$, $L'_E \leftarrow L'_E \cup ((u, v), \rho)$.

Now we will present a brief calculation of the theoretical complexity of our algorithm based on n the total number of vertices in the graph and k the number of vertices possible to group. Algorithm 1 is a polynomial time complexity and proceeds in an iterative way. Specifically, as long as there are adjacent vertices having the same values that represent two scenarios is the same subject or the same object in which the loop is executed k times and the for each loop runs at worst n times.

The function Aj has a constant complexity O (1) and the complexity in the worst case of running Algorithm 1 is $O (k (n)^2)$. This complexity is improved in algorithm 2, wherein a complexity of O is obtained (k.log2 (n)) by parallelizing the processing and by fragmenting the set of vertices with the use of MapReduce. During the execution of the distributed algorithm on a parallel platform, including the volume of data exchanged with the degree of parallelism such as the number of workers. After the end of this step we apply a valuation for graph to obtain a weighted graph. The valuation is to associate a weight w to each vertex vi ∈ V of graph G’ which represents the number of resources contained in the node in other words, it represents the number of data graph fusion such as shown in the algorithm.

Mapping. Since it is difficult to work with triples such as the subjects, predicates, and objects that are in string format, we intend to convert them into a set of unique identifier IDs in this phase to facilitate processing of n triples and to reduce the size of the graph to be stored, and the time course of this graph. The mapping is a binary relation that associates each element of the graph with an item called identifier ID. In other words, it is a generator of float identifier ID that replaces the triples of data in the reduced graph. The IDs are used throughout the partitioning process of the graph. After the obtaining of all the partitions, the IDs are replaced with the values of the subjects, predicates and objects corresponding.

Similarly, Triad [8] uses the mapping before partitioning graph and [9] uses a dictionary to intermediate mapping labels of nodes and edges of string IDs which takes a lot of memory compared to float [9].

For Mapping Algorithm 4 (Fig. 4), we use the function Mp who consists to instantiate the four elements of the quadruplet, which consists the list graph. Mp function accepts four parameters v, (u v), ρ, and ((u, v), ρ), and executes first this instruction $v = \{v\}$, then fills the sets V′, E′, Σ′, and L′ by running the following instructions: $V' \leftarrow V' \cup \{v\}$, $E' \leftarrow E' \cup \{u,v\}$, $\Sigma'_E \leftarrow \Sigma'_E \cup \{\rho\}$, $L'_E \leftarrow L'_E \cup ((u, v), \rho)$.

We plan to implement the mapping phase using JDBC library and Red is for storing the adjacency list, which now contains the IDs generated by the Algorithm 4.

Now it becomes easy to process data because they consume less memory. We use the identifiers throughout the graph partitioning process and finally after we get all the partitions, we will replace them with their corresponding values in the reduced graph.

```
Algorithm3 : Mapping
Input: G0 = (V0, E0, ΣE0, LE0)
Output: G = (V', E', Σ'E, L'E)
V'← Ø, E'← Ø, Σ'E← Ø, L'E← Ø
    for i=0..| V0|do vi← (float) i
                    ui← (float) i+1
                    V'← V' ∪ {vi}
                    E'← E' ∪ {ui,vi}
                    L'E← L'E ∪ ((ui, vi), )
    endfor
    for i=|V0|..|V0|+|Σ'E|do ρi← (float) i+1
                    E'← E' ∪ {ui,vi}
                    L'E← L'E ∪ ((ui-|V0|, vi-|V0|), ρi)
                    Σ'E← Σ'E ∪ { ρi }
    endfor
end
```

```
Algorithm4 : Mapping (MapReduce)
Input: G0 = (V0, E0, ΣE0, LE0)
Output: G = (V', E', Σ'E, L'E)
V'← Ø, E'← Ø, Σ'E← Ø, L'E← Ø
Fonction Map(v0, u0, ρ0)
for i=0.. | V0| do
        Output ((float) i, (float) i+1, (float) i+|Σ'E|))
endfor
endMap
Fonction Reduce (v, u,ρ)
        Mp(({v} ,{u}), (u, v), (ρ), ((u,v), ρi)
endReduce
end
```

Fig. 4 Mapping algorithm

3 Graph Partitioning

The graph partitioning consists to allocate the nodes of the graph to different partitions containing sub graphs so that the size of a partition is larger than a certain size [10]. The graph partitioning can distribute processing and reduce the time of traversing a graph especially for large graphs. For graph partitioning, we use METIS [2] because it provides optimal graph partitioning through the multi-level algorithm (Fig. 5) by reducing the number of edges of the cuts that represents all edges that is in the graph and that are cut as a result of graph partitioning. In our case, it receives as input the list provided by mapping which will allow it to apply multi-level algorithm by reducing the size of the graph by aggregating float, partitioning the reduced graph, gradually Returning to the initial size (mapping output).

```
Algorithme MULTILEVEL(G, k)
        G1 = G
        i = 1
        reapet
              i ← i + 1
              Gi = Coarsening (Gi-1)
        until Coarsening low ou Gi+1 small enough
              P_k^i = partitioning (Gi,k)
        for j = i -1 à 1 faire
              Pk   = projection (P_k^{j+1},Gj )
              P_k^j= Uncoarsening e(Pk)
        endfor
        return P_k^1
end
```

Fig. 5 Multi-level algorithm

After running METIS we reproduce the edges cut into each partition to not lose the results, as in [7] which propose to reproduce the triples at the end of the partitions in the neighboring partitions for data replication. This allows performing certain classes of more complex queries locally in each partition without the need to join another through the partition and no communication.

4 Related Work

There are several RDF data processing systems (Rapid, CliqueSquare, HadoopRDF, Jena-HBase, H_2RDF…) [11], which the principle is to partition the RDF triples on multiple machines, and to parallelize access to these machines at the time of request. To improve this concept, these systems offer the use of MapReduce to distribute the workload among machines during processing SPARQL queries, storing RDF triples, the calculation of the closing of RDF graph, and processing fuzzy RDF data by providing solutions that reduce the cost of communication between machines and the storage costs and input/output.

We present in [11], an overview of the various RDF data processing systems (about twenty systems) using tools to improve the operation and performance of MapReduce. Thus, we present an overview and a classification of these systems based on the impact of the use of MapReduce in their operation.

5 Conclusion

In this paper, we propose to optimize the operation of the RDF data processing system with a configuration based on grouping data using MapReduce. Our goal is to provide a compact format with RDF data representation that effectively supports processing on the basis of triples groups to improve the high cost of input-output and communication due to map and reduce phases. We aim to increase the degree of parallelism to meet the growing needs of data processing on sets of extremely large data. We justify the choice of METIS by its partitioning scheme based on multi-level algorithm that efficiently distributes data on the number of machines available. METIS provides close to an optimal graph partitioning and it runs in O (V) time where V is the number of vertices in the graph, also METIS generates almost uniform vertex partitions. For the rest of our work, we will focus on query optimization and generalization of sub queries, and the formalization of request for graph to build query plans that can be easily parallelized to take advantage of the parallel processing infrastructure.

References

1. Iannella, R.: An idiot's guide to the resource description framework. New Rev. Inf. Network. **4**(1), 181–188 (1998)
2. Pan, J.Z., Horrocks, I.: RDFS(FA) and RDF MT: two semantics for RDFS. In: Proceedings of the 2nd International Semantic Web Conference (ISWC2003), Sanibel Island, pp. 30–46 (2003)
3. Antoniou, G., van Harmelen, F.: Web Ontology Language: OWL. In: S. Staab, R. Studer (Eds.) Handbook on Ontologies in Information Systems. Springer (2003)
4. Pérez, J., Arenas, M., Gutierrez, C.: Semantics and complexity of SPARQL. ACM Trans. Database Syst. **34**(3) (2009)
5. Karypis, G., Kumar, V.: METIS, a Software Package for Partitioning Unstructured Graphs, partitioning meshes, and computing fill-reducing orderings of sparse matrices. Technical Report, Computer Science Department, University of Minnesota1, Minneapolis, MN (1999)
6. Dean, J., Ghemawat, S.: MapReduce: simplified data processing on large clusters. In: 6th Symposium on Operating Systems Design and Implementation (2004)
7. Huang, J., Abadi, D.J., Ren, K.: Scalable SPARQL querying of large RDF graphs. Proc. VLDB Endowment **4**(11), 1123–1134 (2011)
8. Gurajada, S., Seufert, S., Miliaraki, I., Theobald, M.: TriAD: a distributed shared-nothing rdf engine based on asynchronous message passing. In: SIGMOD (2014)
9. Slavov, V., Katib, A., Rao, P., Paturi, S., Barenkala, D.: Fast processing of SPARQL queries on RDF quadruples. In: Proceedings of the 17th International Workshop on the Web and Databases, WebDB 2014, Snowbird, UT (2014)
10. Jindal, A., Rawlani, P., Wu, E., Madden, S., Deshpande, A., Stonebraker, M.: Vertexica: your relational friend for graph analytics. PVLDB **7**(13), 1669–1672 (2014)
11. Benhamed, S., Nait-Bahloul, S.: Classification of knowledge processing by MapReduce. In: 4th International Symposium ISKO-Maghreb (2014)

Developing a Dashboard for Evaluating Higher Education Indicators (in Morocco)

Kamal Zouhri, Saad Elouardirhi and Abdellah Youssfi

Abstract This article is dedicated to the Moroccan university system which raises many questions, especially in terms of evaluation. Moreover, the fact that Morocco started adopting an education sector assessment policy provides several indicators of interest that we intend to highlight in this paper. In this article, we will briefly outline some existing assessment indicators prevailing in open access higher education institutions. We will tackle their disadvantages in the LMD system in order to implement a dashboard by developing an application with the Java language integrated to a SQL server database via the JDBC driver (Java DataBase Connectivity).

Keywords LMD system · Assessment · Indicators · Dashboard · Java language-JDBC

1 Introduction

In the Early 21st Century, the issue of professionalizing Moroccan universities has become a necessity concretized by the introduction of several higher education reforms [1].

These reforms aimed to redesign academic institutions on educational and organizational bases have aroused great interest in the involved areas. Creating effective and innovative universities, which can adapt to the new economic requirements and solving social problems was then a major challenge for Morocco.

K. Zouhri (✉) · S. Elouardirhi · A. Youssfi
ERADIASS Research Team, Faculty of Economics, Law, and Social Sciences
Souissi-Rabat, University Mohammed V, Rabat, Morocco
e-mail: zouhrisage@gmail.com

S. Elouardirhi
e-mail: saad.elouardirhi@yahoo.fr

A. Youssfi
e-mail: yousfi240ma@yahoo.fr

Á. Rocha et al. (eds.), *Europe and MENA Cooperation Advances in Information and Communication Technologies*, Advances in Intelligent Systems and Computing 520, DOI 10.1007/978-3-319-46568-5_26

In 1996, a special Royal Commission was created in order to find out the best way to organize and streamline the university system, as well as other levels of education. It was then in 2003–2004 that the Moroccan university has seen the coming into effect of the new university reform (LMD system) [2, 3].

These reforms have resulted in the multiplication of quantitative indicators in terms of new usages (clear objectives assignment, comparison and evaluation of performance, good governance definition and enforcement criteria) [4].

This article will examine the leading indicators evaluate teaching internationally (number of students per semesters, efficiency ratio, educational wastage rate, and others). This will be carried out by analyzing their advantages and disadvantages with regards to the LMD system, then, we will work on refining and adapting those indicators the new system.

In the end, we will attempt to prepare a Scoreboard gathering those indicators through developing a program with the Java language integrated to a SQL server database via the JDBC pilot (Java DataBase Connectivity).

2 General Question

Improvements in the new LMD system, which is a modular system based on validating modules over successive semesters have caused many problems especially regarding the assessment of this system and more specifically concerning the existing indicators reliability level.

In other words, are the current higher education evaluation indicators suitable for the LMD? And if not, how to adapt them to this new system? Then what are the practical IT tools to monitor this system effectiveness?

3 Indicators for Evaluating Effectiveness

The effectiveness could be measured depending on the results obtained during and at the end of the university course: the number of students enrolled that reach the last grade, the degree of achievement in meeting the set goals, the achievement of the operational program objectives [5],

3.1 *Number of Semesters-Students Devoted to Training Graduates*

It is the total amount of semesters devoted to training each student until graduation or dropouts without a diploma [6].

Formula:

$$NS_{dip} = \sum_{j=n}^{n+k} D_{g,j} \times j + \sum_{j=1}^{n+k} A_{g,j} \times j. \quad (1)$$

where,

$D_{g,j}$: Number of graduates in the cohort g after j semesters
$A_{g,j}$: Number of students (in the cohort g) dropping out after j semesters
g: Cohort of students
n: Number of regular study semesters
k: Number of authorized repetition

3.2 Number of Semesters-Students Consummated by Every Graduate

It is the number of semesters-students taken in order that a student from the cohort could graduate [7].

Formula:

$$NSD_g = \frac{NS_{dip}}{\sum_{j=n}^{n+k} D_{g,j}}. \quad (2)$$

Remark 1. Regarding the previous indicators, we have kept the same formula as per the old indicators, except that we replaced the number of years by the number of semesters.

3.3 Current Efficiency Ratio (CE_s)

The efficiency ratio is the ratio for the optimal number of semesters-students needed (that is to say in the absence of repetition and dropout), in order That a number of students belonging to a given cohort obtain their diplomas, it is the number of semesters-students devoted to training students.

It is calculated using the following formula [7]:

$$CE_s = \frac{\sum_{j=n}^{n+k} D_{g,j} \times n}{NS_{dip}} \times 100. \quad (3)$$

It is clear that this indicator is mainly based on the number of semesters taken, i.e. that each semester consists of four modules, and that in the final grades, one can have six months containing less than four modules, which creates a real problem in relation to the reliability of this indicator.

However, if we consider two students belonging to the same cohort, and who have graduated during the same academic year, after 7 semesters (for example in a 6-semester system with a semester of reserve) for both students and that the first student had 2 modules during the last semester, and that had 3 modules in the final. Obviously the second student has cost more than the first one, and we know well that the more we reduce the cost, more the efficiency ratio increases. Yet with the previous formula that is built on the number of semesters (i.e. on time spent on for graduation), one must consider that both students took the entire last semester, which is untrue. Thus this formula does not apply for the LMD system which is a modular system requiring the introduction of a new formula based upon the number of modules and not the number of semesters.

4 Adjustment of Current Indicators

4.1 The Number of Modules-Students

It is the total of registrations made in each module by every student for studying until graduation or dropping without a diploma.

Formula:

$$NM_{dip} = \sum_{j=n}^{n+k} D_{g,j} \times m_j + \sum_{j=1}^{n+k} A_{g,j} \times m_j. \quad (4)$$

where,

m_j: Total of enrolments in each module for every student after j semester

4.2 The Number of Students-Modules Consummated by a Graduate

It is the number of student-modules consummated by a graduate of the cohort.

Formula:

$$NSD_g = \frac{NM_{dip}}{\sum_{j=n}^{n+k} D_{g,j}}. \quad (5)$$

4.3 Efficiency Ratio (CE_m)

To correct the problem of the current reliability ratio efficiency, we recommend improving it by the following formula:

$$CE_g = \frac{\sum_{j=n}^{n+k} D_{g,j} \times m}{NM_{dip}} \times 100. \quad (6)$$

In that case, the wastage rate is hereby given:

$$CD_g = \frac{1}{CE_g} \times 100. \quad (7)$$

Should be noted that:

More this ratio is close to 1 and sure the system is more efficient and vice versa [7, 8].

5 IT Dashboard Calculating Those Indicators

We got the above dashboard through an application that was carried out by Java connect to SQL Server using the JDBC driver (Java DataBase Connectivity), and that throughout the following steps (Fig. 1):

As a first step, we imported an actual database which shows the journey of a students' cohort, precisely the cohort of 2009 saved in an Excel file into the SQL Server software. Afterwards through the latter, we created functions that show values of the indicators mentioned in our article [9]. Following this step, we prepared a Java program that allowed us to connect to our database via the JDBC driver (Java DataBase Connectivity) [10, 11, 12]. Final step, we created a Java program which displayed the interface of our dashboard in order to obtain the results above mentioned [10, 13].

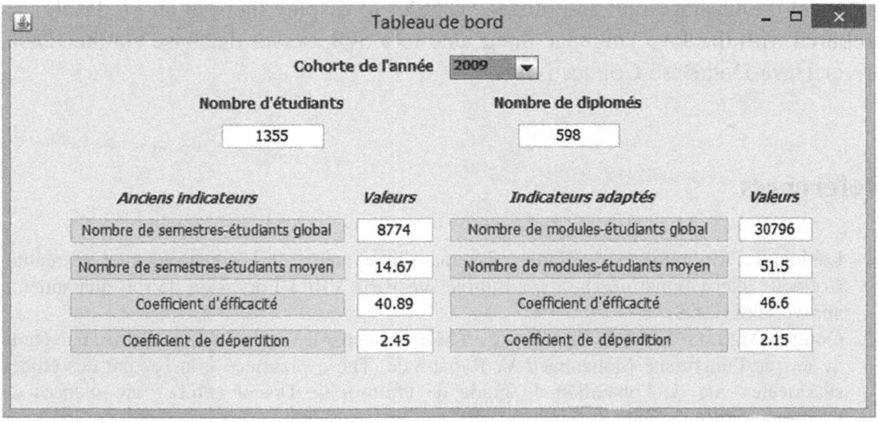

Fig. 1 Dashboard of the 2009 cohort year

At last, our dashboard enables us to properly assess our education system, in particular the example we treated in this article by analysing the results shown above. Furthermore, this dashboard has helped us tracking our system education, just simply by clicking the button presented above the table (cohort year), and selecting the required year.

In our example, although one sees there is a difference between the current indicators and the improvements we made to adapt them, for example regarding the efficiency ratio, one notices that there is a difference of 5,71 %. Through this improvement, one can conclude that wastage (due to repetition and dropout) is actually enormous ($CE_m = 46,6\%$, and $CD_m = 2,15$ very far away from 1).

6 Conclusion

Assessing the efficiency of open access higher education institutions in Morocco, is taking an important place for researchers and experts in this field, and even at a political level. Yet every assessment requires means of measurement that can provide a clear and accurate picture of the field; these means are usually considered as indicators.

The Moroccan university has always undergone changes; the latest being the introduction of the new reform 2003–2004 (LMD). This reform is considered as a drastic shift within the Moroccan university system.

The LMD system brought several changes to the old system, including the fact it is a system based on the concepts of semester and module in the place of year, which presents a real obstacle in relation to the assessment effectiveness. In fact, most current indicators are based on the time spent for forming a graduate (number of semesters student) [7, 6], which is not the case in the LMD system.

In this article, we attempted to fix this adjustment problem by introducing the number of modules-students consummated by the graduates, such as, the efficiency ratio and the wastage rate and others, in order to display them in an IT dashboard prepared with the Java language integrated to a SQL server database via the JDBC driver (Java DataBase Connectivity).

References

1. Lahlou, L.: La réforme de l'université marocaine entre idéal organisationnel et réalité. Colloque international organisé par l'université Paris VIII à l'occasion de son quarantième anniversaire (2009)
2. Gougou, M.: La réforme de l'université au Maroc vue par les auteurs universitaires: une étude de cas de l'université Mohammed V, Rabat-Salé. Thèse présentée à la Faculté des études supérieures vue de l'obtention du grade de Philosophie Doctor (Ph.D.), en sciences de l'éducation à l'université de Montréal (2011)

3. Royaume du Maroc, Ministère de l'Enseignement Supérieur, de la Recherche Scientifique et de la Formation des Cadres (MESRSFC): Système d'enseignement supérieur marocain (2014)
4. Bouras, I., Boudier, M., Canals, V., Calmand, J., Guegnard, C., Menard, B., Murdoch, J.: Evaluer les universités. Analyse critique des indicateurs d'établissement et méthodologie des enquêtes auprès des recruteurs. Groupe de travail sur l'enseignement supérieur (GTES). Relief.47. Echange du Céreq (2014)
5. Loua, S.: Efficacité interne de l'enseignement supérieur malien. Thèse de doctorat en sciences de l'éducation. Institut des Sciences et Pratiques d'Education et de Formation (I.S.P.E.F). Université Lumière LYON 2 (2012)
6. Sall, M.: Efficacité et équité de l'enseignement supérieur, Quels étudiants réussissent à l'Université de Dakar? Tome 1: cadre théorique et méthodologie. Thèse de doctorat d'Etat en sciences de l'éducation. Université Cheikh Anta Diop (Sénégal), Dakar (1996)
7. UNESCO: Indicateurs de l'éducation; directives techniques (2009)
8. Bourne, J., King, M., Albright, P., Ndong-Jatta, A.: Guide méthodologique pour l'analyse sectorielle de l'éducation. Analyse sur l'ensemble du système, avec un accent sur les enseignements primaire et secondaire, vol 1 (2014)
9. Archambeau, T.: Cours SQL: Base du langage SQL et des bases de données (2014), http://sql.sh
10. Tutoriel: Apprenez à programmer en Java, http://sdz.tdct.org/sdz/apprenez-a-programmer-en-java.htmlBh
11. Cariou, E.: Introduction à JDBC: Accès aux bases de données en Java. Université de Pau et des Pays de l'Adour, Département Informatique
12. Genoud, P.: JDBC: Java DataBase Connectivity (2014)
13. Divay, M.: La programmation objet en Java. Dunod, Paris (2006)

First Adaptation of Hunting Search Algorithm for the Quadratic Assignment Problem

Amine Agharghor and Mohammed Essaid Riffi

Abstract Quadratic assignment problem is one of the basic combinatorial optimization problems NP-hard. This paper aims to present a new algorithm for solving the quadratic assignment problem: Hunting Search algorithm. It is a metaheuristic inspired by the method of group hunting of predatory animals. To show the performance of this algorithm, it has been checked on a set of seventeen instances of QAPLib and it gave good results.

Keywords Hunting search algorithm · Quadratic assignment problem · Combinatorial optimization · Meta-heuristic · Evolutionary algorithm

1 Introduction

The quadratic assignment problem (QAP) [1] was introduced in the context of locating "indivisible economic activities". The problem is to assign a set of facilities to a set of locations in order to minimize the total of assignment cost. It is a problem of the class NP-hard, which means that no method can solve it in a polynomial time except for the very small instances (instances of size $n < 20$).

The QAP has several applications in combinatorial optimization problems such as in electronic [2] and in logistic [3]. Several methods have been proposed to solve the problem such as Particle Swarm Optimization [4], Bee Algorithm [5] or Genetic Algorithm [6].

Hunting Search (HuS) [7] is a continuous optimization method adapted as a combinatorial optimization method for solving the TSP [8]. It is a meta-heuristic method inspired by group hunting of some animals.

A. Agharghor (✉) · M.E. Riffi
Laboratory of LAROSERIE, Department of Computer Science,
Faculty of Sciences, Chouaïb Doukkali University, El Jadida, Morocco
e-mail: amine.agharghor@gmail.com

M.E. Riffi
e-mail: saidriffi2@gmail.com

Á. Rocha et al. (eds.), *Europe and MENA Cooperation Advances in Information and Communication Technologies*, Advances in Intelligent Systems and Computing 520, DOI 10.1007/978-3-319-46568-5_27

The present paper proposes the use of HuS for solving QAP. It is structured into five main sections; the first is an introduction of the given problem and method. The second section provides more detailed and description of the QAP. The third section presents the adaptation of HuS for solving QAP. However, numerical results obtained from the use of HuS on the QAPLib instances [9] are presented in the fourth section. The last section concludes the whole work.

2 Quadratic Assignment Problem

Given a set of facilities to assign to a set of locations, and given that between every two facilities there is a required flow, and between every two locations there is a required distance, the problem is to find the best total cost of flows and distances needed to assign the facilities to the locations.

Mathematically, QAP is to find the best solution defined by an objective function from a subset of the set of the feasible solutions.

As an optimization problem, one cannot explore all of the feasible solutions, so one try to find the best solution just from a subset of the feasible solutions.

Let **E** be the set of the feasible solutions, **S** is the subset of **E**, $g: S \rightarrow \mathbf{R}$ is the objective function. The problem is to find:

$$\mathbf{min}\{g(s): s \in S\} \tag{1}$$

s is a solution from **S**, it is a permutation of indexes that represents an assignment of facilities to locations.

Given two matrices of equal dimension, $(f_{ij}) \in \mathbf{R}$ and $(d_{ij}) \in \mathbf{R}$. (f_{ij}) is the square matrix that represents the required flow between facilities i and j. (d_{ij}) is the square matrix that represents the distance between locations i and j.

The objective function gives the cost of the assignment defined as follows:

$$g(s) = \sum_{i=1}^{n} \sum_{j=1}^{n} f_{ij} \times d_{s(i)s(j)} \tag{2}$$

Or

$$g(s) = \sum_{i=1}^{n} \sum_{j=1}^{n} f_{s(i)s(j)} \times d_{ij}$$

Such as s ∈ ∈ **S**, $s(i)$ is the location to which facility i is assigned and n is the dimension of the two matrices.

3 Hunting Search Algorithm

HuS is a metaheuristic for solving continuous optimization problems; it uses the techniques of group hunting of some predatory animals such as wolves. The hunting group represents a set of solutions where each hunter represents one of these solutions. The best hunter named the leader; he represents the best solution. A hunter is characterized by its position that defines the distance between him and the other hunters.

Metaheuristics are the best-used optimization algorithms for solving hard optimization problems.

Evolutionary Algorithms are the metaheuristics inspired by the mechanisms of biological evolution of individuals in a population to find better place or way to live.

HuS is an evolutionary algorithm since it evolves a population of individuals (hunters) via operator selection and variation, in a manner that is similar to the evolution of living beings.

During the hunt process, hunters change their positions to better encircle their prey by movements toward the leader or by correcting their position relatively to each other. Finally, if they are much closed to each other or stuck, they have to be reorganized.

HuS algorithm is as follows:

```
Initialize the parameters
Initialize the Hunting Group (HG)
Make a loop of NE iterations
    Make a loop of IE iterations
        Move toward the leader
        Correct the positions
        If the distance between the best and the worst
      hunter<EPS
               Leave the loop iterations of IE to
               the reorganization
        End if
      End loop iterations of IE
      Reorganization hunters
End loop iteration of NE
```

where

HG (**H**unting **G**roup) is the set of the initial solution.
NE (**N**umber of **E**pochs) is the number of times to run the IE loop during the search.
IE (**I**teration per **E**poch) is the number of times to move toward the leader and correct the hunters' position.
EPS (**Eps**ilon) is the minimum distance between the leader and the worst hunter.

The four main operations of HuS are:

(1) **Initialize the Hunting Group**: generate a set of randomly solutions for the QAP from the QAPLib instance, each solution is going to be presented by a hunter who is implemented in the algorithm as an array, each index of the array represents a location, and each value of the array represents a facilitie.
(2) **Move toward the leader**: every hunter moves toward the leader by copying a part from the best solution; that lets him stay closer to the leader until finding better solution and becomes the leader.
(3) **Correct the positions**: every hunter moves toward the other hunters by copying a part from their solution.
(4) **Reorganization hunters**: when all the hunters becomes very close to each other or stuck, which means that they will represent the same solution, they have to be regenerated except the leader.

4 Experimental Results

This section presents the performance of **HuS** algorithm on instances of QAPLib. The tests were performed on a computer processor Intel(R) Core(TM) i5-4300 CPU @ 1.9 GHz @ 2.50 GHz and 4 GB of RAM. 10 times tested for each instance.

Table 1 Numerical results obtained by HuS applied to some QAP instances of QAPLIB

Instance	Size	Opt	Best Sol	Worst Sol	Suc. (%)	Err. (%)	Time (sec)
Bur26a	26	5426670	5426670	5426670	100	0	0.45
Bur26b	26	3817852	3817852	3817852	100	0	0.51
Bur26c	26	5426795	5426795	5426795	100	0	0.19
Bur26d	26	3821225	3821225	3821225	100	0	0.46
Bur26e	26	5386879	5386879	5386879	100	0	0.35
Bur26f	26	3782044	3782044	3782044	100	0	0.57
Bur26 g	26	10117172	10117172	10117172	100	0	0.17
Bur26 h	26	7098658	7098658	7098658	100	0	0.35
Chr25a	25	3796	3796	3796	100	0	1.1
Esc16a	16	68	68	68	100	0	0
Esc32a	32	130	130	134	80	3.07	1.68
Kra30a	30	88900	88900	90920	20	2.27	0.63
Lipa30a	30	13178	13178	13376	80	1.50	0.84
Lipa40a	40	31538	31538	31923	50	1.22	7.78
Tai12a	12	224416	224416	224416	100	0	0.00
Tai15a	15	388214	388214	388214	100	0	0.01
Tai64c	64	1855928	1855928	1855928	100	0	5.04

Table 1 shows the results obtained. The first column contains the name of the instance. The second column contains the size of the instance. The third column contains the optimal solution in QAPLib. The fourth column contains the findings of the best solution. The fifth column contains the findings of the worst solution. The sixth column contains the percentage of success in getting the optimum in ten tests. The seventh column contains the percentage of error in obtaining the wrong solution. The eighth column contains the best run time of the program while getting the best solution, a maximum run time is fixed at 3600 s. The error percentage is calculated as follows:

$$\boldsymbol{Err} = \frac{\textbf{Best Sol} - \textbf{Opt}}{\textbf{Opt}} \times 100$$

5 Conclusion

In this paper, a first adaptation of HuS algorithm to solve QAP is proposed. The majority optimums of the QAPLib instances that have a size less than 31 are obtained in less than one second, the optimums of the QAPLib instances that have a size less than 65 are obtained in less than 8 s.

Further work is steel needed to solve more complex instances of QAPLib and compare the performance of HuS with the recent methods.

References

1. Lawler, E.L.: The quadratic assignment problem. Manage. Sci. **9**(4), 586–599 (1963)
2. Laporte, G., Mercure, H.: Balancing hydraulic turbine runners: A quadratic assignment problem. Eur. J. Oper. Res. **35**(3), 378–381 (1988)
3. Haghani, A., Chen, M.-C.: Optimizing gate assignments at airport terminals. Transp. Res. Part A: Policy Pract. **32**(6), 437–454 (1998)
4. Szwed, P. et al.: OpenCL implementation of PSO algorithm for the quadratic assignment problem. In: International Conference on Artificial Intelligence and Soft Computing. pp. 223–234. Springer (2015)
5. Chmiel, W., Szwed, P.: Bees algorithm for the quadratic assignment problem on CUDA platform. In: Man–Machine Interactions 4. pp. 615–625. Springer (2016)
6. Ahmed, Z.H.: A multi-parent genetic algorithm for the quadratic assignment problem. OPSEARCH **52**(4), 714–732 (2015)
7. Oftadeh, R., et al.: A novel meta-heuristic optimization algorithm inspired by group hunting of animals: hunting search. Comput. Math Appl. **60**(7), 2087–2098 (2010)
8. Agharghor, A., Riffi, M.E.: Hunting search algorithm to solve the traveling salesman problem. J. Theor. Appl. Inf. Technol. **74**, 1 (2015)
9. Burkard, R.E., et al.: QAPLIB–a quadratic assignment problem library. J. Global Optim. **10**(4), 391–403 (1997)

A New Hybrid Face Recognition System via Local Gradient Probabilistic Pattern (LGPP) and 2D-DWT

Abdellatif Dahmouni, Nabil Aharrane, Karim El Moutaouakil and Khalid Satori

Abstract In last years, the facial biometry is coming back into education field; it is proposed to control student activities. In this paper, we propose a new face recognition system based on our Local Gradient Probabilistic Pattern (LGPP) and 2D-DWT. Firstly, the almost homogeneous and the picks areas are separate according to LGPP confidence interval. Secondly, the obtained images are decomposed using 2D-DWT in "LL, LH, HL and HH" sub-bands. Thereafter, we extract the features vector using 2D-PCA method applied on the approximation (LL-band). In classification phase, we compare between MLP, Bayesian Networks, SVM and KNN classifiers at features vector. The experimental results show that proposed system improves the recognition rate. Indeed, we reach a rate of 97 % for ORL and 98.8 % for Yale.

Keywords LBP · LGP · LGPP · 2D-DWT · 2D-PCA · SVM · Confidence interval

---

Please note that the LNCS Editorial assumes that all authors have used the western naming convention, with given names preceding surnames. This determines the structure of the names in the running heads and the author index.

A. Dahmouni (✉) · N. Aharrane · K. Satori
LIIAN, Department of Mathematics and Computer Sciences, Faculty of Sciences
Dhar-Mahraz, Sidi Mohamed Ben Abdellah University, B.P 1796 Atlas, Fez, Morocco
e-mail: Abdellatifdahmouni@Gmail.com

N. Aharrane
e-mail: Aharranenabil@Gmail.com

K. Satori
e-mail: Khalidsatorim3i@Yahoo.fr

K.E. Moutaouakil
ENSAH, National School Applied of Sciences, Mohammed I University,
BP 03, Al-Hoceima, Morocco
e-mail: Karimmoutaouakil@Yahoo.fr

Á. Rocha et al. (eds.), *Europe and MENA Cooperation Advances in Information and Communication Technologies*, Advances in Intelligent Systems and Computing 520, DOI 10.1007/978-3-319-46568-5_28

1 Introduction

Face recognition is a biometric modality widely used in different sector where the person's identity check is an essential task such as computer applications access, banks access, countries frontiers crossing, students schools control and others sensitive services access. Any face recognition hybrid system has three stages. The first stage consists in principal face components extracting using local methods such as Scale-Invariant Feature Transform (SIFT) [1], Speeded-Up Robust Features (SURF) [2], Weber Local Descriptor (WLD) [3] and some Local Binary Pattern (LBP) varieties [4]; see Table 1. The second stage consists in features vectors extracting by reducing the data redundancy using the subspace techniques such as Principal Component Analysis (PCA) [5, 6], Linear Discrimination Analysis (LDA) [7, 8], Discrete Wavelet Transforms (DWT) [9, 10], Discrete Cosine Transforms (DCT) [11] and Artificial Neural Networks (ANN) [12]. The last stage consists to classify the features vector dataset using the adequate classifiers; see Fig. 1.

In a previous work, we have proposed a new face representation based on Local Gradient Probabilistic Pattern (LGPP) [13]. To evaluate the current pixel the LGPP uses the confidence interval concept calculated according to the probabilistic low in the current neighborhood. In this paper, we propose to hybrid the LGPP, the 2D-DWT, the 2D-PCA and the adequate faces classifiers.

Table 1 Some local binary pattern varieties

Varieties	Description	Authors	References
CLBP	Based on sign and magnitude information	Guo et al.	(2010) [16]
LTP	Based on three bits transformation	Tan et al.	(2010) [17]
LDP	Based on kernel of Kirsch edge pattern	Zhang et al.	(2010) [18]
LGP	Based on local gradient transformation	Jun et al.	(2013) [19]
LBPP	Based on confidence interval of grayscale	Dahmouni et al.	(2015) [20]
LGPP	Based on confidence interval of gradient	Dahmouni et al.	(2015) [13]

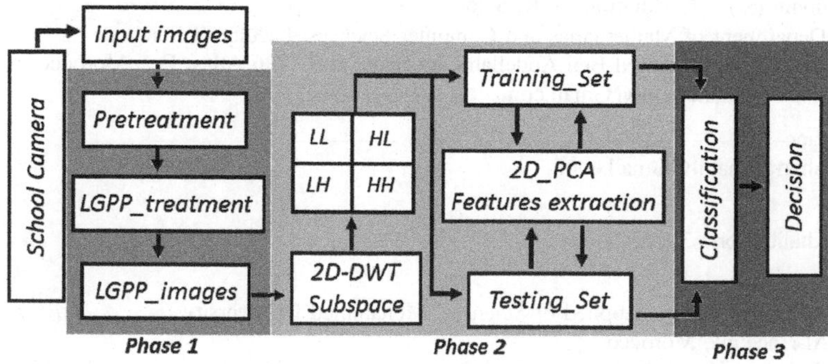

Fig. 1 Diagram of proposed face recognition system

The remainder of the paper consists of three sections followed by a conclusion. Section 2 presents in detail the Local Gradient Probabilistic Pattern (LGPP) descriptor. Section 3 presents in addition to 2D-DWT and 2D-PCA subspace algorithms some of data learning algorithms. Section 4 evaluates and analyzes the experimental results.

2 Local Gradient Probabilistic Pattern (LGPP)

The principal idea of the LGPP is to define a confidence interval associated to the gradient values knowing the confidence interval associated grayscale values [13]. In this context we propose to generate the grayscale values confidence interval $[a_1, a_2]$ based on statistical moments in the current neighborhood of (N = 5 * 5).

$$\text{Variation coefficient: } \delta = \frac{\sigma}{\mu} \quad \text{Symmetry coefficient: } S = \sum_{n=1}^{N} \frac{(i_n - \mu)^3}{\sigma^3} \tag{1}$$

To separate between the almost homogeneous areas and the peaks areas, we associated to any grayscale value a random variable X. Indeed, the pixel whose the neighborhood is governed by a distribution that is approximately a normal law can be considered as homogeneous pixel. As it is proved in [13] the grayscale confidence interval is given by:

$$\begin{aligned} &\text{If } (S = 0 \text{ or } \delta < \beta) \text{ then,} \\ &[\alpha_1, \alpha_2] = [\mu - k\sigma, \mu + k\sigma] \quad \text{Where,} \quad 0.1 < \beta < 0.2 \end{aligned} \tag{2}$$

$$\begin{aligned} &\textit{else}, \quad [\alpha_1, \alpha_2] = [\mu - K, \mu + K] \quad \text{Where,} \\ &K = \begin{cases} \sigma & \text{if } \beta < \delta < \beta + 0.1 \\ (\beta + 0.1)\,\mu & \text{other} \end{cases} \end{aligned} \tag{3}$$

Likewise, the LGPP confidence interval is given by:

$$\begin{aligned} &[\gamma_1, \gamma_2] = [\max(0, -K + g_m), K + g_m] \quad \text{Where,} \\ &\max(0, -K + g_m) \le g_n \le K + g_m \quad \text{and} \quad g_m = |\mu - i_c|. \end{aligned} \tag{4}$$

However, based on statistical moments in the neighborhood centered on current pixel, each of eight neighboring pixels having a gradient value located inside the LGPP confidence interval is coded by 1, the others by 0, the obtained byte is converted into new gray level. The LGPP value is expressed by following equation:

Fig. 2 Obtained ORL images by: original, LBP and LGPP (k = 1, 2, 3, 4) descriptors

$$\mathrm{LGP}_{\mathrm{P,R,N}}(\mathrm{x_c}, \mathrm{y_c}) = \sum_{\mathrm{n}=0}^{\mathrm{P}-1} \mathrm{s}(\mathrm{g_n}) * 2^{\mathrm{n}} \quad \text{Where: } \mathrm{s}(\mathrm{x}) = \begin{cases} 1, & \text{if } \gamma_1(N) \le x \le \gamma_2(N) \\ 0, & \text{others} \end{cases} \tag{5}$$

Contrary to most LBP varieties, which are the deterministic models, LGPP considers the pixel on the face distributions as a law of probability. We thus pass to a probabilistic description where the notion of the confidence interval allows assigning values near 255 at all pixels of almost homogeneous areas, and the values near 0 at all pixels of peaks areas. The nose, the eyes and the mouth which having strong curves are easily localized, see Fig. 2. To establish a face recognition system that uses this property we propose to fuse the LGPP-k = 4 and different features extraction methods.

3 Features Extraction Using the Subspace Algorithms

Generally, the subspace methods build the features vectors by a linear transformation of the data. There are spatial methods such as PCA, LDA, 2D-PCA and 2D-LDA, and frequency methods such as DWT, DCT, 2D-DWT and 2D-DCT. In this work, we use the 2D-DWT and 2D-PCA as features vectors extraction methods. Equations should be punctuated in the same way as ordinary text but with a small space before the end punctuation mark.

3.1 2D-DWT Frequency Algorithm

Discrete Wavelet Transform (2D-DWT) uses the low pass filter and high pass filter to decompose the original image into succession of approximation component (LL band) and details component (HL, LH and HH bands) following the level decomposition. For each level of decomposition, the low pass filtering in horizontal direction and high pass filtering in vertical direction to generate four (LL, LH, HL and HH) sub bands (Fig. 3). The LL band, which store the maximum of information and represents the approximate coefficients, can be used to engender the

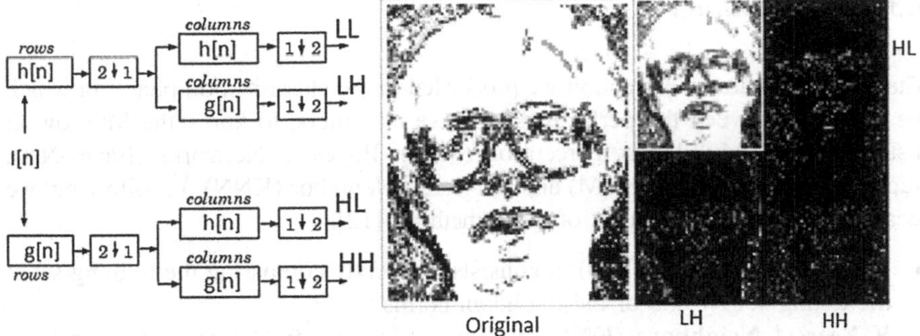

Fig. 3 2D-DWT frequency decomposition

next level decomposition. The sub bands LH, HL and HH represent the spatial orientations features. In this paper, we use the Haar wavelet into one level to produces the four (LL, LH, HL and HH) sub bands. The approximation (LL) is used as direct input to 2D-PCA spatial subspace methods to extract the reduce face features vector.

3.2 2D-PCA Subspace Algorithm

In contrast to conventional PCA subspace algorithm that represent the image by simple 1D_vector, 2D-PCA keeps the 2D structure of the image. However, the covariance matrix will be of small size, (n, n) instead of (n^2, n^2), it will be easily evaluated, less time is necessary to define the eigenvectors basis. Let **R** a vector of dimension n and **Γ** a matrix of size (m, n). The linear transformation $\mathbf{Y} = \mathbf{\Gamma} \times \mathbf{R}$ is a projection of **Γ** on **R**; the projected vector **Y** of dimension m represents the features vector of **Γ**. For M training images [Γ_1, …,Γ_M] we define the 2D-PCA covariance matrix **C** by:

$$\mathbf{C} = \frac{1}{M}\sum_{j=1}^{M}(\Gamma_j - \psi_{mean})^t(\Gamma_j - \psi_{mean}) \quad \text{Where:} \quad \psi_{mean} = \frac{1}{M}\sum_{j=1}^{M}\Gamma_j \tag{6}$$

The optimal projection matrix $\mathbf{R}_{opt} = \text{argmax}(J(\mathbf{R}))$ is obtained by maximization of generalized variance criterion: $J(\mathbf{R}) = \mathbf{R}^t\mathbf{C}\mathbf{R}$. R_{opt} is composed by the Eigen-vectors R = [R1… Rd] corresponding to, d, greater Eigen values of the covariance matrix C. Then we obtain for each image Γ a characteristic matrix of dimension m × d: $\mathbf{Y} = [Y_1 \ldots Y_d]$, whose components $\mathbf{Y}_k = \Gamma\,\mathbf{R}_k (k = 1 \ldots d)$. To classify the Eigen extracted features we use most known classification algorithms are used.

3.3 Classification

The After the features extraction we proceeded to the classification phase, in which we compare between the performance of five classifiers, to know the Murkowski distance (MD), Multilayers Perceptron (MLP), Bayesian Networks (Bayes-Net), Support Vector Machine (SVM) and K-Nearest Neighbor (KNN). In following, we recall the important principals of these methods [12].

- **Murkowski distance (MD)**: it consists in directly comparing the training set at the testing set by one of the Euclidian norms.
- **K-Nearest Neighbors (KNN)**: it consists in classify the new instance by a majority vote of its neighbors; the new instance is allocated to the class most common among its k nearest neighbors.
- **Multilayers Perceptron (MLP)**: it is a most popular neural networks used in machine learning. It is a multilayers oriented artificial neural network, with supervised learning such as the back propagation of gradient.
- **Bayesian Networks (Bayes-Net)**: it consists in represent the set of random variables and their conditional dependencies via a directed acyclic graph. It represents a probability distribution on a set of random variables X, which admits the following joint distribution: $P(X_1, X_2, \ldots, X_n) = \pi\ (P(X_i/P_a(X_i)))$.
- **Support Vector Machine (SVM)**: it is a universal constructive learning procedure based on the statistical learning theory. It consists in seeks an optimal separating hyper-plan, where the margin is maximal. An important SVM characteristic is that the solution is based only on those data points, which are at the margin. These points are called support vectors. The linear SVM can be extended to nonlinear version using the kernel functions.

It should be noted that several varieties of these methods are proposed in the literature.

4 Experimentations

To evaluate the performances of the proposed approach, we have used ORL and Yale databases. ORL database is made of 40 subjects having each one 10 different views [14]; the images in gray levels have the same size (92 * 112) pixels and are taken under various conditions. YALE database is made up of 165 images of 15 subjects representing 11 various conditions of size (320 * 243) [15], and scaled to size (112 * 92) pixels. One pre-treatment with histogram equalization is necessary to harmonize the gray levels distribution.

In our previous work [13], we have proposed the LGPP + 2D-PCA as recognition system. To improve this vision we introduce the 2D-DWT method in the features extraction phase. For a random number of 5 training (the rest are testing) images per class of ORL and Yale databases, Table 2 shows that LGPP(k > 2)

Table 2 Recognition rate of different size of confidence interval on ORL and Yale databases

	Databases	Methods	LGPP k = 1	LGPP k = 2	LGPP k = 3	LGPP k = 4
Rate (%)	ORL(5/5)	2DPCA + 2DDWT	96	96.5	97	97
	Yale(5/6)	2DPCA + 2DDWT	97.78	98.67	98.89	98.89

gives the best recognition rate, we reach 97 % for ORL and 98.89 % for Yale. Table 3 shows that the proposed approach improves the recognition rate with the significant reduction in computation time compared to the old system.

In each pattern recognition system, the classification phase is decisive. In fact, the choice of the classification tools infects the recognition rate. As it is impossible to compare theoretically the non-parametric MLP, KNN, Bayes-Net and SVM classifiers, we propose to compare them experimentally, see Table 5. Some machine learning parameters are defined: True Positive (TP), True Negative (TN), False Positive (FP), False Negative (FN), Precision (p), Recall (r), F-Measure and Accuracy, see Table 4.

Table 5 shows the global results of all used classifiers with the specific configuration for each one. We noted that, SVM and MLP give the best results they reach 98 % on ORL, the KNN is dominant on Yale it reaches 100 %. Hence, in the

Table 3 Recognition rate of some face recognition methods on ORL and Yale databases

Methods	LGPP + 2DDWT + 2DPCA		LGPP + 2DPCA		LBP + 2DDWT + 2DPCA	
Data	Rate%	Time (s)	Rate%	Time (s)	Rate%	Time (s)
ORL (5/5)	97	1.58	96	1.94	89	1.6
Yale (5/6)	98.67	0.68	96	0.78	91.11	0.68

Table 4 Machine learning parameters

	Prediction +	Prediction -	
Truth +	TP	TN	Precision = TP/(TP + FP) Recall = TP/(TP + FN)
Truth -	FP	FN	F-Measure = 2pr/(p + r) Accuracy = (TP + TN)/Ntotal

Table 5 Comparison of different used classifiers on ORL and Yale databases

Databases	Classifiers	Precision (%)	Recall (%)	F-measure (%)	Accuracy (%)
ORL	SVM	0.981	0.98	0.98	98
	KNN	0.972	0.97	0.97	97
	Bayes-Net	0.92	0.92	0.916	92
	MLP	0.981	0.98	0.98	98
Yale	SVM	0.994	0.994	0.994	99.39
	KNN	1	1	1	100
	Bayes-Net	0.983	0.982	0.982	98.19
	MLP	0.994	0.994	0.994	99.39

confusion matrix, the sum of diagonal elements represent the number of correctly classified instances, all others are incorrectly classified. We note that, almost all instances are correctly classified the SVM classifier is the most qualified to be used in this work; see Table 6.

Table 6 Different confusion matrix of Yale database

<table>
<tr><td>MLP classifier</td><td>SVM classifier</td></tr>
<tr><td>a b c d e f g h i j k l m n o ← classified as</td><td>a b c d e f g h i j k l m n o ← classified as</td></tr>
<tr><td>11 0 0 0 0 0 0 0 0 0 0 0 0 0 0 | a = 1</td><td>11 0 0 0 0 0 0 0 0 0 0 0 0 0 0 | a = 1</td></tr>
<tr><td>0 11 0 0 0 0 0 0 0 0 0 0 0 0 0 | b = 2</td><td>0 11 0 0 0 0 0 0 0 0 0 0 0 0 0 | b = 2</td></tr>
<tr><td>0 0 11 0 0 0 0 0 0 0 0 0 0 0 0 | c = 3</td><td>0 0 11 0 0 0 0 0 0 0 0 0 0 0 0 | c = 3</td></tr>
<tr><td>0 0 0 10 1 0 0 0 0 0 0 0 0 0 0 | d = 4</td><td>0 0 0 10 1 0 0 0 0 0 0 0 0 0 0 | d = 4</td></tr>
<tr><td>0 0 0 0 11 0 0 0 0 0 0 0 0 0 0 | e = 5</td><td>0 0 0 0 11 0 0 0 0 0 0 0 0 0 0 | e = 5</td></tr>
<tr><td>0 0 0 0 0 11 0 0 0 0 0 0 0 0 0 | f = 6</td><td>0 0 0 0 0 11 0 0 0 0 0 0 0 0 0 | f = 6</td></tr>
<tr><td>0 0 0 0 0 0 11 0 0 0 0 0 0 0 0 | g = 7</td><td>0 0 0 0 0 0 11 0 0 0 0 0 0 0 0 | g = 7</td></tr>
<tr><td>0 0 0 0 0 0 0 11 0 0 0 0 0 0 0 | h = 8</td><td>0 0 0 0 0 0 0 11 0 0 0 0 0 0 0 | h = 8</td></tr>
<tr><td>0 0 0 0 0 0 0 0 11 0 0 0 0 0 0 | i = 9</td><td>0 0 0 0 0 0 0 0 11 0 0 0 0 0 0 | i = 9</td></tr>
<tr><td>0 0 0 0 0 0 0 0 0 11 0 0 0 0 0 | j = 10</td><td>0 0 0 0 0 0 0 0 0 11 0 0 0 0 0 | j = 10</td></tr>
<tr><td>0 0 0 0 0 0 0 0 0 0 11 0 0 0 0 | k = 11</td><td>0 0 0 0 0 0 0 0 0 0 11 0 0 0 0 | k = 11</td></tr>
<tr><td>0 0 0 0 0 0 0 0 0 0 0 11 0 0 0 | l = 12</td><td>0 0 0 0 0 0 0 0 0 0 0 11 0 0 0 | l = 12</td></tr>
<tr><td>0 0 0 0 0 0 0 0 0 0 0 0 11 0 0 | m = 13</td><td>0 0 0 0 0 0 0 0 0 0 0 0 11 0 0 | m = 13</td></tr>
<tr><td>0 0 0 0 0 0 0 0 0 0 0 0 0 11 0 | n = 14</td><td>0 0 0 0 0 0 0 0 0 0 0 0 0 11 0 | n = 14</td></tr>
<tr><td>0 0 0 0 0 0 0 0 0 0 0 0 0 0 11 | o = 15</td><td>0 0 0 0 0 0 0 0 0 0 0 0 0 0 11 | o = 15</td></tr>
<tr><td>KNN classifier</td><td>Bayes-Net classifier</td></tr>
<tr><td>a b c d e f g h i j k l m n o ← classified as</td><td>a b c d e f g h i j k l m n o ← classified as</td></tr>
<tr><td>11 0 0 0 0 0 0 0 0 0 0 0 0 0 0 | a = 1</td><td>11 0 0 0 0 0 0 0 0 0 0 0 0 0 0 | a = 1</td></tr>
<tr><td>0 11 0 0 0 0 0 0 0 0 0 0 0 0 0 | b = 2</td><td>0 11 0 0 0 0 0 0 0 0 0 0 0 0 0 | b = 2</td></tr>
<tr><td>0 0 11 0 0 0 0 0 0 0 0 0 0 0 0 | c = 3</td><td>0 0 11 0 0 0 0 0 0 0 0 0 0 0 0 | c = 3</td></tr>
<tr><td>0 0 0 11 0 0 0 0 0 0 0 0 0 0 0 | d = 4</td><td>0 0 0 10 1 0 0 0 0 0 0 0 0 0 0 | d = 4</td></tr>
<tr><td>0 0 0 0 11 0 0 0 0 0 0 0 0 0 0 | e = 5</td><td>0 0 0 0 11 0 0 0 0 0 0 0 0 0 0 | e = 5</td></tr>
<tr><td>0 0 0 0 0 11 0 0 0 0 0 0 0 0 0 | f = 6</td><td>0 0 0 0 0 11 0 0 0 0 0 0 0 0 0 | f = 6</td></tr>
<tr><td>0 0 0 0 0 0 11 0 0 0 0 0 0 0 0 | g = 7</td><td>0 0 0 0 0 0 11 0 0 0 0 0 0 0 0 | g = 7</td></tr>
<tr><td>0 0 0 0 0 0 0 11 0 0 0 0 0 0 0 | h = 8</td><td>0 0 0 0 0 0 0 11 0 0 0 0 0 0 0 | h = 8</td></tr>
<tr><td>0 0 0 0 0 0 0 0 11 0 0 0 0 0 0 | i = 9</td><td>0 0 0 0 0 0 0 0 10 0 0 0 0 1 0 | i = 9</td></tr>
<tr><td>0 0 0 0 0 0 0 0 0 11 0 0 0 0 0 | j = 10</td><td>0 0 0 0 0 0 0 0 0 11 0 0 0 0 0 | j = 10</td></tr>
<tr><td>0 0 0 0 0 0 0 0 0 0 11 0 0 0 0 | k = 11</td><td>0 0 0 0 0 0 0 0 0 0 11 0 0 0 0 | k = 11</td></tr>
<tr><td>0 0 0 0 0 0 0 0 0 0 0 11 0 0 0 | l = 12</td><td>0 0 0 0 0 0 0 0 0 0 0 11 0 0 0 | l = 12</td></tr>
<tr><td>0 0 0 0 0 0 0 0 0 0 0 0 11 0 0 | m = 13</td><td>0 0 0 0 0 0 0 0 0 0 0 0 11 0 0 | m = 13</td></tr>
<tr><td>0 0 0 0 0 0 0 0 0 0 0 0 0 11 0 | n = 14</td><td>0 0 0 0 0 0 0 0 0 0 0 0 0 11 0 | n = 14</td></tr>
<tr><td>0 0 0 0 0 0 0 0 0 0 0 0 0 0 11 | o = 15</td><td>0 0 0 0 0 0 0 0 0 0 0 0 0 0 11 | o = 15</td></tr>
</table>

5 Conclusion

In this paper, we have proposed a new hybrid system based on our Local Gradient Probabilistic Pattern (LGPP) and 2D-DWT frequency method. The 2D-PCA subspace method are used to extract the face features vector, which serve as inputs for the different classification algorithms. According to the experimental results, our system reaches it maximum performance (98 %) with SVM and MLP classifiers on ORL database. The KNN is best on Yale database it reaches 100 %.

View the performance of our system, we will use it, in the future, in the education field to analyze the behavior of the student record their absences and equalize their fairness of examinations.

References

1. Lowe, D.G.: Distinctive image features from scale-invariant key points. Int. J. Comput. Vision **60**, 91–110 (2004)
2. Bay, J.M., Ess, A., Tuytelaars, T., Gool, L.V.: Surf: Speeded up robust features. Comput. Vis. Image Underst. **110**, 346–359 (2008)
3. Pal, A., Das, N., Sarkar, S., Gangopadhyay, D., Nasipuri, M.: A new rotation invariant weber local descriptor for recognition of skin diseases. In: Morocco Recognition and Machine Intelligence, pp. 355–360. Springer, Heidelberg (2013)
4. Ahonen, T., Hadid, A., Pietikäinen, M.: Face description with local binary patterns application to face recognition. IEEE Trans. Pattern Anal. Mach. Intell. **28**, 2037–2041 (2006)
5. Yektaii, M., Bhattacharya, P.: A criterion for measuring the separability of clusters and its applications to principal component analysis. SIViP **5**, 93–104 (2011)
6. Zhu, Q., Xu, Y.: Multi-directional two-dimensional PCA with matching score level fusion for face recognition. Neural Comput. Appl. **23**, 169–174 (2013)
7. Huang, S., Yang, D., Zhou, J., Zhang, X.: Graph regularized linear discriminant analysis and its generalization. Pattern Anal. Appl. **18**, 639–650 (2015)
8. Forczmański, P., Łabędź, P.: Improving the recognition of occluded faces by means of two-dimensional orthogonal projection into local subspaces. In: Image Analysis and Recognition, pp. 229–238. Springer International Publishing (2015)
9. Anbarjafari, G., Izadpanahi, S., Demirel, H.: Video resolution enhancement by using discrete and stationary wavelet transforms with illumination compensation. SIViP **9**, 87–92 (2015)
10. Huang, Z.H., Li, W.J., Shang, J., Wang, J., Zhang, T.: Non-uniform patch based face recognition via 2D-DWT. Image Vis. Comput. **37**, 12–19 (2015)
11. Gmira, F., Hraoui, S., Saaidi, A., Oulidi, A.J., Satori, K.: Securing the architecture of the JPEG compression by an dynamic encryption. In: (ISCV) IEEE (2015)
12. Aharrane, N., El Moutaouakil, K., Satori, K.: A comparison of supervised classification methods for a statistical set of features: Application: Amazigh OCR. In: (ISCV) IEEE (2015)
13. Dahmouni, A., El moutaouakil, K., Satori, K.: Robust face recognition using local gradient probabilistic pattern (LGPP). In: Proceedings of the MedICT2015, vol. 380, pp. 277–286. Springer, International Publishing (2016)
14. The ORL face database at the AT&T. http://www.uk.research.att.com/facedatabase.html
15. The Yale Face Database. http://cvc.yale.edu/proiects/yalefaces/yalefaces.html
16. Guo, Z., Zhang, L., Zhang, D.: A completed modeling of local binary pattern operator for texture classification. IEEE Trans. Image Process. **19**, 1657–1663 (2010)

17. Tan, X., Triggs, B.: Enhanced local texture feature sets for face recognition under difficult lighting conditions. IEEE Trans. Image Process. **19**, 1635–1650 (2007)
18. Jabid, T., Kabir, M.H., Chae, O.S.: Local directional pattern for face recognition. In: Proceeding of the IEEE International Conference of Consumer Electronics (2010)
19. Jun, B., Kim, D.: Robust face detection using local gradient patterns and evidence accumulation. Pattern Recogn. **45**, 3304–3316 (2012)
20. Dahmouni, A., El moutaouakil, K., Satori, K.: A new local binary probabilistic pattern (LBPP) and subspace methods for face recognition. Wseas. Trans. Comput. **14**, 588–597 (2015)

The Detection of Smell in Spoiled Meat by TGS822 Gas Sensor for an Electronic Nose Used in Rotten Food

Nihad Benabdellah, Mohammed Bourhaleb, Naima Benazzi, M'barek Nasri and Sanae Dahbi

Abstract The rotten food is one of the serious problems for health. In our subject we will focus on the design of an electronic nose that can detect and define the rotten food before the human nose. Their heart is the array sensors. In this paper the TGS822 gas sensor one of these sensors is studied to detect toxic gas Acetone and Ethanol exist in the spoiled meat in the early time.

Keywords Rotten food · Meat · Gas sensor · Electronic nose · Acetone gas · Ethanol gas

1 Introduction

The types of spoiled food can't be defined by the human nose because they have the same smell. Our subject is the design of an electronic nose used in the fridge which can define and detect the rotten food in early time before human. Some concentration of gases increase in rotten food for example: acetone, ethanol, Ammonia (NH3), Hydrogen Sulfide (H2S)…are detected by the array sensors which are the heart of our nose [1] Those sensors are overlapping selectively along with a pattern reorganization component that identifies the smell [2].

An electronic nose is made up of a mechanism for chemical detection. It is an intelligent sensing device simulates the human olfactory system [3]. Nowadays, the electronic noses have supplied external advantage to several of commercial industries, environment, cosmetics, biomedical, food, water and various scientific research fields. The electronic nose can detect the dangerous molecules for human [4].

N. Benabdellah (✉) · M. Bourhaleb · N. Benazzi · M. Nasri · S. Dahbi
Laboratory of Electrical Engineering and Maintenance, Higher School of Technology, Oujda, Morocco
e-mail: benabdellah.nihad@gmail.com

Á. Rocha et al. (eds.), *Europe and MENA Cooperation Advances in Information and Communication Technologies*, Advances in Intelligent Systems and Computing 520, DOI 10.1007/978-3-319-46568-5_29

In this paper, the proposed conception of the electronic nose for the rotten food will be presented. In the second part, the gas sensor TGS822 one of array sensors is used to detect the gases toxic exist in the spoiled meat. Finally, sensitivity for TGS822 will be calculated each day during 5 days and deduce that our sensor can determine the rotten meat in the first day before the human nose.

2 The Rotten Meat

Meat is a complex niche that has chemical and physical properties which allow a variety of microorganisms that can be colonized and developed. Many factors can influence on microbes that are presented on certain meat. After slaughtering, meat can be contaminated by bacteria from water, air, and soil as well as from the workers and the equipments involved during the manufacturing process [5].

2.1 Microbes Present in Rotten Meat

Brochothrix Thermosphacta is a microorganism of which meat is considered an ecological niche. This microbe aerobic metabolism of glucose produces foul-smelling odor such as acetone and acetic acid [6]. Clostridium is a rod-shaped cell with a gram-positive membrane. These microbes are anaerobes and some are toxin-producing pathogens. Some of them produce acetone, butanol, ethanol, isopropanol, and organic acids [7]. Leuconostoc is one of the lactic acid bacteria; it produces D-lactate and ethanol. This group of microbe is responsible for the discoloration, gas production, and buttery smell of spoiled meat [8]. The foul smell is the result of toxic gas which has a tart odor.

3 The Proposed Electronic Nose for the Rotten Food

Block diagram in Fig. 1 Further illustrates the sensing concept of electronic nose. The sample are prepared in the odor handling system that the Static headspace (SHS) [9] method is used after that the volatile component heading in the system when the odor well be detected by the array sensors consists of: the gases sensor such us TGS822 and the humidity and temperature sensor. To the output of the array sensor we have the footprint odor. The pattern recognition system is used to identify this foot print and compare to pattern data base.

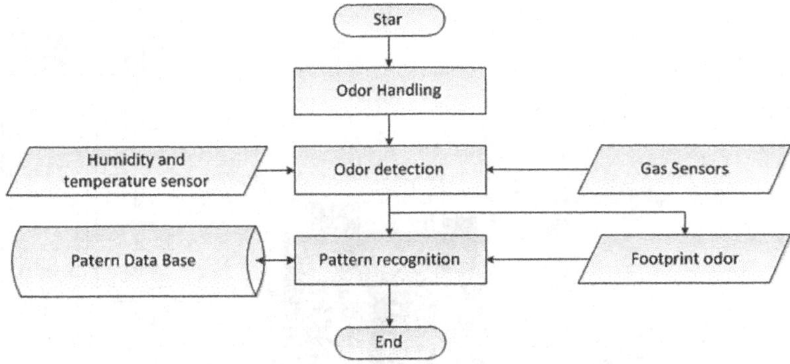

Fig. 1 Block diagram of Electronic nose

4 Gas Sensor TGS822

The TGS822 semiconductor gas sensor is used to detect organic solvent vapors such as ethanol and acetone gases present in the rotten meat. This sensor is one of the array sensors of our electronic nose. This sensor is the type Figaro. It works like a resistor, if the concentration of gas increases the resistance decreases. It has 6 pins, the two middle pins are the power to the internal heater and the other are the resistor connections. It is very important that the power provided to the internal heater is exactly 5 V.

4.1 The Output Voltage for TGS822 in the Presence of Gas Toxic

The design and simulation of TGS822 with arduino in ISIS proteus (Fig. 2):

The output voltage in the presence of Ethanol and Acetone: ISIS proteus is used to design the circuit of TGS822 gas sensor with arduino en presence of ethanol and acetone gas in the temperature of room 25 °C. The designed sensor circuit is being tested on breadboard in the presence of ethanol and acetone gas.

The presence of Ethanol:

In the presence of Acetone:

The recorded results in Figs. 3 and 4 illustrate that whenever the concentration of Ethanol gas and Acetone gas, increases in the environment the output voltage for our sensor increases so the presence of gas are determined according to the increase of the output voltage.

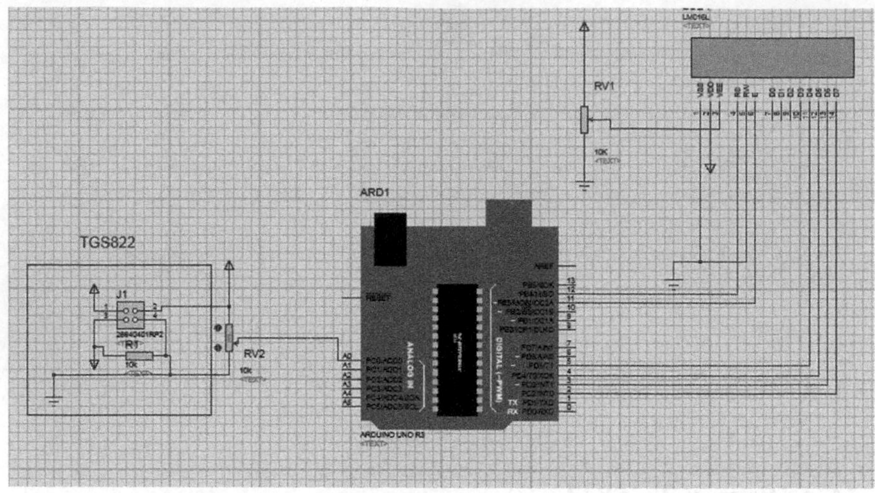

Fig. 2 Block diagram of Electronic nose

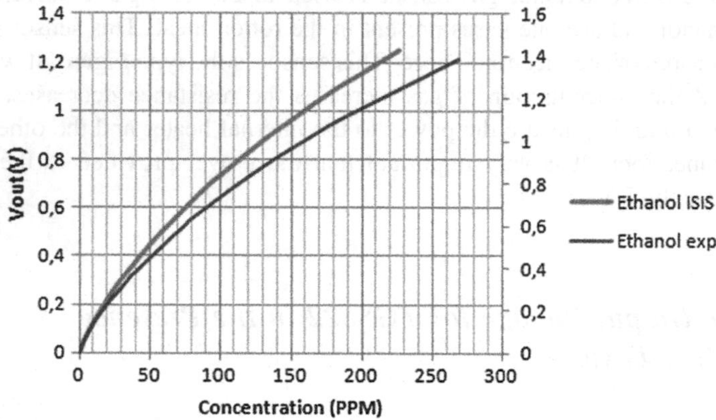

Fig. 3 Electrical output voltage (V) proportional to concentration of ethanol gas in ISIS proteus and in breadboard

4.2 The Sensitivity for TGS822 in the Presence of Toxic Gas

Sensitivity of a sensor is defined as the change in output of the sensor per unit change in the parameter being measured. The factor may be constant over the range of the sensor (linear), or it may vary (nonlinear). Sensitivity of TGS822 formula is:

$$S = \frac{\Delta R}{\Delta C} \quad \text{With} \quad R = \frac{Rs}{R0} \tag{1}$$

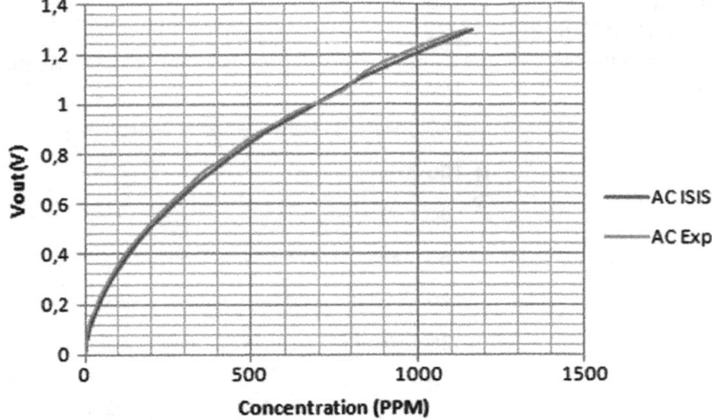

Fig. 4 Electrical output voltage (V) proportional to concentration of Acetone gas in ISIS proteus and in breadboard

Defined as follows: R0 is Sensor resistance in 300 ppm ethanol and Rs is Sensor resistance of displayed gases at various concentrations, C is the concentration of gas.

The sensitivity for TGS822 in the presence of Ethanol gas:

Calculating the sensitivity of TGS822 in the presence of ethanol gas from Fig. 5 and formula (1): Sexp = 0.191 ppm^{-1} and Sisis = 0.143 ppm^{-1}.

The sensitivity for TGS822 in the presence of Acetone gas:

Calculating the sensitivity of TGS822 in the presence of acetone gas from Fig. 6 and formula (1): Sexp = 0.335 ppm^{-1} and Sisis = 0.367 ppm^{-1}.

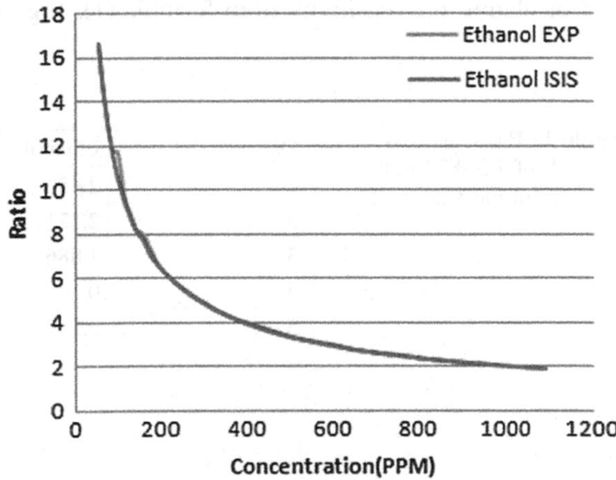

Fig. 5 Sensitivity characteristic for TGS822 in the presence of Ethanol gas in ISIS proteus and in breadboard

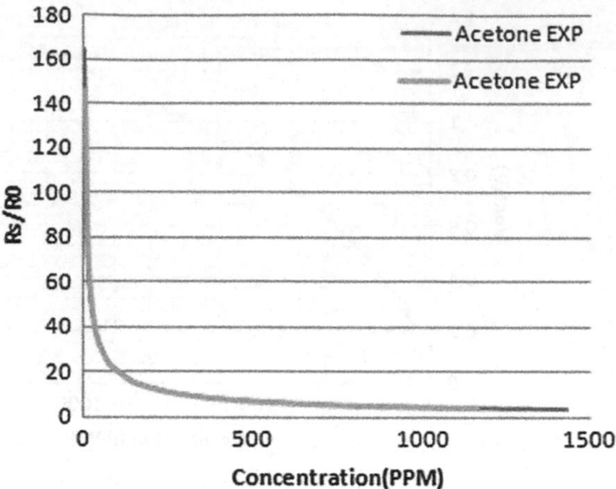

Fig. 6 Sensitivity characteristic for TGS822 in the presence of Acetone gas in ISIS proteus and in breadboard

The recorded results in Figs. 5 and 6 Illustrate that whenever the concentrations of Ethanol gas, and Acetone gas, increase in our environment the resistance of our sensor TGS822 decreases.

4.3 The Detection of Toxic Gas Present in Rotten Meat

We have experienced a 40 g of meat for 5 days with TGS822 gas sensor in the room temperature 25 °C, the Ethanol and Acetone gas will be detected when this meat become spoiled (Table 1).

The detection of Ethanol gas in spoiled meat for 5 days (Fig. 7):

The detection of Acetone gas in spoiled meat for 5 days:

Calculating the sensitivity from formula (1), Fig. 8:

Table 1 The sensitivity (ppm^{-1}) of TGS822 in the presence of meat during 5 days

Days	Ethanol	Acetone
1	2.45	0.888
2	2.121	0.393
3	1.886	0.38
4	0.195	0.06
5	0.147	0.023

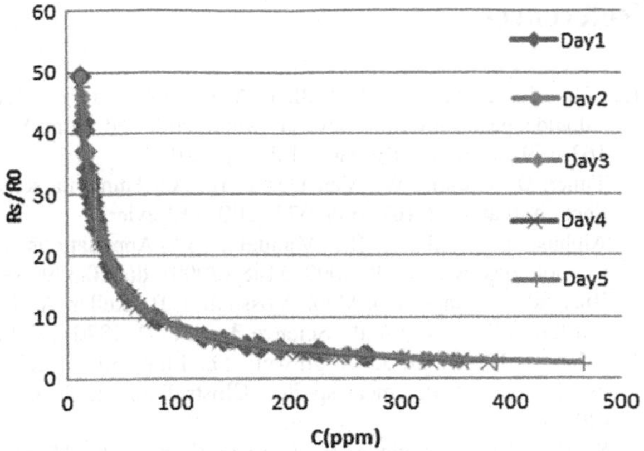

Fig. 7 Sensitivity characteristic for TGS822 of 5 days and detection of ethanol in spoiled meat

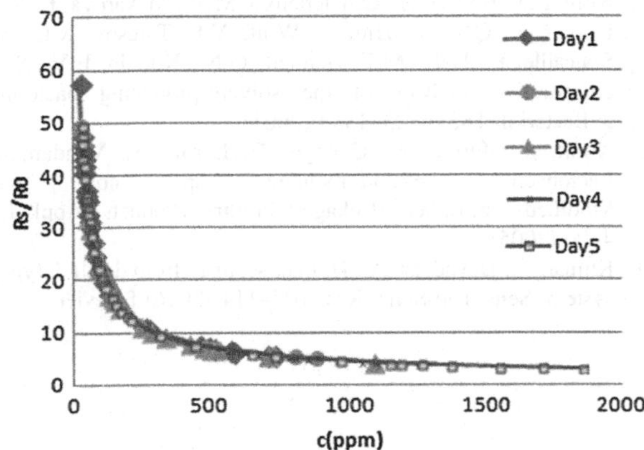

Fig. 8 Sensitivity characteristic for TGS822 of 5 days and detection of Acetone in spoiled meat

5 Conclusion

The sensor sensitivity increases when we have detected a small variation of ethanol and acetone gas concentration, so we deduce that we have a toxic gas which means that the meat is rotten but when its smell is detected by the human sensibility for our sensor tends to zero because our sensor detect the rotten meat in the early time before human. The paper has presented the detection of toxic gas in spoiled meat in the early time using TGS822 Metal Oxide Semiconductor gas sensor from Figaro one of our array sensors. We deduce that the sensor detect spoiled meat in the early time before the human nose. The subject is to detect many rotten foods by our system of detection and determine for the design of an electronic nose used in the fridge detect the rotten food.

References

1. Lohitesh, K., Alok, K.B., Allen, A.A., Suneetha, V.: Detection and removal of hydrogen sulphide gas from food sewage water collected from Vellore. Der Pharmacia Lettre **5**(3), 163–1692, Scholars Research Library (2013)
2. Timer, B., Olthuis, W., Van Den Berg, A.: Ammonia sensors and this application-a review. Sens. Actuators B **107**, 666–677 (2005) Elsevier
3. Alphus, D., Wilson, B., Manuela, B.: Applications and advances in electronic-nose technologies. Sensors **9**, 5099–5148 (2009). doi:10.3390/s90705099 (USA)
4. Bushdid, C., Magnasco, M.O., Vosshall, L.B., Keller, A.: Humans can discriminate more than 1 trillion olfactory stimuli. Science **343**(6177), 1370–1372 (2014)
5. Adam, K.H., Brunt, J., Brightwell, G., Flint, S.H., Peck, M.W.: Spore germination of the psychrotolerant, red meat spoiler. Clostridium frigidicarnis. Lett. App. Microb. **53**, 92–974 (2011)
6. Sperber, W.H., Doyle M.P.: Compendium of the Microbiological Spoilage of Foods and Beverages. Food Microbiol. Food Saf. (2009). doi:10.1007/978-1-4419-0826-1_1 (Springer)
7. Nolling, J., Breton, G., Omelchenko, M.V., Makarova, K. S., Zeng, Q., Gibson, R., Lee, H.M., Dubois, J., Qui, D., Hitti J., Wolf, Y.I., Tatusov, R.L., Sabathe, F., Doucette-Stamm, L., Soucaille, P., Daly, M.J., Bennett, G.N., Koonin, E.V., Smith D.R.: Genome sequence and comparative analysis of the solvent-producing bacterium Clostridium acetobutylicum. J. Bacteriol. **183**, 4823–4838 (2001)
8. Koort, J., Murros, A., Coenye, T., Eerola, S., Vandamme, P., Sukura, A., Bjorkroth, J.: Lactobacillus oligofermentans sp. nov., Associated with Spoilage Modified-Atmosphere-Packaged Poultry Products. Appl. Environ. Microbiol. **71**(8), 4400–4406 (2005)
9. Kumar, V., Devadoss, A.: Human security from death defying gases using an intelligent sensor system. Sens. Bio-Sens. Res. 107–114 (2016) Elsevier

New Image Steganography Method Based on Haar Discrete Wavelet Transform

Youssef Taouil, El Bachir Ameur and Moulay Taib Belghiti

Abstract Steganography is a technique of dissimulating information in digital media. In contrast to cryptography, it is not just to keep the non intended recipients from knowing the hidden information but it is to keep in secret even the existence of a hidden information. In this paper we propose a new image steganography method based on Haar discrete wavelet transform, data is hidden in the frequency domain since it is the most robust area. The embedding is performed in the integer part of the coefficients of the transform to prevent the loss of data coming from the floating point; in such a way that increases both the imperceptibility and the capacity of hiding. To test the performance of the proposed method, experiments on variety of images were accomplished. The results show better image quality and a high imperceptibility and payload in comparison with prior works. The security was strengthened using a random key that chooses randomly the location where to hide data.

Keywords Steganography · Data hiding · Haar discrete wavelet transform · Random key

E.B. Ameur
Faculty of Science, Research Team MSISI, LaRIT Laboratory,
Ibn Tofail University, 14000 Kénitra, Morocco
e-mail: ameurelbachir@yahoo.fr

Y. Taouil (✉) · M.T. Belghiti
EECOMAS Laboratory, ENSAK, Ibn Tofail University, 14000 Kénitra, Morocco
e-mail: taouilysf@gmail.com

M.T. Belghiti
e-mail: belghititaib@yahoo.fr

Á. Rocha et al. (eds.), *Europe and MENA Cooperation Advances in Information and Communication Technologies*, Advances in Intelligent Systems and Computing 520, DOI 10.1007/978-3-319-46568-5_30

1 Introduction

Today, the confidentiality of information has become a big concern in the field of communication since the technological development is a major threat. Cryptography protects data by coding its content and making it incomprehensible to anyone aside the intended recipient; at this point, Steganography presents stronger protection by making secret even the existence of data through concealing it in digital files.

Steganography comes from the Greek word "steganos-graphos" meaning covered writing; it is the art of hiding a message within a digital file called cover so that the presence of the hidden message is indiscernible. The key concept behind steganography is that the message to be transmitted is not detectable to the casual eye, and people who are not intended to be the recipients of the message should not even suspect the existence of a hidden message.

The first approaches of Steganography were based on the spatial domain, especially the technique of hiding in the Least Significant Bit (LSB) of the image pixels [1]. Hong and Chen [2] proposed a reversible data hiding method based on image interpolation; data is embedded into interpolation errors using the histogram shifting technique. A semi reversible data hiding method that utilizes interpolation and the least significant substitution technique is proposed in [3]. Hu in [4] propose a high payload image steganographic scheme based on an extended interpolating method. The prominent disadvantage in spatial domain techniques is the vulnerability to the slight attacks. Martins et al. proposed in [5] an energy-sufficient steganographic method to secure wireless sensor networks; data is hidden in the MAC layer of the 802.15.4 protocol.

In the frequency domain approaches, the message is embedded after applying a transformation on the image, this way the hidden message resides in more robust areas providing a better resistance against statistical attacks. Singla and Siyal [6] proposed Steganographic methods based on Discrete Cosinus Transform (DCT). In Wavelet-based steganographic methods, the message is hidden in the wavelet transform of the image and then the stego image is obtained by applying the inverse transform [7]. A high payload steganographic method based on pixel value differencing (PVD) technique is proposed in [8].

In Taouil et al. [9] the proposed methods have a high imperceptibility but a satisfying value of capacity. In order to ameliorate this criterion while maintaining the imperceptibility at an equivalent level, we developed, in this paper, a new image steganographic method based on Haar discrete wavelet transform. The binary sequence of the secret message is divided into packets of five bits and hidden in the integer part of the coefficients H, V and D, 2 bits are embedded in the 1st and 2ndLSB of the coefficients of two sub bands, and the last bit is embedded in the 2ndLSB of the coefficient of the remaining sub band. The security is fortified by a key that chooses randomly the coefficients where to hide data. The experiments show a high capacity, and a high level of imperceptibility evaluated by the Peak Signal to Noise Ration and the Mean Square Error.

In Sect. 2 the Haar discrete wavelet transform with the algorithms of decomposition and reconstruction is presented. In Sect. 3 we explain the proposed method. Section 4 discusses the experiment, the comparison of the results of the method proposed in [8] and [9] with our method. In Sect. 5 we conclude this paper and we discuss our future works.

2 Multiresolution Analysis and Haar Discrete Wavelet

Wavelets theory and its applications are rapidly developing fields in applied mathematics and signal analysis. Multi-resolution Analysis was formulated based on the study of orthonormal and compactly supported wavelet bases; it is the main theory in wavelets that analyzes a signal in frequency domain in detail. In this transform, spatial domain is passing through low pass and high pass filters to extract low and high frequencies respectively. Applying one level 2D wavelet transform on image decomposes the cover image into four non overlapping sub bands by namely A, H, V and D as shown in Fig. 1.

The sub bands A include the low pass coefficients and presents a soft approximation of image. The other three sub bands show respectively horizontal (H), vertical (V) and diagonal (D) details. In this paper Haar wavelet was chosen as a case study. The decomposition algorithm of Haar Discrete Wavelet Transform (DWT) is given by the following equations:

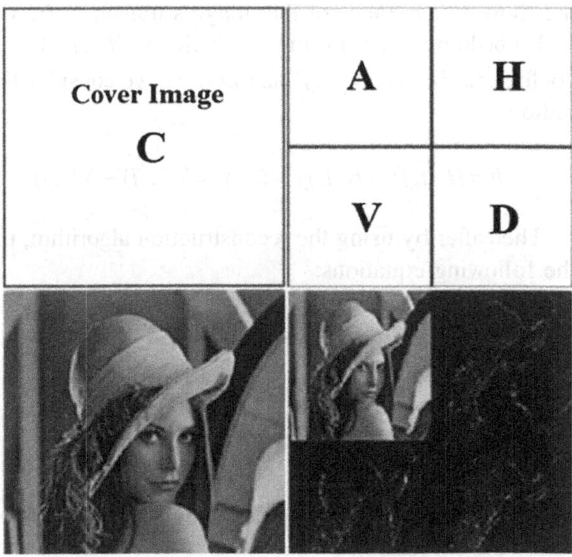

Fig. 1 Decomposition of Haar DWT

$$\begin{cases} A(i,j) = \frac{C(2i-1,2j-1)+C(2i-1,2j)+C(2i,2j-1)+C(2i,2j)}{2} \\ H(i,j) = \frac{C(2i-1,2j-1)+C(2i-1,2j)-C(2i,2j-1)-C(2i,2j)}{2} \\ V(i,j) = \frac{C(2i-1,2j-1)-C(2i-1,2j)+C(2i,2j-1)-C(2i,2j)}{2} \\ D(i,j) = \frac{C(2i-1,2j-1)-C(2i-1,2j)-C(2i,2j-1)+C(2i,2j)}{2} \end{cases}$$

The reconstruction algorithm or inverse Haar Discrete Wavelet Transform (IDWT) is given by the following equations:

$$\begin{cases} C(2i-1,2j-1) = \frac{A(i,j)+H(i,j)+V(i,j)+D(i,j)}{2} \\ C(2i-1,2j) = \frac{A(i,j)+H(i,j)-V(i,j)-D(i,j)}{2} \\ C(2i,2j-1) = \frac{A(i,j)-H(i,j)+V(i,j)-D(i,j)}{2} \\ C(2i,2j) = \frac{A(i,j)-H(i,j)-V(i,j)+D(i,j)}{2} \end{cases}$$

3 Proposed Work

The aim of this work is to develop a new steganography method to dissimulate texts in images with a high level of imperceptibility and a high capacity of hiding. The method is based on the transform domain, using Haar discrete wavelet, to provide a strong resistance against attacks. Human eyes are much sensitive to the low frequency part (A sub-image); A is the most important component in the decomposition process and is to not be used in the embedding. Therefore, data is hidden in the most robust zones of the image's transform H, V and D.

Embedding data in the coefficients $H(i,j), V(i,j)$ and $D(i,j)$ produce new coefficients $H^{'}(i,j), V^{'}(i,j)$ and $D^{'}(i,j)$; we consider the difference coming from the hiding:

$$h = H^{'}(i,j) - H(i,j) \quad ; \quad v = V^{'}(i,j) - V(i,j) \quad ; \quad d = D^{'}(i,j) - D(i,j)$$

Thereafter by using the reconstruction algorithm, the stego-image is obtained by the following equations:

$$\begin{cases} S(2i-1,2j-1) = \frac{A(i,j)+(H(i,j)+h)+(V(i,j)+v)+(D(i,j)+d)}{2} \\ S(2i-1,2j) = \frac{A(i,j)+(H(i,j)+h)-(V(i,j)+v)-(D(i,j)+d)}{2} \\ S(2i,2j-1) = \frac{A(i,j)-(H(i,j)+h)+(V(i,j)+v)-(D(i,j)+d)}{2} \\ S(2i,2j) = \frac{A(i,j)-(H(i,j)+h)-(V(i,j)+v)+(D(i,j)+d)}{2} \end{cases}$$

Then the relation between the stego image and the cover image is given by the following expressions:

$$\begin{cases} S(2i-1,2j-1)=C(2i-1,2j-1)+e_1 & with \quad e_1=\frac{h+v+d}{2} \\ S(2i-1,2j) \quad =C(2i-1,2j)+e_2 & with \quad e_2=\frac{h-v-d}{2} \\ S(2i,2j-1) \quad =C(2i,2j-1)+e_3 & with \quad e_3=\frac{v-h-d}{2} \\ S(2i,2j) \quad\quad =C(2i,2j)+e_4 & with \quad e_4=\frac{d-v-h}{2} \end{cases}$$

Since $S(i,j)$ and $C(i,j)$ are integers then e_k must also be integers, and knowing that $h=e_1+e_2$, then h is necessarily an integer, the same goes for v and d; we conclude from this that the hiding must be performed in the integer part of the coefficients $H(i,j), V(i,j)$ and $D(i,j)$.

To increase the payload of the method we insert data in two bits of the integer part of the coefficients $H(i,j), V(i,j)$ and $D(i,j)$ in such a way that guaranties that the pixels of the stego image are integers.

The imperceptibility is measured in function of the Mean Square Error (MSE): $MSE=\frac{1}{MN}\sum_{i=1}^{M}\sum_{j=1}^{N}(S(i,j)-C(i,j))^2$,We remark that:

$$MSE=\frac{1}{MN}\sum_{i=1}^{M/2}\sum_{j=1}^{N/2}E_{ij} \; while \; E_{ij}=e_1^2+e_2^2+e_3^2+e_4^2=h^2+v^2+d^2$$

To have a high level of imperceptibility we must minimize the MSE, and this requires minimizing the absolute value of h, v and d. Since we are dissimulating 2 bits of data in the integer part of the coefficients, the hiding must be done in the 1st and 2nd least significant bits to minimize the MSE.

$$H(i,j)=\pm\sum_{r=-1}^{8}a_r 2^r with \, a_r=0 \, or \, 1$$

When a bit of data $''b''$ is hidden in the coefficient bit of weight $''p''$, the resultant coefficient is:

$$H'(i,j)=\pm\left(\sum_{r=-1,r\neq p}^{8}a_r 2^r+b2^p\right);$$

Then the difference relative to the weight $''p''$ is:

$$h_p=H'(i,j)-H(i,j)=\pm(b-a_p)2^p \; then \, h_p\in\{-2^p,0,2^p\}$$

We can decompose the added values h, v and d as follows:

$$h=h_0+h_1 \quad ; \quad v=v_0+v_1 \quad ; \quad d=d_0+d_1$$

The expression of e_1(the same for e_2, e_3 and e_4) becomes:

$$e_1 = e_{10} + e_{11} \; while \; e_{1r} = \frac{h_r + v_r + d_r}{2}$$

The parameters $h_1, v_1, d_1 \in \{-2, 0, 2\}$ which makes $e_{11}, e_{21}, e_{31}, e_{41}$ integers, but $h_0, v_0, d_0 \in \{-1, 0, 1\}$. In this case e_{p0} can be non-integers; to solve this problem we choose just 2 parameters h_0 and v_0 to dissimulate data and the third one d_0 we deduce its value, accordingly to h_0 and v_0, to have e_{p0} integers and to minimize $|d|$.

The relation between h_0, v_0 and d_0 is given by the following equation:

$$d_0 = h_0 \oplus v_0; \; while \oplus is \; the \; xor \; bit \; operator.$$

The message is divided into packets of five bits $(b_0, b_1, b_2, b_3, b_4)$, we hide (b_0, b_1) in the 1st and 2nd LSB of the coefficient $H(i,j)$, (b_2, b_3) in the 1st and 2nd LSB of $V(i,j)$ and (b_4) in the 2nd LSB of $D(i,j)$ after adding d_0 to $D(i,j)$. This makes the capacity of hiding equal to $\left(\frac{3K-1}{4}\right)$ bit per pixel (bpp).

The values of $h; v; d$ are in the set $\{-3, -2, -1, 0, 1, 2, 3\}$. They can be reduced when they are equal to 3 or –3, this improves the imperceptibility.

After hiding (b_0, b_1) in $H(i, j)$:

$$H(i,j) = \pm \sum_{k=-1}^{8} a_k 2^k$$

$$H'(i,j) = \pm\left(\sum_{k=2}^{8} a_k 2^k + \sum_{k=0}^{1} b_k 2^k + a_{-1} 2^{-1}\right)$$

$$then \; H'(i,j) - H(i,j) = \pm \sum_{k=0}^{1} (b_k - a_k) 2^k = 3s; \; while \; s = \pm 1$$

Now we construct a new coefficient $H''(i, j)$ carrying the same embedded data in $H'(i, j)$ and reducing the difference from 3s to a smaller value; b_1, b_0 are to be kept unchanged since they carry the hidden data, and a_{-1} also cannot be changed otherwise the stego pixel maybe non integer; hence the expression of $H''(i, j)$:

$H''(i,j) = \sum_{k=2}^{8} \beta_k 2^k + \sum_{k=0}^{1} b_k 2^k + a_{-1} 2^{-1}$, We can easily verify that:

$$\sum_{k=2}^{8} (\beta_k - a_k) 2^k = \left(H''(i,j) - H(i,j)\right) - 3s$$

By remarking that $\sum_{k=2}^{8} (\beta_k - a_k) 2^k$ is divisible by 4, we can deduce that among $\{\pm 2s, \pm 1s, 0\}$ the number $H''(i, j) - H(i, j)$ can only be equal to $-1s$:

$$then \sum_{k=2}^{8} (\beta_k - a_k)2^k = -4s, \text{Hence:}$$

$$H''(i,j) = \sum_{k=2}^{8} a_k 2^k - 4s + \sum_{k=0}^{1} b_k 2^k + a_{-1}2^{-1} = H'(i,j) - 4s$$

$$H''(i,j) - H(i,j) = H'(i,j) - H(i,j) - 4s = -s = \pm 1$$

This scheme requires that while adding $-4s$ the bits b_0 and b_1 must be not change, and it is not the case hence $H(i,j) = 0$ or ± 0.5.

The maximum absolute value of e_p is 4, so if a pixel of the cover image is $< 4\, or > 251$ then after the hiding process the stego pixel can fall off $[0, 255]$. To prevent this problem we keep, before embedding data, the pixels of the cover that are going to be used in the embedding in $[4, 251]$:

$$\begin{cases} C(i,j) = 4 & \text{if } C(i,j) < 4 \\ C(i,j) = 251 & \text{if } C(i,j) > 251 \end{cases} \quad (5)$$

The key: To fortify the security of the proposed method we use a random key that scrambles the hiding location, the order of selection of the pixels where to dissimilate data is given randomly.

Embedding algorithm:

- Step 1: Read the cover image as two dimensional file (matrix).
- Step 2: Transform the text into a binary sequence appending to it the key and the length of the text.
- Step 3: Apply the key to select the coefficients where the text is going to be embedded.
- Step 4: Adjust the values of pixels selected by the key to keep them between 4 and 251.
- Step 5: Perform the 2-D Haar discrete wavelet transform.
- Step 6: Divide the message into packets of five bits, and hide two bits in the 1st and 2nd LSB of the horizontal coefficients, two bits in the 1st and 2nd LSB of the associated vertical coefficient, and the remaining bit in the 2nd LSB of the associated diagonal coefficient.
- Step 7: Apply the Haar inverse discrete wavelet transform to obtain the stego image.

Extracting algorithm:

- Step 1: Read the stego image as two dimensional file (matrix).
- Step 2: Apply the Haar discrete wavelet transform to the stego image.
- Step 3: Extract the key to select the coefficients where the message is hidden.
- Step 4: Extract the length of the message and the content of the message from the 1st and 2nd LSB of the horizontal and vertical coefficients and the 2nd LSB of the diagonal coefficient.
- Step 5: Regroup the message extracted by blocks of 8 bits to obtain the hidden text.

4 Experiment Results

In this section, experiments were carried out to test the performance of the proposed method. We used a variety of images containing some well known ones as "Baboon", "Peppers" and "Barbara", where we dissimulated a text of 1000 bytes. The proposed work is compared with the method developed by Abu et al. [8] and the methods of Taouil et al. [9] based on the following criterions:

Imperceptibility: A steganography method is said imperceptible when human eye is unable to distinguish between the cover image and the stego image. This parameter is measured by the Mean Square Error (MSE) and Peak Signal to Noise Ratio (PSNR).

$$PSNR = 10\,Log\left(\frac{255^2}{MSE}\right)$$

Capacity: it refers to the quantity of information that can be dissimulated without degrading the quality of the cover image.

Security: it is attached to the random nature of the secret message and its independence from the cover image [10], and defined in terms of undetectability which is assured when the statistical tests cannot distinguish between the cover and the stego images, the security can be measured by the entropy between the probability distribution of the cover image and the stego image, the method is said ε-secure if the entropy is smaller than ε [11].

$$D(PDc//PDs) = \int PDc\,Log\left(\frac{PDc}{PDs}\right) \leq \varepsilon$$

The results shown in tables prove that the proposed method has a high level of imperceptibility knowing that a steganographic process is considered imperceptible when the PSNR is beyond 36 dB (Tables 1, 2, 3 and 4).

Figure 2 also illustrates that the proposed work is more imperceptible than the method of Azman [8], the difference in PSNR is greater than 15 dB. The PSNR values of the proposed method are between those of the two methods of Taouil [9], which indicates that the imperceptibility was maintained as almost the same level while increasing the capacity of embedding from 50 and 75 % bit per pixel to 125 % bit per pixel.

Table 1 MSE, PSNR and security for the method of Azman [8]

Method of Azman [8]				
Image	MSE	PSNR	Entropy	Capacity (%)
Baboon	2.437	44.262	11.841	75
Pepper	0.717	49.571	7.0897	75
Barbara	1.714	45.790	9.6325	75
Boat	2.782	43.685	3.9953	75
Flower	0.827	48.952	5.5758	75
Bridge	1.981	45.161	7.2474	75

Table 2 MSE, PSNR AND security for the 1st method of Taouil [9]

First method of Taouil [9]				
Image	MSE	PSNR	Entropy	Capacity (%)
Baboon	0.030	63.301	2.0567	75
Pepper	0.029	63.365	2.8249	75
Barbara	0.030	63.295	8.8921	75
Boat	0.028	63.613	2.5243	75
Flower	0.016	66.025	0.5565	75
Bridge	0.041	61.991	4.3243	75

Table 3 MSE, PSNR AND security for the 2nd method of Taouil [9]

Second method of Taouil [9]				
Image	MSE	PSNR	Entropy	Capacity (%)
Baboon	0.011	67.491	2.0567	50
Pepper	0.012	67.474	2.8249	50
Barbara	0.011	67.542	7.3928	50
Boat	0.010	67.827	2.5243	50
Flower	0.006	70.180	0.5565	50
Bridge	0.015	66.182	4.3243	50

Table 4 MSE, PSNR AND security for the proposed method

Proposed method				
Image	MSE	PSNR	Entropy	Capacity (%)
Baboon	0.014	66.524	1.2635	125
Pepper	0.015	66.135	1.2609	125
Barbara	0.015	66.092	5.0706	125
Boat	0.016	65.899	1.8856	125
Flower	0.008	68.616	0.3980	125
Bridge	0.021	64.828	2.6631	125

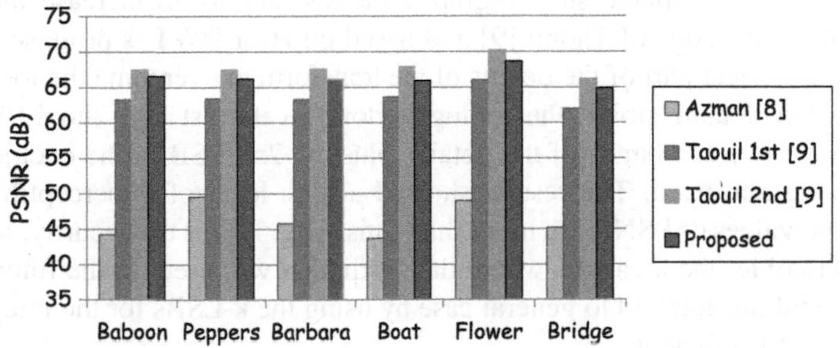

Fig. 2 Comparison of imperceptibility with Azman and Taouil

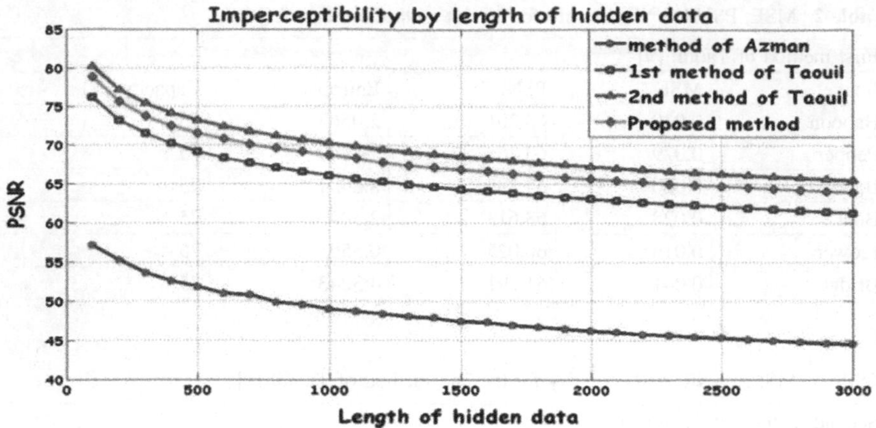

Fig. 3 Comparison of imperceptibility with Azman and Taouil

Figure 3 shows the results of PSNR while embedding a variable quantity of data from 100 bytes to 3000 bytes and adding 100 bytes each time, the values of the proposed work are between the methods of Taouil [9] and greater than those of the method developed by Azman [8] showing the high level of imperceptibility. The capacity of the method of Azman [8] is variable depending on a threshold; in the experiment this threshold was fixed in a value such as the capacity is 75 %. The first method of Taouil [9] has a capacity of 75 % while the second one's is 50 %; in the proposed method we use four coefficients to hide five bits of the message, this makes 125 % bits per pixel. The amelioration of the capacity of the methods of Taouil [9] was successful; it went up from 50 and 75 % to 125 % while the imperceptibility is on a better level than the 1stmethod and almost at the same level of the 2nd method.

5 Conclusion

In this paper, a new image steganography method aiming to increase the hiding capacity of the methods of Taouil [9] and based on Haar DWT is proposed, data is hidden in the integer part of the details of the transform to overcome the loss of data caused by the floating point, the hiding is done in the 1st and 2nd LSB of the coefficients of two sub bands of the details, plus the 2nd LSB of the coefficients of the remaining sub band. The results showed a high level of imperceptibility and capacity, the values of PSNR are more than satisfying. As for the security, a random key that scrambles the locations where data is hidden was used. In the future works we will extend this method to general case by using the k-LSBs for the integer parts of the wavelet coefficients.

References

1. Channalli, S., Sinhgad, A.J.: Steganography, an art of hiding data. Int. J. Comput. Sci. Eng. **1**(3), 137–141 (2009)
2. Hong, W., Chen, T.-S.: Reversible data embedding for high quality images using interpolation and reference pixel distribution mechanism. J. Vis. Commun. Image R. **22**, 131–140 (2011)
3. Jung, K.-H., Yoo, K.-Y.: Steganographic method based on interpolation and LSB substitution of digital images. Multimed. Tools Appl. doi:10.1007/s11042-013-1832-y
4. Hu, J., Li, T.: Reversible steganography using extended image interpolation technique. Comput. Electr. Eng. **46**, 447–455 (2015)
5. Martins, D., Guyennet, H.: Steganography in MAC Layers of 802.15.4 Protocolfor securing wireless sensor networks. doi:10.1109/MINES 2010.175.Source:IEEE Xplore (2010)
6. Singla, D., Syal, R.: Data security using LSB & DCT steganography in images. IJCER **2**(2), 359–364 (2012)
7. Yang, B., Deng, B.: Steganography in gray images using wavelet. In: Proceedings of ISCCSP (2006)
8. Abu, N.A., Adi, P.W., Mohd, O.: Robust digital image steganography within coefficient difference on integer haar wavelet transform. Int. J. Video Image Process. Netw. Secur. IJVIPNS-IJENS **14**(02) (2014)
9. Taouil, Y., El Bachir, A., Belghiti, M.T., El Harraj, A., Souhar, A.: High imperceptibility image steganography methods based on Haar DWT. Int. J. Comput. Appl. (0975–8887) **138**(10) (2016)
10. Zollner, J., Federrath, H., Klimant, H., Pitzman, A., Piotraschke, R., Westfeld, A., Wicke, G., Wolf, G.: Modeling the security of steganographic systems. In: Information Hiding Workshop, pp. 345–355, Portland, USA (1998)
11. Cachin, C.: An information theoretic model for steganography. Inf. Comput. **192**(1), 41–56 (2004)

Building HMM Independent Isolated Speech Recognizer System for Amazigh Language

Safâa El Ouahabi, Mohamed Atounti and Mohamed Bellouki

Abstract This paper describes the implementation of Hidden Markov Model based speaker independent spoken digits and letters speech recognition system for Amazigh language which is an official language in Morocco. The system is developed using HTK. The system is trained on 33 Amazigh alphabets and 10 first digits by collecting data from 60 speakers and is tested using data collected from another 20 speakers. This document details the experiment by discussing the implementation using the HTK Toolkit. Performance was measured using combinations of HMM 8-states and various number of Gaussian mixture distribution. The experimental results show that the system have given better recognition rate 85.95 % with 4 Gaussian Mixture. The results obtained are improved in comparison with our previous work.

Keywords Speech recognition system · Hidden Markov model · Gaussian mixture distribution · Hidden Markov model toolkit (HTK) · Amazigh letters and digits

1 Introduction

Speech Recognition is a technology that allows a computer to identify the words that a person speaks into a microphone or telephone. It is the process of converting an acoustic signal into a sequence of words. To recognize the underlying symbol sequence given a spoken utterance, the continuous speech waveform is first converted into a sequence of equally spaced feature vectors. The task of the recognizer is to map between sequences of feature vectors and the wanted underlying symbol sequences. Automatic speech recognition (ASR) is one of the fastest growing areas

S. El Ouahabi (✉) · M. Atounti · M. Bellouki
Faculty Polydisciplinary of Nador, Laboratory of Applied Mathematics and Information System, Mohammed First University, Oujda, Morocco
e-mail: safaa.elouahabi@gmail.com

Á. Rocha et al. (eds.), *Europe and MENA Cooperation Advances in Information and Communication Technologies*, Advances in Intelligent Systems and Computing 520, DOI 10.1007/978-3-319-46568-5_31

of engineering and technology [1] and there is lot of systems which are developed for English and other major languages spoken in developed countries [2–7], in the other hand, spoken alphabets and digits for different languages were targeted by ASR researchers [8, 9], as for the Amazigh language, many ASR were developed for digits and letters [10, 11]. Automatic speech recognition systems have been implemented using various toolkits and software, the most commonly used amongst them are the Hidden Markov Model ToolKit, Sphinx Toolkit, ISIP Production System, Julius Open-Source Large Vocabulary CSR Engine, HMM Toolbox for Matlab etc, among all these tools the HTK toolkit is the most popularly used tool to design ASR systems, since it is used in building and manipulating hidden Markov Models, it has applications in other research areas as well. HTK is well documented and provides guided tutorials for its use. The work presented in this paper aims to build a HMM independent isolated speech recognizer system for Amazigh language based on HMM toolkit (HTK). The system is trained on 33 Amazigh alphabets and 10 first digits by collecting data from 60 speakers and it is tested using data collected from another 20 speakers including both males and females. This paper details the experiment by discussing the implementation using the HTK Toolkit, performance was measured using combinations of HMM 8-states and various number of Gaussian mixture distribution. The article is organized as follows: Sect. 2 presents a brief description of the Amazigh language. In Sect. 3 we describe the hidden markov models. While Sect. 4 emphasizes on the description of HTK toolkit. In Sect. 5, we describe the Amazigh ASR developed, experiments and results. Finally, Sect. 6 concludes this paper.

2 Amazigh Language

The Amazigh language, known as Berber or Tamazight, stretches over an area that is so vast in Africa, from the Canary Islands to the Siwa Oasis in the North, and from the Mediterranean coast to Niger, Mali and Burkina-Fasso in the South. Historically, the Amazigh language has been autochthonous and was exclusively reserved for familial and informal domains [12]. In Morocco, we may distinguish three big regional varieties, depending on the area and the communities: Tarifiyt in the Northern, Tamazight in the Center and Tashlhiyt in the South-west and the High Atlas of the country. As regards the Amazigh varieties, there are 2.5 million Tachelhit-speakers in the south of Morocco known as Sus, 3 million Tamazight-speakers in the Atlas Mountains, and about 1.7 million Tarifiyt-speakers in the Riff [13]. In Morocco, the status of Amazigh has achieved the most advanced level, especially by the foundation of the Royal Institute of Amazigh Culture (IRCAM), in the Dahir on October 17th 2001, which stipulates that the Amazigh culture is a "national matter" and, this being the case, is a concern of all the citizens. Like any language passes throw oral to a written mode, the Amazigh language has been in need of a graphic system. In Morocco, the choice ultimately fell on Tifinaghe for

Arabic correspondance	Tifinaghe Transcription [IRCAM]	Arabic correspondance	Tifinaghe Transcription [IRCAM]	Arabic correspondance	Tifinaghe Transcription [IRCAM]	Arabic correspondance	Tifinaghe Transcription [IRCAM]	Arabic correspon dance	Tifinaghe Transcription [IRCAM]
صفر	ⴰⵎⵢⴰ	ا	ⴰ	ه	ⵀ	-	ⵓ	ي	ⵢ
واحد	ⵢⴰⵏ	ب	ⴱ	ح	ⵃ	ر	ⵔ	ز	ⵣ
اثنان	ⵙⵉⵏ	-	ⴳ	ع	ⵄ	-	ⵕ	-	ⵥ
ثلاثة	ⴽⵕⴰⴹ	-	ⴳⵯ	خ	ⵅ	غ	ⵖ		
اربعة	ⴽⴽⵓⵥ	د	ⴷ	ق	ⵇ	س	ⵙ		
خمسة	ⵙⵎⵎⵓⵙ	ض	ⴹ	ي	ⵉ	ص	ⵚ		
ستة	ⵚⴹⵉⵚ	ڥ	ⴻ	ج	ⵊ	ش	ⵛ		
سبعة	ⵙⴰ	ف	ⴼ	ل	ⵍ	ت	ⵜ		
ثمانية	ⵜⴰⵎ	ڭ	ⴽ	م	ⵎ	ط	ⵟ		
تسعة	ⵜⵥⴰ	-	ⴽⵯ	ن	ⵏ	و	ⵡ		

Fig. 1 Table of the 10 Amazigh digits and 33 alphabets

technical, historical and symbolic reasons. Since the Royal declaration on February 11th 2003, Tifinaghe has become an official graphic system for writing Amazigh, particularly in schools. Thus, the IRCAM has developed an alphabet system called Tifinaghe-IRCAM. This alphabet is based on a graphic system towards phonological tendency. This system does not retain all the phonetic realizations produced, but only those that are functional [14–16].

Figure 1 shows the ten Amazigh digits and the 33 Amazigh letters, along with their transcription in Tifinagh and Arabic correspondence.

3 Hidden Markov Model

Hidden Markov Model (HMM) is very powerful mathematical tool for modeling time series. It provides efficient algorithms for state and parameter estimation, and it automatically performs dynamic time warping of signals that are locally stretched. Hidden Markov models are based on the well-known chains from probability theory that can be used to model a sequence of events in time. Markov chain is deterministically an observable event. The most likely word with the largest probability is produced as the result of the given speech waveform. A natural extension of Markov chain is Hidden Markov Model (HMM), the extension where the internal states are hidden and any state produces observable symbols or observable evidences [17].

Mathematically Hidden Markov Model contains five elements.

1. Internal States: These states are hidden and give the flexibility to model different applications. Although they are hidden, usually there is some kind of relation between the physical significance to hidden states.
2. Output: $O = \{O_1, O_2, O_3, \ldots, O_n\}$ an output observation alphabet.
3. Transition Probability Distribution: $A = a_{ij}$ is a matrix. The matrix defines what the probability to transition from one state to another is.
4. Output Observation: Probability Distribution $B = b_i(k)$ is probability of generating observation symbol $o(k)$ while entering to state i is entered.
5. The initial state distribution ($\pi = \{ \pi_i \}$) is the distribution of states before jumping into any state.

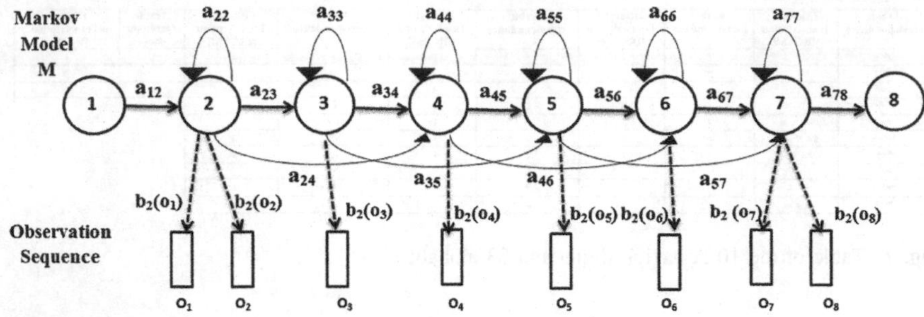

Fig. 2 Hidden markov model (HMM)—8-states

Here all three symbols represents probability distributions i.e. A, B and π. The probability distributions A, B and π are usually written in HMM as a compact form denoted by lambda as $\lambda = (A, B, \pi)$.

In our work, we are using the same kind of HMM. We are using a file called 'proto' which contains all necessary information and specifications. This file has been taken as it is from the HTK book [17]. This prototype 'proto' has to be generated for each digit and letter in the dictionary. Same topology is used for all the HMMs and the defined topology consists of 6 active states (observation functions) and two non-emitting states (initial and the last state with no observation function), See Fig. 2.

4 HTK Toolkit

Hidden Markov Model Toolkit i.e., HTK is a portable toolkit developed by the Cambridge University Engineering Department (CUED) freely accessible for download after registration at the URL http://htk.eng.cam.ac.uk. It consists of several library module and C program code and with good documentation (HTK Book [17]). Precompiled binary versions are also available for download (for Unix/Linux and Windows operating systems). Apart from speech recognition it has been applied to character recognition, speech synthesis, DNA sequencing etc. The toolkit provides tools for data preparation, training, testing and analysis, see Fig. 3.

5 Amazigh Speech Recognition System Using HTK

This section presents the different phases of the development of our system which are: data preparation, training, test and the analysis of results. It highlights on the system performance based on HTK.

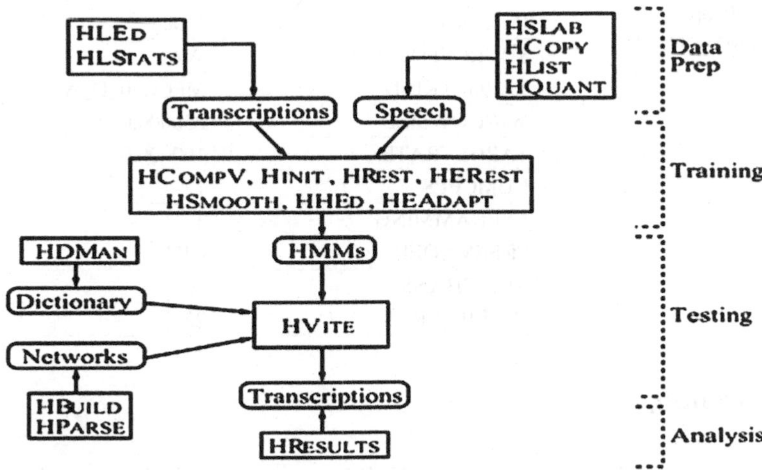

Fig. 3 HTK tools

5.1 Data Preparation

This phase consists of recording the speech signal. Firstly the data is recorded with the help of a microphone using a recording tool wavesurfer [18] in .wav format. The sampling rate used for recording is 16 kHz. The corpus consists of 10 Amazigh spoken digits (0–9) and 33 Amazigh alphabets each digits and alphabet is uttered 10 times in separated data files, each file containing 1 utterance and so the 10 distinct digits and 33 alphabets spoken by 80 person results in 19,500 file. Each utterance was visualized back to ensure that the entire word was included in the recorded signal, wrongly pronounced utterances were ignored and only correct utterances are kept in the database.

Before the corpus can be used to train HMMs it must be converted to an appropriate parametric form and the transcriptions must be converted to the correct format and use the required labels. Each wav file in corpus was labeled with the Wavesurfer tools, which results in to 19600 file (.lab). Due to the significant amount of files, a long time was required to validate this step of labeling signals.

To parameterize audio HCopy is used. In this work, we have used the acoustic parameters type MFCC. The parameters necessary for acoustic analysis such as format of input speech files, features to be extracted, window size, window function, number of cepstral coefficients, pre-emphasis coefficients, number of filter bank channels and length of cepstral filtering is provided to the HCopy in a configuration parametrisation.conf. The values to these parameters used for our experiment are given in Table 1.

Table 1 HCopy configuration parameters

Parameter	Value
SOURCEFORMAT	WAV
TARGETKIND	MFCC_0_D_A
WINDOWSIZE	250000.0
TARGETRATE	100000.0
NUMCEPS	12
USEHAMMING	T
PREEMCOEF	0.97
NUMCHANS	26
CEPLIFTER	22

5.2 Training

Before starting the learning phase, the HMM parameters must be properly initialized. The Hinit command in HTK tool is used for initializing the HMM by using the time alignment algorithm Viterbi from prototypes, and the training data in their MFCC form and their associated labeled file. This prototype has been generated for each word in the dictionary. Same topology is used for all the HMMs and the defined topology consists of 6 active states (observation functions) and two non-emitting states (initial and the last state with no observation function), see Fig. 2. This topology is used for all the HMMs. various numbers of Gaussian distributions with diagonal matrices are used as observation functions and these are described by a mean vector and variance vector in a text description file known as prototype. After initialization of models, Hcompv and HRest tool are applied in several iterations to estimate simultaneously all models on all sequences of acoustic vectors not labeled. The resulting models are improved by increasing the number of Gaussians for estimating the observation emission probability in a state. The models have been re-estimated by Hrest and Herest.

5.3 Testing

Once the models are trained, they are tested using the HTK command called HVite. We have created a dictionary and the word network before executing HVite. HVite uses the Virterbi algorithm to test the models. As input, HVite requires a network describing the allowable word sequences, a dictionary defining how each word is pronounced and a set of HMMs in addition to the result of decoding phase (.mfcc). HVite will convert the word network to a phone network and attach the appropriate HMM definition to each phone instance, after which recognition can be performed on direct audio input or on a list of stored speech files.

5.4 Analysis

Analysing an HMM-based recognizer's performance is done by the tool HResults. It uses dynamic programming to align the transcriptions output and correct reference transcriptions. The recognition results of the test signals are compared with the reference labels by a dynamic tracking performed by HResult.

5.5 Results

The evaluation of the performance of the speech recognition system was done by using HTK tool HResults. In HMM training, The system was tested on Baum Welch algorithm and Viterbi algorithm. Performance was measured using combinations of HMM 8-states and various number of Gaussian mixture distribution (1, 2, 4, 8, 16, 32), the state was modeled using Left Right (LR) HMM topologies. After initialization of models (Single Gaussian Mixture), Hcompv and HRest tool are applied in several iterations to estimate simultaneously all models on all sequences of acoustic vectors not labeled. The resulting models are improved by increasing the number of Gaussians for estimating the observation emission probability in a state. The models have been re-estimated by Hrest and Herest. In each mixture number, evaluation of the performance of the speech recognition system was measured by using HResults. The system is relatively successful, as it can identify the spoken digit and alphabets at an accuracy of 85.95% for speaker independent approach when the system trained using 8 states-HMM and 4 Gaussian Mixture, see Table 2. There is possibility that by using more robust model, we can increase the accuracy further.

Table 2 Results of test

Number of mixture	Number of state	Word correction rate (%)
1	8	82.66
2	8	84.14
4	8	**85.95**
8	8	85.83
16	8	85.30
32	8	84.25

6 Conclusion and Future Works

The system was tested using testing corpus data and the system scored up to 85.95 % word recognition for speaker independent approach. The work has catered for only an Isolated Digit-Letters Speech data. As much as it has created a basis for research, it can be expanded to cater for more extensive language models and larger vocabularies. The system can be enhanced to a larger vocabulary including commonly used words; it can be made robust by using larger database for training.

References

1. Anusuya, M.A., Katt, S.K.: Speech recognition by machine: a review. Int. J. Comput. Sci. Inf. Secur. **6**(3) (2009)
2. Kimutai, S.K., Milgo, E., Milgo, D.: Isolated Swahili words recognition using Sphinx4. Int. J. Emerg. Sci. Eng. **2**(2) (2013). ISSN:2319–6378
3. Ananthi, S., Dhanalakshmi, P.: Speech recognition system and isolated word recognition based on Hidden markov model (HMM) for Hearing Impaired. Int. J. Comput. Appl. **73**(20), 30–34 (2013)
4. Kumar, K., Jain, A., Aggarwal, R.K.: A Hindi speech recognition system for connected words using HTK. Int. J. Comput. Syst. Eng. **1**(1), 25–32 (2012)
5. Sameti, H., Veisi, H., Bahrani, M., Babaali, B., Hosseinzadeh, K.: A large vocabulary continuous speech recognition system for Persian language. EURASIP J. Audio Speech Music Process. (2011)
6. Abushariah, M.A., Ainon, M.R.N., Elshafei, R.M., Khalifa, O.O.: Natural speaker-independent Arabic speech recognition system based on Hidden Markov Models using Sphinx tools. In: International Conference Computer and Communication Engineering (ICCCE), Kuala Lumpur, Malayzia, doi:10.1109/ICCCE.2010.5556829 (2010)
7. Gales, M.J.F., Diehl, F., Raut, C.K., Tomalin, M., Woodland, P.C., Yu. K.: Development of a phonetic system for large vocabulary Arabic speech recognition. In: IEEE Workshop on Automatic Speech Recognition & Understanding (ASRU), Kyuto, Japan, pp. 24–29, doi:10.1109/ASRU.2007.4430078, 9–13 Dec 2007
8. Alotaibi, Y.A.: Investigating spoken Arabic digitsin speech recognition setting. Inf. Sci. **173**, 115–139 (2005)
9. Chapaneri, S.V.: Spoken digits recognition using weighted MFCC and improved features for dynamic time warping. Int. J. Comput. Appl. (0975–8887) **40**(3) (2012)
10. EL Ghazi, A., Daoui, C., Idrissi, N.: Automatic speech recognition for Tamazight enchained digits. World J. Control Sci. Eng. **2**(1), 1–5 (2014)
11. Satori, H., El Haoussi, F.: Investigation amazigh speech recognition using CMU tools. Int. J. Speech Technol. **17**(3), 235–243 (2014). doi:10.1007/s10772-014-9223-y
12. Boukous, A.: Société, langues et cultures au Maroc: Enjeux symboliques. Najah El Jadida, Casablanca, Maroc (1995)
13. Moustaoui, A.: The Amazigh language within Morocco's language policy, Dossier 14, University of Autònoma de Madrid (2003)
14. Boukous, A.: Phonologie de l'amazighe. Institut Royal de la Culture Amazighe, Rabat (2009)
15. Outahajala, M., Zenkouar, L., Rosso, P.: Building an annotated corpus for Amazighe. In: 4th International Conference on Amazigh and ICT, Rabat, Morocco (2011)

16. Fadoua, A., Siham, B.: Natural language processing forAmazigh language: Challenges and future directions. Language Technology for Normalisation of Less-Resourced Languages, (2012)
17. Young, S., Evermann, G., Hain, T., Kershaw, D., Moore, G., Odell, J., Ollason, D., Povey, D., Valtchev, V., Woodland, P.: The HTK Book (2002). http://htk.eng.cam.ac.uk
18. http://sourceforge.net/projects/wavesurfer/

Part III
Network & Web Technologies Applications

Performance Evaluation of LT Codes for Wireless Body Area Network

Nabila Samouni, Abdelillah Jilbab and Driss Aboutajdine

Abstract In recent years the wireless body area network (WBAN) technology has appeared as a subcategory of wireless sensor network (WSN) to facilitate and improve the quality of medical care. The reliability of transmissions and energy efficiency in WBAN are studied in this paper by using the Luby Transform (LT) code as an effective solution, for this we propose to evaluate its performances and adapt it according to the constraints imposed by WBAN. The validation results show that this code has not only the best performance but also the best energy efficiency.

Keywords WBANs · LT code · ARQ · Energy efficiency · Reliability

1 Introduction

In recent years the wireless body area network (WBAN) technology has appeared as a subcategory of wireless sensor networks (WSNs) to facilitate and improve the quality of medical care and remote monitoring. WBAN consist of small, portable, autonomous and intelligent electronic devices called sensor nodes, which are placed or implanted in the human body and connected together wirelessly. Their objective is to measure the patient's vital signs electrocardiogram (ECG), photoplethysmogram (PPG), electroencephalography (EEG), pulse rate, blood flow, pressure and temperature (see Fig. 1) and to transfer them to the Personal Device (PD) (named also sink).

N. Samouni (✉) · D. Aboutajdine
Faculty of Sciences, LRIT Associated Unit with CNRST,
Mohammed V University in Rabat, Rabat, Morocco
e-mail: nabila.samouni@gmail.com

D. Aboutajdine
e-mail: aboutaj@hotmail.com

A. Jilbab
ENSET, Mohammed V University in Rabat, Rabat, Morocco
e-mail: a_jilbab@yahoo.fr

Á. Rocha et al. (eds.), *Europe and MENA Cooperation Advances in Information and Communication Technologies*, Advances in Intelligent Systems and Computing 520, DOI 10.1007/978-3-319-46568-5_32

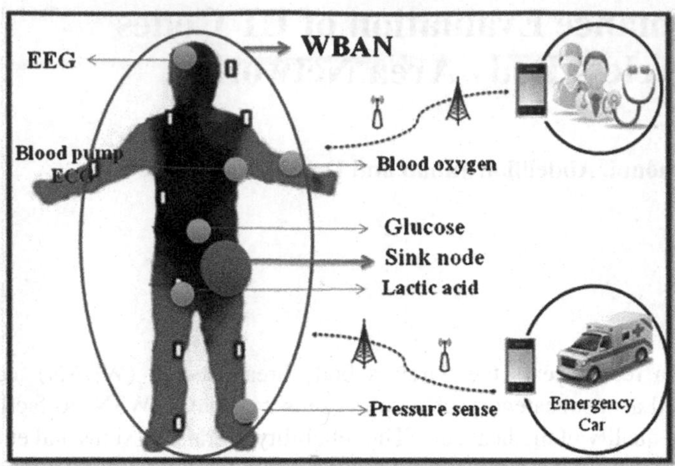

Fig. 1 A wireless body area network system

The objective of this last is collecting all information received from sensors node and transmitted it to interested doctor and provide him a real-time reporting [1].

Because of the specific constraints of the medical sensor networks, it is difficult to use the existing norms of traditional wireless sensors networks such as IEEE 802.15.4 (Zigbee) and IEEE 802.15.1. As a result, group of IEEE was launched in November 2007 known as IEEE 802.15.6 [1]. IEEE 802.15.6 define an optimized communication standard for WBANs which is a short range, low power and highly reliable data communication.

Despite the great success of WBANs, many problems still remain open. Take for example the problem related to the transmission channel, indeed its reliability is affected by noise and interference phenomena. A multiple reflections and distortions are highly present, while the patient requires continuous monitoring via a network that must deliver reports and alerts about their health in an effective manner. Whereas the reliability is the fundamental requisite of medical application. It is in this context that is part of our work, which aims to offer a reliable communication in WBAN. In traditional WSNs, to provide reliable data communication various encoding schemes [2], including Fountain codes, turbo codes and LDPC (Low Density Parity Check) codes etc. The most of them are good but just for some scenarios and applications in WSNs, also cannot be adopted for all fields. They were complicated to implement in energy-limited WBANs and very expensive in term of energy consumption that is one of the area requiring more attention in WBAN. Consequently, the data transmission must consider both low energy consumption and high delivery reliability. The reliability of transmissions and energy efficiency in WBAN are studied in this paper by using the Luby Transform (LT) code that is a subcategory of Fountain code as an effective solution, for this we propose to evaluate its performances by comparing it with Automatic Repeat reQuest (ARQ) in terms of energy consumption and bit error rate (BER).

After this introduction, we organize the rest of this paper as follows: Sect. 2, concentrates on a brief discussion and some existing works on error correction techniques especially in LT codes. Section 3 presents and compares the BER performances and energy consumption of ARQ and Fountain code in WBAN. Finally, in Sect. 4, we present a conclusion of this study.

2 Methodology

The requirements described in the preceding section indicate that WBAN must support the combination of low energy consumption and high delivery reliability. In this section a description of the two major error control techniques will be given. Further the principle and some exiting works about these techniques especially about LT code are discussed.

In view of reliable communication, several solutions have been proposed in the literature for WSN, and they can be grouped into two major error control modes one is Automatic Repeat reQuest (ARQ) and the other one is Forward error correction (FEC). The main idea of ARQ is to retransmit the packets received in error by the transmitter using an error detection codes to the data. In [3, 4], the energy efficiency has compared between ARQ and other error correcting such as BCH (Bose, Chaudhuri, and Hocquenghem) and convolutional codes for wireless sensor network, the results show that the ARQ has the best energy efficiency.

FEC employs error correcting codes which lets the receiver node to detect and correct errors in data packet if it existed, by adding parity bits (redundancy) to information packets before they are transmitted. Several codes have been investigated for error correction but the choice of error control schemes is very critical in the case of WBAN.

In this paper, the Fountain code is implemented, because it has numerous advantages. Firstly, this class can be implemented with far less complicated encoding and decoding algorithms, making such codes easy to be employed in modern communication systems. Secondly, it can automatically adapted with all channels, reduces the use of the return channel and it has an infinite yield [5]. In [6], it is shown that by applying Fountain code in WSN, wireless transmission efficiency and reliability can be dramatically improved. In [7, 8], the impact of Fountain code on power consumption in wireless sensor network was studied. The results show that this code not only has the best performance but also has the best energy efficiency. Among the known Fountain codes, three categories stand out: Random Linear Fountain (RLF), Luby Transform (LT), and Raptor codes. In the following we consider LT code in favor of its lower encoding/decoding complexity and we compared it with ARQ.

2.1 LT Codes

LT codes are the first practical realization of Fountain codes proposed by Luby [2] in 1998. They are rateless, i.e., the number of generated encoded packets are potentially limitless, and encoded symbols are generated on the fly.

Encoding LT

The principle encoding LT is to combine randomly fragments from K fragments to be transmitted. For this in a first step, we divide the information transmitted in K fragments of the same size. In a second step, we select randomly a degree $d_m \in \{1, \ldots, k\}$ according to the distribution $\Omega(x)$ and we assign the XOR of the chosen d information symbols to the encoding symbol.

Decoding LT

The process decoding is done iteratively by using the Belief Propagation algorithm, that is based on the fact that the degree of packet 1 may be considered decoded. Thus, using the previously decoded packet, the decoder iteratively reduces the amount of encoded packets, until all of the fragments are decoded.

3 Implementation and Results

In this section, we propose to evaluate the performances and energy efficiency provided by Luby Transform code and ARQ within Wireless Body Area Networks in order to determine the best performing and most suitable for medical application. The following simulation parameters were considered (Table 1).

Simulation experiments are performed using IEEE 802.15.6 standard and the performance evaluation holds in terms of energy consumption and BER. In the following, we present the effects of two systems peer-to-peer, one with channel coding using LT code and other without coding (ARQ) see Fig. 2.

Table 1 Simulation parameters

Parameter	Type or value
P_t Transmit power	11 dBm
R Transmission rate	20 kbit/s
F Frequency carrier	868 MHz
N_b number of bits per packet	200 octets (ARQ) 250 octets (LT code)
E_{Ele}	50 nJ/bit
E_{amp}	10 pJ/bit/m^2 or 0.0013 pJ/bit/m^4

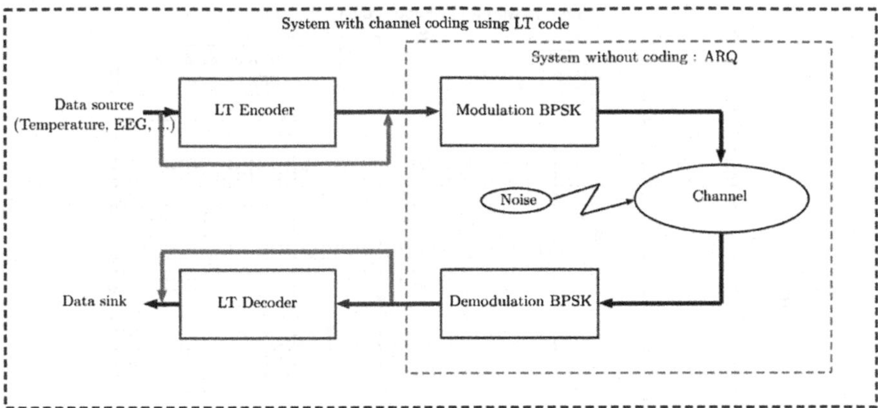

Fig. 2 Transmissions with and without coding

3.1 BER Test

This section was conducted to analyze the performances of LT code and ARQ in terms of BER. An AWGN (Additive White Gausian Noise) channel was assumed with variance N0/2 ($N_0 = -111$ dBm/Hz) and zero mean, using BPSK modulation. The results obtained are shown in Fig. 3, it can be observed that the BER decreases with increasing Signal-to-Noise Ratio (SNR), and on other hand, that the coding channel of the LT codes performs better than ARQ.

3.2 Energy Consumption

In this section, we first carried out a theoretical comparison of the energy consumed for LT code and ARQ. Figure 3 shows an example of the transmission scenario for two systems : with LT codes and without (ARQ) (Fig. 4).

With the ARQ mechanism, source (S) encapsulates each fragment in a data packet (DATA), and passes one after the other at node Destination (D). After sending each packet, it waits for a specific time (time out) to receive an acknowledgment. If S has not received the acknowledgment (ACK) in a given time interval, it retransmits the packet until it receives the acknowledgment. Note that the acknowledgment process is performed for each data packet individually and the process keeps repeating until the transmitter receives an ACK, or a specific number of retransmission is reached. In contrast, in case with LT code, each packet transmitted by S is encoded by the linear combinations XOR of a number of moieties selected from K. S sends consecutively an unlimited number of packets encoded to D without waiting for the acknowledgment for each packet Data (DATA) sent. When the recipient receives enough packets (about $K + \epsilon$) to decode correctly, it does send only a single packet to acknowledge

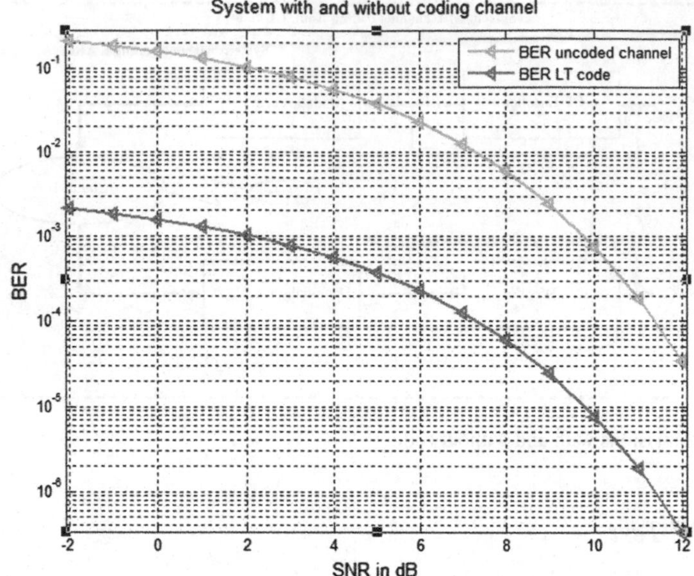

Fig. 3 Performance in terms of BER versus SNR for coding channel with LT code

Fig. 4 Comparison acknowledgment mechanisms for ARQ and LT code

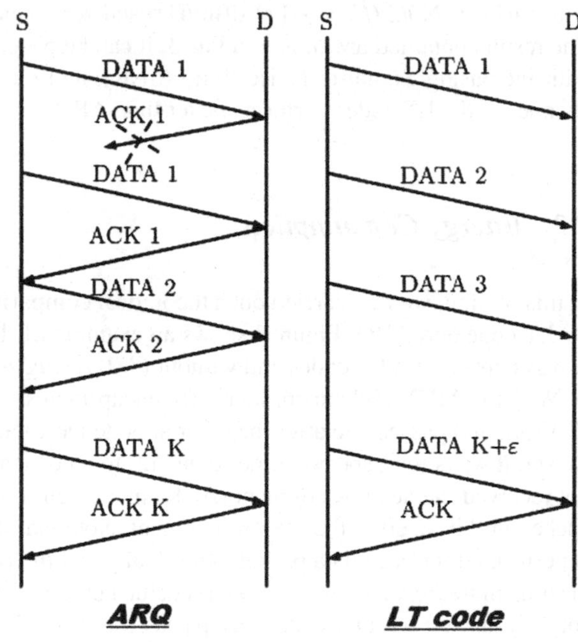

receipt of K fragments. The number of acknowledgment transmission attempts is reduced from K to 1 in relation to the previous case.

Energy Model for ARQ and LT Code

In general, the energy of communication represents the largest portion of the energy consumed by a sensor node and it is determined by the quantity of data (s) and transmission distance (d). According to [9], the energy consumed for the transmission, of one packet can be decomposed as:

$$E_p = E_T(s,d) + E_R(s) + E_{ack} \tag{1}$$

where E_T, E_R and E_{ack} is the energy consumed in transmission, receiver and acknowledgement respectively see Fig. 5, where E_{elec} and E_{amp} represent the energy of electronic transmission and amplification respectively.

The total energy consumed in the sensor node for communication with LT code to transmit K fragments of information is given by:

$$E_{LT} = (K+\varepsilon-1).\frac{1}{\gamma_{dt}}.\left(E_{T_{LT}}+E_{R_{LT}}\right)+\frac{1}{\gamma_{ack}}.\left(\frac{1}{\gamma_{dt}}.\left(E_{T_{LT}}+E_{R_{LT}}\right)+E_{ack}\right) \tag{2}$$

with γ_{dt} and γ_{ack} is the transmission probability and acknowledgement probability respectively.

$E_{T_{LT}} = E_T + E_{enc}$ and $E_{R_{LT}} = E_R + E_{dec}$ is the energy consumed in transmission and receiver in case with LT code respectively. E_{enc} and E_{dec} represent the energy used to encode and decode LT respectively LT. Consumption to the calculation process of these energies is small and can be neglected compared to the energy consumption required for the transmission radio.

For uncoded channel (ARQ), the total energy used to transmit K fragments of information can be expressed, in the form:

$$E_{T_{ARQ}} = K.\frac{1}{\gamma_{ack}}.\left(\left(\frac{1}{\gamma_{dt}}.\left(E_T+E_R\right)\right)+E_{ack}\right) \tag{3}$$

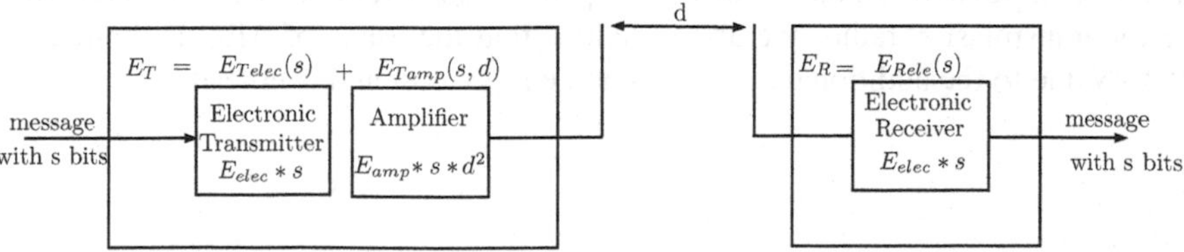

Fig. 5 Radio energy model

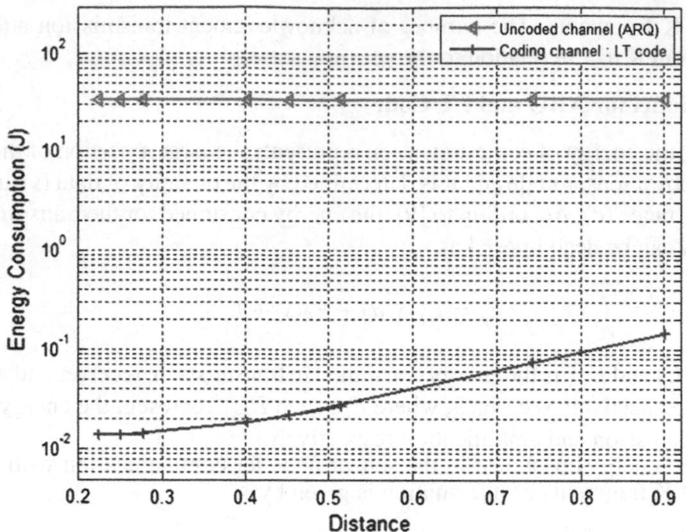

Fig. 6 Energy consumption as a function of the distance

To validate this theoretical comparison, we consider a WBAN deployed on the body of a person with height 1.6 m. The network is composed of 7 nodes (left palm, ECG, glucose, right palm, temperature, lactic acid, and EEG). The energy consumption of ARQ and LT codes, as a function of the distance with K = 1000 packets and the number of packets sent plotted in Figs. 6 and 7 respectively.

From Fig. 6 the results clearly show that the system using LT code has better energy efficiency than that using ARQ and consumes less resources. Also, the results displayed in Fig. 7 founds that the transmission with the LT code is less gourmand in energy.

Despite the number of data packets transmitted in case using LT code, the simulations results and theoretical comparison show that this code provides not only a good performance but also good energy efficiency. This happens because the sends of acknowledgment packets can occur for each packet, while for LT, it is just for the last packet which can be transmitted, as well as consumption to the calculation process is small and can be neglected compared to the energy consumption required for the transmission radio. We also conclude that the usage of ARQ is limited for WBAN due to the additional retransmission energy cost and overhead.

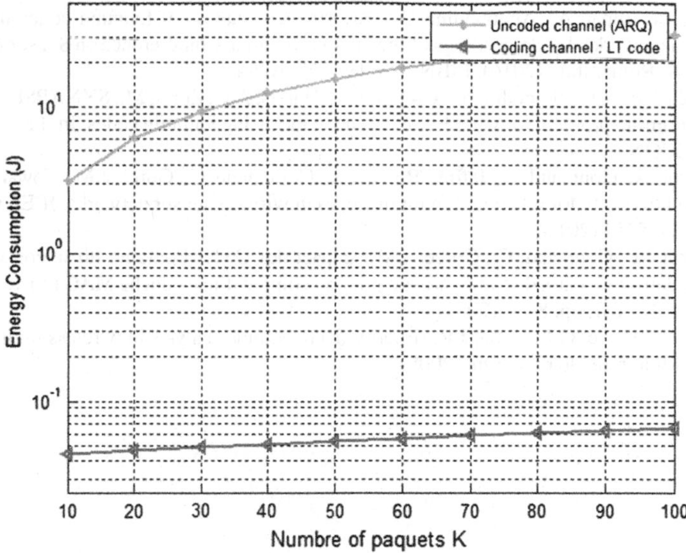

Fig. 7 Energy consumption as a function of the number of packets sent

4 Conclusion

It is well known that the reliability of transmission for application medical is very important because the reports and patient health alerts must delivered in a perfect manner. This paper has studied the reliability of transmissions and energy efficiency by using the LT code as a best solution, and compared it with ARQ scheme. We simulated and verified that LT code has the best performance not only in BER, but also in resources and energy consumption.

Due to the cooperative characteristic of the WBANs, our future work will be devoted to the study of cooperation between nodes to improve system performance.

References

1. Movassaghi, S., Abolhasan, M., Lipman, J., Smith, D., Jamalipour, A.: Wireless body area networks: a survey. IEEE Commun. Surv. Tutor. (2014)
2. Alrajeh, N.A., Marwat, A., Shams, B., Shah, S.S.H.: Error correcting codes in wireless sensor networks : an energy perspective. Appl. Math. **9**(2), 809–818 (2015)
3. Tian, T., Yuan, D.F., Liang, Q.Q.: Energy efficiency analysis of error control schemes in wireless sensor networks. In: International Wireless Communications and Mobile Computing Conference, IWCMC'08, pp. 401–405. IEEE (2008)
4. Shih, E., Cho, S.H., Ickes, N., Min, R., Sinha, A., Wang, A., Chandrakasan, A.: Physical layer driven protocol and algorithm design for energy-efficient wireless sensor networks. In: Proceedings of the ACM Mobicom, pp. 272–286. Rome, Italy (2001)

5. Samouni N., Jilbab A., Aboutajdine D.: Performance analysis of fountain codes in wireless body area networks. In: Proceedings of the 5th International Confererence on Sensor Networks, pp. 45–49. Rome, Italy (2016). ISBN:978-989-758-169-4
6. Rossi, M., Zanca, G., Stabellini, L., Crepaldi, R., Harris, A.F., Zorzi, M.: SYNAPSE: a network reprogramming protocol for wireless sensor networks using fountain codes. In: Proceedings of IEEE SECON (2008)
7. Apavatjrut, A., Goursaud, C., Jaffres-Runser, K., Comaniciu, C., Gorce, J.M.: Toward increasing packet diversity for relaying lt fountain codes in wireless sensor networks. IEEE Commun. Lett. **15**(1), 5254 (2011)
8. Apavatjrut, A., Goursaud, C., Gorce, J.M.: Impact des Codes Fontaine sur la consommation d'nergie dans les rseaux de capteurs avec prise en compte d'une couche MAC raliste. In XXIIe colloque GRETSI, pp. 8–11 (2009)
9. Ruifeng, Z., Marie, G.J., Katia, J.R.: Energy-delay bounds analysis in wireless multi-hop networks with unreliable radio link (2008)

Wireless Sensor Networks in Biomedical: Wireless Body Area Networks

Bahae Abidi, Abdelillah Jilbab and Mohamed E.L. Haziti

Abstract A wireless sensor networks can be used in several applications to monitor certain parameters like smart homes, agriculture field monitoring and healt care monitoring. In this work, we propose a routing protocol for wireless body area network, to transfer data with minimum energy consumption and longer network life time, we use a multi-hop topology and we evaluate the performance of our approach by simulations.

Keywords Wireless sensor networks · Wireless body area networks · Multi-hop · Routing · Clustering · Stability · Energy consumption

1 Introduction

A wireless Sensor Network (WSN) [1] refers to a group of dispersed nodes dedicated for monitoring and recording the physical conditions of the environment and organizing the collected data at a central location. Every nodes is connected to one or sometimes to several sensors depending of the topology of the network, each of which is small, lightweight and portable. A sensor network has typically several parts.

The four parts which forms the sensor node are sensing unit, processing unit, communication unit and power supply unit. The deployment of the sensor node can be random, regular or mobile to produce a high-quality of information in WSN. To send the captured data from differents nodes at the base station, the network use

B. Abidi (✉)
Faculty of Sciences, LRIT Associated Unit with CNRST (URAC No 29),
Mohammed V University in Rabat, Rabat, Morocco
e-mail: bahae.abidi@gmail.com

A. Jilbab
ENSET, Mohammed V University in Rabat, Rabat, Morocco

M.E. Haziti
Higher School of Technology, Salé, Morocco

Á. Rocha et al. (eds.), *Europe and MENA Cooperation Advances in Information and Communication Technologies*, Advances in Intelligent Systems and Computing 520, DOI 10.1007/978-3-319-46568-5_33

a routing protocol. Routing [2] is forwarding data from source to destination, the route between both extremities is determined by many techniques relatively to the application field. Many routing protocol developed for wireless sensor networks, it's a system of digital rules for exchanging data between all the nodes in the network and achieves it to the base station with an energy efficient manner.

The available types of sensors [3], their small sizes and the decrease in cost of micro-sensors have expanded the domains of application of sensor networks. The potential application of sensor networks include differents domains: industrial automation, automated and smart homes, video surveillance, traffic monitoring, air traffic control, robot control, Medical device monitoring.

The application in a medical context requirer operation at low power consumption. This is one of the most important constraints, because each node must operate at reasonable temperatures. Another constraint is related to the quality of information sent by the network, indeed the patient's state is primordial and any error can be harmful. Biomedical sensors are used solely in health monitoring applications, this would have a radical impact on the quality of life of patients and treatment success rates.

Wireless body area networks (WBANs) are a new generation of wireless sensor networks, they are new emerging sub-field of wireless sensor network, where wireless sensors are placed on the human body, the use of wireless body area network technology reduces the expenditures of patient in hospital.

The paper is organized in following order. In Sect. 2, we provide an introduction of wireless body area network, while Sect. 3 deals with our motivation, Sect. 4 describes the proposed protocol. Performance and simulation results are presented in Sect. 5. Finally, Sect. 6 gives conclusion.

2 Wireless Body Area Network

Wireless Body Area Network (WBAN) [4], a new generation of wireless sensor networks. These networks are composed of tiny biomedical nodes, dedicated to ensure a continous monitoring of vital parameters of patients. A Wireless body area network [5] consists of low power devices operating on, in or around the human body, to serve a variety of application including medical. Although WSNs and WBANs share many difficulties, such as miniaturization. The main challenges in terms of research remain in the biomedical and healthcare monitoring application. Indeed, the evolution of WBANs [6] should follow the increasing development in the medical domain. Its main objective being to ensure constant monitoring of patient at home or at work.

The WBAN [7] can be medical or non- medical. Medical WBAN can be relegated as wearable WBAN and implantable WBAN, and we have three types of nodes used:

- Implant node: Implantable nodes are those nodes which are placed inside the human body just below the sink.
- Body surface node: Body surface node is those nodes which are placed on the surface of the human skin.
- External node: External nodes do not have any contact with the human skin.

Multiple sensors communicate with a mobile phone using wireless interfaces forming a WBAN, these sensors are used to sense, process and transmit the vital sign of the human body and biological informations to the sink node in real time such as, pressure, temperature, electromyogramme (EMG), Glycmie and electrocardiogram (ECG). Doctors can then access in real time to the data collected by the sink node and do the treatment according to the requirement (Fig. 1).

The WBAN use the IEEE 802.15.6 [9]. The IEEE 802.15.6 working group had a number of success stories in the realization of the international standarization for WBAN, it established the first draft of the communication standard of WBAN, optimized for low power on or in body nodes.

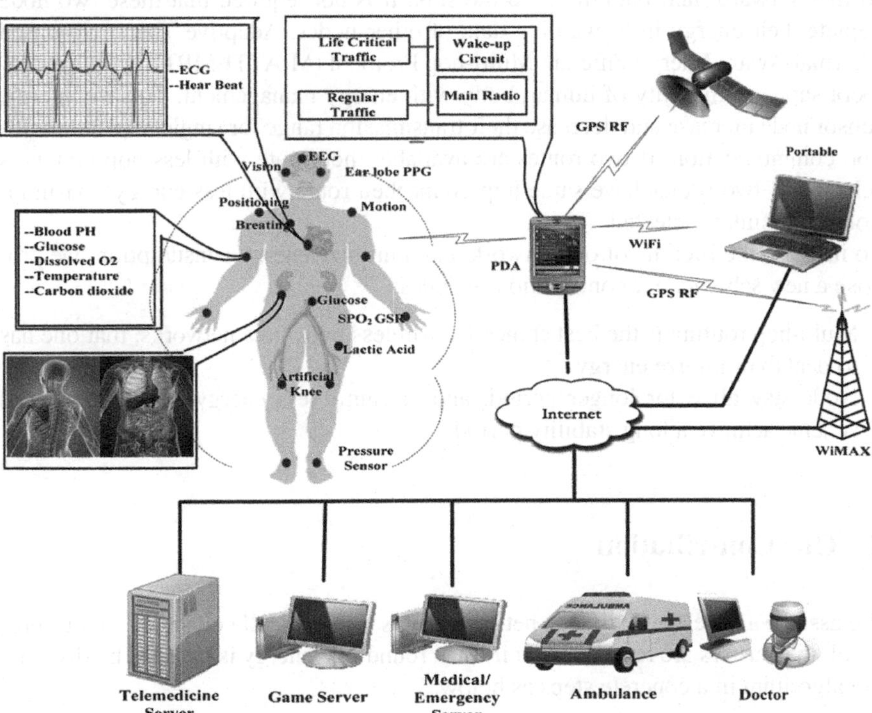

Fig. 1 WBAN: wireless body area network

3 Motivation

The Energy consumption in WBANs is crucial [8], especially in implanted biosensors, since they are inaccessible and difficult to replace. So, different energy efficient routing schemes are used to forward data from body sensor to medical server. In thermal aware routing protocol, where each node selects the minimum hop rout to the sink is based on thermal heat nodes, which can change another optimal route. The Cascading Information retrieval by Controlling Access with Distributed slot Assignment (CICADA) routing protocol employs a spanning tree structure [10]. Another clustering based routing protocol known as Anybody [11] has features to restrict the sensor nodes to transmit directly to the sink, which affects the CHs selection mechanism . Also Quwaider and Biswas [12] proposed a delay tolerant protocol. Stable Increased- throughput Multi-hop protocol for Link Efficiency in wireless body area network (SIMPLE) [13] is a novel routing protocol for WBANs, he use a cost function to select appropriate route to sink. Cost function is calculated based on the residual energy of nodes and their distance from sink. Nodes with less value of cost function are elected as parent node. Other nodes become the children of that parent node and forward their data to parent node. Two others nodes has critical and important medical data (ECG and Glucose monitoring), they are placed near to the sink, so they forward their data direct to the sink. It is not required that these two node deplete their energy in forwarding data of other nodes. Adaptive Threshold-based Thermal-Aware Energy-efficient Multi-hop Protocol (M-ATTEMPT) [14], this protocol supports mobility of human body with energy management. To save energy, sensor node increase and decrease their transmission range for single hop and multi-hop communication. If two routes are available then route with less hop-counts is selected, if two routes have same hop-count then route with less energy consumption to the sink is selected.
To increase the lifetime of our network and minimize energy consumption, we propose a new scheme, our contribution includes:

- Multi-hop routing is the best choice for wireless body area networks, that one has to deal to minimize energy.
- Node stay alive for longer period, and consume less energy, so our proposed scheme achieve a long stability period.

4 Our Contribution

We assume a wireless body area network that is hierarchically clustered. In our protocol, the clusters are reestablished in each round, the energy is reduced by dividing the algorithm in a concrete steps as below:

4.1 Deployment of Sensor

The first step is plotting the sensor node in the network randomly, we have 8 sensor nodes implanted in the human body with an initial energy Eo.

4.2 Gateway Node Election

The nodes are dispersed randomly in the human body, Eo is the initial energy of normal node. A gateway node, is a node who routes data to base station. The node chance of being a gateway node is evaluated to it's physical characteristics, such as the internal energy and the distance between the node and base station. The selection of gateway node consists, firstly to calculate the distance between all nodes and base station, store them in a specific array, choose the appropriate one according to mentioned features.

4.3 Clustering Process

The first step in clustering process, is to elect a cluster head, then the cluster formation. All nodes except the gateway node will participate in the formation of clusters. In every round, each sensor node generate a random number between 0 and 1 to determinate if node will become cluster head or not, then compares this number with the value of the threshold given by Eq. (1), if the number chosen by each node is less than the threshold value, the node become a cluster head for current round.

$$T(n) = \begin{cases} \frac{P}{1-P\times(r \bmod \frac{1}{P})} & n \in G \\ 0 & \text{otherwise} \end{cases} \qquad \text{equation} \quad (1)$$

where P = K/n, k is the expected number of CHs in the round, n is the number of nodes in the network. The value r is the round number. G is the set of nodes that have not been a CH in the last (r mod(N/k)) rounds. After election of cluster head each normal nodes should to connect with the nearest cluster-head, to construct the clusters. The cluster-head aggregate the received data and send it to the appropriate gateway node.

4.4 Connects Each CH with the Appropriate Gateway Node

Knowing that the gateway node will route all data to base station, each cluster-head (CH) must to connect to the appropriate gateway node. In order to consume less energy, each cluster-head try to connect to the nearest gateway node. So, the distance between the cluster-head and gateway node are calculated, if the distance between gateway node and cluster-head is smaller then the distance between Cluster-head and base station, the gateway node will be adopted.

4.5 Collection and Transmission

To reduce energy consumption, a multi-hop communication was adopted, the data was send to the sink through the gateway node, the key idea is to combine data from differents sensors to eliminate redundant transmissions and reduce the number of data transmission, and this according to our topologie, because a large part of the energy is lost during the transmission and reception.

In our concept, node will transmit data, if and only if sensed value is greater than the current value. So, nodes will transmit data, if difference between currently sensed value and the value stored is equal or greater then the value stored in the node. According to this concept, cluster members nodes collect data, transmits to their cluster-head nodes. Each cluster head fuse the receiving data from all its cluster members, then send it to its gateway node. The main task for gateway node is to send this receiving data to base station. Each node will check these energetics remains after accomplish its transmission assignment, if energy is above a 0, the iterations continue and all alive nodes can execute the step above again.

5 Simulation and Results

The performance of our routing protocol are evaluated by comparing these results with that of SIMPLE and ATTEMPT, protocols cited in Sect. 3. The performance evaluation holds in terms of energy consumption via simulation using IEEE 802.15.6 Standard, with 8 sensor nodes implanted in the human body. The following simulation parameters were considered (Table 1).

To compare these protocols, the performance below are taken into consideration:

5.1 No of Dead Nodes

The network life time represents the total network operation time till the last node die, dead node is the node in the network which has an energy remaining. For that we use the first node die per round to measure the lifetime of our network. The simulation show that the first node die for SIMPLE protocol die in round 4437, for ATTEMPT these nodes die early, on the other side, for our proposed protocol the first node die untill the round 5596 (Fig. 2).

Table 1 Simulation parameters

Simulation parameters	
Parameter	Value
Size of network	1 × 0.6
Number of nodes	8
Deployment	Randomly
Sink location	1 × 0.25
Initial energy	0.5
Number of round	10000

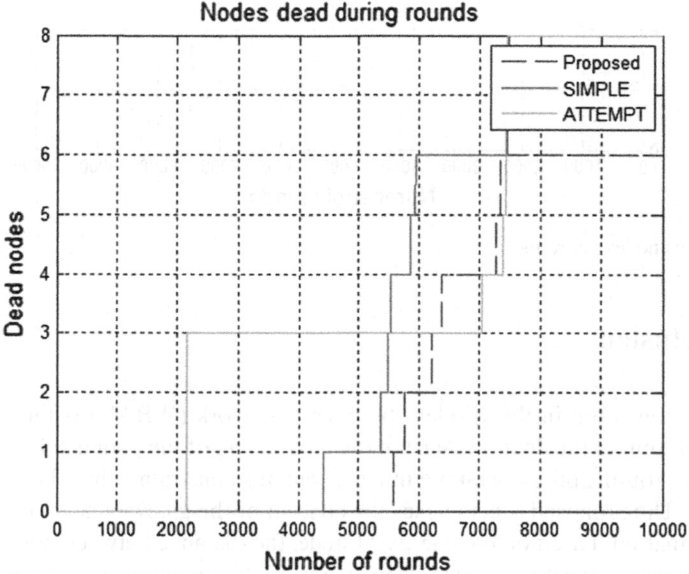

Fig. 2 First dead nodes per rounds

5.2 *No of Alive Nodes*

The second metric adopted is the number of alives nodes, the stability period is the time in which all node in network are alives and by analyzing the number of alives node, we decide the network life time. Figure 3 show the number of alives nodes per rounds for the 3 protocols. Our protocol and untill the round 5500 keep all the 8 nodes alives, then in round 5596 the first node die. Therefor based on Fig. 3, our protocol is better compared to others protocols.

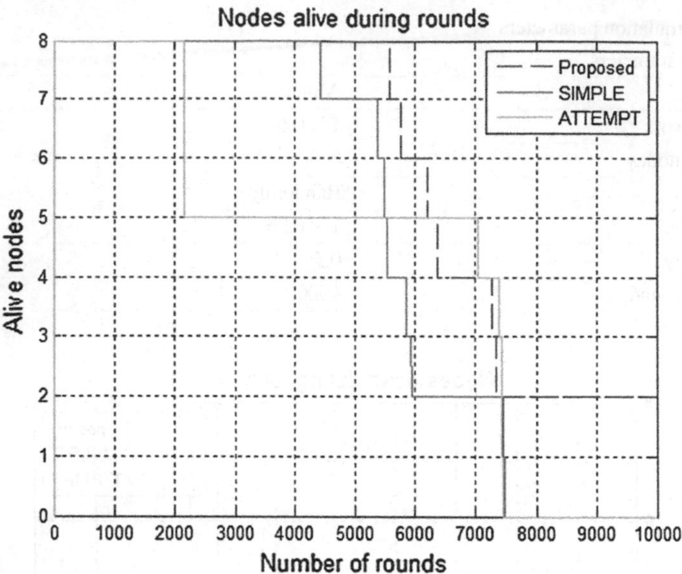

Fig. 3 Alives nodes per rounds

6 Conclusion

The major constraint in the wireless body area network (WBAN) is the quality of information sent to the doctor to ensure the monitoring of the patients. In this paper, we present a routing protocol taking into account this constraint when data are routed in WBAN. The proposed scheme uses the concept of the gateway node to route data to the destination, based on the energy of node, the distance between node and also with the easier aggregation used to keep energy, maintain quality of information and to avoid a redundant data. Our simulation results shows that the proposed routing protocol improve the network stability and lifetime. It can be concluded that our protocol will perform in large as well as in small sized networks.

References

1. Singh, S., Chauhan, A.K., Raghav, S., Tyagi, V., Johri, S.: Heterogeneous protocols for increasing the life time of wireless sensor networks. J. Global Res. Comput. Sci. **2**(4), 172–176 (2011)
2. Almobaideen, W., Hushaidan, K., Sleit, A.: and M. Qatawneh : A cluster based approach for supporting qos in mobile adhoc networks. Int. J. Digit. Content Technol. Appl. **5**(1), 1–9 (2011)
3. Liu, X.: A survey on clustering routing protocols in wireless sensor networks. Sensors (2012). ISSN:1424-8220

4. Quwaider, M., Biswas, S.: DTN routing in body sensor networks with dynamic postural partitioning. Ad Hoc Netw. **8** (2010)
5. Ullah, S., Higgins, H., Braem, B., Latre, B., Blondia, C., Moerman, I., Saleem, S., Rahman, Z., Kwak, K.: A comprehensive survey of wireless body area networks. J. Med. Syst., 130 (2010)
6. Chen, M., Gonzalez, S., Vasilakos, A., Cao, H., Leung, V.: Body area networks: a survey. Mobile Netw. Appl. **16**(2), 171193 (2011)
7. Latr, B., Braem, B., Moerman, I., Blondia, C., Demeester, P.: A survey on wireless body area networks. Wirel. Netw. **17**, 118 (2011)
8. Movassaghi, S., Abolhasan, M.: Wireless body area networks: a survey. IEEE Commun. Surv. Tutor. **16.3**, 1658–1686 (2014)
9. Kwak, K.S., Ullah, S., Ullah, N.: An overview of IEEE 802.15.6 standard. In: International Symposium on Applied Sciences in Biomedical and Communication Technologies (ISABEL), 3rd edn, pp. 1–6 (2010)
10. Latre, B., Braem, B., Moerman, I., Blondia, C., Reusens, E., Joseph, W., Demeester, P.: A low-delay protocol for multihop wireless body area networks. In: Proceedings of the 4th Annual International Conference on Mobile and Ubiquitous Systems: Networking and Services, Philadelphia, pp. 1–8, 6–10 Aug 2007
11. Watteyne, T., Auge-Blum, I., Dohler, M., Barthel, D.: Anybody: A self-organization protocol for body area networks. In: Proceedings of the ICST 2nd International Conference on Body Area Networks, pp: 1–6, Belgium, 11–13 June 2007
12. Tsouri, G.R., Sapio, A., Wilczewsk, J.: An investigation into relaying of creeping waves for reliable low-power body sensor networking. In: IEEE Trans. Biomed. Circuits Syst. **5**, 307–319 (2011)
13. Javaid, N., Abbas, Z., Fareed, M., Khan, Z., Alrajeh, N.: M-attempt: a new energy-efficient routing protocol for wireless body area sensor networks. In: The 4th International Conference on Ambient Systems, Networks and Technologies, vol. 19, pp. 224–231 (2013)
14. Javaid, N., Abbas, Z., Fareed, M.S., Khan, Z.A., Alrajeh, N.: M-ATTEMPT: a new energy-efficicent routing protocol for wireless body area sensor networks. In: The 4th International Conference on Ambient Systems, Networks and Technologies (2013)

Information Technology Governance in Public Sector Organizations

Amine Laita and Mustapha Belaissaoui

Abstract There is a huge paucity of empirical researches on Information Technology Governance (ITG) in public sector organizations even with the noticeable growing importance of IT in the public sector through several IT projects. ITG happens to be one of the most important levels of governance. It is an important approach to structure process through relational mechanisms. There is little evidence that success in implementing ITG reforms leads to morerapid and inclusive economics. This paper aims to study the importance of ITG in public sector organizations. The public sector is considered to be a set of organizations with the mission to serve citizens. Furthermore, this paper is going to come up with the answer to an essential question which is if ITG should rely on stable functions of a public sector organization, or should it consider broader and more evolving objectives touching the whole government?

Keywords IT · Governance · Public sector · Framework · Standards

1 Introduction

This work finds its motivation in the scarcity of empirical research concerning ITG in the public sector. There is a noticeable presence of information system projects in the public sector and in other projects dedicated in this sense. Additionally governance happens to be an important variable that structures the process through relational mechanisms. ITG is an important is a part that starts to gain more and more important place in governance pyramid.

ITG can be defines from 3 point of view:

A. Laita (✉) · M. Belaissaoui
SIAD Laboratory, ENCG, Hassan I University, Settat, Morocco
e-mail: Amine.laita.sg@gmail.com

M. Belaissaoui
e-mail: mbelaissaoui@gmail.com

Á. Rocha et al. (eds.), *Europe and MENA Cooperation Advances in Information and Communication Technologies*, Advances in Intelligent Systems and Computing 520, DOI 10.1007/978-3-319-46568-5_34

- Entreprise "an integral part of enterprise governance and consists of the leadership and organizational structures and processes that ensure the enterprise's IT sustains and extends the organization's strategies and objectives" [1].
- IT and business fusion "IT governance is the organizational capacity exercised by the board, executive management and IT management to control the formulation and implementation of IT strategy and in this way ensuring the fusion of business and IT" [2].
- Decision-making "IT governance is defined as specifying the decision rights and accountability frameworks to encourage desirable behavior in using IT" [3].

The purpose of this paper is to discuss the importance of ITG in the public sector. Nonetheless public sector is gathering of organizations aiming to serve citizens. In this sense, this paper is going to answer a question of huge importance to see if ITG should rely on stable functions of a public sector organization, or should it consider broader and more evolving objectives touching the whole government.

2 Research Gap

Which ITG framework to be implemented for an effective ITG in public sector organizations.

Should ITG rely on stable functions of a public sector organization, or should it consider broader and more evolving objectives touching the whole government?

- What is the level of involvement of ITG in public sector organizations?
- What is the effective approach to be used in public sector organizations (Local or global)?
- At what level the IT objectives should be aligned with public sector objectives?
- What added value does IT bring to public sector?
- At what level external relations influence decision making mechanisms?

3 Literature Review

IT governance (also termed Information Systems (IS) governance) consists of the leadership, organisational structures and processes that ensure that an organisation's IT sustains and extends its strategies and objectives [4]. IT governance aims to ensure that the expectations and achievements from IT are matched, and that the risks associated with IT are controlled. In particular IT governance focuses on the strategic alignment between an organisation's use of IT and achievement of its business goals and objectives, an issue which is also important in public sector organizations. As IS is positioned within organisational settings and involves people, IT (IS) governance considers much broader issues than technology. These issues include policy, planning, culture, training and change management. As it is now well accepted that poor IT governance is the major explanation for failure to achieve the goals from IT-related projects.

In a survey conducted by the IT governance Institute [5] it was found that the top ten management problems include inadequate view on how well IT is performing, non-alignment between IT and business strategies and the higher cost of IT with low return on investment.

The need for effective IT governance is also becoming essential in the public sector, in which increasingly we see a variety of fragmented IT initiatives and activities, with loss of synergies and exploitation of economies of scale. The need is equally amplified by, alongside resources constraints, relatively lower IT literacy, culture and leadership, basic citizens competing needs and priorities. Also on the fact that the sector is complex and its effectiveness is characterized by a need for intra and inter-organizational synergies that call for common and effective strategies, services, communication, collaboration and accountability to multiple stakeholders [3]. These problematic governance related issues in the sector stem from IT strategic planning and implementation to management, support and monitoring [1, 5]. As a result effective integration of ICT and indeed ICT-enabled transformational government in these environments needs consistent strategic alignment of IT and business goals in order to increase efficiency in public service delivery and meet stakeholders' expectations [4] (Fig. 1).

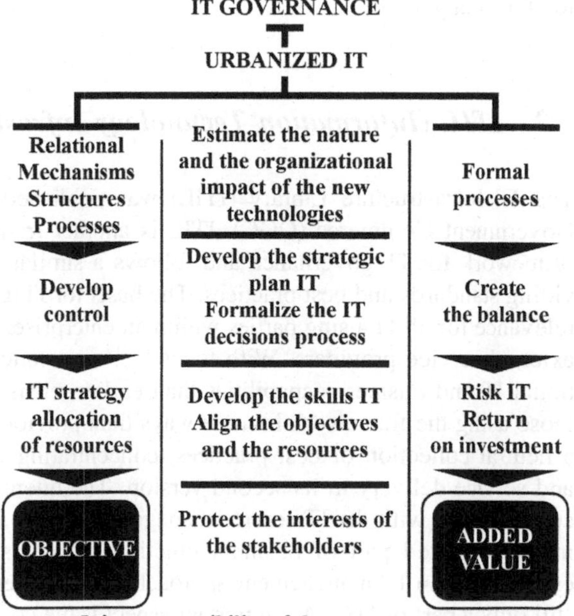

Fig. 1 Concepts and definition of the ITG. Laita (2016)

4 ITG Standards and Frameworks

4.1 *COBIT (Control Objectives for Information and Technology)*

COBIT is one of the most commonly accepted systematic approaches for ITG. The framework provides process descriptions and resents activities in a manageable and clear structure. Its mission is "…to research, develop, publicize and promote an authoritative, up-to-date, internationally accepted IT governance control framework for adoption by enterprises and day-to-day use by business managers, IT professionals and assurance professionals." [1] Johannsen and Goeken justify the strong and increasing public interest in COBIT with its ability to bridge the well known gap between a company's business interests and its IT.

The COBIT framework has been developed by the Information Systems Audit and Control Association (ISACA), and the ITGI and was published in its first version in 1996, followed by the second edition in 1998, and the third edition in 2000, and the fourth edition in May 2007. The latest version 5 was released in June 2012.

IT is not surprising, when historically looking at the development of COBIT by the audit and control association ISACA, that COBIT clearly focuses on IT controls. One of the primary purposes is o help defining goals for strategic alignment and performance measurement, but also to provide metrics an practices for risk management and performance measurement. The ITGI defines resource management as a separate domain, which is an integral part of the Value Creation domain for this analysis [6].

4.2 *ITIL (Information Technology Infrastructure Library)*

The IT Infrastructure Library (ITIL) was published by the UK-based Office of Government Commerce (OGC), ITIL is another frequently in literature discussed framework for IT governance and follows a similar concept like COBIT by providing standards and best practices. The basis for ITIL V3's success is its operating relevance for all IT-using parties within an enterprise, from small IT departments to external service providers. With focus "… on a much broader range of organizational IT and business capabilities than earlier releases, this new version will help those using the framework in more ways than previously. Historically seen, ITIL is a neutral collection of best practices, concentrating especially on service support and service delivery in its second version. The intention of ITIL is to enhance the compatibility with the IT service management norm ISO. IT service has become a more integrated part of business function, which is why ITILV3 now supports establishing an IT management approach. Nevertheless, IT service Management is still only a part of ITG. An important aspect is the consideration of IT as a business itself, rather than treating it as a separate function. Some goals and valuable highlights of this de facto-standard for IT service management are:"

- Alignment exclusively towards business usage;
- Primary focus on the Service Life Cycle processes as a second priority;
- Support for the fulfillment of the compliance requirements (SOX, Basel ….)
- Basis for Balanced Scorecard;
- Learning organization at the centre of interest;
- Coordination with the ISO/IEC 20000 standard;
- Agile and adjustable Service Design;
- Assistance with the management of Service Providers;
- Improved measurability and traceability of real added values.

4.3 ISO/IEC 17799 (International Organization for Standardization and the International Electrotechnical Commission)

The ISO 27002 standard is titled "Information Technology—Security Techniques-Code of Practice for information security management" and has replaced the former ISO 17799 standard in july 2005. The name has changed due to its international acceptance while contents remained the same. For the following proceeding both labeling will be used with the same contextual meaning.

ISO 27002 belongs to the family of the ISO 27000 series of standards and is closely related to the ISO 27001 standard. The latter provides a specification for an information security management system that intents" … to serve as a single reference point for identifying the range of controls needed for most situations where information systems are used in industry and commerce." It is a code of practice for information security that outlines potential control and control mechanisms. In comparison, the ISO 27002 standard "…established guidelines and general principles for initiating, implementing, maintaining and improving information security management for an organization." This standards defines information security policy as the managerial alignment of security issues by integrating the management for an organization-wide security policy [Müll2003, 3].

It also treats information like a valuable business asset for the organization that constantly calls for protection. Therefore, the main goal is to ensure confidentiality, integrity and availability of critical information. This protected information ensures business continuity, minimized business damage, and maximized return on investments and capitalizing on business opportunities.

4.4 Val IT

The availability of information transforms capital investment decisions into business decisions based on the probability of alternative strategic assumptions. From the opposite perspective, an organization's business decisions depend on investment decisions. Since IT governance has changed the role of IT to an integrated organizational part, measuring and managing IT investments have become a more difficult business matter. In early times of simple IT management, a calculated budget was distributed to an IT project with responsibility for IT investment. In 2006, the lack of investment and management structures has resulted in the Val IT initiative by the ITGI due to company-wide IT integration.

Val IT is based on the COBIT framework and focuses on investment decisions and the realization of its benefits, while COBIT focuses on the implementation of demand processes. It extends and complements COBIT from both the business and the financial perspective with the purpose of creating real business value from IT-enabled investments. Where COBIT provides a framework for the means of creating value, Val IT provides guidance on meeting the end.

IT governance includes leadership and commitment from the top management. In contrast to ITIL, for example, this framework does not focus on operating processes rather than on top level decision making. The Val IT framework specifically provides guidance for executives in order to help understand their roles in business investments. If managed well within effective IT governance, the Val IT initiative provides significant opportunities to create value.

Albeit the *ITGI Global Status Report 2008* still reports occurring problems with applying Val IT, 50 % of the respondents plan to apply this framework, but are not familiar with the brand itself. The major obstacles to adopting the framework's principles include uncertainly regarding the return on investment and lacking and experience [7].

4.5 Discussion About IT Governance Standards

Frameworks like COBIT and ITIL tend to result in descriptions of what to do. In comparison, frameworks with high abstraction levels tend to offer more detailed descriptions on how to execute activities for improving IT governance. From this point of view, mapping two frameworks facilitates and extends solving of problems in certain cluster. If, for example, both frameworks are control-oriented, efficiency could still increase by offering a more detailed description of processes. In other words, none of the criteria are mutually exclusive, if another one fits.

Table 1 A summary of differences between sectors

Attribute/factors	Sector			
	Public		Private	
	Public service	Semi-government	Non-profit	Private
Goals	Multiple and intangible	Multiple and tangible	Multiple	Specific and tangible
product	Provide services and public goods	Sell services	Provide services	Profit
Achievement measured by	Political efficiency and achieving policy mission	Sustainability of service provision	Achieving mission	Financial profitability and efficiency
Environmental	Less incentives for productivity	May have more incentives than government	No incentives, uses volunteers	More incentives
	More legal and formal constraints —red tape	Less formal constraints	No red tape	No red tape
	Political influences	Some political and market influences	Free of influences	Market influences
Proprietary versus shared IT	Shares IT resources, applications and technical help	IT is proprietary to give an edge	Lacks in sharing of resources	Treats IT as proprietary to stay ahead and competitive

Sources Caudle et al. [3], Kraemer and Dedrick (1996), Dawes et al. (2004), DCITA (2005)

5 Difference Between Public and Private Sector

See Table 1.

6 IT Governance Versus IT Management

See Fig. 2.

Fig. 2 Governance versus management

Governance	Management
External and internal focus	Internal focus
Whole-of-organization	Departments & individuals
Future	Present
Strategic	Operations & projects
Benefit realization	Cost & quality
Wise investment	Budget accountability
Delegation	Hands-on

7 ITG Conceptual Framework for Public Sector Organizations

Be in the public or private sector, IT governance can be deployed using a combination of processes, structures and relational mechanisms. Processes could be monitoring, decision-making, service level agreements (SLAs), balanced IT scorecards; structures may include IT councils, committees (like IT strategy committee, IT steering committee); while mechanisms could be business partnerships, shared learning, stakeholder participation and collaboration between functional areas or workgroups. framework. Each aspect is indispensable to successful IT governance [8–10] (Fig. 3)

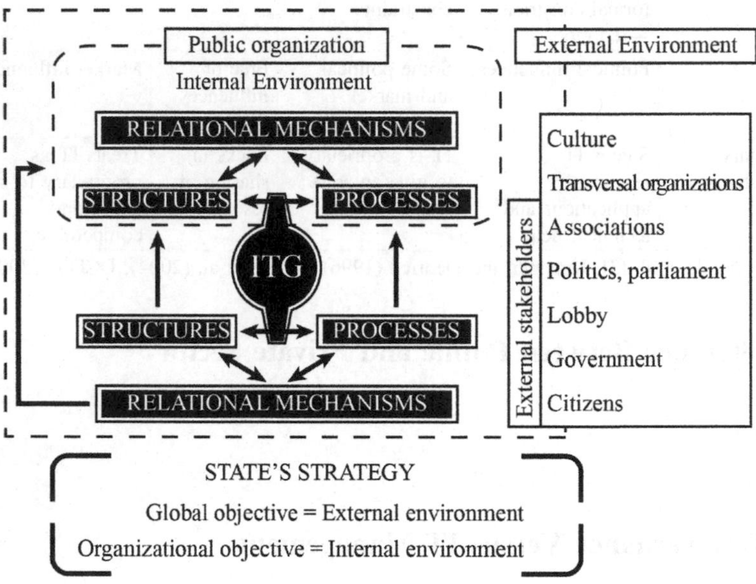

Fig. 3 Conceptuel framework: IT governance in public sector. Laita (2016)

8 Conclusion

Bozeman and Bretschneider (1986) first hypothesized that the differences between the public and private sectors require different principles in the fundamental management of organizational information systems. It is imperative that further research is conducted to capture and better understand these fundamental differences, even as they relate to IT governance. Evidently, a 'one-size-fits-all' approach is not appropriate when studying the two sectors, and failure to address the differences will be 'a mistake' [11] when studying IT governance. Acknowledging the scarcity of empirical research done in this area, further studies are clearly needed to establish the IT governance approaches that work best in a public sector context and whether the adopted approach depends on the functions of a government agency. A study on the contribution of IT governance to service delivery in government will be another important area to investigate, as well as the extent to which IT is aligned with the objectives of different government agencies. Another possible area of research could be investigating what threat is posed by shrinking IT funds to IT governance in the public sector, the influence this might have on service delivery and possibly how it can be avoided.

It is also necessary to examine organizational activities and the mechanisms necessary for effective implementation of IT governance in the public sector. Subsequent research could replicate prior studies from the private sector in the public sector, and thereby provide empirical evidence for the differences between IT governance in the two sectors as discussed in this paper. Also, an investigation could reframe the underlying IT governance theories and develop alternatives to a public service organization. It is hoped that this paper has highlighted some of the significant differences between the public and private sector, which are pertinent to consider when addressing IT governance. Hopefully, the issues raised will provide motivation for empirical research to examine what is currently an under researched area in IT governance.

References

1. IT Governance Institute. http://www.itgi.org
2. Van Grembergen, W.: Introduction to the minitrack IT Governance and its Mechansims, Proceedings of the 35th Hawaii International Conference on System Sciences (HICSS) (2002)
3. Weill, P., Woodham, R.: Don't Just Lead, Govern: Implementing Effective IT Governance, CISR Work. P. no. 326 (2002)
4. Van Grembergen, W., De Haes, S., Guldentops, E.: Structures, processes and relational mechanisms for information technology governance: theories and practices, in Strategies for Information Technology Governance. In: Van Grembergen (ed.) Idea Group Publ. (2004)
5. IT Governance Institute and PricewaterhouseCoopers 2006 IT Governance in Practice, on-line available at http://www.pwc.com

6. Entreprise Governance of Informtion Technology : Achieving Alignment and Value, Featuring COBIT 5, Steven De Haes and Wim Van Grembergen, Springer (2015) ISBN 978-3-319-14546-4
7. Christian Häfner : Building a Framework for an efficient IT Governance. Grin Verlag gmbh (2008) ISBN : 978-3-640-61282-6
8. Campbell, J., McDonald, C., Sethibe, T.: Public and private sector IT governance identifying contextual differences Faculty of Information Sciences and Engineering University of Canberra, Bruce ACT 2601 (2009)
9. Weill, P., Woodham, R.: 'Don't just lead, govern: Implementing effective IT governance', CISR Working Paper, April 2002, http://dspace.mit.edu/handle/1721.1/1846 (2002)
10. De Haes, S., Van Grembergen, W.: 'IT governance and its mechanisms', Information Systems Control Journal, vol. 1. (2004)
11. Khalfan, A., Gough, T.G.: 'Comparative analysis between the public and private sectors on the IS/IT outsourcing practices in a developing country: a field study', Logistics Information Management (2002)

Toward a Big Data Platform to Get Public Opinion from French Content on the Web/CMS

Abdelkader Rhouati, El Hassane Ettifouri, Mohammed Ghaouth Belkasmi and Toumi Bouchentouf

Abstract Public Opinion is a very important criterion of making political decision. However, obtaining it is a difficult task. So far the only efficient way is using surveys, on the net or by asking people directly. In this work we introduce the implementation of a new platform based on Big Data approach and named POK—abbreviation of Public Opinion Knowledge, which is a solution to get the public opinion from French content published on the Web/CMS, and also presented as an alternative of surveys.

Keywords Big data · Web · CMS · Content · Data · Public opinion · Hadoop · NoSQL · Web mining · Survey · Jsoup · Casandra · OpenNLP

1 Introduction

The arrival of the web 2.0 on the last decade has changed the habits of the world. All things are passing on the net. The world now is expressing his ideas and his convictions on the net by articles and comments. We talk about Public Opinion on the internet. In fact, the web 2.0 also make possible to created a new approaches and strategies to know how people are thinking and what they really want? Be able to answer this lasts two questions are very important, especially to the political class, which must take the appropriate decisions.

A. Rhouati (✉) · E.H. Ettifouri · M.G. Belkasmi · T. Bouchentouf
Team SIQL, Laboratory LSEII, ENSAO, Mohammed First University, BP 60000 Oujda, Morocco
e-mail: abdelkader.rhouati@gmail.com

E.H. Ettifouri
e-mail: h.ettifouri@gmail.com

M.G. Belkasmi
e-mail: ghaouth@gmail.com

T. Bouchentouf
e-mail: tbouchentouf@gmail.com

Á. Rocha et al. (eds.), *Europe and MENA Cooperation Advances in Information and Communication Technologies*, Advances in Intelligent Systems and Computing 520, DOI 10.1007/978-3-319-46568-5_35

Till now the unique way to get an idea about Public Opinion is to use surveys. In our last Work [1] we propose an alternative to this method, that fills its disadvantages which can be summarized in reduced and insignificant specimen of people and the huge time in aggregating and processing of responses. Our new approach is based on a Big Data System, that use articles and comments from the internet to get the Public Opinion about a given subject. This new approach contains basically 4 steps: data extraction, data management, modeling and processing of data and display of results.

This work is carried out to present a platform that implement this new approach. The platform is called "Public Opinion Knowledge (POK)".

This article is organized in different sections. Section 2 presents the general context of our Work, and brief reminder of our approach. Sections 3, 4, 5 and 6 are respectively present the detail of the four steps of the approach. Finally, a conclusion and future works will be presented in Sect. 7.

2 Context General: From the Approach to the Platform

2.1 The Impact of Web2.0 on the Emergence of Public Opinion

The public opinion is a major factor of the political and economic decisions of all governmental organisms. In fact, a respectful politician can't ignore the public opinion on any action. More his actions are consistent with the public opinion more they will be appreciated and supported. However, the public opinion is becoming more complex to measure. On one hand, because it's continuous evolution depending on events and news of every day. And on other hand it is measurable almost using a single and basic method: "the Surveys", which does not evaluate in the last decade, even if it was developed in a technical way using surveys diffused on web2.0: web sites, blogs, newsletter and social networks. So we need a new way to measure a public opinion that must be fast, simple and on real time.

With the explosion of the Internet and the emergence of Web 2.0, the public opinion has changed. It is no more only a discussion on Radio or TV, at conferences or even over coffees. The public opinion can be now an article on blogs, a comment on social networks, etc. Briefly any content on the Web represents a part of public opinion. It presents a huge unexploited data.

In fact, the goal of POK Platform is studying this data published by users on the web to get out a positive or negative public opinion on a given topic.

2.2 *New Approach Based on Big Data to Get Public Opinion from Web Content*

Our last work [1] was an introduction to new approach to get Public Opinion from web Content based on Big Data. This approach is based on 4 steps:

- Data source: consists on the extraction of data from several web sources.
- Data management: consists in modeling data and proceed to store it on a NoSQL storage platform.
- Modeling: consists in using a Web mining process to analyze data.
- Result: consists in visualizing the results and distinguish the positive and negative opinion of people.

This approach is explaining on the following figure (Fig. 1):

2.3 *The Objective Behind the Realization of "POK"*

The functional design of the platform is based on distributed computing in order to addressing the problem of massiveness of data to process. After the step of extracting data (articles and comments) from blogs on the web and directly from database of CMS, then saving this data in a Big Data database, we will apply an algorithm of Web Mining in order to deduce the public opinion from all stored data. In order to optimize the Web Mining treatment, a distributed system of several machines will be used. The following figure explains the 4 steps (Fig. 2).

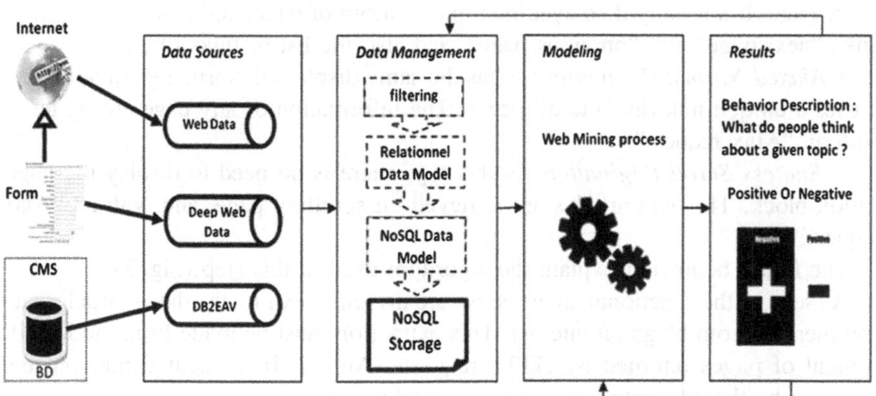

Fig. 1 A new approach to get Public Opinion from web content based on Big Data

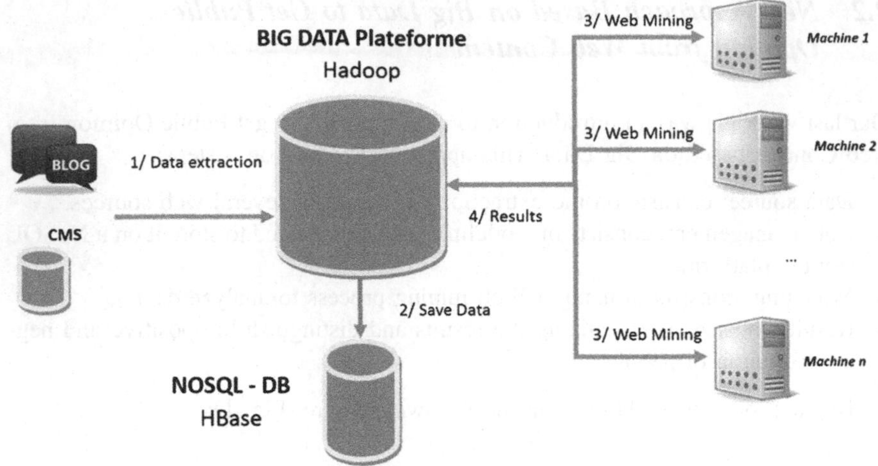

Fig. 2 The functional design of POK platform using a distributed system

3 Step 1: Data Extraction

3.1 Data Extraction from Blogs

The extraction of data from blogs is based on HTML code of web pages. The extraction of every single information, for example title or author, is done using a CSS Selector configuration.

+ *Normal Pagination:* The normal pagination is based on a single link (href) for every page. It's usually displayed on top or bottom of page, and contains two main links "next page" and "previous page", and also the list of links of pages.

+ *Ajaxed Normal Pagination:* It has the same display of Normal Pagination, but it uses a unique link (href) to all pages. The information of any page are retrieved with an "Ajax request".

+ *Endless Scroll Pagination:* In this type there is no need to display the pagination block. The next articles are retrieved on scrolling page, and with an "Ajax request".

The figure bellowing explain the algorithm used in this step (Fig. 3):

As seen in the functional analysis, we are driven to extract the data—articles and comments—from blogs on internet. This extraction must be made from the HTML content of pages returned by HTTP requests. An HTML content extraction tool seems to be the adequate solution to our problem.

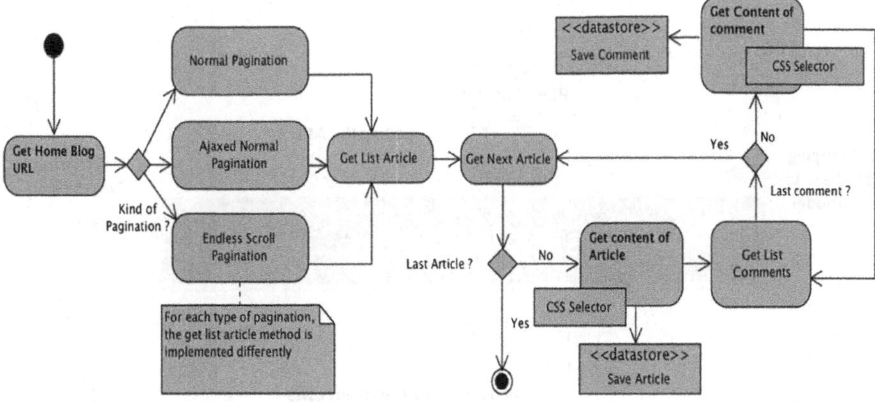

Fig. 3 The algorithm of extraction data from Blogs

Before embarking on a technical choice we conducted a comparative study on three existing solutions, the following table shows the result of the comparison:

Tools	Prog. Lang	License	Last update	Parse html	Clean html	Update html
Jsoup	Java	MIT (a)	08/2015	Yes	Yes	Yes
Jaunt API	Java	Jaunt (b)	01/2016	Yes	Yes	No
Html Cleaner BSD	Java	BSD (c)	12/2015	Yes	Yes	No

a. MIT License is permissive free software license

b. Jaunt license is a specific license of product that contains many constraints

c. BSD License is a permissive free software licenses, imposing minimal restrictions on the redistribution of covered software

The final choice is the use of JSOUP API [2, 3], firstly because that it can change the HTML code, which is required to have a valid code processing and secondly because of its MIT license which is no restrictive license. So, Jsoup is very convenient to retrieve and manipulate data, using the best of DOM, CSS selectors, and several methods inspired by Jquery. Jsoup implements the WHATWG HTML5 [4] specification, and builds the DOM arborescence in a similar way that modern browsers.

3.2 Data Extraction from CMS

Data extraction from CMS sources is an option when we can offer a direct access to database. For this type of extraction, we will use DB2EAV API [5], that make possible to retrieve data from any databases which are based on EAV schema. So using the DB2EAV API we fetch all existing articles and comments of several Web Site, as explain on Fig. 4.

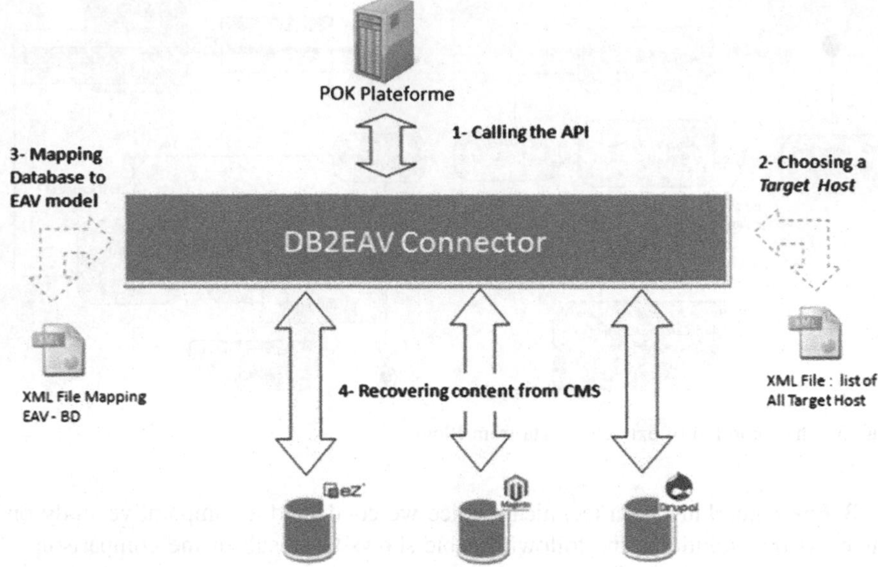

Fig. 4 The approach of extraction data from directly access to CMS's databases

4 Step 2: Data Saving on NoSQL Database

The basic need to NoSQL [6] is the performance. To meet this need corporate developers such as Google and Amazon have made compromises on the ACID [7] properties of relational Database. These compromises on relational concept enabled RDBMS to release their brakes to horizontal scalability. Another important aspect of NoSQL is that it caters to the CAP [8] theorem which is more suitable for distributed systems.

To choose the NoSQL database type to will work on it, you must initially make a comparison between the various existing kinds. A different types of NoSQL database existing as explain on [9]. As we have already mentioned in previous sections our solution is supposed to use homogeneous data (as textual content), and these data are distinguished between two types of content, Articles and comments, the ideal choice for our case is to opt for a Column-Oriented database.

At first we focus our comparison on the criterion of the type of recording data supported by Database system. The following table illustrate a comparative of most popular NoSQL System.

Data base system	Type of data	Our decision
Casandra (http://cassandra.apache.org/)	Columns	**Ok**
MongoDB (https://www.mongodb.org/)	BSon (Json Binary Format)	Ko

(continued)

(continued)

Data base system	Type of data	Our decision
CouchDB (http://couchdb.apache.org/)	Documents, Json	Ko
HBase (https://hbase.apache.org/)	Columns	**OK**
Riak (http://basho.com/products/riak-ts/)	Key-value	Ko

Finally, we decided to use Hbase because it is built on top of Hadoop HDFS (the physical architecture that wants to establish), which makes it a strategic choice for our case, but also because it is in coherence with CAP theorem. Hbase, an open source version of Bigtable [10], is a data management system for a distributed environment and coupled to the Hadoop platform, indeed Base is an infrastructure over Hadoop. Hbase can store data by columns, allowing it to greatly reduce information search time. In Hbase, data is managed in large tables, called htable and composed of rows and column families. The column families are constructed from several columns, called also qualifier. The rows are identified by a unique value (equivalent to the primary key in the relational database). The data, or the value, is stored in a row and a specific column. But several entries can be made to one location. This is called version data at different time, stored on timestamp format.

For better management of the memory space, data is still stored as bytes in Hbase. Hbase is written in Java. So a Java API is available to communicate with it. Whatever other means of communication with Hbase have been implemented: Hbase shell [11], Hive [12], Pig [13], API JRuby [14], Cascading API [15].

5 Step 3: Web Mining

5.1 Overview of Web Mining Constraints Applied to a French Text

Data-Mining, also known as Web-Mining [16] when data is taken from the web, is a knowledge discovery process. It aims at extracting knowledge from large amounts of data, by automatic or semi-automatic methods. The concept of data mining is born of the need to search in very large databases in order to extract relevant information to be used to make decision on a very large number of domains. WebMining have a standard process known by a several steps [17].

Investigate an opinion from a given French text is the act to determine if the opinion is positive or negative about certain subject. However, all texts are characterized by a several aspects of natural languages, which is very complexes. Indeed, the identification of the presence of a word which expresses a positive opinion, as “belle” (beautiful), “super” (super), “facile” (easy), or a negative opinion, as “mal” (evil), “Hanteux” (shameful), “villain” (bad), is not enough to classify the entire text. For example, the sentence “Belle voiture” (as beautiful car) expresses a positive point of view identified by the presence of the adjective "belle" (as beautiful). But, the affirmation "pas belle voiture" still contains the word "belle" (beautiful) but the opinion of the sentence was reversed by the presence of the particle of negation "pas" (not). Thus, implementing an algorithm that represents the subtleties of the French language is a complex task that requires an advanced set of linguistic knowledge.

- *Lexical Level*: Basically, it defines the lexical errors as the manifestation of a lack of language proficiency. However, the lexical errors in our case are also due to various possible orthographic forms generated by the use of Tchat language or/and by the use of abbreviations to not exceed the limit of text characters on blogs.
- *Syntactic Level*: The data that we manage is in the form of free text, the analysis may be confronted with various Syntactic forms. These do not always meet the grammatical standards of the French language. Thus, the language used by some users is spontaneous. Words are not always used in their original form, for example "jolieee" (beautiful) with serval "e" to express the superlative form.
- *Semantic level*: The semantic level has a set of difficulties. The first one is in the multiple meanings of words, which can make an ambiguous meaning and create misunderstandings. The second one is in the negation. Usually the sentence negations are recognizable by the use of particle "NE", N″, NI″, plus verb and the particle "Pas, AUCUN, Plus…". But in some cases we can use Only one particle "Pas or NE". The third one is in the use of some tenses in French, as conditional, can change the meaning. The use of the conditional in French can expresses a negative opinion, while the used verb expresses a positive feeling.
- *Pragmatic level:* This linguistic level implies a general background knowledge of the situation. This often includes elements outside the language, different personal information (age, sex, social status). However, an automatic analyzer can not have contextual knowledge. So this level will never be handled with POK Platform.

5.2 "OpenNLP" as Solution of WebMining

OpenNLP [18] is a JAVA library that allow an automatically treatment of text. It takes care on tasks of grammatical libeling and named entities extraction. The main features and characteristics of OpenNLP library can be resumed on [19].

Before any analyze of text, it's must be standardized. We decided in the beginning to use an existing tool which allow the normalization of rules applied on sentences. However, those tools do not respond perfectly to our need. Indeed, we have specific and determined actions to implements on the goal to cleaning text. For this, we decide to conceive and develop our self preprocessing model based on java classes java. lang.String java.util.regex.Pattern and java.util.regex.Matcher that contain pretty advanced support for regular expression. The following table explains the actions of cleaning.

The analysis algorithm based on OpenNLP to scoring every sentences and texts:

1. Divide the text on list of sentences using sentence Detector
2. Divide each sentence of the list into words (token) through tokenization OpenNLP
3. Mark each word.
4. Develop a data dictionary. This is a set of files (key, value) whose key is a word (positive, negative, negative particle, adverb) and value is one of the following: Positive, Negative, AdverbePlus (eg: trop, très, plus…), AdverbeMoins (eg: Moins, peu…) or Negation
5. Tag every word based on all dictionary files, as it takes as tag: Positive, Negative, Particle denial or Amount of adverb placed before or after a token of positive or negative.
6. Making a scoring algorithm of sentences with following operations:

   + If the word is positive the score of sentence is incremented by 1.

   + If the word is positive but it is preceded or followed by a negative particle score is reduced by 1.

   + If the word is positive and it is preceded by an adverb (très, trop) the score increments of 2.

   + If the word is negative score is reduced by 1.

   + If the word is negative but it is preceded or followed by a negative particle score increments by 1.

   + If the word is negative and it is preceded by an adverb (très, trop) score is reduced by 2.

7. Then the score of a text is the sum of the score of all sentences which construct it.

6 Step 4: Display the Results—Implementing the Web Mining in Hadoop Framework

Hadoop [20] is an open source framework developed in Java, it's also a project of the Apache Foundation since 2009. It is intended to facilitate the development of distributed applications, enabling the management of thousands of nodes and petabytes of data. Hadoop is based on two products, the first is the "Hadoop Distributed File System". The second is the MapReduce programming model. In Hadoop, the different types of data (structured or not) are stored in the HDFS [21].

There given a high-performance data access in distributed clusters. When HDF collects a given data, the system divides the information into several blocks and distributes them on multiple nodes in the cluster, which then allows parallel processing. The file system copies each data brick several times (three times by default) and distributes copies of each of the nodes, placing at least one copy on a separate server in the cluster.

MapReduce is a design pattern developed in Java and invented by Google, in which calculations are performed in parallel, and often distributed to potentially very large data, typically greater than 1 petabyte. The principle is to break down a task into several smaller tasks, or more precisely cut a task involving very large volumes of data in identical tasks on subsets of the data. The tasks (and their data) are then dispatched to different servers, and the results are collected and consolidated after. The upstream phase of task decomposition, is the part map, while the downstream phase, the consolidation of the results is the reduce part.

In order to implement the module of the processing on the data stored in HBase harnessing the power of Hadoop MapReduce model, we have developed three classes necessary to run a Hadoop MapReduce program. So, we proceeded as follows:

- *The first class "Driver class":* The driver class is the main class of the program. It informs Hadoop about the different types and classes used, as Mapper class and Reducer class, the input and output files… etc.
- *The second class "Mapper Class"*: The Mapper class is responsible for implementing the method of the map program.
- *The third class "Reducer class":* The Reducer class is similar to the Mapper class. It implements the reduce method of map/reduce program.

The following diagram explains the approach of a MapReduce operation within our algorithm of Web-Mining (Fig. 5):

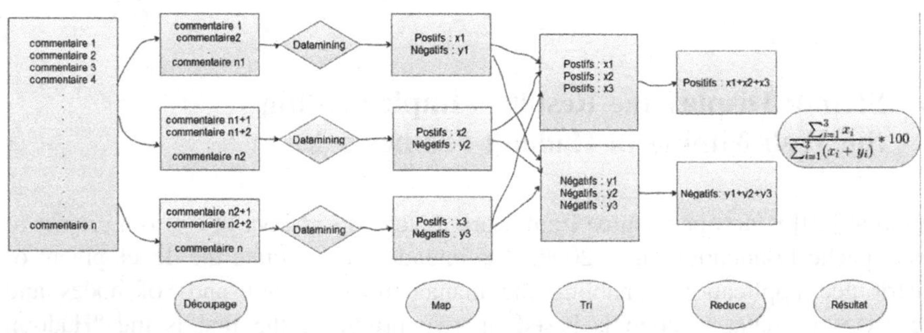

Fig. 5 Implementation of data mining algorithm in Hadoop MapReduce

7 Conclusion and Future Works

In this paper we presented the POK platform, as Public Opinion Knowledge based on the tool Hadoop Big Data, and allowing getting a public opinion from French content extracted from the web/CMS. We also provide detailed engineering design of the platform and the implementation of each layer. POK platform is currently in its final recipe phase. This will lead us to our next work which we will communicate about a real use case with concrete results. Then we would like to discover other Web-Mining analysis algorithm and tools, such as GATE Framework [22], then we will to implement to the POK Platform and carry out a comparison with the API OpenNLP.

References

1. Rhouati, A., Ettifouri, H., Belkasmi, M.G., Bouchentouf, T.: Get the public opinion from content published on the Web/CSM: new approach based on Big Data. In: Proceedings of the Mediterranean Conference on Information and Communication Technologies 2015. ISBN 978-3-319-30296-6
2. Jsoup (2016). https://github.com/jhy/jsoup
3. Jsoup (2016). http://jsoup.org/
4. https://html.spec.whatwg.org/multipage/
5. Rhouati, A., Ettifouri, H., Belkasmi, M.G., Bouchentouf, T.: The DB2EAV API of mapping database to EAV model as solution of data interoperability between Content Management Systems (CMS). In: The 2nd World Conference on Complex Systems, Agadir, Morocco, 10–12 Nov 2014. ISBN: 978-1-4799-4648-8
6. Schram, A., Anderson, K.M.: MySQL to NoSQL: data modeling challenges in supporting scalability. In: Proceedings of the 3rd annual conference on Systems, programming, and applications: software for humanity (SPLASH'12). ACM, New York, NY, USA, 191–202
7. Cattell, R.: Scalable SQL and NoSQL data stores. ACM SIGMOD Rec. **39**(4) (2010)
8. Frank, L., Pedersen, R.U., Frank, C.H., Larsson, N.J.: The CAP theorem versus databases with relaxed ACID properties. In: Proceedings of the 8th International Conference on Ubiquitous Information Management and Communication (ICUIMC'14). ACM, New York, NY, USA, Article 78
9. Leavitt, N.: Will NoSQL databases live up to their promise? Computer **43**(2), 12–14 (2010)
10. Chang, F., Dean, J., Ghemawat, S., Hsieh, W.C., Wallach, D.A., Burrows, M., Chandra, T., Fikes, A., Gruber, R.E.: Bigtable: A Distributed Storage System for Structured Data, pp. 15–15 2006
11. Hbase shell (2016). http://hbase.apache.org/book.html#shell
12. Hive (2016). https://hive.apache.org/
13. Pig (2016). https://pig.apache.org/
14. API JRuby (2016). https://github.com/junegunn/hbase-jruby
15. Cascading API (2016). https://hbase.apache.org/book.html#cascading
16. Cooley, R., Mobasher, B., Srivastava, J.: Web mining: information and pattern discovery on the World Wide Web. In: Ninth IEEE International Conference on Tools with Artificial Intelligence. Proceedings, Newport Beach, CA, pp. 558–567 (1997)
17. Malarvizhi, R., Saraswathi, K.: Web content mining techniques tools and algorithms—a comprehensive study. Int. J. Comput. Trends Technol. **4**(8), 2940–2945 August Issue (2013). ISSN 2231-2803. www.ijcttjournal.org

18. The Apache Software Foundation. Apache OpenNLP developer documentation: Written and maintained by the apache OpenNLP development community. http://opennlp.apache.org/documentation/1.6.0/manual/opennlp.html
19. Open NLP: Manuel (2016). https://opennlp.apache.org/documentation/1.6.0/manual/opennlp.html
20. Apache foundation: Hadoop (2016). http://hadoop.apache.org/
21. Ghazi, M.R., Gangodkar, D.: Hadoop, MapReduce and HDFS: a develjopers perspective. Procedia Comput. Sci. **48**, 45–50 (2015). ISSN 1877-0509
22. Cunningham, H., Maynard, D., Bontcheva, K., Tablan, V.: GATE: A framework and graphical development environment for Robust NLP tools and applications. In: Proceedings of the 40th Anniversary Meeting of the Association for Computational Linguistics, ACL'02, Philadelphia, 2002

Parallel Implementation of the Multi Capacity VRP on GPU

Abdelhamid Benaini, Achraf Berrajaa and El Mostafa Daoudi

Abstract We present a parallel implementation of an heuristic for the multi capacity vehicle routing problem on GPU. This algorithm involves two kinds of decision: the selection of a mix of vehicles among the available vehicle types and the routing of the selected vehicles. The proposed algorithm computes in parallel an initial solution (tours), and then calculates in parallel all the possible cases to obtain the more suitable vehicles to be used. Finally an improved procedure of the cost of all pairs of neighboring tours on GPU, is developed. In order to highlight the performance of our approach, Ochi (in Parallel and distributed processing, 216–224 [11]) and Karagul (in GU J Sci 27(3):979–986 [7]) test problems and random problems are used. Obtained experimental results on GPU outperform other implementations in execution times and quality of solutions. This means that our algorithm is well suited to the computational power of the GPU and our implementation exploits efficiently the power of the GPU.

Keywords VRP · Multi capacity · Parallel algorithm · GPU · CUDA

1 Introduction

The Vehicle Routing Problem (VRP) is an NP-Hard combinatorial optimization problem. Several variants of this problem are well studied because of their importance in the fields of logistics and transport. Among them the classical VRP, the VRP with time windows, with fleet size and mix vehicles, with split delivery,

A. Benaini (✉) · A. Berrajaa
Normandie UNIV, UNIHAVRE, LMAH, Le Havre, France
e-mail: abdelhamid.benaini@univ-lehavre.fr

A. Berrajaa
e-mail: berrajaa.achraf@gmail.com

A. Berrajaa · E.M. Daoudi
University Mohammed I, LaRI, Oujda, Morocco
e-mail: m.daoudi@fso.ump.ma

Á. Rocha et al. (eds.), *Europe and MENA Cooperation Advances in Information and Communication Technologies*, Advances in Intelligent Systems and Computing 520, DOI 10.1007/978-3-319-46568-5_36

and recently the searches are focused on electric and green VRP. The literature on the methods that solve VRP is vast and large including exact heuristics and meta-heuristics methods. In this paper, we are interested by VRP with multi capacity [2, 4, 5, 7, 11, 12]. This last can be modeled by directed graph G = (V, A), where V is the set of n + 1 nodes, node 0 represents the depot, while the remaining nodes 1, …, n corresponds to the n cities (or customers). The cost of the arc between cities i and j is noted c_{ij} and represents the distance between i and j. The demand of city i is noted d_i. The fleet is composed of m different types of vehicle. Each vehicle of type of k has a capacity C_k and is associated with a fixed cost F_k which can be seen as rental costs [7, 11, 16].

The multi capacity VRP is clearly NP-Hard problem as it reduces for m = 1 to the classical VRP.

A feasible tour $R^k = (i_1, i_2, \ldots, i_t)$, with $i_1 = i_t = 0$ and $i_1, \ldots, i_t$ is a simple circuit in G and k is the type of the vehicle used to realize this tour, that is the total demand of the customers visited in R^k does not exceed the vehicle capacity C_k. i.e., $\sum_{h=2}^{t-1} d_{ih} \le C_k$. The cost of the R^k tour is the sum of the costs of the arcs forming the tour (traveling distance) plus the fixed cost of the vehicle associated with it i.e., $\cos t(R^k) = F_k + \sum_{h=0}^{t-1} c_{i_h i_{h+1}}$. As in Karagul [7] and in Ochi [11] we assume that the number of vehicles of each type is unlimited and the vehicle capacity and cost types satisfy $C_1 \le C_2 \le \ldots \le C_m$, and $F_1 \le F_2 \le \ldots \le F_m$.

Several algorithms are proposed to solve this problem. The first one was proposed by Golden et al. [16]. They suggested five adaptations of Clarke and Wright's algorithm to the multi capacity VRP, among them the combined savings that extends the concept of saving to include fixed vehicle cost. Tabu search based procedures are presented in Gendreau et al. [17]. Renaud and Bactor [12] proposed a sweep-based heuristic for the multi capacity VRP which generates a number of tours that are serviced by one or two vehicles (1-petal or 2-petal). The selection of tours and vehicles to be used is then made by solving a set-partitioning problem having a special structure. Dell' Amico et al. [4] develop a constructive insertion heuristic for the multi capacity VRP with time windows.

In this paper, we propose a parallel algorithm in three steps that solves this problem and we describe its implementation on the GPU (Graphics Processing Unit). To our knowledge, few parallel and/or GPU implementations that solve the VRP are proposed in the literature. Li et al. [9] propose a parallel simulated annealing algorithm based on GPU-acceleration to solve a large scale VRP with time windows. Uthayopass [15] a fast software for solving the pickup and delivery problem with time windows using GPU cluster. Lekaez et al. [8] present an implementation of genetic algorithm to find a solution of TSP on GPU. Szymon and Dominik [13] solve the multi criteria discrete optimization of distance constrained VRP using parallel tabu search on GPU. Genetic methods on GPU are presented by Talbi [14], other GPU implementations are local search for TSP [6], Constraint-based local search [1], Pickup and Delivery Problem with Time

Windows using GPU Cluster [15], Solving Multi-criteria VRP by Parallel Tabu Search on GPU [13], we have proposed a GPU implementation of an algorithm based on the Clarke and Wright algorithm that solve the multi depot VPR [3].

The remainder of this paper is organized as follows. In Sect. 2, we present the outline of the proposed algorithm and their three principal steps: partition in sectors, construction of the initial solution and finding the more suitable type of vehicle for each tour of the initial solution and improving the obtained solution by using the 2-Opt that we adapted to GPU. Section 3 is devoted to the implementation of our algorithm on GPU. In Sect. 4, we present the experimental results and we compare them to those of Ochi [11] and of Karagul [7] and also we discuss the efficiency of our implementation.

2 The Proposed Algorithm

To find a feasible solution to the VRP with multi capacity having an efficient cost, we must conceive an algorithm that tests the maximum possible solutions in reasonable time execution. The concept and power of GPU can achieve this aim. So we propose an algorithm that tests a maximum of feasible solutions in parallel. The three principal steps of our parallel algorithm are:

- Step 1: Partition the multi capacity VPR into Ns independent sub-problems (Ns sectors). Each of them is considered as a Traveling Salesman Problem (TSP) and follows the construction of an initial solution in parallel using any algorithm that solves the TSP. For this, we use the Clarke and Wright algorithm, noted CW.
- Step 2: Compute all feasible solutions in parallel for each sector using all types of vehicles. i.e. calculate all costs of routes to deserve the sector using all types of vehicle and then select the vehicle with the minimum cost to deserve this sector.
- Step 3: Improve the result using the 2-Opt principle that we adapt to GPU concept.

In the following, we will detail each of these steps:

2.1 Partition of the VRP into Ns Sectors

As in [3], each city is defined by its polar coordinates (θ_i, ρ_i) with the depot at (0, 0). The cities are numbered by increasing angle $\theta_i \leq \theta_{i+1}$, for each $1 \leq i \leq n$. A more general study of the partition in sectors can be found in [10]. In the example of Fig. 1, cities are numbered so that the city 1 is the city whose θ angle (θ_1) is the smallest. If two cities have the same θ angle then they are numbered in increasing radius ρ. We partition the set of cities into sectors such that the demand of each

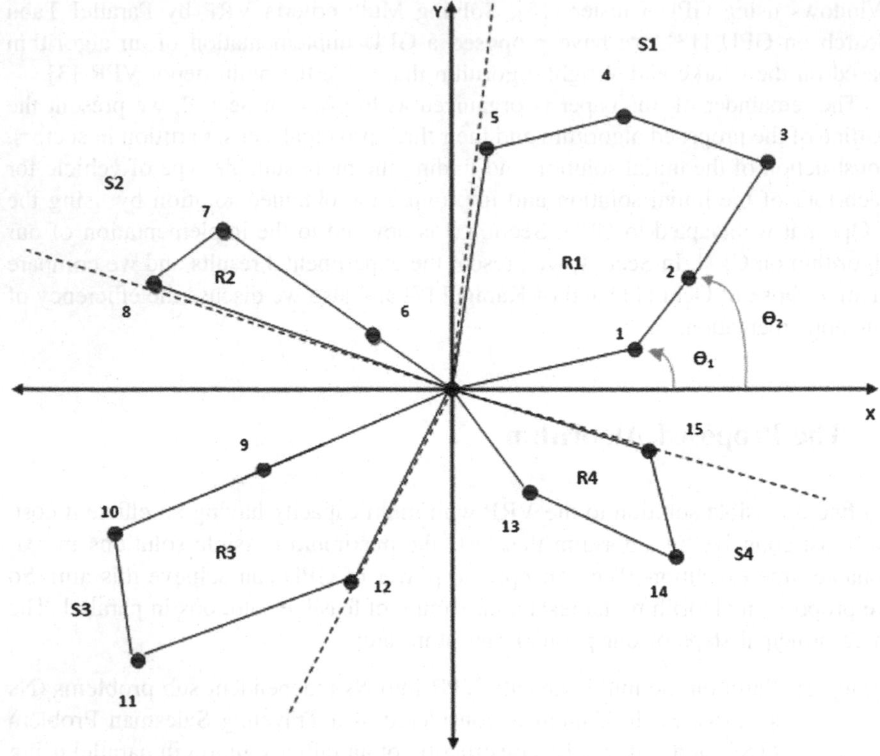

Fig. 1 Partitioning the cities in sectors

sector is less than the maximal capacity C_m. The choice of this capacity is not random indeed with this capacity we will make sure that the sector is saturated with the maximum of cities.

Let $S_1, \ldots, S_{Ns}$ these sectors and $S_i = \{r_{i-1} + 1, \ldots, r_i\}$, with $r_0 = 0$, and r_i is the unique integer satisfying: $\sum_{k=r_{i-1}+1}^{r_i} d_k \le C_m < \sum_{k=r_{i-1}+1}^{r_i+1} d_k$ for $1 \le i < Ns$, and $\sum_{k=r_{Ns-1}+1}^{r_{Ns}} d_k \le C_m$ for $i = Ns$. Here, $h_i = ri - 1$ is the number of cities in S_i. Each S_i can be served in one tour with vehicle of capacity C_m and S_i U $\{r_i + 1\}$ cannot be served in one tour by any vehicle. Hence, we can use any TSP algorithm to construct a tour that serves S_i in minimum cost. In our case we use CW algorithm which we adapted to GPU implementation.

Figure 1 shows an example of partitioning cities into 4 sectors: $S_1 = \{1, 2, 3, 4, 5\}$, $S_2 = \{6, 7, 8\}$, $S_3 = \{9, 10, 11, 12\}$, $S_4 = \{13, 14, 15\}$. For this example, we have assumed that:

$d_1 + \cdots + d_5 \le C_m < d_1 + \cdots + d_6$, $d_6 + \cdots + d_8 \le C_m < d_6 + \cdots + d_9$,
$d_9 + \cdots + d_{12} \le C_m < d_9 + \cdots + d_{13}$, $d_{13} + \cdots + d_{15} \le C_m$.

Construction of the Tour for each Sector

Each sector is a set of cities where the sum of their demand does not exceeds the maximum vehicles capacity C_m. So any TSP algorithm can be used to construct a tour that serves S_i. We have used the CW algorithm and implemented it on GPU to construct these tours. Therefore CW will be called in parallel, for $S = S_1, \ldots, S_{Ns}$ to construct Ns initial tours $R_1, \ldots, R_{Ns}$. Recall that the CW algorithm is based on the computation of saving for combining two cities into the same tour. Initially, each city is considered to be on separate tour the parallel GPU implementation of CW is given in [3], many techniques that improve this algorithm are proposed where the definition of saving is modified [10]. Each sector is considered as an independent TSP solved by a single tour with vehicle of type m. We construct these tours $R_i^m = CW(S_i)$ for $i = 1, \ldots, Ns$, in parallel.

2.2 *Finding the More Suitable Vehicle for Each Sector*

In the previous section, we partitioned the problem into Ns-TSP using the capacity C_m, and then used the CW algorithm in parallel to obtain an initial solution.

Now we will calculate for each sector all possibilities when using all vehicle types. We do this in parallel exploiting the parallelism of the GPU. Indeed, each R_i^m is treated independently of other tours and hence the Ns tours will be treated in parallel. Precisely, let the tour $R_i^m = (c_1, \ldots, c_{ti})$ then for each type k of vehicle, we partition R_i^m, starting with c_1 to c_{ti} and conversely starting with c_{ti} down to c_1, into sub-tours, …, that can be served by vehicles of type k. Finally we select the type k* that minimize the $cost(R_i^k) + F_k$.

Algorithm that find the more suitable vehicle for $R_i^m = \{c_1, \ldots, c_{ti}\}$

```
for all k, 1≤ k ≤ m do in parallel
    capacity  = 0;   sub-tour = empty ;
      for j = 1 to ti or for j= ti  down to 1
       if capacity  +demand(cj) ≤ Ck then
          add cj to the current sub-tour;
          capacity = capacity  + demand(cj);
       else
          create a new tour and add cj to this tour.
          capacity = demand(cj);
      costk=∑(cost(sub-tour of Rk)+ Fk
      find k* such that costk* = mink costk
```

At end of this step, the tour R_i^m is partitioned into a set of sub-tours $R_{i1}^{k*}, \ldots, R_{it}^{k*}$ all served with vehicle of type k*. We find in parallel the best type of vehicle for each R_i^m, for all $1 \leq i \leq Ns$. To simplify the notations, let us note $R_1^{k1}, \ldots, R_t^{kt}$ the obtained tours from the initial solution $R_1^m, \ldots, R_{Ns}^m$. (Note that $Ns \leq t$).

2.3 *Improvement of the Initial Solution*

The k-opt concept [4, 5] is applied to sets of k tours by removing cities from some tours and inserting them into another one for a saving in travel distance. In our case, we do this between pairs of adjacent tours R_i and R_{i+1} and we adapted the move() and swap() to the multi capacity case. A move of cities from R_i to R_{i+1} or from R_{i+1} to R_i could reduce the cost of these two tours. Similarly, swaps of cities between R_i and R_{i+1} could reduce the cost of these two tours. This process of moves/swaps of cities between pairs of adjacent tours is iterated in order to minimize the global cost of the multi capacity VRP. These operations are done between pairs (R_1, R_2), (R_3, R_4), …, (R_{t-2}, R_{t-1}) in parallel, and between (R_2, R_3), (R_4, R_5), … , (R_{t-1}, R_t) since these pairs are also neighbor tours. As well as for the odd even sort steps, we apply alternatively these two steps. Call Opt(R, R′) the function that consists in move/swap cities between R and R′ tours. The Improve of the solution (R_1, …, R_t), is the following:

```
Improve(R1,…Rt)
 If t is odd than:
   repeat
     1.do in parallel Opt(R1,R2), … Opt(Rt-2,Rt-1)
     2.do in parallel Opt(R2,R3), … Opt(Rt-1,Rt)
   while the global cost is improved.
 If t is even then:
   repeat
     1.do in parallel Opt(R1,R2), … Opt(Rt-1,Rt)
     2.do in parallel Opt(R2,R3), … Opt(Rt-2,Rt-1),Opt(Rt,R1)
   while the global cost is improved
```

In the following we precise these two operations (*move()* and *swap()* functions):

2.3.1 Move a City C from R^k to $R'^{k'}$

The city c can be moved from R^k to $R'^{k'}$ if there is k1 (type of vehicle) such that demand(R′ + c) = demand (R′) + demand (c) $\leq C_{k1}$ and if the operation is gainful that is if the total cost of the new tours is less than that the old one i.e.:

$$\mathrm{cost}(R-c)+F_{k2}+\mathrm{cost}(R'+c)+F_{k1}<\mathrm{cost}(R)+F_k+\mathrm{cost}\left(R'\right)+F_{k'}$$

where k2 is the type of vehicle used to serve the tout (R – c) necessary k2 $\leq$ k. The tour (R – c) is obtained from R by replacing the path i – c – j by i – j in R and

$$\mathrm{cost}(R-c)=\mathrm{cost}(R)-c_{ic}-c_{cj}+c_{ij}.$$

The tour (R′ + c) is obtained from R by inserting c in R′ at the location that minimizes the obtained cost. That is, if c is inserted between i′ and j′ then the edge i′ − j′ of R′ will be replaced with the path i′ − c − j′ and the

$$\mathrm{cost}(R' + c) = \mathrm{cost}(R') - c_{i'j'} + c_{i'c} + c_{cj'} = \min_{i,j \text{ in } R'} \mathrm{cost}(R') - c_{ij} + c_{ic} + c_{cj}$$

2.3.2 Swap Two Cities C and C′ Between R^k and $R'^{k'}$

The swap of a cities c and c′ between R and R′ is done if there exist two vehicle types k1 and k2 such that demand(R − c + c′) $\leq C_{k1}$ and the demand (R′ − c′ + c) $\leq C_{k2}$, and if the operation is gainful that is if the total cost of the new tours is less than that the old one i.e.

$$\mathrm{cost}(R - c + c') + F_{k1} + \mathrm{cost}(R' - c' + c) + F_{k2} < \mathrm{cost}(R) + F_k + \mathrm{cost}(R') + F_{k'}$$

c is replaced by c′ in R and c′ is replaced by c in R′ (c and c′ can be inserted in R and in R′ in more appropriate places).

3 The GPU Implementation

Now we explain the GPU implementation of the proposed algorithm, and the CUDA C syntax, the blocks, grid and the kernel for each part of our algorithm on GPU. We first recall the GPU architecture.

3.1 The GPU Architecture

Graphics Processing Units GPU is actually available in most of personal computers. They are used to accelerate the execution times of variety of problems. The GPU consists of a large number of lightweight stream processors that are capable of performing many simple calculations in parallel. The smallest unit in GPU that can be executed is called thread. Threads (all executing the same code and can be synchronized) are organized into blocks of equally sizes and blocks are organized in grid that represent one kernel. Blocks are independent and cannot be synchronized.

A simplified GPU architecture is given in [3]. The memory hierarchy of the GPU consists of three levels: (1) the global memory that is accessible by all threads. (2) the shared memory accessible by all threads of a block and the local memory (register) accessible by a thread. Shared memory has a low latency (2 cycles) and is of limited size (48 Kb for the GeForce). Global memory has a high latency (400 cycles) and is of large size (4 GB for the quadro). So, each class of memory has a different data access time. Thus a carful consideration must be taken in programming a GPU.

An entire block is assigned to a single SM (Stream Multiprocessor). Several blocks can run on the same SM. Each block is divided into Warps (32 threads by Warp) that are executed in parallel. The programmer must control the block sizes, the number on Warps and the memories access. GPU provide massive computation when the needed operations are properly parallelized.

The following steps summarize the proposed GPU implementation of our parallel algorithm for the multi capacity VRP.

3.2 Outline of the Implementation

Our implementation includes the following five steps:

1. DataCopy_to_GPU(); //copy to GPU D the distance matrix //D(0:n,0:n),the demands d(1:n), the capacity C(1:m) and the fixed costs F(1:m).
2. Partition_into_sectors(); //Partitioning the VRP into Ns Sector (Ns-TSP).
3. Find_Initial_solution(); //Solving the Ns TSP by CW algorithm
4. Find_Best_vehicle(); //Find the more suitable type of vehicles for each TSP
5. Improve_solution(); //Improve iteratively the solution using 2-Opt().

In the following we explain each of these steps.

1. We use the classical CUDA function cudaMemcpy () to transfer the data from CPU to GPU. However parallel programming with CUDA requires a precise methodology and careful consideration must be taken in utilizing the different memories of the GPU [3, 14]. Here the D matrix is distributed to the shared memory of each block of threads as we will see latter. The demands, the capacity and the fixed costs are duplicated in each shared memory.
2. The construction of sector is intrinsically sequential and is done by the CPU.
3. We use CW to find in parallel the Ns tours for the Ns sectors $R_i = CW(S_i)$ $i = 1, \ldots, Ns$. Each R_i is computed on a block of $h_i \times h_i$ threads; where h_i is the number of cities in the sector. Since, in CUDA we can define only uniform size's blocks of threads, we use blocks of hxh threads with $h = \max(h_i)$. The needed data for computing R_i are the columns $r_i, \ldots, h_i + r_i$ and the capacity $C_1, \ldots, C_m$ and the fixed costs $F_1, \ldots, F_m$. Using CUDA syntax, the blocks, grid and the kernel CW() are defined as follows:

```
dim3 dimBlock(h,h); dim3 dimGrid(Ns,1);
CW < ≪dimGrid,dimBlock ≫>(Si);
```

4. Each tour R_i is treated independently of the other tours on a block of hxh threads. Hence, the Ns tours are treated in parallel on Ns blocks. Precisely, let $R_i = (c_1, \ldots, c_{ti})$ then for each type k of vehicle we partition R_i, starting with c_1 to c_{ti} and/or starting with c_{ti} down to c_1, into sub-tours $R_{i1}, \ldots, R_{is}$ that can be

served by vehicles of type k. Finally we select the more suitable type k^* that minimize the $cost(R_{i1}) + F_{k*} + \cdots + cost(R_{is}) + F_{k*}$. A block of m thread is used to find k^* for each tour R_i. Since Ns blocks each of m threads are used to find the type of vehicle that minimizes the cost for Ns tours R_i. Hence, with CUDA syntax, the blocks, grid and the kernel Find_Best_vehicle() are defined as follows:

```
dim3 dimBlsock(m,1);dim3 dimGrid(Ns,1);
Find_Best_vehicule< ≪dimGrid,dimBlock ≫>(Si);
```

5. Recall that the Opt(R, R′) procedures which are based on the move/swap operations are done between adjacent pairs of tours R and R′. The move() operation is launched on bloc of h x m threads, each thread treats one city of the route R which can be moved to the neighboring tour R′. We use the second dimension of block to find the suitable type of vehicles for the tours after the move if any. Similarly the swap operation is launched on 3 dimensional of $h \times h \times m$ threads. A thread of coordinal (i, j, k) deals with city i of R and city j of R′ and a vehicle of type k after the swap(i, j) if any. The Opt(R, R′) executes in parallel and calculates the gain of each move/swap operations and choose the greatest profit. Since both operations require the same data we will start each tour in a block and each operation in dimension of block (move in the first dimension and swap in the second dimension). Hence, this treatment is done on a two-dimensional grid where each dimension of the block launches a set of threads in the form of three-dimensional however every operation uses only the number of threads that needs to do its treatment.

```
dim3 dimBlock(h, h, m); dim3 dimGrid(Ns,2);
Opt< ≪ dimGrid, dimBlock ≫>(tours);
```

As we have already seen each $Opt(R_i, R_{i+1})$, $1 \leq i \leq Ns$, is executed on bloc with shared memory and requires all distances between cities c and c′ of these two tours which are stored in blocs D_i and D_{i+1} of the matrix D as shown in Fig. 2. So D_i and D_{i+1} must be stored in the shared memory M_i of the bloc B_i. Now, if c is moved from R_i to R_{i+1} then we must update the shared memories M_{i-1}, M_i and M_{i+1} as shown on Fig. 3a. Similarly, a swap of c and c′ between R_i and R_{i+1} needs to update M_{i-1}, M_i and M_{i+1} as shown in Fig. 3b.

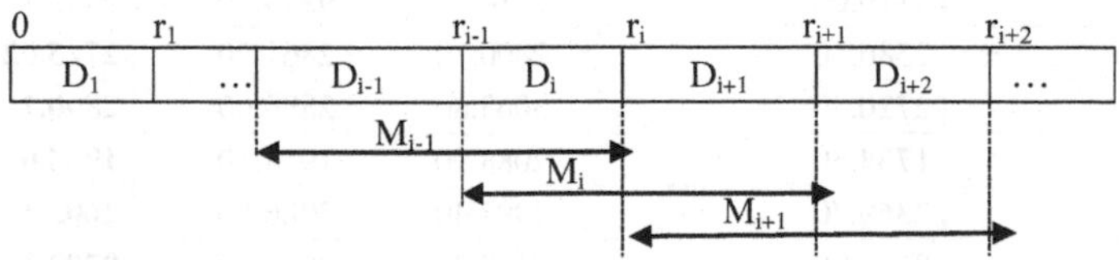

Fig. 2 Data repartition on the shared memory

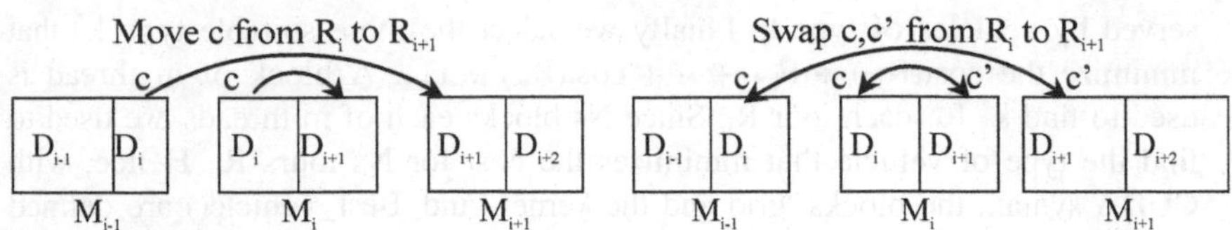

Fig. 3 Updating the shared memory in the move/swap

4 Experimental Results

The proposed method is tested under the 12 sample problems given in by Golden et al. [16] and theses results are compared with those of Ochi [11] and Karagul [7] (the well known benchmarks for this problem). As shown in the table below, our algorithm gives better solutions (in term of costs) for the most of these problems (Table 1).

In the following we compare the execution times of the CPU versus GPU of our GPU implementation. We have used Nvidia GeForce, 1 Gb, with 48 CUDA cores (1,17 GHz) and an i3 with 4 core (2,4 GHz) as CPU. So the optimal acceleration, for this hardware, cannot exceed 8. In Fig. 4 we compare the CPU versus GPU execution times for each problem of Golden. The obtained results show that the execution times are reduced on GPU for all studied problems and the acceleration factor reaches 8 for problem number 18.

The Fig. 5 gives the execution times of the CPU versus GPU for random problems. In these problems the coordinate of customers is randomly generated and the number of vehicles types is chosen as follow. The problems with a number of customers less than 50 use 3 types of vehicles. The problems with a number of customers between 50 and 80 use 7 types of the vehicles and the problems with a number of customers greater than 80 use 9 types of the vehicles.

Table 1 The qualities of the results in term of costs

Problem number	Best know solution	Ochi	Karagul	Our algorithm
3	961.03	1088.70	999.20	**983.51**
4	6437.30	7324.70	7324.7	7408.5
5	1007.10	1153.00	1097.4	**1053.03**
6	6516.50	7031.40	7031.40	7112.1
13	2406.40	2670.70	2680.20	**2539.6**
14	9119.00	9214.40	9214.40	9423.1
15	2586.40	2800.10	2861.20	**2745.02**
16	2720.40	3063.80	2899.00	**2890.7**
17	1734.50	2088.90	1954.10	**1934.0**
18	2369.70	2992.40	2986.50	**2603.1**
19	8661.80	9599.20	9824.80	**9702.1**
20	4039.50	4459.10	4498.90	**4375.2**

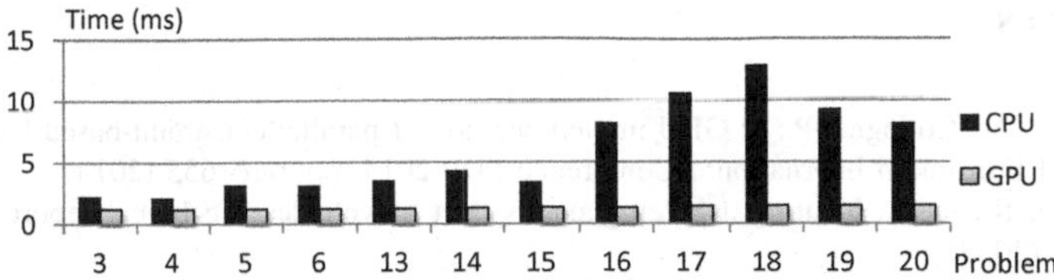

Fig. 4 GPU versus CPU of algorithm implementation for Golden's 12 test problems

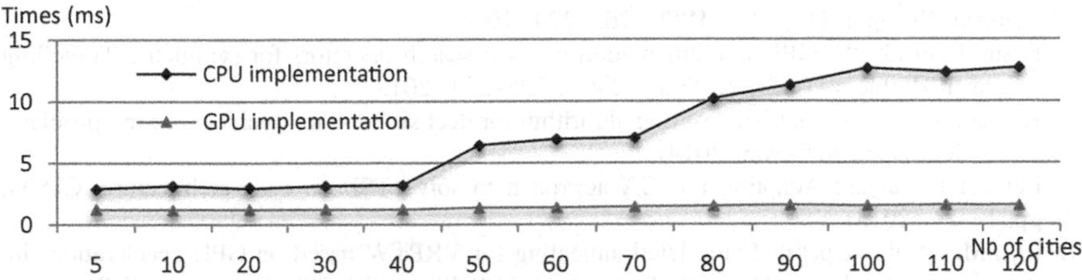

Fig. 5 GPU versus CPU of algorithm implementation for random problems

Here the acceleration factor is greater than 8 for n = 120 and this means that our algorithm is adapted meadows with the power of the GPU. Clearly, the increase of data size (number of cities and number of types of vehicle) might increase the acceleration factor.

5 Conclusion

In this paper, we have presented an algorithm and its GPU implementation for solving the multi capacity VPR. This algorithm exploits efficiently the parallelism of the GPU. Indeed, our solution computes in parallel an initial solution (tours) in one step and then improves this initial solution by selecting the best vehicles that minimize the cost. We adapted the move() and swap() procedures to the multi capacity case and we used them to improve in parallel the costs of all pairs of neighbor tours.

Our parallel algorithm gives feasible solutions with a best cost and in a very interesting time compared to the known others proposed solutions [7, 11]. To our knowledge, this is the first implementation on GPU of such class of heuristics that solve the VRP with multi capacity. Actually, we work to extend this approach (heuristic and GPU implementation) to the electric and green VRP and to the dynamic VRP.

Acknowledgments We thank Pr. K. Karagul who provided us the instances test.

References

1. Arbelaez, A., Codogne, P.: A GPU implementation of parallel constraint-based local search. In: 22nd Euromicro International Conference PDP'2014, pp. 648–655 (2014)
2. Baldacci, R., et al.: Routing a heterogeneous fleet of vehicles. Technical report DEIS OR INGCE (2007)
3. Benaini, A., Berrajaa, A., Daoudi, E.M.: GPU implementation of the multi depot vehicle routing problem. In: IEEE AICCSA 2015
4. Dell'Amico, M., et al.: Heuristic approach for the fleet size and mix vehicle routing problem with time windows. Transp. Sci. **41**(4), 516–526 (2007)
5. Desrochers, M., Verhoog, T.: A new heuristic for the fleet size and mix vehicle routing problem. Comput. Ops. Res. **18**(3), 263–274 (1991)
6. Fosin, J., et al.: "A GPU implementation of local search operators for symmetric Travelling Salesman Problem. Traffic & Trans. **25**(3), 225–234 (2013)
7. Karagul, K.: A new heuristic routing algorithm for fleet size and mix vehicle routing problem. GU J. Sci. **27**(3), 979–986 (2014)
8. Lekaez, U., et al.: Adapting the GA approach to solve TSP on cuda architecture. CINTI, pp. 19–21 (2013)
9. Li, J.M., et al.: A parallel simulated annealing for VRPTW based on GPU acceleration. In: Advances in Intelligent Decision Technologies SIST, pp. 201–208. Springer (2010)
10. Meesuptweekoon, K., Chaovalitwongse, P.: Dynamic vehicle routing problem with multiple depots. Eng. J. **18**(4), 135–149 (2014)
11. Ochi, L., et al.: A parallel evolutionary algorithm for the VRP with heterogeneous fleet. In: Parallel and Distributed Processing, pp. 216–224 (1998)
12. Renaud, J., Boctor, F.: A sweep-based algorithm for the fleet size and mix vehicle routing problem. European J. Oper. Res. **140**, 618–628 (2002)
13. Szymon, J., Dominik, Z.: Solving multi-criteria VRP by parallel tabu search on GPU. Proc. Comput. Sci. **18**, 2529–2532 (2013)
14. Talbi, E.G., Hasle, G.: Metaheuristics on GPUs. J. Parallel Distrib. Comput. **73**(1), 1–3 (2013)
15. Uthayopas, P., et al.: Speeding up the pickup and delivery problem with time windows using GPU cluster. Int. J. Eng. Ind. **4**(2), 53–61 (2013)
16. Golden, B., Assad, A., Levy, L., Gheysens, F.: The fleet size and mix vehicle routing problem. Comput. Oper. Res. **11**(1), 49–66 (1984)
17. Gendreau, M., Laporte, G., Musaraganyi, C., Taillard, É.D.: A tabu search heuristic for the heterogeneous fleet vehicle routing problem. Comput. Oper. Res. **26**(12), 1153–1173 (1999)

Backslide Based Correction for MAC Layer Misbehavior in Wireless Networks

Mohamed El Fissaoui, Abderrahim Beni-hssane and Mostafa Saadi

Abstract In this paper, we present an enhanced scheme to handle MAC layer misbehavior in wireless networks. It's known that IEEE 802.11 wireless LAN medium access control (MAC) provides transmission fairness to nodes in a network. But some nodes can take advantage and modify its MAC to improve its own performance at the expense of other nodes. To handle these misbehaving nodes, we use a backslide strategy to react against the misbehaving nodes, which is based on the number of times the misbehaving nodes they misbehave. Extensive simulation results are presented to validate our proposed strategy by improving honest nodes throughput and then the performance of the network.

Keywords Mac layer misbehavior · Wireless networks · IEEE 802.11

1 Introduction

Wireless ad hoc networks also known as IBSS—Independent Basic Service Set are networks formed by connecting computers using a wireless medium without an infrastructure. Ad hoc networks follow the IEEE 802.11 standard [1] for wireless networks.

The main advantage of The IEEE 802.11 standard for wireless networks is its fairness method to access to the medium for all nodes in the network. But some

---

M. El Fissaoui (✉) · A. Beni-hssane
LAROSERI Laboratory, Computer Science Department Sciences,
Chouaïb Doukkali University, El Jadida, Morocco
e-mail: mohamed.el.fissaoui@gmail.com

A. Beni-hssane
e-mail: abenihssane@yahoo.fr

M. Saadi
Département Informatique & Télécoms Ecole Nationale des Sciences
Appliquées (ENSA), Khouribga Université Hassan 1er, Settat, Morocco
e-mail: saadi_mo@yahoo.fr

Á. Rocha et al. (eds.), *Europe and MENA Cooperation Advances in Information and Communication Technologies*, Advances in Intelligent Systems and Computing 520, DOI 10.1007/978-3-319-46568-5_37

nodes can deviate from the standard and modify its MAC (Medium Access Control) to increase its access to the medium at expense of other nodes in the network.

In this paper a novel backslide strategy to penalize misbehaving node(s) and to improve honest nodes throughput is proposed and validated by simulation.

The rest of this paper is organized as follow: In Sect. 2 we briefly present IEEE 802.11 standard for wireless networks. In Sect. 3 we discuss the MAC layer misbehavior. In Sect. 4 we give an overview on related work. In Sect. 5 we present our proposed scheme. Simulation results are presented in Sect. 6, while Sect. 7 concludes the paper.

2 IEEE 802.11 Standard for Wireless Networks

In IEEE 802.11 standard for wireless networks there are two modes that can be used by nodes to access the medium: Distributed Coordination Function (DCF) or Point Coordination Function (PCF). The IEEE 802.11 DCF is the main access method of IEEE 802.11 MAC, and it uses carrier sense multiple access with collision avoidance (CSMA/CA) and random backoff time. The IEEE 802.11 PCF uses a master node that polls devices in the network to send and receive data. Binary Exponential Backoff of the IEEE 802.11 standard assures randomness, which is guarantees fair throughput, for participants' nodes accessing the medium.

3 MAC Layer Misbehavior

In IEEE 802.11 standard, fair access to the shared medium is provided by BEB algorithm. Some nodes may use the shared medium more than other nodes by modifying its MAC parameter. This kind of behavior is named MAC Layer Misbehavior, and these nodes are referred to as misbehaving nodes or selfish nodes.

Nodes that follow the IEEE 802.11 standard are referred to as genuine nodes or honest nodes. In this work two types of misbehaviors [2], are used: α-misbehavior and CWfix-misbehavior. In α-Misbehavior, the misbehaving node chooses a backoff at random from [0, α (CW-1)] where $0 < \alpha < 1$ and use the shared medium more often than other nodes. In the Fixed contention window, instead of choosing a random backoff from [0, CW-1]. The misbehaving node modifies its contention window to a fix value and always chooses its backoff interval randomly from [0, CWfix].

4 Related Work

Researchers have proposed many solutions to detect and react against MAC Layer misbehavior. In this section we present the literature related to MAC misbehavior. In this work we are interested just in reaction scheme.

4.1 Detection Schemes

In [3] Radosavac et al. monitor backoff intervals of node neighbors' and apply Sequential Probability Ratio Test on the used backoffs to detect the misbehavior. Rong et al. [4] uses the Sequential Probability Ratio on packet inter arrival time distribution for detecting misbehavior.

DOMINO [5] System for Detection Of greedy behavior in the MAC layer of IEEE 802.11 public Networks, has been proposed by Raya et al. Domino, it's deployed in access points to detect misbehavior.

In [6] centralized detection scheme is used to detect Selfish nodes. Ming Li et al. [7] propose a realtime selfish misbehavior detection scheme through honest nodes' observations.

4.2 Reaction Schemes

Modifications to the IEEE 802.11 protocol have been proposed by Kyasanur et al. [8] to detect the misbehavior and to penalize misbehaving nodes. Concept of receiver-assigned backoff in order to detect misbehavior is presented by authors. DREAM [9] system for detection and reaction to a timeout MAC-layer misbehavior is a detection and reaction scheme in the same time.

Cardenas et al. [2] evaluate the theoretical performance of DOMINO [5] and SPRT [3] and by simulations. Guang et al. [10] propose modifications to the BEB algorithm to facilitate the misbehavior detection and decrease misbehaving sender performance. In [11] Jaggi et al. use throughput degradation to detect misbehavior. Alocious et al. [12]. Use centralized approach; it enables the access point to allocate backoff values for nodes.

The reaction of previous work, against misbehaving nodes is not adequate enough, because the same nodes can misbehave many times without being penalized. That is why; we propose a penalizing strategy to prevent misbehaving nodes to use the shared medium, based on the number of times they misbehave. Simulation results show the efficiency of our proposed scheme to penalize misbehaving nodes and to improve honest nodes throughput.

5 Our Proposed Scheme

The basic idea of our proposed scheme is to react against misbehaving nodes by using backslide correction strategy. If the node(s) misbehave for the first time the honest nodes use a collective strategy to react against misbehaving nodes in way for misbehaving node(s) to behave like honest nodes, but if the misbehaving node misbehaves for the second time, the honest nodes react aggressively against misbehaving nodes(s). Third misbehaving action will not be tolerated; the honest nodes will react in a way to prevent the misbehaving node to use the network. To our knowledge this the first work that use backslide correction strategy to penalize the misbehaving node(s).

Misbehaving nodes use α-misbehavior to misbehave, in the other side honest nodes uses fixed contention window to react. When nodes misbehave many times, the value of contention window become smaller and smaller until the misbehaving node is prevented from using the shared medium. Algorithm 1 shows our backslide correction scheme.

In Algorithm 1 when honest nodes detect the misbehaving node for the first time, they use a fixed contention window misbehavior to react, but when the misbehavior node appear the second time, honest nodes act aggressively by reducing the size of the contention window, In the third misbehaving, honest nodes use smaller contention window in way to prevent misbehaving node to use the shared medium.

Algorithm 1: Backslide correction strategy.

```
 If Misbehavior AND MisbehaveCount = 0 then
         MisbehaviorType = CWfix-Misbehavior
         Misbehavevalue  = CWfix
         MisbehaveCount  = MisbehavueCount + 1
else if Misbehavior AND MisbehaveCount = 1 then
         MisbehaviorType = CWfix-Misbehavior
         Misbehavevalue  = CWfix/2
         MisbehaveCount  = MisbehavueCount + 1
else
         MisbehaviorType = CWfix-Misbehavior
         Misbehavevalue  = CWfix/4
  end if
```

6 Simulation and Performance Evaluation

We use NS-2 for the simulation. NS2 is discrete event simulator written in OTCL and C++. NS-2 is free and open source and it supports a variety of protocols. Most of the wireless network research all over the world is simulated in NS-2. The simulation parameter is presented in Table 1.

Throughput metric are used to evaluate the effectiveness of our proposed scheme.

In Fig. 1 When node misbehave, the throughput of honest nodes fall about 35 Kbps in comparison to its previous throughput when there is no misbehaving node. This is due to misbehaving node presence. After the first reaction the throughput of the honest nodes starts to increases again and the throughput of the

Table 1 Simulation parameter

Simulation parameter	Value
Field size	100 × 100 m
Data rate	2 Mb/s
Simulation time	6000 s
MAC	802.11.b
Packet size	512 Bytes

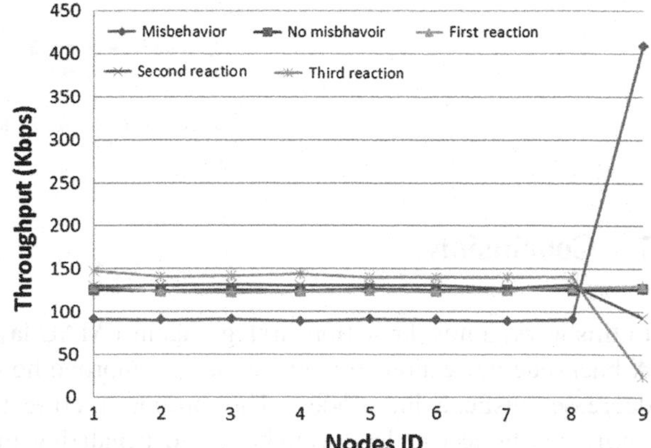

Fig. 1 Throughput with 10 nodes

misbehaving node decrease to the same as the honest nodes. When the node misbehaves for the second time, the honest nodes react aggressively to decrease misbehaving node throughput (about 40 Kbps) and increase its own throughput. But when the node misbehaves the third time, honest nodes react in way to prevent misbehaving node to use the network as shown in the Fig. 1.

In Fig. 2, when misbehaving node start to misbehave its throughput increase about 300 Kbps and the honest nodes throughput decrease dramatically in comparison to its throughput before the misbehavior. But when honest nodes react the first time its throughput start to recover and misbehaving node is forced to react normally as the other honest nodes. In the second reaction misbehaving node throughput decrease about 40 %. When the misbehavior is detected for the third time, the misbehaving node is prevented to use the shared medium as shown in Fig. 2.

With 20 nodes, the impact of misbehavior becomes more harmful on honest nodes throughput as shown in Fig. 2. But when honest nodes use the backslide correction strategy, its throughput improves significantly.

By using backslide correction strategy; the misbehaving node(s) is handled in an appropriate way, by counting the number of times it misbehaves and penalizing it by preventing it to use the shared medium when the misbehavior occurs for the third time.

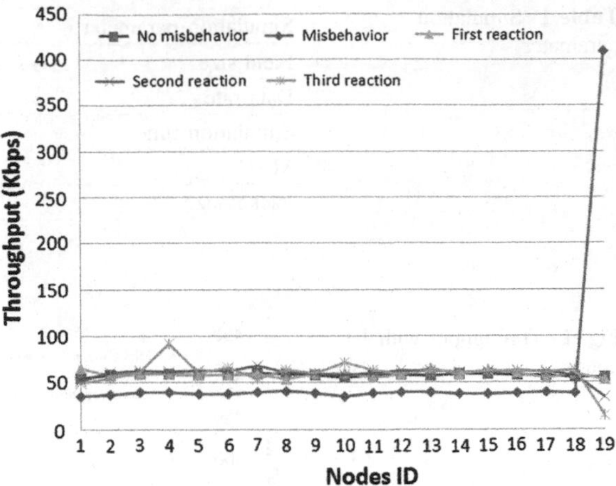

Fig. 2 Throughput with 20 nodes

7 Conclusion

In this work, a novel reaction strategy against MAC layer misbehavior is presented. A backslide correction strategy is used to improve honest nodes throughput and to decrease misbehaving nodes. The proposed correction scheme is effective in improving honest nodes throughput and penalizing misbehaving nodes. Our proposed backslide correction strategy is shown to be effective in handling MAC Layer Misbehavior, even if there is more than one misbehaving nodes, by counting the number of time the nodes misbehave and attach it the misbehaving node ID.

As part of our future work, we will try to minimize the time between detection and reaction of misbehavior in way that, the reaction take place the same time the misbehavior is detected in.

References

1. IEEE standard for information technology- telecommunications and information exchange between systems- local and metropolitan area networks specific requirements- part 11: Wireless LAN medium access control (mac) and physical layer (Phy) specifications. ANSI/IEEE Std 802.11, 1999 Edition (R2003) (2003)
2. Cardenas, A.A., Radosavac, S., Baras, J.S.: Evaluation of detection algorithms for mac layer misbehavior: Theory and experiments. IEEE/ACM Trans. Netw. (2009)
3. Radosavac, S., Baras, J.S., Koutsopoulos, I.: A framework for mac protocol misbehavior detection in wireless networks. In: Proceedings of the 4th ACM Workshop on Wireless Security, WiSe 05 (2005)
4. Rong, Y., Lee, S.K., Choi, H.A.: Detecting stations cheating on backoff rules in 802.11 networks using sequential analysis. In: Proceedings of IEEE INFOCOM 2006. 25TH IEEE International Conference on Computer Communications (2006)
5. Raya, M., Aad, I., Hubaux, J.P., El Fawal, A.: Domino: Detecting mac layer greedy behavior in IEEE 802.11 hotspots. IEEE Trans. Mobile Comput. **5**, (2006)

6. Choi, Min, A.W., Shin, K.G.: A lightweight passive online detection method pinpointing misbehavior in wlans. IEEE Trans. Mobile Comput. (2011)
7. Li, M., Salinas, S., Li, P., Sun, P., Huang, X.: Mac-layer selfish misbehavior in IEEE 802.11 ad hoc networks: Detection and defense. IEEE Trans. Mobile Comput. (2015)
8. Kyasanur, P., Vaidya, N.H.: Selfish mac layer misbehavior in wireless networks. IEEE Trans. Mobile Comput. (2005)
9. Dream: A system for detection and reaction against MAC layer misbehavior in ad hoc networks. Comput. Commun. (2007)
10. Guang, L., Assi, C., Benslimane, A.: Modeling and analysis of predictable random backoff in selfish environments. In: Proceedings of the 9th ACM International Symposium on Modeling Analysis and Simulation of Wireless and Mobile Systems, MSWiM '06 (2006)
11. Jaggi, N., Giri, V.R., Namboodiri, V.: Distributed reaction mechanisms to prevent selfish misbehavior in wireless ad hoc networks. In: Global Telecommunications Conference (GLOBECOM 2011). IEEE (2011)
12. Alocious, C., Xiao, H., Christianson, B., Malcolm, J.: Evaluation and prevention of mac layer misbehaviours in public wireless hotspots. In: 2015 IEEE International Conference on Computer and Information Technology; (CIT/IUCC/DASC/PICOM) (2015)

Impact of Location Data Freshness on Routing in Wireless Sensor Networks

Rania Khadim, Ansam Ennaciri, Mohammed Erritali and Abdelhakime Maaden

Abstract The most critical and important parameter to study is the Location Cache Maximum Age (LCMA). This parameter defines the maximum age at which location information is not considered fresh enough to be used when routing packets. Instead of this, a location request is launched to catch a more fresh location. In other words, it defines the freshness of the stored position in the location cache memory. In this paper, we propose a combination of GPSR (Greedy Perimeter Stateless Routing) and Location-based Services (Grid and Hierarchical Location Services (GLS/HLS)). In order to implement this concept, we have proposed a patch over the NS-2 simulator which mixes GPSR, GLS and HLS according to our proposal. We have conducted various simulations and we have varied the freshness of the destination location data. The obtained results are promising in terms of latency, packet delivery rate and overhead.

Keywords Wsns · Location-based services · Routing protocols · LCMA

1 Introduction

WSNs (Wireless Sensor Networks) are a special case of MANETs (Mobile Ad hoc Networks). Usual topology based routing protocols have limited performances in such networks due to the high network dynamics. Geographic routing protocols

---

R. Khadim (✉) · A. Maaden
Laboratory of Mathematics and Applications, Faculty of Sciences and Technics,
Sultan Moulay Slimane University, B.P: 523, Beni Mellal, Morocco
e-mail: khadimrania@gmail.com

A. Ennaciri · M. Erritali
TIAD Laboratory, Department of Computer Sciences, Faculty of Sciences
and Technics, Sultan Moulay Slimane University, B.P: 523, Beni Mellal, Morocco
e-mail: ennaciri.ansam@gmail.com

M. Erritali
e-mail: m.erritali@usms.ma

Á. Rocha et al. (eds.), *Europe and MENA Cooperation Advances in Information and Communication Technologies*, Advances in Intelligent Systems and Computing 520, DOI 10.1007/978-3-319-46568-5_38

were designed to provide better performances for such networks. Each node has to care about its actual geographic position and the position of the targeted node to reach. The paradigm position-to-position is used. The Location-based Services is required to catch the destination position. The combination of this service with routing is quite natural in order to ensure better performances. When a sender needs to send a packet to a destination, the sender looks for the destination position in its records. If it is fresh enough, the packet is then sent immediately. If not, the sender sends an HLS or GLS request in order to get the actual destination position. When this later is received, the packet is then sent.

For this purpose, we have proposed a patch over the NS-2 simulator which mixes GPSR, GLS and HLS according to our proposal. We have undertaken a set of experimentations and we have considered the freshness of the destination location data. On one hand, we have presented our results in terms of latency, packet delivery rate and overhead and on the other hand the freshness of the location data has a real impact on the network performances.

The paper is organized as follows. Section 2 is dedicated to related works. Section 3 details our combination algorithm about GPSR, GLS and HLS. Section 4 talks about our experimentations and the obtained results. Finally, Sect. 5 concludes the study and gives some hints about future works.

2 Related Works

2.1 Geographic Routing Protocols

Routing protocols algorithms must choose some criteria to make routing decisions, for instance the number of hops, latency, transmission power, bandwidth, etc. The Topology-based Routing Protocols suffer from heavy discovery and maintenance phases, which lead to scalability problems. This is due to the high mobility, which generates frequent topology changes and short links. This is why Geographic Routing Protocols are suitable for large scale wireless sensor networks.

The first routing protocol using the geographic information is the Location-Aided Routing (LAR) [1]. The LAR [1] is a reactive routing protocol that uses the geographic information in the route discovery, to limit the propagation of route request packets RREQs to the geographic region where it is most probable the destination is located. The route discovery is initiated in a Request Zone. If the request doesn't succeed, it initiates another request with a larger Request Zone and the decision is made on a routing table.

The first real geographic routing protocol is the Greedy Perimeter Stateless Routing (GPSR) [2]. It is a reactive and efficient routing protocol designed and adapted for mobile ad hoc networks and sensor networks. It forwards the packet to the target's nearest neighbor (Greedy Forwarding approach) until reaching the destination. Therefore, it scales better than the topology based protocols, because it uses the geographical position of the nodes for routing packets.

Another geographic routing protocol is the Distance Routing Effect Algorithm for Mobility (DREAM) [3]. DREAM is a proactive routing protocol where each node maintains a location table that contains the geographical coordinates of all the destinations obtained by a positioning system such as GPS. The use of this location information allows calculating the direction of each destination and the distance to each. Each node broadcast packets containing its current position within the network, the frequency of the broadcast is determined by considering:

- Distance effect: More important is the distance between two nodes, slower they move relative to each other. Therefore, remote nodes share their location information with each other less frequently than close nodes.
- Node mobility rate: A node with a high mobility must frequently inform the other nodes of its location and vice versa [4].

To forward a packet, a node looks up the location of the destination in its local location table, and forwards the packet to all its neighbors in the direction of the destination. DREAM can deliver exponentially many copies of a packet to the destination, which ensures better reliability [5].

We have used the Greedy Perimeter Stateless Routing (GPSR) as the geographic routing protocol for the combination.

2.2 Location-Based Services

Geographic Routing Protocols use location information when they need to route packets. Obviously, location information is maintained by Location-based Services provided by network nodes in a distributed way. Routing and location services are very related but are considered separately in the literature. The Location-based Services (in the context of WSNs) is a distributed service without infrastructure in general. It has to answer to a location query such as: "Where is the node X?".

There are mainly two types of Location services according to Fig. 1: Flooding-based and Rendezvous-based location services. An example of flooding-based location services is Reactive Location Service (RLS) [6].

In this service, every node floods its location request through the whole network. Once this request has reached a node with information, it responds directly. The location response is sent after receiving a location request.

The two major Hierarchical Rendezvous-based location services are, the Grid Location Service (GLS) [7] and the Hierarchical Location Service (HLS) [8].

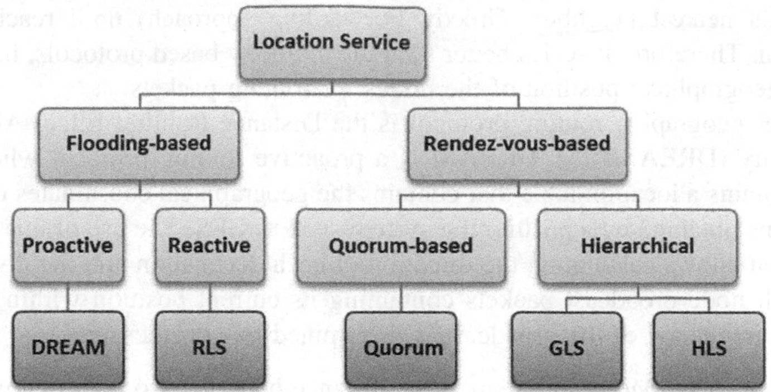

Fig. 1 Location-based services taxonomy

The network is divided into several levels. At each level, a node recruits location servers. The location query is forwarded up and down in the hierarchy. This limits the number of forwarded packets and avoids flooding.

Nodes in Hierarchical Location Services must elect location servers in different levels to keep updating their positions with those servers. When another node needs this position, it sends a request to those servers which reply with a location response packet indicating location information received from the last update.

In GLS (Fig. 2a), each node sends a periodic update of its position to all location servers at all levels. While in HLS (Fig. 2b), to update its location information, a node frequently sends packets to the responsible cell (the cell where a node must select its location servers) at the first level and only the level i responsible cell's location is sent to the level i + 1.

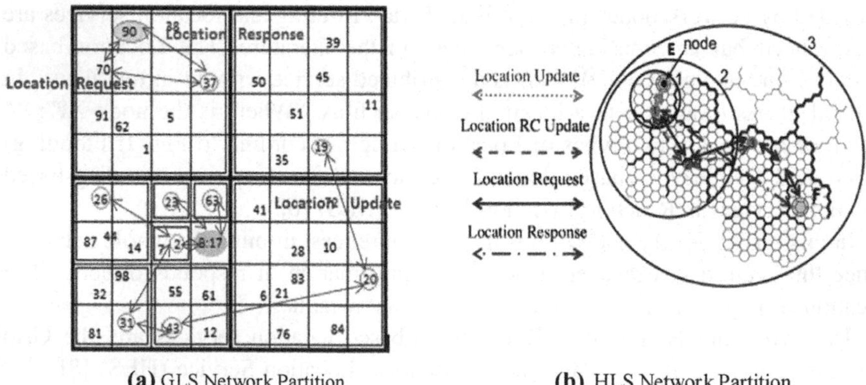

(a) GLS Network Partition **(b)** HLS Network Partition

Fig. 2 Hierarchical network partition

3 GPSR and Location-Based Services Combination

3.1 Description

GPSR takes care of routing packets and GLS or HLS are called to get destination position when the target node position is unknown or is not fresh enough. When a destination is quite far away from the sender, the exact position of the target is first calculated.

The GLS and HLS location algorithm is presented in the Fig. 3. In GLS and HLS if a node needs to send a packet to another node, it looks for its position. If it has a fresh location, it uses it to send data until reaching the destination. Otherwise, it sends a location request to catch the new position. After receiving the response, it begins sending data.

3.2 Algorithm

The function Location (Algorithm 1) handles the destination position queries; it looks into the local cache memory of the current node and updates the packet information with the destination position.

Algorithm 1 GLS or HLS :: Location

```
cacheThreshold ← LocationCacheMaxAge ;
procedure LOCATION(packetToSend)
if (Destination Location age < cacheThreshold) then
PreparePacket (packetToSend);
ChooseNextBestHop(packetToSend);
ForwardPacket(packetToSend);
else
LaunchPositionQuery(destination);
StickToBuffer(packetToSend);
end if
end procedure
```

4 Simulations

4.1 Working Environment for Experimentations

The simulations were performed using the NS-2 Simulator 2.35 [9]. The geographic routing protocol used is Greedy Perimeter Stateless Routing (GPSR) [2]. The selected area is a 2 * 2 km^2. The Media Access Control (MAC) layer used is the

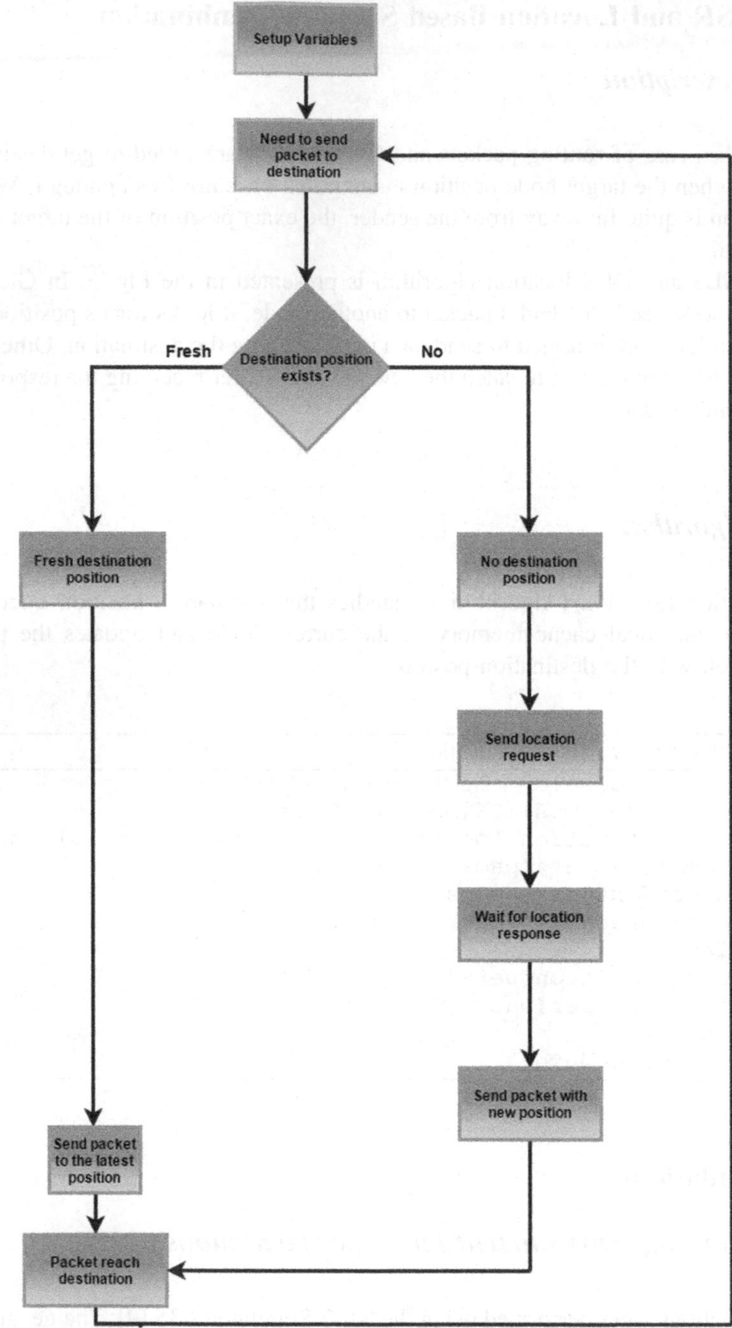

Fig. 3 Schematic presentation of routing and grid/hierarchical location based combination algorithms

Table 1 The simulation parameters

Parameters	Value
Channel type	Channel/WirelessChannel
Propagation model	Propagation/TwoRayGroud
Network interface	Phy/WirelessPhyExt
MAC layer	802.11p
Interface queue type	Queue/DropTail/PriQueue
Link layer	LL
Antenna model	Antenna/OmniAntenna
Interface queue length	512 packets
Ad-hoc routing protocol	GPSR
Location-based service	GLS or HLS
Location cache maximum age	4, 8, 12, 16 and 22 s
Number of nodes	100
Area	2 * 2 km^2
Simulation time	150 s
GPSR beacon interval	0.5 s
CBR traffic	4 * 20 packets/node
CBR packet size	128 B
CBR send interval	1 s

IEEE 802.11p. The simulation is run 10 times and the results are averaged for more accuracy. The parameters used in the simulations in NS-2 are summarized in Table 1.

4.2 Experimentation Results

Our experimentations have provided the following figures. On each figure, we show the network behavior with HLS and GLS.

In Fig. 4, we observe that when the LCMA decreases, the number of sent location requests also decreases and the HLS scheme presents the lowest values on comparison with the GLS scheme. This is due to the efficiency of the HLS, which reduces the location requests.

Figure 5 shows that the average overhead in the HLS scheme is lower to the average overhead of the GLS scheme. The difference comes mainly from location updates, because location requests are not undergoing change.

Figure 6 presents the Packet Delivery Rate (PDR). The PDR is defined as the ratio between the received and emitted CBR packets. This Figure shows that the PDR in the HLS scheme is greater than the PDR of the GLS scheme. Regardless of the value of the LCMA, the PDR is always improved.

The latency is measured as the time between the moment of dispatch and the moment of receiving CBR packets. In Fig. 7, we observe that the average latency in the HLS scheme is lower than the average latency of the GLS scheme.

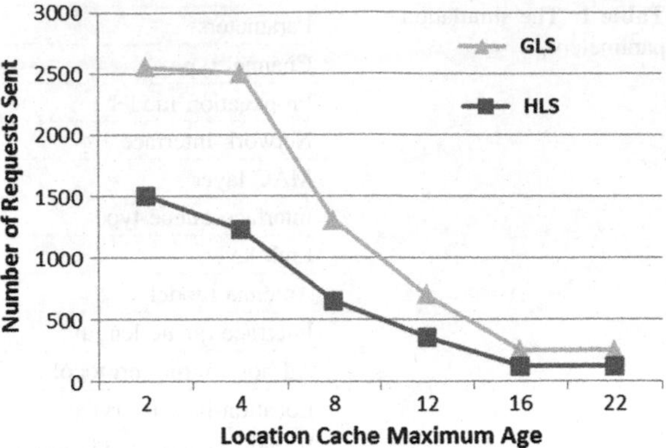

Fig. 4 Location requests number average

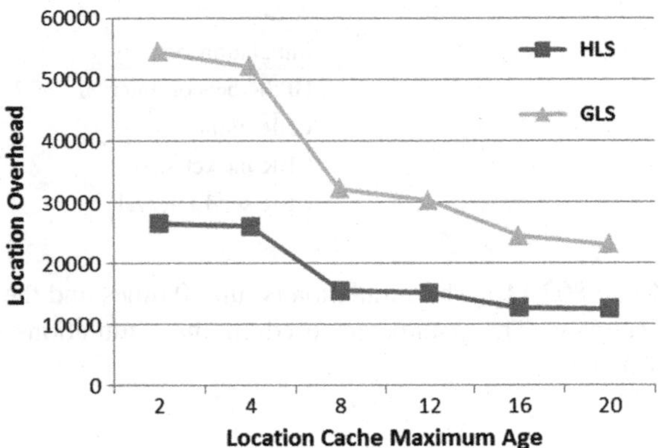

Fig. 5 Overhead average

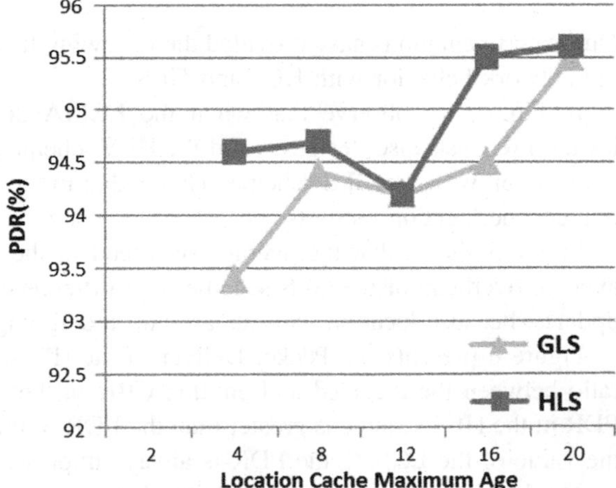

Fig. 6 Packet delivery rate

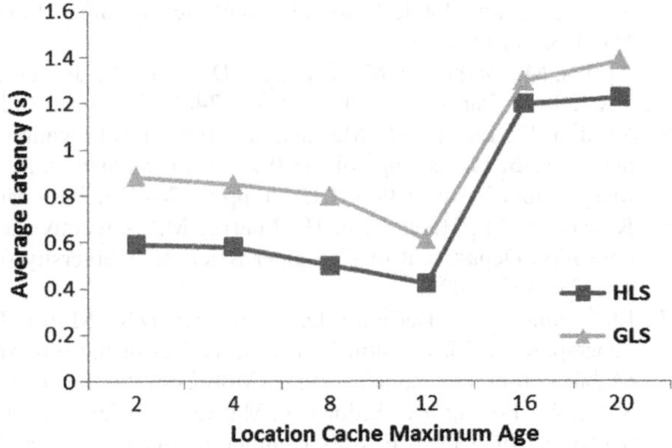

Fig. 7 Latency average

We finally conclude that the results observed on the HLS scheme are better than the GLS scheme, in term of LCMA. This is due to the fact that in HLS location requests are not launched consistently, but the packet is sent directly with old location information. This reduces the location cost like the location requests and then the overhead without affecting the network performances such as the latency and the packet delivery rate, instead it even improves them.

5 Conclusion and Future Works

We have proposed a combination approach to handle location based service with routing protocols in WSNs. We implemented it in the NS-2 framework within a patch. We have conducted many experimentations with this patch in order to observe network performances.

We have shown in this paper that a smart combination of GPSR with Location-based services (GLS or HLS) could provide better results in terms of network performances in particular for the packet delivery rate, the latency and the overhead. As a future work, we intend to check the network behavior for larger networks with higher density.

References

1. Ko, Y.-B., Vaidya, N.H.: Location-aided routing (lar) in mobile ad hoc networks. Wirel. Netw. **6**(4), 307–321 (2000)
2. Karp, B., Kung, H.T.: Gpsr greedy perimeter stateless routing for wireless networks. In: Proceedings of the 6th Annual International Conference on Mobile Computing and Networking (MobiCom'00), pp. 243–254. NewYork, NY, USA (2000)
3. Basagni, S., Chlamtac, I., Syrotiuk, V.R., Woodward, B.A.: A Distance Routing Effect Algorithm for Mobility (DREAM). In: Proceedings of the 4th Annual ACM/IEEE International

Conference on Mobile Computing and Networking (MobiCom'98), pp. 76–84. New York, NY, USA (1998)
4. Ayaida, M., Barhoumi, M., Yacine, G.D., Afilal, L.: Joint routing and location-based service in VANETs. J. Parallel Distrib. Comput. **74**(2), 2077–2087 (2014)
5. Khadim, R., Erritali, M., Maaden, A.: Hierarchical location-based services for wireless sensor networks: In Proceedings of the IEEE 13th International Conference on Computer Graphics, Imaging and Visualization (CGIV), pp. 457—463. Beni-Mellal, Morocco, March 2016
6. Kasemann, M., Hartenstein, H., Mauve, M.: A reactive location service for mobile ad hoc networks, Department of Computer Sciences, University of Mannheim TechRep, TR02014, pp. 121–133 (2002)
7. Li, J., Jannotti, J., DeCouto, D.S.J., Karger, D.R., Morris, R.: A scalable location service for geographical ad hoc routing. In: Proceedings of the 6th Annual International Conference on Mobile computing and networking (MobiCom'00), pp. 120–130. New York, NY, USA (2000)
8. Kiess, W., Fussler, H., Widmer, J., Mauve, M.: Hierarchical location service for mobile ad-hoc networks. SIGMOBILE Mob. Comput. Commun. Rev. **8**, 47–58 (2004)
9. http://nsnam.isi.edu/nsnam/

Detecting Text Similarity Using MapReduce Framework

Marouane Birjali, Abderrahim Beni-Hssane, Mohammed Erritali and Youness Madani

Abstract The evaluation of similarities between textual documents was regarded as a subject of research strongly recommended in various domains. There are many of documents in a large amount of corpus. Most of them are required to check the similarity for validation. In this paper, we propose a new MapReduce algorithm of document similarity measures. Then we study the state of the art of different approaches for computing the similarity of amount documents to choose the approach that will be used in our MapReduce algorithm. Therefore, we present how the similarity between terms is used in the assessment of the similarity between documents. Simulation results, on Hadoop framework, show that our MapReduce algorithm outperforms classical ones in term of running time.

Keywords Hadoop cluster · Document similarity · MapReduce programming model · Similarity measure

M. Birjali (✉) · A. Beni-Hssane
LAROSERI Laboratory, Faculty of Sciences, Department of Computer Sciences, University of Chouaib Doukkali, El Jadida, Morocco
e-mail: birjali.marouane@gmail.com

A. Beni-Hssane
e-mail: abenihssane@yahoo.fr

M. Erritali · Y. Madani
TIAD Laboratory, Faculty of Sciences and Technologies, Department of Computer Sciences, University of Sultan Moulay Slimane, Béni Mellal, Morocco
e-mail: m.erritali@usms.ma

Y. Madani
e-mail: younesmadani9@gmail.com

Á. Rocha et al. (eds.), *Europe and MENA Cooperation Advances in Information and Communication Technologies*, Advances in Intelligent Systems and Computing 520, DOI 10.1007/978-3-319-46568-5_39

1 Introduction

In our days, we are witnessing an unceasing development of information technologies. These new technologies have enabled an exponential increase in the volume of data by online contents like blogs, posts, social networking.... Every day 2.5 quintillion bytes of data are created and a very large amount so the 90 % of data in the world are created in last 2 years [1]. This rapid increase in the volume of information has created the problem of how to find the pertinent information that interests us in this great mass of data. To overcome this problem a discipline as a whole is born. This discipline is called the search for relevant information in a Big Data environment.

In an Information Retrieval System (IRS), each document is represented by an intermediate representation. It describes the contents of the document by descriptors. These descriptors are significant units in the document. In our context, to find the relevant documents by comparison with a document query, the ISR compares the representation of this query to the representation of each document.

In this work, we specifically focus on text corpus to find the relevant documents based on similarity measure. The identification of the similarity between documents resulting from the indexing and the similarity measure that is a fundamental phase in our work.

The remainder of this paper is structured as follows:

In the next section, we proposed our MapReduce algorithm of document similarity measures. We then study in Sect. 4 the state of the art of different approaches of the similarity measure. Then the analysis and simulation of our MapReduce algorithm in Sect. 5, before concluding and present the perspectives of this work.

2 Related Work

Many studies have been presented on detecting document similarity in recent years for facilitating the search for information in complex information systems. Kumar et al. [2] and Chowdhury [3] surveyed duplicate or near duplicate data detection algorithms. Related work on text similarity detection can be mainly classified into two categories: traditional method and parallel method.

For the traditional methods, Lyon et al. [4] proposed a characteristic distribution defines the trigrams of words, and actions to find the similarity based on theoretical principles using the MD5 hash function or report the similarity between the words based on the overlapping ratio of a similar value or maximum common subsequence. Matveeva [5] presented a vector space model (VSM) algorithm to compute the similarity between the text document using the cosine measure the vector. Yih [6] explored different approaches, based on the weight of TF-IDF to study the long-term weight function.

For methods based in parallel, most methods centered model MapReduce. Zhang et al. [7] presented a sequence-based method to detect the relevent similarity of the web page using MapReduce, which is composed of two tasks that the sentence level near duplicate detection and the corresponding sequence. However, a manual procedure is impossible with a large corpus. Many of applications using the similarity detection, such as the similarity of recommendation [8].

3 Our Proposed MapReduce Algorithm

In this section, we propose our new MapReduce algorithm for computing the similarity measure of documents and queries. The Map phase is for each term of document, the mapper emits the document ID as the key, and his words as the value. The second phase, named Reduce, will take the output of the Map function and computes the similarity relation between each collection of values of each document and the query. This similarity measure computed by our algorithm with the use of the weight of the words and one of the approaches already existed in the next section.

To compute the similarity between documents and query we apply the following algorithm:

```
Class Mapper
    Method Map(Docid, term)
        RemoveStopWord&removePunctuation (term)
        For each element ∈ (Docid,term)
            Write(Docid,term)
        End for
Class Reducer
    Method Reduce(Docid,List(term))
        List(q) = indexing(Query)
        S=0
          X ← 0
          Y ← 0
        For each n ∈ List(term)
            F=calculateoccurence(n)
            For each e ∈ List(q)
                R= calculateoccurence(e)
            X←X+F×R×Sim(n, e)
                Y←Y+F×R
            End for
        End for
        S←X/Y
        Write(Docid, S)
```

Our similarity measure is presented by the following formula:

$$\mathrm{Sim}(q,d) = \frac{\sum_{i=1}^{n}\sum_{j=1}^{m} q_i \times d_j \times \mathrm{Sim}(i,j)}{\sum_{i=1}^{n}\sum_{j=1}^{m} q_i \times d_j} \quad (1)$$

i: represents the terms of the query q
j: represents the terms of the document d
q_i: is the frequency of the term i in query q
d_j : is the frequency of the term j in document d
Sim (*i*, *j*) : is the similarity measure between the two terms i and j

4 Similarity Measures

4.1 Measurement Approaches Similarity

Wu and Palmer Measurement: The principle of similarity measurement is based on the distances (N1 and N2) which separate the X and Y nodes from the root node and the distance (N) which separates the Subsuming Concept (SC) by this formula:

$$\mathrm{Sim}(X,Y) = \frac{2 \times N}{N1 + N2} \quad (2)$$

Resnik measure: The notion of the Informational Contents (IC) was initially introduced by [9]. The informational contents are obtained by computing the term frequency in the corpus. The formula of this measure is:

$$\mathrm{Sim}(c_1,c_2) = \mathrm{Max}[\mathrm{CS}(c_1,c_2)] = \mathrm{Max}[-\log P(\mathrm{CS}(c_1,c_2))] \quad (3)$$

$\mathrm{CS}(c_1,c_2)$: represents the most concept specific between the concept c_1 and c_2. *Jiang and Conrath Measure*: To cure the problem presented to the level of the Resnik measurement, Jiac [10] brought a new formula which consists in combining the Informational Contents of the specific concept to those of the concepts which we seeks the similarity. This approach is computed by the formula following:

$$\mathrm{Sim}(X,Y) = \frac{1}{\mathrm{distance}(X,Y)} \quad (4)$$

Leacock and Chodorow: The proposed measure combines between counting of the arcs method and the informational contents method. This technique is defined by the formula:

$$\mathrm{Sim_{lc}(X, Y)} = -\log\left(\frac{\mathrm{cd(X, Y)}}{2\times\mathrm{M}}\right) \quad (5)$$

M is the longest way length, which separates the concept root, of the concept more in bottom. We indicate that $cd(X, Y)$ is the shortest way length between X and Y.

4.2 *Evaluation of Measurement Approaches Similarity*

Our contribution propose a new algorithm for computing the similarity between text documents and the query based on MapReduce. To evaluate the comparison between these different approaches, we compute the similarity measure with the same document. The results show that the similarity changes for each approach, it is very important in the Leacock and Chodorow with good execution time. Based on this evaluation, our MapReduce algorithm will be based on Leacock and Chodorow approach. This table shows the computed similarity with the running time for each approach (Table 1).

5 Experimentation and Discussion of Our MapReduce Algorithm

Table 2 represents the processing time of our MapReduce algorithm based on Leacock and Chodorow approach for the same document to see how change similarity measure. We note that the similarity measure given by our proposed algorithm is two times of the similarity given by Leacock and Chadorow approach. However the running time is also twice but with the use of the main property of Hadoop, which is for processing and managing large data sets with a parallel, distributed algorithm, it decreases the running time in large number of documents as shown in the Fig. 1.

Table 1 Comparing of ranning time and similarity measure of deferent approaches

Approaches	Similarity measure	Running time (msec)
Leacock and Chodorow	0.18	1137
Wu and Palmer	0.15	1216
Resnik	0.09	1384
Jiang Conrath	0.11	1362

Table 2 Evaluation of running time and similarity measure for our approaches

Approach	Similarity	Time (msec)
Leacock and Chodorow	0.14	1016
Our MapReduce algorithm	0.26	2329

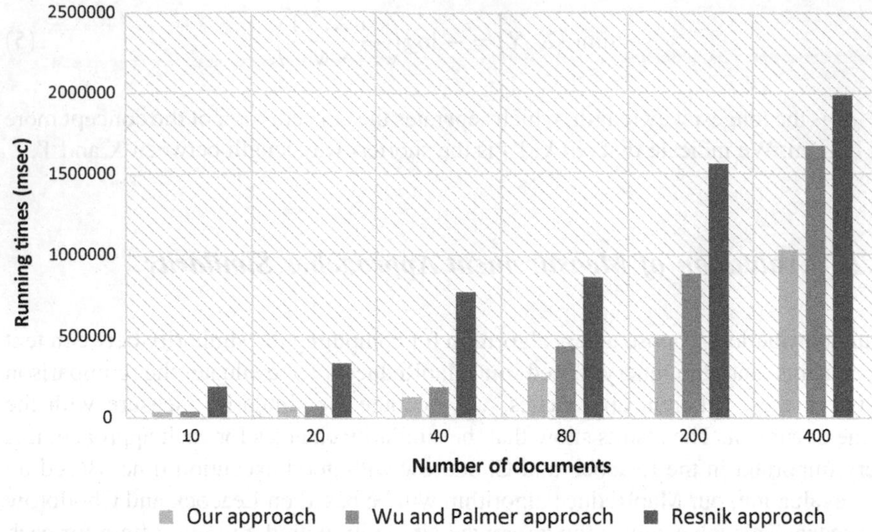

Fig. 1 Comparison between performance of running times in our approach and the state of the art ones

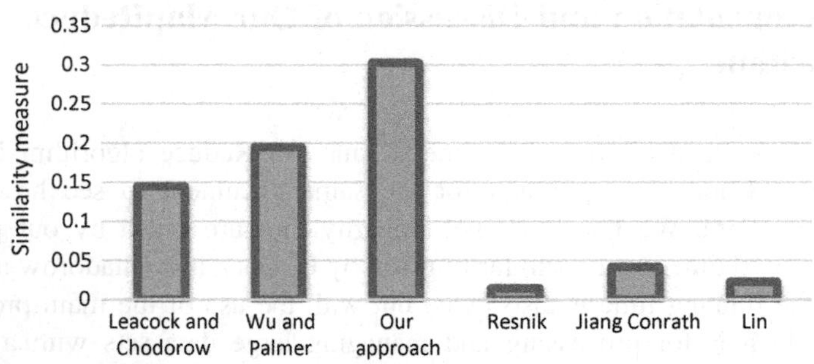

Fig. 2 Comparison between the semantic similarity measure of our approach and other approaches

To evaluate the efficiency of our MapReduce algorithm based on Leacock and Chodorow, we compared it with the approaches presented in Sect. 4.

Figure 2 shows the results of the similarity mesures between documents using our approach and the approaches already exist in the literature.

Our approach provides the double of the similarity measure compared to other approaches because it is based on the Learock and Chodorow approach for the method Cosine and the vector representation of the relations between the the concepts.

6 Conclusion and Future Research

The paper discussed our MapReduce algorithm to compute the similarity between a query and the documents and find the most pertinent documents. The simulation results, on Hadoop, that our Mapreduce algorithm outperforms the state of the art ones on running time performance and increases the measurement of the similarity. The future research involves integrate the notion of the semantic similarity measure and the recall and precision of our Mapreduce algorithm.

References

1. Improving Decision Making in the World of Big Data: http://www.forbes.com/sites/christopherfrank/2012/03/25/improving-decision-making-in-the-world-of-big-data/#7a89869c4b4d
2. Kumar, J.P., Govindarajulu, P.: Duplicate and near duplicate documents detection: a review. Eur. J. Sci. Res. **32**, 514–527 (2009)
3. Chowdhury, A.: Duplicate data detection. Accessed http://ir.iit.edu/~abdur/Research/Duplicate.html (2004)
4. Lyon, C., Malcolm, J., Dickerson, B.: Detecting short passages of similar text in large document collections. In: Processing of EMNLP (2001)
5. Matveeva, I.: Document representation and multilevel measures of document similarity. In: Proceedings of ACLHLT (2006)
6. Yih, W.: Learning term-weighting functions for similarity measures. In: Proceedings of EMNLP (2009)
7. Zhang, Q., Zhang, Y.Z., Hao, Yu, Huang, X.: Efficient partial-duplicate detection based on sequence matching. In: Proceedings of SIGIR (2010)
8. Almohsen, K.A., HudaaAl-Jobori, A.: Recommendera Systems in a light of big data. Int. J. Electr. Comput. Eng. (IJECE), vol. 5, no. 6, December 2015, pp. 1553–1563 (2015)
9. Lin, D.: An Information-Theoretic Definition of asimilarity. In: Proceedings of the Fifteenth International Conference on Machine Learning (ICML'98). Morgan-Kaufmann, Madison (1998)
10. Jiang, J., Conrath, D.: Semantic similarity based on corpus statistics and lexical taxonomy. In: Proceedings of International Conference on Research in Computational Linguistics, Taiwan (1997)

CIMP: Cloud Integration and Management Platform

Omar Sefraoui, Mohammed Aissaoui and Mohsine Eleuldj

Abstract Cloud Computing has emerged as a model, where all the essential components of a computer system such as software applications, platforms for software development or physical infrastructure are considered as services. Nowadays, the Cloud services offered is very diverse in terms of hardware and software, providing a wide range of configurations, and great flexibility. In this paper we propose a Cloud Integration and Management Platform (CIMP), aimed to act as an intermediate between users and Cloud solutions and offers additional components that enhance their functionalities. In addition, we propose an experimental validation showing the feasibility of our approach.

Keywords Cloud solutions · OpenStack · Dynamic reconfigurable component · Cloud migration

1 Introduction

Today's information system (IS) becomes increasingly complex making it difficult to follow the complex growing demands of information technology (IT) resources requirements. This leads to study and suggest new approaches to rationalize and optimize these resources.

In this way, a new concept of Cloud Computing has emerged as a technical and economic model for the use of IT resources [1]. This is the new trend of computing where IT resources are dynamically scalable, virtualized and exposed as a service

O. Sefraoui (✉) · M. Aissaoui
National School of Applied Sciences, Mohammed First University, Oujda, Morocco
e-mail: osefraoui@gmail.com

M. Aissaoui
e-mail: aissaoui@ensa.ump.ma

M. Eleuldj
Mohammadia School of Engineers, Mohammad V University, Rabat, Morocco
e-mail: eleuldj@emi.ac.ma

Á. Rocha et al. (eds.), *Europe and MENA Cooperation Advances in Information and Communication Technologies*, Advances in Intelligent Systems and Computing 520, DOI 10.1007/978-3-319-46568-5_40

on the Internet [2]. Cloud Computing is often associated with the supply of new mechanisms that allow providers to give users access to a virtually unlimited number of resources (Resource Outsourcing) [3].

In the Cloud, the equipment is provided in the form of virtual machines running by a hypervisor software. Each virtual machine is characterized by a set of hardware resources, consisting essentially of CPU, memory and external storage network, software and platform. The provisioning of virtual machines is on-demand and dynamically allocated to users. In Cloud system, the clients order resources in the form of a lease, but in general they use fewer resources than requested. These unused resources are a loss for the users as well for the provider.

Different solutions exist for the deployment of open source Clouds. Among these, Nimbus [4], Eucalyptus [5], OpenStack [6], OpenNebula [7] and CloudStack [8].

We present in this paper, a Cloud Integration and Management Platform (CIMP) that aimed to:

- Integrate Cloud solutions for Cloud provider.
- Capture user specifications and translate them into commands that will help generate the Cloud settings.
- Provide additional components that enhance the functionality of Cloud manager or provide missing functionality.

In the remaining of this paper, a review of the related work is presented. Then a presentation of CIMP architecture is giving. We present the CIMP Interface and components. Then, the result of the implementation is shown using OpenStack.

2 Related Works

Several Cloud solutions are available nowadays:

Eucalyptus [5] is a solution that allows the installation of a private and hybrid Cloud infrastructure, with a main storage controller walrus and controllers on each node.

OpenStack [6] is an open source Cloud Computing solutions. OpenStack controls large pools of compute, storage, and networking resources.

OpenNebula [7] is an open source project aimed at building the industry standard opensource Cloud Computing tool to manage the complexity and heterogeneity of large and distributed infrastructures.

Apache CloudStack [8] is an opensource solution designed to deploy and manage large networks of virtual machines as Infrastructure-as-a-Service (IaaS).

Recent studies have addressed the use of managing Cloud environments.

In [9], the authors present current efforts to develop an opensource Cloud Application Management Framework (CAMF) based on the Eclipse Rich Client Platform.

The paper [10] present a CloudTUI-FTS tool able to interact with different Cloud platforms, the user can perform both basic tasks (e.g., start-up/shut-down a service) and advanced tasks (e.g., create policies and mechanisms to prevent faults and to provide service scalability).

The different Cloud solutions contain some common components and modules like: Compute node, Network, Dashboard, Virtualization, VM.

We focus mainly in this paper on integration and management platform for the optimization of Cloud management. We discuss the development and the implementation of CIMP and component for helping tenant to select an objective parameters initialization. It can be done with Interface and Components.

3 Cloud Integration and Management Platform

The Cloud solutions are comprised of a set of components aimed to manage Cloud resources. As shown in Fig. 1, the KPI measuring, is used to collect measurements data, and is transferred to the component for apply policy.

In order to integrate a many different Cloud solutions, we propose the CIMP platform as shown in the Fig. 2, where the CIMP act as an intermediary between users and Cloud Computing solutions, different users (U1, U2, ... Un) can exploit the CIMP that supports multiple Cloud deployment solutions (CS1, CS2, ... CSm).

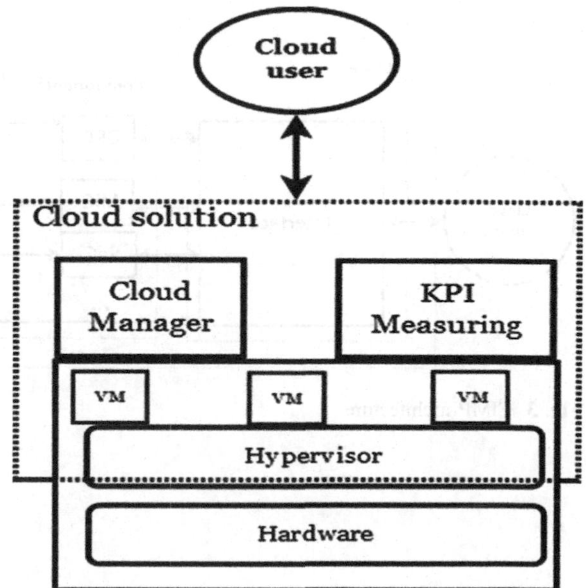

Fig. 1 Components of Cloud Solution

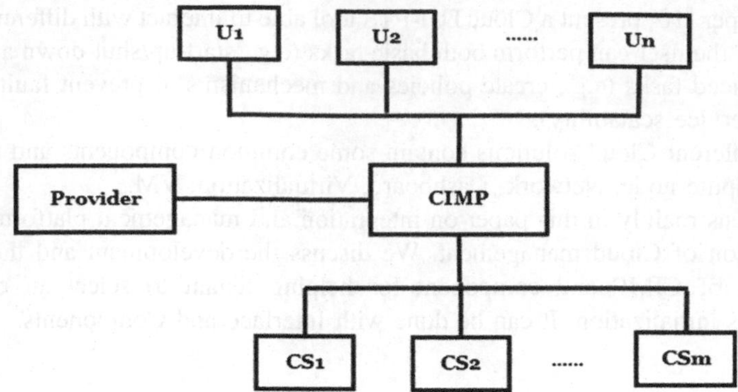

Fig. 2 Integration of Cloud solutions in CIMP

3.1 CIMP Architecture

The CIMP allows capturing user requirements and translating them into commands for the Cloud parameter settings [11]. As shown in Fig. 3, the CIMP is composed of Interface and Components.

The CIMP consists of several components; each plays a different role and defines the user and the provider profits to meet the different needs of optimization, performance, security, QoS and cost. Among these components as shown in Fig. 3, we find:

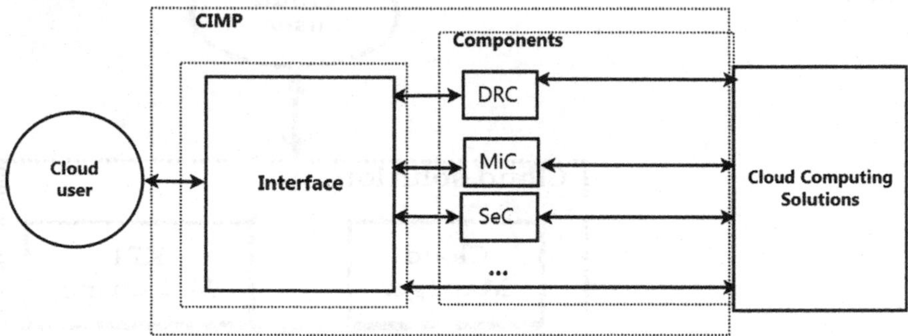

Fig. 3 CIMP architecture

- Dynamic Reconfigurable Component (DRC): The DRC component optimizes the use of Cloud resources and enables dynamic resource allocation. This component determines the optimal solutions depending on policy and strategy defined by users.
- Migration component (MiC): The MiC component allows the users to select the services it wishes to move; outsource the software, platform or infrastructure services.
- Security component (SeC): The SeC component can improve the security mechanism offered by Cloud providers, deploying their own security policies and could be developed as a perspective of this paper.

When new features are missing or insufficient, the CIMP uses additional components to overcome this deficiency and add missing functionality to the Cloud solution.

In the remaining, the CIMP Interface and MiC component are presented in the next sections.

3.2 *CIMP Interface*

The CIMP interface allows the tenant to select the configuration functions. These parameters reflect the actual needs of the user. Among these parameters, we find:

- User Activity (e-commerce, web blog …)
- User resource need (small, medium, huge, wide)
- Dynamic behavior (yes, no)
- Migration Services (SaaS, PaaS, IaaS)
- Security level (low, medium, strong)

The CIMP interface is used to evaluate the user requirement parameters, and to determine and propose configuration commands as shown in Fig. 4. The Requirements module allows the user to specify the required resources. The

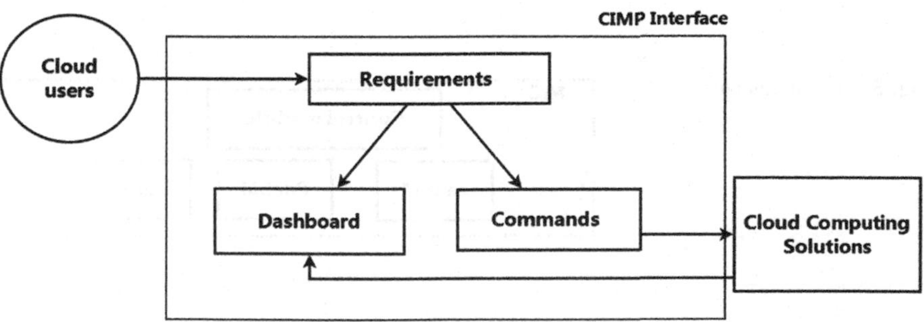

Fig. 4 CIMP interface

Dashboard module provides users a graphical interface to access, provision, and management of Cloud resources. The Commands module allows adding new parameters to the Cloud resources.

The commands are deducted from the requirement parameters in two ways:

1. The parameters of the user activity allow determining the characteristics of Cloud resources, and therefore the configurations of the virtual machines.
2. All remaining parameters are translated to the adjustment of these resources. Depending on the type of user activity, the level of resource requirements and the choice to enable or disable the dynamic behavior of specific requirements selected by the user, the CIMP generates the appropriate commands.

Finally, using these commands, the CIMP interface generates various parameters and Cloud adjustment instructions, necessary for the control and management of Cloud. The commands can directly control components or modify Cloud settings.

3.3 MiC Component

Generally, the migration process involves porting of applications, the software development platform and the entire IS infrastructure in the Cloud. The organization maintains that connection and consultation tools.

The MiC component allows the user to select the services it wishes to move. Among these services, we find:

IaaSM: IaaS Migration
PaaSM: PaaS Migration
SaaSM: SaaS Migration

With MiC component, the user will choose the service they desire to outsource. The component directs it to the relevant service of the Cloud environment to start the migration process as shown in Fig. 5.

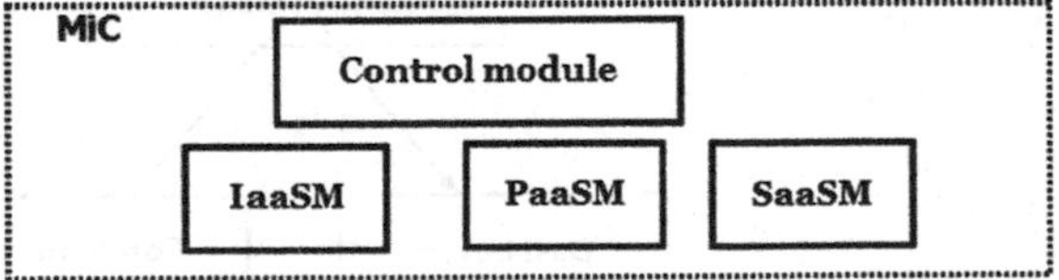

Fig. 5 MiC architecture

4 Experimental Results

To demonstrate the feasibility of the CIMP, we used a testbed based on the OpenStack Cloud [6]. OpenStack architecture is built using many components such OpenStack Compute, Block and Object Storage, Networking, Identity Service, Dashboard, Ceilometer, Orchestration, Database and many others projects [6]. The other Cloud solutions can also be integrated due to the interoperability of the CIMP.

The CIMP is composed by a CIMP interface and a MiC component module. The CIMP interact with the underlying Cloud environment via RESTful API to guarantee the independence and Cloud solutions interoperability.

To demonstrate the efficiency of MiC, and to show the result provided, we propose to test our approach. The classical architecture of an organization's information systems represents some basic elements for the operation of a number of services. We consider that the organization needs to accommodate for example, a website, a database (DB), a set of machines connected together through a local area network (LAN) that is connected to the outside (Internet), a development environment and application of the organization's activities. Users of the organization (U1, U2, U3) are connected to the Internet through the network equipment as the gateway and router.

Migration service PaaS platform as shown in Fig. 6, will allow the organization to leave the infrastructure management and the development environment to the Cloud provider.

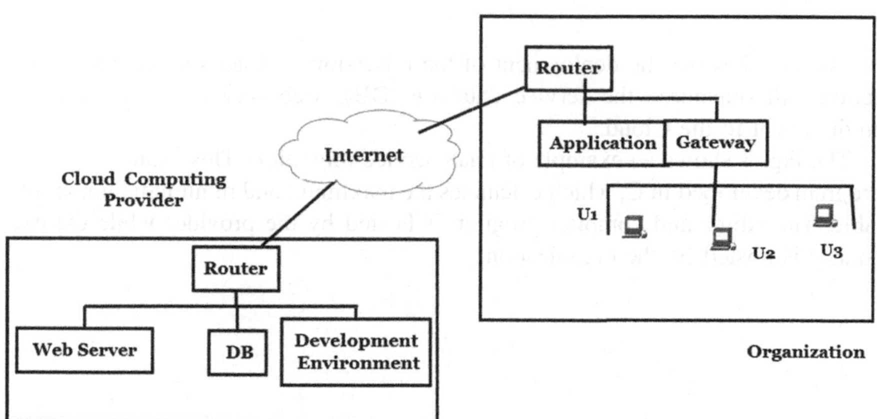

Fig. 6 Migration service PaaS

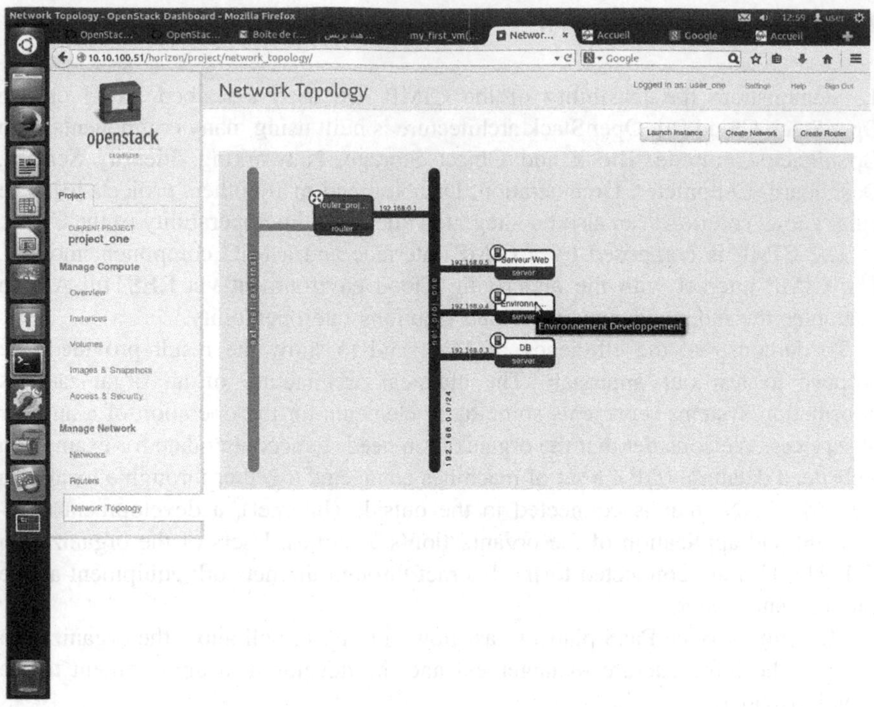

Fig. 7 OpenStack Deployment—PaaS

The Fig. 7 show the deployment of the migration to PaaS service, the organization will outsource the service database (DB), web server and development environment to the Cloud.

The Fig. 8 shows an example of PaaS service migration. This example shows a program developed in C, which calculates the maximum and minimum number of a table. The editor and compiler program is hosted by the provider while the executable is hosted by the organization.

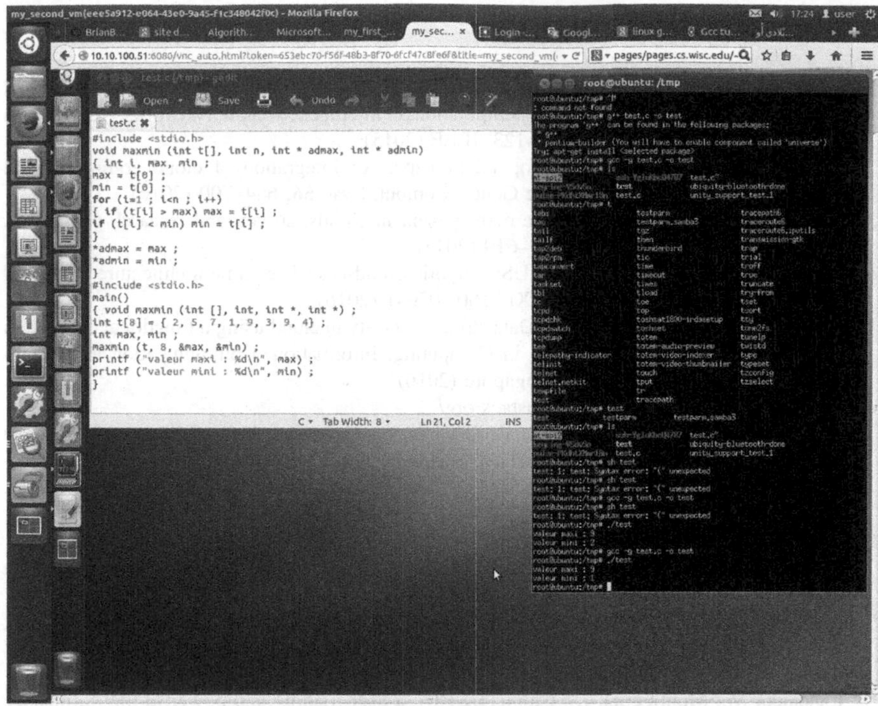

Fig. 8 Example of PaaS migration service

5 Conclusions

In this paper we introduced the Cloud Integration and Management Platform (CIMP). The CIMP is an open source, scalable, interoperable and extensible.

The CIMP is composed of an interface and some components such as DRC and the MiC. The implementation of these is proposed to act on the Cloud resources required to provide integration of Cloud solutions and an optimized environment to the needs defined by users, and offer them the opportunity to choose and to out-source their services to the Cloud. These components could be useful for the Cloud user as well as for the provider.

In continuity of this work and to improve the functionality of the CIMP, the concept of artificial intelligence could be used to develop an intelligent CIMP. In addition, other components may be integrated in the platform.

References

1. Puthal, D., Sahoo, B.P.S., Mishra, S., Swain, S.: Cloud computing features, issues, and challenges: a big picture. In: Computational Intelligence and Networks (CINE) on International Conference, pp. 116–123. IEEE (2015)
2. Botta, A., de Donato, W., Persico, V., Pescapé, A.: Integration of cloud computing and internet of things: a survey. Future Gener. Comput. Syst. **56**, 684–700 (2016)
3. Jennings, B., Stadler, R.: Resource management in clouds: survey and research challenges. J. Netw. Syst. Manage. **23**(3), 567–619 (2015)
4. DePalma, N., Breazeal, C.: NIMBUS: a hybrid cloud-crowd realtime architecture for visual learning in interactive domains. arXiv:1602.07641 (2016)
5. Bharati, P.V., Mahalakshmi, T.S.: Data storage security in cloud using a functional encryption algorithm. In: Emerging Research in Computing, Information, Communication and Applications, pp. 201–212. Springer Singapore (2016)
6. Openstack project. http://www.opnstack.org/
7. Vogel, A., Griebler, D., Maron, C. A., Schepke, C., Fernandes, L.G.: Private IaaS clouds: a comparative analysis of OpenNebula, CloudStack and OpenStack. In: Euromicro International Conference on Parallel, Distributed, and Network-Based Processing, pp. 672–679. IEEE (2016)
8. Nanig, T.T.: Private cloud deployment model for academic environment using CloudStack. In: Genetic and Evolutionary Computing, pp. 155–164. Springer International Publishing (2016)
9. Loulloudes, N., Sofokleous, C., Trihinas, D., Dikaiakos, M.D., Pallis, G.: Enabling interoperable cloud application management through an open source ecosystem. In: IEEE Internet Computing, pp. 54–59 (2015)
10. Canonico, M., Monfrecola, D.: CloudTUI-FTS: a user-friendly and powerful tool to manage Cloud Computing Platforms. In: Proceedings of the 9th EAI International Conference on Performance Evaluation Methodologies and Tools, pp. 220–223. ICST (Institute for Computer Sciences, Social-Informatics and Telecommunications Engineering) (2016)
11. Sefraoui, O.: Towards management platform Cloud solutions. Doctoral dissertation, UMP (2015)

Improvement of the HEFT Algorithm by Lookahead Technique for Heterogeneous Resource Environments

Soukaina Khaldoune and Mohsine Eleuldj

Abstract Task scheduling is a major issue in heterogeneous resource environments and particularly in multiprocessor system and Cloud. It means selection of the best suitable resources for task execution. Among the scheduling heuristics proposed in the literature, the Heterogeneous Earliest Finish Time (HEFT) heuristic is known by giving a good schedules in short time. In this paper, we propose three improvement versions of HEFT, where generally if certain conditions were verified, decisions made by the heuristic do not accounting a task being in scheduling only, but also take into account the impact of this decision to the children of the task. Thus, the heuristic can allocate this task and one of its children on the processor that can minimize execution time of the two tasks at the same time. Simulation results indicate that the new versions of HEFT can effectively reduce the schedule length in most cases without making the execution time prohibitively high otherwise.

Keywords HEFT algorithm • Task scheduling

1 Introduction

The scheduling problems are varied. We can meet them in many fields: industrial production systems, computer systems, administrative systems, transportation systems, etc. The reason that they have done and continued so as to be a subject of many researches. In a computer system, the scheduling problem consists to organize in time the tasks subject to constraints of time and resources, while satisfying as the best one or more objectives.

S. Khaldoune (✉)
National School of Applied Sciences, Mohammed I University, Oujda, Morocco
e-mail: s.khaldoune@ump.ac.ma

M. Eleuldj
Mohammadia School of Engineers, Mohammed V University, Rabat, Morocco
e-mail: eleuldj@emi.ac.ma

Á. Rocha et al. (eds.), *Europe and MENA Cooperation Advances in Information and Communication Technologies*, Advances in Intelligent Systems and Computing 520, DOI 10.1007/978-3-319-46568-5_41

Scheduling is an important issue that greatly influences the performance of the heterogeneous resource environment such as multiprocessor systems and Cloud Computing.

Task scheduling problem is a well-known combinatorial optimization problem. The problem of finding an optimal schedule is NP-complete [1]. There are various methods to solve optimization problems that are classified into two main categories: exact and approximate methods [2]. The exact methods ensure to find the global optimum. To find an optimum, the only way is often to do an exhaustive research on the set of feasible solutions, but this method requires very time consuming calculation for NP-complete problems. The development of the complexity theory of algorithms clarifies the difficulty of the problem. So, we turn to another way to tackle these problems is to find a suboptimal solution but in a reasonable time using approximate methods.

The task scheduling algorithm is classified into two major categories, namely static scheduling and dynamic scheduling [3]. In the static category, all information about the number of tasks, number of processors, execution time of each task on the processors, and communication time between the tasks that are supposed to be known by the scheduling algorithm at compiled time. It is assumed that task execution time can be estimated and does not change during the course of execution. But, in the dynamic category, all the above information is known at run time.

As a result, a number of heuristics have been developed in recent years. Heuristic algorithms, class of approximate methods, are easy to implement. They attempt to strike a good balance between running time, complexity and schedule quality and find good solutions with relatively small computational effort [3]. The problem considered in this paper is the static scheduling of an application on a heterogeneous system.

An overview of the different heuristics proposed in the literature to solve the scheduling task problem can be found in [4–7]. Among the heuristics most widespread and having a better performance complexity reports is Heterogeneous Earliest Finish Time HEFT algorithm [8]. It is one of the most used and cited. This prompted us to detail the HEFT algorithm and use it in our contributions.

Thus, the contribution is a proposal of three improvement versions for HEFT algorithm in order to enhance the scheduling tasks, considering information about the children of a task. We made a comparison between the versions of improvement and HEFT based on performance metrics. The metrics used in the performance analysis are scheduling length ratio, speedup and efficiency.

The rest of the paper is organized as follows. Section 2 present the task scheduling problem and define some basic attributes of DAG scheduling. Section 3 present the proposed improvements of HEFT algorithm, which are then evaluated with the comparison metrics in Sect. 4. Finally, Sect. 5 concludes the paper.

2 Task Scheduling Problem

A scheduling system model consists of an application, a target Computing environment and performance criteria for scheduling [8].

The application is be represented by a Directed Acyclic Graph (DAG), $G=(N, E)$. $N=\{n_i, \quad i=1\ldots n\}$ is the set of n tasks and $E=\{e,\ e=(i,j)\}$ is the set of communication edges between tasks. Each edge represent the precedence constraint such that task n_i can not start its execution before receiving all data from its parents. Tasks and edges are weighed for computation cost and communication cost respectively. An entry task has no predecessor tasks, and an exit task has no successor tasks in a DAG.

The target computing platform is defined as $P=\{p_j, \quad j=1\ldots p\}$, a set of p heterogeneous processors available in the system. They are connected in a fully connected network topology.

W is a $n \times p$ matrix of computation cost in which each $w_{i,j}$ represents the execution time to complete task n_i on processor p_j. The average execution cost of a task is defined as:

$$\overline{w_i} = \sum_{j=1}^{p} w_{i,j}/p \tag{1}$$

Performance criteria are another component in workflow scheduling system modeling. It is used to evaluate the effectiveness of the scheduling strategy.

Makespan criterion is the most used performance measure in the literature. The makespan, also referred to as schedule length, is the time difference between workflow start and its completion.

So, our objective of the proposed algorithm is to map each task of given DAG to a processor from the set P of processors and finding the minimum scheduling length.

Among the heuristics most used and having a better performance complexity reports is Heterogeneous Earliest Finish Time HEFT algorithm [8].

The HEFT algorithm determines a task scheduling graph of a heterogeneous resource environment. HEFT algorithm consists of two main phases: A prioritizing tasks phase and a processor selection phase.

In the first phase, each task to be scheduled is given a priority. Tasks are ordered by their scheduling priorities that are based on upward ranking defined as follows:

$$rank_u(n_i) = \overline{w_i} + \max_{n_j \in succ(n_i)} \left\{ \overline{c_{i,j}} + rank_u(n_j) \right\}. \tag{2}$$

where $succ(n_i)$ is the set of immediate successors of task n_i, $\overline{w_i}$ is the average computation cost of task n_i, and $\overline{c_{i,j}}$ is the average communication cost of the edge from task n_i to task n_j. For the exit task n_{exit}, upward rank value is equal to:

$$rank_u(n_{exit}) = \overline{w_{exit}}. \quad (3)$$

It is also necessary to define another attribute downward rank which will be used in the proposed scheduling algorithms.

$$rank_d(n_i) = \max_{n_j \in pred(n_i)} \left\{ rank_d(n_j) + \overline{w_j} + \overline{c_{i,j}} \right\}. \quad (4)$$

where $pred(n_i)$ is the set of immediate predecessors of task n_i. For the entry task n_{entry}, the downward rank value is equal to zero.

In the second phase tasks are assigned to processors. Each task, in order of its priority, is scheduled on the processor which will result in the earliest finish time of that task.

HEFT algorithm is among the heuristics having a better performance complexity reports, but it has some limitations. For example, in the processor selection phase, the task being scheduled is allocated to the processor which gives it the minimal earliest finish time (EFT). The earliest finish time of the task in each processor is calculated based on the time when that processor becomes available and when the predecessors terminate their execution ignoring the impact of this decision to the task's successors. Thus, our objective is trying to minimize the estimated finish time of the task being scheduled and their children at the same time and also taking into consideration the communication cost between two tasks successors.

3 Proposed Algorithm

We present three different approaches to improve HEFT with look ahead strategy.

The new approaches are two phases. A task prioritizing phase for computing the priorities of all tasks and a processor selection phase for scheduling, each task selecting in the order of its priority, on a processor which minimizes the task's finish

Task Prioritizing Phase. For assigning priority to a task, we use above mentioned attributes $rank_u$ and $rank_d$ for each task. After computing the attribute of each task of a given DAG, we have to sort the tasks according to the attribute value. As a result, tasks in the same level can be executed at the same time.

Using the attribute $rank_u$ and $rank_d$, two different task lists are generated by sorting the tasks decreasing order of $rank_u$ and increasing order of $rank_d$, respectively. Then, we try the two different priority lists, and we use the priority task list which can give a better schedule length.

Resource Selection Phase. In this phase, tasks are assigned to the processors according to three strategies. Each approach proposes a strategy for assigning tasks on a processor.

(1) *Algorithm 1.* First, the task selected is allocated on a processor that minimizes the completion time. Next, if this task has successors where it is the only predecessor of them, we calculate the earliest finish time of each successor on each processor.

Algorithm 1

```
Compute rank_u and rank_d for all tasks.
priorityListUpward ← Task list sorted by decreasing order of rank_u value.
priorityListDownward ← Task list sorted by increasing order of rank_d value.
priorityList ← priorityListUpward AND priorityListDownward are tested, and the
priority task list used is that can minimize EFT of the exit task.
n_0 ← priorityList[0] : task being scheduled
for j ← 1 à p do
    EFT(n_0, p_j) ← w_{n_0,j}
endfor
Assign the task n_0 to the processor p_j which minimizes the EFT(n_0, p_j).
Update avail[j].
while there are unscheduled tasks do
    n_m ← priorityList[m]: task being scheduled.
    for j ← 1 à p do
        EFT(n_m, p_j) ← w_{n_m,j} + max{avail[j], max_{n_h ∈ pred(n_m)}(EFT(n_h, proc(n_h)) + c_{n_h,n_m})}
    endfor
    Assign task n_m to the processor p_j which minimizes the EFT(n_m, p_j).
    Update avail[j].
    for i ← 1 à n do
        if n_m is the only predecessor for n_i then
            for j ← 1 à p do
                EFT(n_i, p_j) ← w_{n_i,j} + max{avail[j], (EFT(n_m, proc(n_m)) + c_{n_m,n_i})}
            endfor
        endif
    endfor
    Select n_i which resulted in the smallest EFT(n_i, p_j).
    Assign task n_i to the processor p_j that minimizes EFT(n_i, p_j).
    Update avail[j].
endwhile
```

Each successor will have then a minimum completion time on a processor. After that, we assign only the one successor that has a minimum completion time.

In other words, if the successors have a single predecessor (the task selected), we can assign the task and its successor also having the minimum completion time according to where the task selected is assigned, successively and simultaneously. The details of the proposed algorithm are given in Algorithm 1.

(2) *Algorithm 2*. This version works in a similar way to Algorithm 1, but with a difference in the condition. The idea behind this version is to try the task selected to be scheduled on every processor and affect him on the processor that minimizes the completion time. Next, we calculate the estimated finish time of its children according to where the task selected is assigned. We only consider the children or the successors where the task selected to be scheduled is the only predecessor of them or it is the only predecessor which is still not affected (we cannot assign a task as if its predecessors have not yet completed their execution). After that, we assign also the successor on a processor having the minimum completion time simultaneously.
(3) *Algorithm 3*. This version works in a similar way to Algorithm 2, but with a difference in calculation of the earliest finish time of the selected task and their successors. First, if the selected task has successors where it is the only predecessor of them or it is the only predecessor which is still not affected, we calculate the earliest finish time of that selected task and their successors on each processor in all possible combinations. After that, we assign the selected task and only one of its successors on the processors that can minimize the completion time. In other words, if the successors have a single predecessor or they have only one predecessor which is still not affected, we can assign the task and its successor also based on the minimum completion time of its successors and not on the task itself, successively and simultaneously.

4 Simulation Results and Discussion

This section presents a performance comparison of the proposed algorithms with HEFT algorithm, which is followed by the simulation results.

4.1 Comparison Metrics

Scheduling Length Ratio. The main performance measure of a scheduling algorithm is the schedule length (makespan). Since a large set of task graphs with different properties is used, it is necessary to normalize the schedule length to the lower bound, which is called the schedule length ratio (SLR). SLR is defined as follows:

$$SLR = \frac{makespan}{\sum_{n_i \in CP_{MIN}} \min_{p_j \in p} \{ w_{i,j} \}}. \quad (5)$$

where CP_{MIN} is the minimum computation cost of the tasks on a critical path. The task scheduling algorithm that gives the lowest SLR is the best algorithm with respect to performance.

Speedup. The speedup value is calculated by dividing the sequential execution time by the parallel execution time. The sequential execution time is computed by assigning all tasks to a single processor that minimizes the computation costs. Speedup is defined as follows:

$$Speedup = \frac{\min_{p_j \in p} \sum_{n_i \in V} w_{i,j}}{makespan}. \quad (6)$$

Efficiency. In the general case, efficiency is calculated by dividing the speedup value by the number of resources used in schedule task graph. Efficiency is defined as:

$$Efficiency = \frac{speedup}{number\ of\ resource}. \quad (7)$$

4.2 Results and Discussion

In order to be consistent and to have the better compare results, the model defined for workflow scheduling in a heterogeneous environment by [8] is the same example used in our simulation with minor revision.

A DAG model is shown in Fig. 1, which represents an application. The table represents the computation cost of each task on a processor and the weight of each edge in the figure represents its average communication cost. If two tasks are allocated on the same processor, their communication time is considered to be zero.

As an example, Fig. 2b shows the schedule obtained by the Algorithm 1, which is equal to 76. Based on the values of the $rank_u$ and $rank_d$, we have two priority lists of tasks with respect to the Algorithm 1 that are $\{n_1, n_3, n_4, n_2, n_5, n_6, n_9, n_7, n_8, n_{10}\}$ and $\{n_1, n_4, n_5, n_3, n_6, n_2, n_7, n_9, n_8, n_{10}\}$ respectively. Using the first priority list, the schedule length is shorter than the schedule length produced by using the second priority list. Thus, the algorithm uses the priority list which gives the minimum schedule length. On the other hand, the algorithm choose, for each task on the priority list, the best resource that gives the smallest EFT which can reduce the EFT of its child or itself since it does not have children.

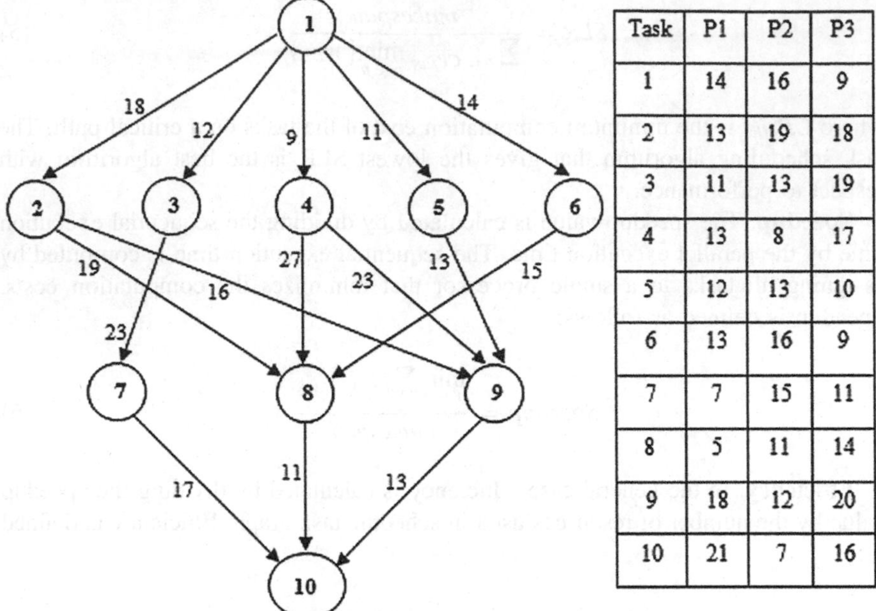

Task	P1	P2	P3
1	14	16	9
2	13	19	18
3	11	13	19
4	13	8	17
5	12	13	10
6	13	16	9
7	7	15	11
8	5	11	14
9	18	12	20
10	21	7	16

Fig. 1 A DAG graph and computation time of the tasks in each processor

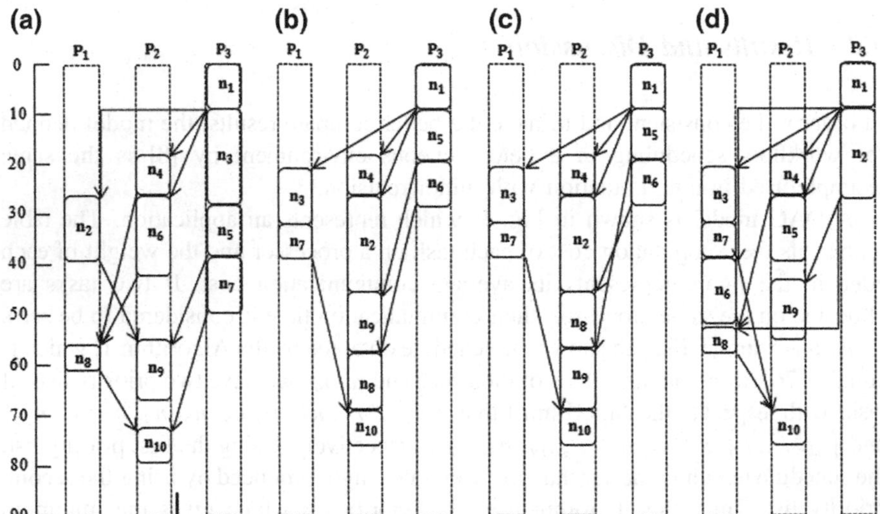

Fig. 2 Scheduling of DAG graph with HEFT algorithm (schedule length = 80) and the improvement algorithms (schedule length = 76)

The proposed algorithms also outperform the HEFT algorithm in terms of SLR, speedup and efficiency. The simulations were run using 2, 3, 5, 7 and 9 heterogeneous resources in the target environment, which are shown in Figs. 3, 4 and 5.

One main advantage of the proposed algorithms that they try two priority lists to choose which priority list can give a better earliest finish time. Then, they try not only to minimize the earliest finish time of the task being scheduled but also the earliest finish time of their children.

Another important point is that the algorithms proposed can significantly improve the schedule length returned by HEFT, especially in cases where the communication cost is higher. The versions proposed take into consideration the computation cost of the task being scheduled and the communication cost between her and their immediate children before deciding which task child will be scheduled right after the parent task.

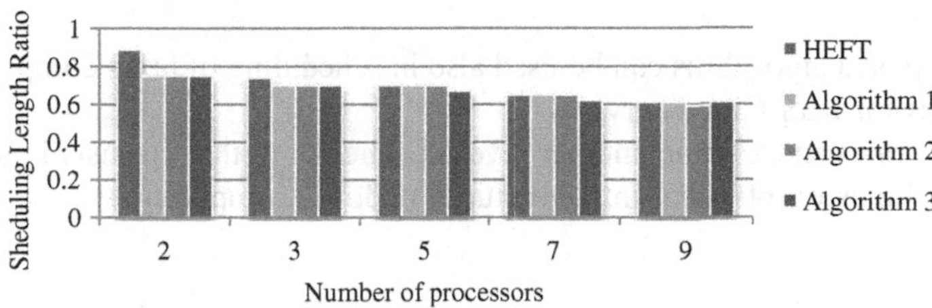

Fig. 3 Scheduling length ratio

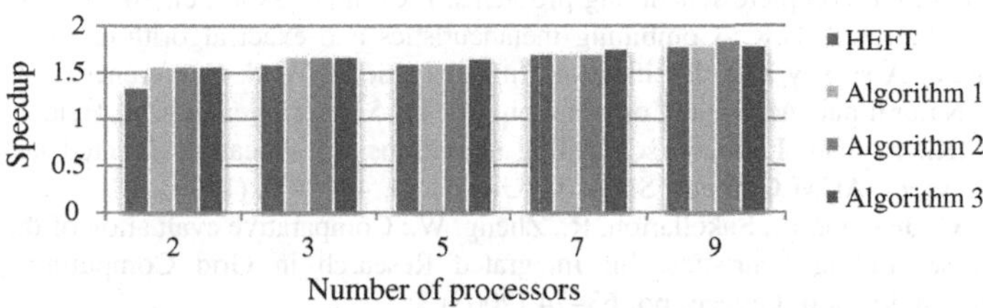

Fig. 4 Speedup

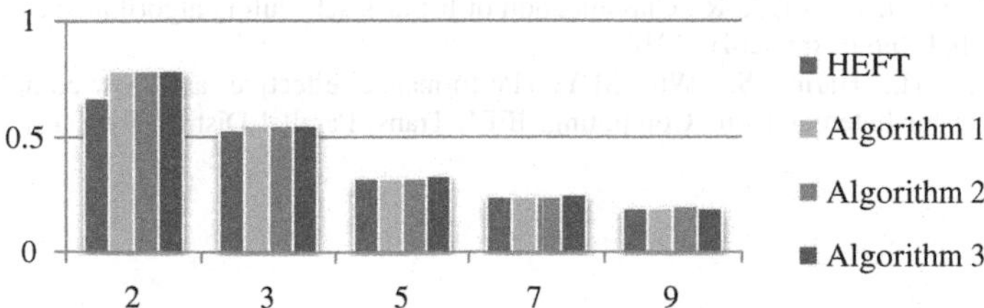

Fig. 5 Efficiency

5 Conclusion

In this paper, we have proposed new approaches to schedule DAG graph in heterogeneous systems. The problem is solved by heuristic algorithms. They are an improvement of the Heterogeneous Earliest Finish Time (HEFT) algorithm.

The new approaches are a two-phase algorithm. In the task prioritizing phase, two different priority lists are tried to choose which priority list can give a better makespan or schedule length. In the resource selection phase, the algorithm assigns the task selected and one of its children. The algorithm computes the earliest finish time for the current task and the child tasks on all processors. Then, the processors selected for task being scheduled and one of its child are the processors that minimize the earliest finish time from the child of task on all resources where task is tried.

Results show that new approaches are efficient compared to the HEFT algorithm and they outperform it in terms of Scheduling Length Ratio, speedup, and efficiency.

The proposed algorithms can be used also in scheduling of tasks composing user applications on IaaS Cloud services.

Future work can examine another level of scheduling that consists in allocation virtual machines on physical infrastructure in Cloud Computing.

References

1. Ullman, J.D.: NP-complete scheduling problems. J. Comput. Syst. Sci. **10**(3), 384–393 (1975)
2. Puchinger, J., Raidl, G.R.: Combining metaheuristics and exact algorithms in combinatorial optimization: A survey and classification. In: International Work-Conference on the Interplay Between Natural and Artificial Computation, pp. 41–53. Springer, Berlin, Heidelberg (2005)
3. Kwok, Y.K., Ahmad, I.: Static scheduling algorithms for allocating directed task graphs to multiprocessors. ACM Comput. Surv. (CSUR) **31**(4), 406–471 (1999)
4. Canon, L.C., Jeannot, E., Sakellariou, R., Zheng, W.: Comparative evaluation of the robustness of DAG scheduling heuristics. In: Integrated Research in Grid Computing, CoreGRID Integration Workshop, Greece, pp. 63–74 (2008)
5. Singh, L., Singh, S.: A survey of workflow scheduling algorithms and research issues. Int. J. Comput. Appl. **74**(15) (2013)
6. Arabnejad, H.: List based task scheduling algorithms on heterogeneous systems—an overview. In: Doctoral Symposium in Informatics Engineering, pp. 93 (2013)
7. Rajak, N., Dixit, A., Rajak, R.: Classification of list task scheduling algorithms: a short review paper. J. Ind. Intell. Inf. **2**(4) (2014)
8. Topcuoglu, H., Hariri, S., Wu, M.Y.: Performance effective and low-complexity task scheduling for heterogeneous Computing. IEEE Trans. Parallel Distrib. Syst. **13**(3), 260–274 (2002)

Smart Intrusion Detection Model for the Cloud Computing

Mostapha Derfouf, Mohsine Eleuldj, Saad Enniari and Ouafaa Diouri

Abstract Nowadays, Cloud computing is turning into a major trend in the field of computer science. It is referred to as a new data hosting technology that became very popular lately thanks to the repayment of costs induced to companies. However, and since this concept is still in its first stages of use, many new security risks start to appear, making its huge benefits fade compared to the security risks it brings with it. This paper proposes a smart intrusion detection model that is based on the principle of collaboration between many IDSs (Intrusion detection systems). These IDSs are host based intrusion detection systems (HIDS) that are customized in order to support data mining and machine learning and we call them SHIDS (Smart host based intrusion detection systems). The SHIDS are deployed on the different virtual machines in the cloud to detect and protect against attacks targeting applications running on virtual machines by exchanging encrypted IDMEF alerts. This approach provides many benefits in terms of portability and costs. That being said, this paper will set forth the different architectures of intrusion detection systems in the cloud, provide a comparison between the major architectures, propose our intrusion detection architecture and validate it with an experiment using Open stack.

Keywords IDS · HIDS · NIDS · Cloud computing · Machine learning · Open stack

M. Derfouf (✉) · M. Eleuldj · S. Enniari · O. Diouri
Department of Computer Science, Mohammadia School of Engineers, Mohammed V University in Rabat, Rabat, Morocco
e-mail: mostaphaderfouf@research.emi.ac.ma

M. Eleuldj
e-mail: saadenniari@research.emi.ac.ma

S. Enniari
e-mail: eleuldj@emi.ac.ma

O. Diouri
e-mail: diouri@emi.ac.ma

Á. Rocha et al. (eds.), *Europe and MENA Cooperation Advances in Information and Communication Technologies*, Advances in Intelligent Systems and Computing 520, DOI 10.1007/978-3-319-46568-5_42

1 Introduction

Cloud computing is a new data hosting technology that stands today as a satisfactory answer to the problem of data storage and computing. It can provide processing and accommodation of digital information via a fully outsourced infrastructure allowing users to benefit from many online services without having to worry about the technical aspects of their uses. Cloud computing appears as a tremendous opportunity for companies but logically raises the question of the security of data when they are hosted by a third party.

We have already dealt with the problem of data storage security on Cloud computing and proposed a solution based on encryption to secure data [1], in this paper, we propose a smart intrusion detection model designed to secure the cloud. First we will give an overview about the different intrusion detection models in the cloud environments then we provide a comparison between the different IDS models finally we propose our own intrusion detection model for the Cloud and provide the experiment.

2 Related Work

In the “Advanced IDS Management Architecture” [2] the authors proposed an IDS which uses an Event Gatherer combined with the Virtual Machine Monitor (VMM). This IDS is composed of many sensors and a central management unit. The Event Gatherer plugin plays the role of Handler, sender, and receiver in order to provide an integration of different sensors. This architecture uses the IDMEF (Intrusion Detection Message Exchange Format). An interface is designed to expose the result reports for users.

The multilevel IDS concept is proposed by Kuzhalisai and Gayathri [3] which deals with effective use of system of resources. The proposed system binds user in different security groups based on degree of anomaly called anomaly level. It consists of AAA module which is responsible for authentication, authorization and accounting. When user tries to access the cloud the AAA checks the authentication of the user and based on it, it gets the recently updated anomaly level. Security is divided into three levels: high, medium and low.

Gul and Hussain [4] have proposed a multi-threaded NIDS designed to work in distributed cloud environment. This multi-threaded NIDS contains three modules: capture and queuing module, analysis/processing module and reporting module. The capture module is responsible of reading the network packets and sending them to the shared queue for analysis.

In [5] the authors proposed a framework that integrates a network intrusion detection system (NIDS) in the Cloud. The proposed NIDS module consists of Snort and signature apriori algorithm that is capable of generating new rules from captured packets. The new rules are added to the Snort configuration file.

Table 1 Synthesis of the related work

IDS/Features	Advanced IDS management architecture	Cloud intrusion detection system	Cloud detection and prevention system	Improved hybrid IDS
Type	Collaborative	Collaborative	Intelligent	Intelligent
The ability to detect unknown attacks	No	No	Could be	Could be
The ability to analyze the content of encrypted streams	Yes	Yes	No	Yes
Encrypting exchanged alerts	No	No	No	No
Diffusion of detected attacks	No	Using alert system message	No	No

In [6] the authors proposed Ajit Kumar Gautam, Vidushi Sharma, Shiv Prakash and Manak Gupta proposed an improved Hybrid IDS. The Improved hybrid IDS is combination of anomaly based detection and honey pot technology with KFSensor and Flowmatrix. The Honey pot is used to attract more and more attackers, the detection obtained can be used to create new signatures and update the database. Finally anomaly can be used to detect unknown attack in the whole network.

The Cloud Detection and Prevention System (CIDPS) architecture [7] is illustrated and presented as a workflow scenario. Sensor inputs or alerts generated while monitoring network, host, platform and applications together with the latest CIDPS challenges and enterprise CIDPS policies and their updates, drive through the CIDPS Trust Management system to be analyzed. Inference Engine (IE) is the logical and main part of IDE.

In CIDS architecture [8] each node has its own analyzer and detector components that are connected to the behavior and knowledge based databases. The individual analysis reduces the complexity and the volume of exchanged data, but at the expense of the node processing overhead. This framework contains CIDS components, cloud system components and NIDS components.

Table 1 presents the different architectures of intrusion detection systems in a comparative table.

3 Virtual Machine with Smart Intrusion Detection Model

The Cloud Computing is based on the principle of virtualization in which the different services and applications are deployed in virtual machines. The virtual machines use various operating systems (OS) and expose different application to the cloud customers, thus any vulnerability in these systems and applications can be remotely exploited by hackers hence the importance of implementing intrusion detection systems in the Cloud at the virtual machine level.

We propose a smart host based intrusion detection system (SHIDS) as security tool to monitor the hypervisor and virtual machines on that hypervisor, detect malicious activities at the VM level and protect against attacks targeting applications running on virtual machines, this approach provides many benefits in terms of portability and costs. The virtual machine is equipped with a SHIDS (Smart host based intrusion detection system) it is a customized HIDS (Host based intrusion detection system) designed to support data mining in order to detect unknown attacks, the proposed SHIDS behaves like a HIDS and controls the state of the virtual machine, since it has access to its stored information, whether in RAM, in the file system, log files or audit trails. In addition the SHIDS can analyze all the activities on the virtual machine hosting the cloud services.

The SHIDS has the ability to analyze the content of encrypted streams since SHIDS has access to the encryption keys and certificates on the machine where it is installed while the NIDS (Network based intrusion detection system) cannot analyze encrypted traffic. The SHIDS can detect easily the "Trojan" attacks while this type of attack is difficult to detect by a NIDS. The HIDS is the suitable solution to secure our virtual machines in the Cloud environment.

3.1 Virtual Machine Components

Figure 1 shows the virtual machines equipped with a SHIDS (Smart Host based IDS).

The proposed virtual machine contains the following components:

- Os software: Is the system software that manages computer hardware and software resources, it can be Linux, Microsoft windows etc.
- User software: It is the application provided by the cloud and designed to be used by the customers of the cloud.
- SHIDS (Smart Host Based Intrusion Detection System): it refers to an intrusion detection system that is placed on a single host system. It is a customized HIDS whose detection engine is modified and improved in order to support and integrate data mining and machine learning modules to detect unknown and new attacks that are not stored in the IDS database.

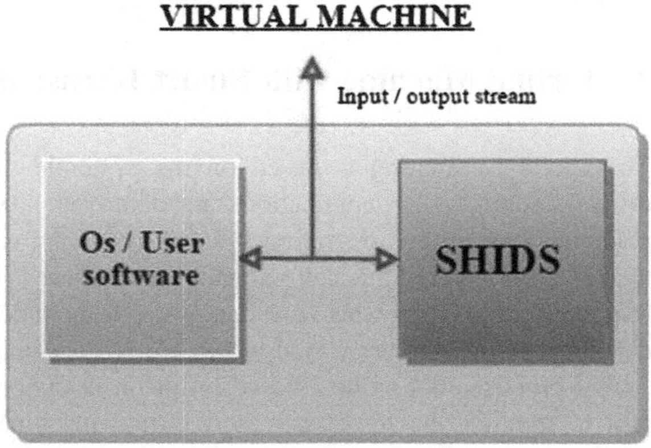

Fig. 1 Virtual machine integrating a smart host based IDS

3.2 SHIDS Architecture

The different components (Fig. 2) composing the SHIDS are:

- Logging and alert system: is the module responsible for logging alerts, their storage in the database and the exchange of alerts with the other intrusion detection systems.
- IDS database: It is a MySQL database that stores the signatures of the various attacks known previously.
- Behavior database: It is a database used to store the past behaviors of individual users.
- Configuration interface: It is the interface used for the configuration and the parameterization of the SHIDS.
- Data Mining Engine: Is a program that allows using data mining techniques such as statistics, frequent pattern mining, clustering and classification to detect the new attacks that are unknown previously.
- knowledge base: It is a database that stores the attacks detected by the data mining engine after the application of different data mining techniques.

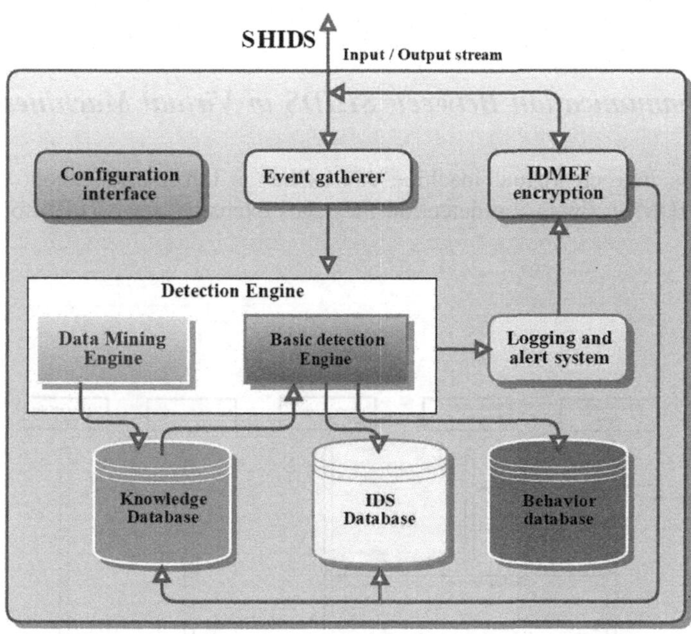

Fig. 2 SHIDS architecture

The SHIDS behaves like a demon or a standard service on a host system that detects suspicious activities. The SHIDS can monitor the machine activities, user activities and malicious activities like worms, virus and Trojans. The event gatherer which collects all the events coming from outside to the virtual machine, send the coming streams to the basic detection engine which is the customized program responsible of analyzing the events and detecting attacks based on the signatures stored in the IDS database. When an input is received the basic detection engine solicits the database to find a matching and in order to improve the SHIDS we added a behavior database using anomaly detection technique based on statistical measures, it focuses on characterizing the past behavior of individual users or groups of users to detect significant deviations. Many metrics are taken in consideration such as the connection time of the user, the number of password failures and CPU and memory usage. If no matching is found this means that the signature is not stored in the database so the basic detection engine will call the data mining engine, this latter applies the data mining techniques such as the clustering, association and fuzzy logic to check if the received signature matches an intrusion, in this case if an intrusion is detected it will be stored in the knowledge base and the basic detection engine will be notified. Finally the logging and alert system sends the detected attack in an IDMEF format to the central ids in order to update the other SHIDS as shown in Fig. 3.

3.3 Communication Between SHIDS in Virtual Machines

The HIDSs in each virtual machine can exchange information about intrusions using the IDMEF (intrusion detection messages exchange format) [9] format.

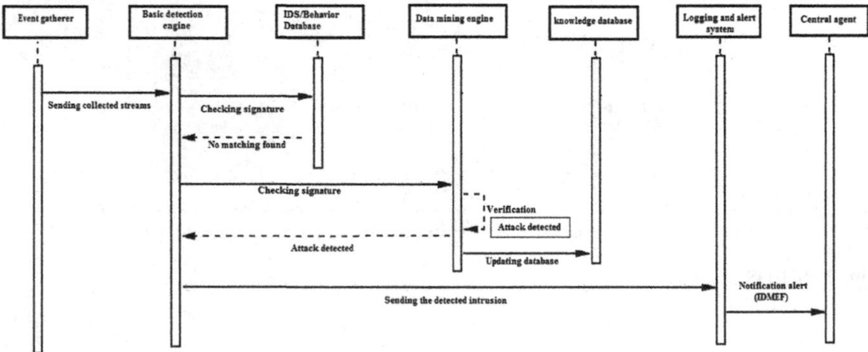

Fig. 3 UML sequence diagram of the SHIDS

```
<?xml version="1.0" encoding="UTF-8"?>
<idmef:IDMEF-Message xmlns:idmef="http://iana.org/idmef" version="1.0">
  <idmef:Alert messageid="abc123456789">
    <idmef:Analyzer analyzerid="sensor01">
      <idmef:Node category="dns">
        <idmef:name>sensor.example.com</idmef:name>
      </idmef:Node>
    </idmef:Analyzer>
    <idmef:CreateTime ntpstamp="0xbc71f4f5.0xef449129">
```


Fig. 4 Example of IDMEF message

The IDMEF data model is an object-oriented representation of the alert sent to central intrusion detection managers by intrusion detection analyzers [9]. IDMEF (intrusion detection messages exchange format) is a data format used to exchange incident reports between intrusion detection systems, intrusion prevention systems and software that must interact with them. IDMEF messages are designed to be easily handled. The details of the format are described in RFC 2007. The RFC 4765 [10] presents an implementation of the XML data model and associated DTD. An example of an IDMEF message is shown in Fig. 4.

The IDMEF format was chosen for its openness and extensibility. IDMEF message can be either an alert or a heartbeat. The IDMEF library is generated by JAXB based on the IDMEF XML schema provided by the RFC [10]. JAXB generates a class and their members based on a specific XML Java mapping. To protect the exchange of IDMEF alerts we used RSA encryption to secure the exchanges between the central agent and the other SHIDSs so that nobody can read the IDMEF alerts on the network, the rng-tools [11] (Hardware Random Generator) was used to prevent users from sniffing traffic from the insiders.

4 The Proposed Architecture

Two scenarios are possible to secure the cloud environment, the first one is to set up a distributed intrusion detection architecture in which each HIDS communicate by exchanging IDMEF alerts, in this case if an intrusion is detected by a SHIDS, it will be communicated to all other SHIDS to update their database as shown in Fig. 5.

The drawback of this architecture is that it will burden the network traffic because there are a lot of exchanges of alerts between SHIDS in addition to that if the number of servers is very large it will generate a lot of alerts which degrades the performance of the SHIDS since the alerts come from several sources.

The second scenario is based on a centralized IDS architecture (Fig. 6) that is based on the principle of collaboration between many SHIDS deployed on the different virtual machines (assuming that we have n virtual machines VM1, VM2 .... VMn and m Physical machines PM1, PM2.... PMm with m # n) in the cloud to detect and protect against attacks targeting applications running on these virtual

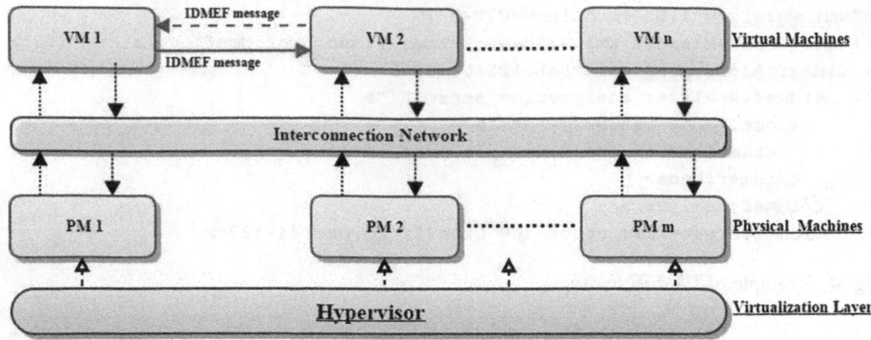

Fig. 5 Distributed intrusion detection architecture

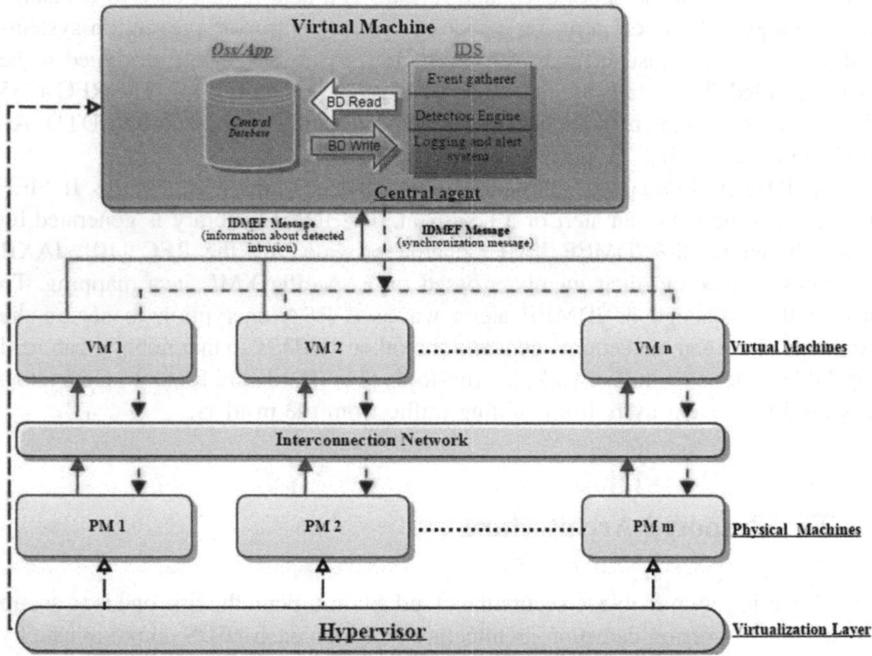

Fig. 6 Centralized intrusion detection architecture

machines, this approach provides many benefits in terms of portability and costs. The concept of this model is that the different SHIDSs are placed in each VM and cooperate with each other by exchanging alerts about detected intrusions. In our model there is a central agent that is responsible for the reception of notifications from the other SHIDS as well as writing and reading from a central database in order to synchronize the local database of each SHIDS. Thanks to this concept the different SHIDS are synchronized and each detected attack is communicated to other neighbors by the central agent. The model is improved by using the data mining and machine learning techniques to detect unknown attacks.

The model proposed in Scenario 2 offers more benefits than the one proposed in Scenario 1, so we adopted the architecture of the second scenario.

5 Experiment

The objective of this experiment is to valid the proposed architecture and test the detection of intrusions that are already stored in the database, in this case we will simulate four types of attacks (remote login with a wrong password, accessing a protected resource, trying to do administrative tasks such as changing the root password, and trying to log in without password) on a server in a virtual machine and verify that the attacks are detected by the SHIDS and sent in an IDMEF format to the central agent finally check that this later will notify all other SHIDS to update their databases. The data mining module of our architecture will be tested in the next paper taking into consideration the performance and the validity of the results.

The experiment was performed by using OpenStack [12] installed on an Ubuntu machine [Intel(R) Core(TM) i3-2370 M CPU @ 2,70 GHz 2,70 GHz, 8 Go de RAM]. We created three virtual machines (VM1, VM2 and VM3). During this experiment we put the focus on the following modules of OpenStack:

- Glance: It is the image Service of OpenStack that we used to perform registration and discovery of virtual machine images.
- Nova: It is the compute service of OpenStack that we used to automate and manage pools of computer resources.
- Horizon: It is the OpenStack Dashboard that provides a graphical interface to access and automate cloud resources.

We used Prelude [13] as a hybrid IDS (IDS) compound of heterogeneous types of sensors:

- NIDS: Network Intrusion Detection System (Snort).
- HIDS: Host based Intrusion Detection System (Samhain) that we customized to support data mining and machine learning techniques.

Prelude is designed in a modular way so as to adapt to any environment, the Manager is the central component that receives events from the different sensors and processes them, the Libdb is the library that provides an abstraction layer for the storage of IDMEF alerts in a database, the LML is the logs processing module and Prewikka is a web based graphical user interface (GUI) for Prelude.

On VM1 we installed Prelude Manager which is the module that processes and centralizes the system alerts, on VM2 and VM3 we installed the SHIDSs, the idea is that if one of these machines (for example VM2) detects an intrusion, it will use the IDEMF format to inform the central IDS (installed on VM1) and this later will

Classification	Source	Target	Analyzer	Time
1 x User login failed with an invalid user (failed) 6 x Remote Login (failed)	::1	::1	sshd (localhost)	06:59:01 - 06:54:16
1 x Credentials Change (failed) 1 x Invalid user in authentication request (failed)	n/a	::1	PAM (localhost) sshd (localhost)	06:58:51 - 06:43:37
Credentials Change (failed)	localhost Userid number: n/a	localhost ::1 127.0.0.1 Userid name: root Process name: sshd (3436)	PAM (localhost)	06:54:15

Fig. 7 Alerts provided by the Prelude sensors displayed on the Prewikka

update the VM3 with the new intrusion detected. The different exchanged alerts are encrypted using RSA standard. In order to be able to persist IDMEF alerts describing the intrusion detected we used Libpreludedb [14] Library that provides an abstraction layer for storing IDMEF alerts. In the experiment we simulated attacks on the VM2 and check if these attacks are transmitted in IDMEF format to the central IDS (VM1). In the central IDS the alerts are stored using Mysql database. To display the different alerts sent by the different sensors we use a web graphical interface called Prewikka. The Fig. 7 shows the alerts provided by the Prelude sensors and displayed on the web interface of the central IDS.

6 Conclusion

To sum up Cloud computing can be considered as a new concept that has brought many benefits but several security threats and vulnerabilities have appeared with this new concept that is why the cloud providers must identify these security issues and try to protect against them.

In this paper we have outlined the different architectures of intrusion detection systems in the cloud and provide a comparative study between them finally we implemented the proposed architecture using Openstack. In the next paper we will improve the architecture through the use of data mining techniques to detect new and unknown intrusions, improve accuracy and speed of intrusion detection and ensure a good adaptive capacity and scalability.

References

1. Derfouf, M., Mimouni, A., Eleuldj, M.: Vulnerabilities and storage security in cloud computing. In: International Conference on Cloud Computing Technologies and Applications —CLOUDTECH 2015, Marrakech, Maroc, 2–4 June 2015
2. Roschke, S., Cheng, F., Meinel, C.: An advanced ids management architecture. J. Inf. Assur. Secur. **5**, 246–255 (2010)
3. Kuzhalisai, M., Gayathri, G.: Enhanced security in cloud with multi-level intrusion detection system. In: IJCCT, vol. 3, Issue 3 (2012)
4. Gul, I., Hussain, M.: Distributed cloud intrusion detection model. Int. J. Adv. Sci. Technol. **34**, 71–82 (2011)

5. Modi, C.N., Patel, D.R., Patel, A., Rajarajan, M.: Integrating signature apriori based network intrusion detection system (nids) in cloud computing. Proc. Technol. **6**, 905–912 (2012)
6. Gautam, A.K., Sharma, V., Prakash, S., Gupta, M.: Improved hybrid intrusion detection system (HIDS): mitigating false alarm in cloud computing. JCT (2012)
7. Patel, A., Taghavi, M., Bakhtiyari, K., Júnior, J.C.: Taxonomy and proposed architecture of intrusion detection and prevention systems for cloud computing. In: Cyberspace Safety and Security, pp. 441–458. Springer (2012)
8. Kholidy, H.A., Baiardi, F.: CIDS: a framework for intrusion detection in cloud systems. In: 2012 Ninth International Conference on Information Technology: New Generations (ITNG), pp. 379–385. IEEE (2012)
9. Debar, H., Curry, D., Feinstein, B.: The intrusion detection message exchange format, internet draft. Technical report, IETF Intrusion Detection Exchange Format Working Group, Jul 2004
10. Debar, H., Curry, D., Feinstein, B.: The intrusion detection message exchange format IDMEF, RFC 4765, IETF, 1 Mar 2007. http://tools.ietf.org/html/rfc4765
11. https://www.gnu.org/software/hurd/user/tlecarrour/rng-tools.html
12. https://www.openstack.org/
13. https://www.prelude-siem.org/
14. https://www.prelude-siem.org/projects/prelude/wiki/InstallingPreludeDbLibrary

Gateways Selection for Integrating Wireless Sensor Networks into Internet of Things

Hassan El Alami and Abdellah Najid

Abstract This paper presents an approach to integrate Wireless Sensor Networks into Internet. In WSN some sensors have two radio components: Zigbee radio which used on all sensors to send data within the WSN and WiMax radio which activated only on subset of sensors, indicated to as gateways, for sending data to Internet. Objective of our approach is to increase lifetime and throughput of WSN where sensors can join the Internet through gateway, to achieve this goal, we use link cost to select gateway. The link cost is the ratio between distance to Internet and energy level of sensor. The sensor has been selected as gateway which has minimum link cost. To ensure minimum link cost can be performed, we divided WSN into grids where, gateways have been selected in each grid. Results justify its accuracy.

Keywords Wireless sensor network · Internet of things · Gateway · Link cost · Throughput · WiMax · Zigbee

1 Introduction

As important applications of the Internet, the Internet of Things (IoT) provides immediate bridge between the physical world and the smart things like sensors or RFID tags in the digital world and can lead to innovative applications and services with high efficiency and productivity. According to Cisco [1], 50 billion smart things will be connected to the IoT in 2020, thus overshadowing the data generated by humans. This is restricted by the birth rate: in 2020, it is expected to have 8 billion people around the world [2]. IoT is receiving vast attention because it allows

H. El Alami (✉) · A. Najid
STRS Laboratories, National Institute of Posts and Telecommunications - INPT, Rabat, Morocco
e-mail: h.elalami.a@ieee.org

A. Najid
e-mail: najid@inpt.ac.ma

Á. Rocha et al. (eds.), *Europe and MENA Cooperation Advances in Information and Communication Technologies*, Advances in Intelligent Systems and Computing 520, DOI 10.1007/978-3-319-46568-5_43

for the pervasive interaction with/between smart sensors leading to an effective integration of information into the digital world. However, IoT, as it stands nowadays, is just at the beginning of a longer travel. The technology of WSNs/IoT has been developed in order to provide a baseline architecture framework and basic functionalities allowing for enormous scale operational distribution. Covering vast application domains, WSNs can play an important role by collecting environment information including monitoring environment, medical treatments, military surveillance, and so on. In particular, WSNs are connecting into IoT [3]. The most usual strategy integrating the WSNs through gateways to Internet in which sensors can employ independent routing protocols. Additionally, each sensor is responsible to communicate with gateway and gateway is responsible to forward the sensed data to the user or the server web through Internet. Thus, Workload on gateway is maximized i.e., the gateway consumes more energy causing the gateway lifespan to decrease, where, transmission from the WSNs to the Internet becomes blocked. Therefore, the gateways selection in WSNs to connect the Internet is one of the challenges in integrating the WSNs into the IoT. Therefore, in this paper, we investigate gateways selection to maximize lifetime and throughput of WSNs.

In this paper, we consider a WSN where sensors transmit sensed data to Internet (BS_WiMax) via gateways which are selected initially. On the other hand, after the sensors are distributed in network area, information sharing among sensors and base station BS_WiMax is performed with the help of HELLO packets broadcast mechanism. BS_WiMax first broadcasts HELLO packet to inform each sensor with the position of BS_WiMax and received signal strength, after receiving HELLO packet, each sensor stores the position of BS_WiMax, and sends a response HELLO packet which contains ID of sensor, distance to BS_WiMax, link status and its energy level status. After this phase, BS_WiMax select sensor as gateway based on minimum link cost for current round. Then, the sensors send sensed data to Internet through gateways. In addition, link cost is the ratio between distance to BS_WiMax and energy level of sensor. Rest of the paper is organized as: in Sect. 2 related works is presented, Sect. 3 provides brief description of our proposed approach, Sect. 4 takes the discussion of simulation results into consideration, and finally Sect. 5 ends paper with conclusion and future work.

2 Related Works

In recent years, the interconnection of sensors to Internet has become increasingly, and sensor networks have already achieved much commercial success in applications [4] like tele-controlling of human physiological data, environmental monitoring, commercial applications, and many more. Thus, WSNs applications request for integration with existing Internet Protocol (IP) networks, especially the IoT. Therefore, WSNs are connecting things to the Internet through a gateway that interfaces the WSNs to the Internet. Unlike traditional networks, WSNs have the particular characteristic of collecting sensed data such as motion, pressure, temperature, fire detection, Current/voltage, etc., and forwarding it to the user through gateways. In general idea of integration WSNs into Internet, many approaches have been proposed.

In [5] authors propose different scenarios to connect WSNs to Internet, the first scenario includes of connecting both independent WSNs and the Internet through a single gateway. The second scenario makes a hybrid network, still composed of independent network, where a small number of dual sensors can access the Internet. The last scenario is inspired from Wireless Local Area Network (WLAN) and forms a dense IEEE 802.15.4 access point network, where multiple sensors can join the Internet in single hope.

From [6], authors present a prototyping implementation of IoT Gateway based on Zigbee-GPRS protocols, which realizes sensed data forwarding, protocol transformation, WSN control and management. Therefore, this proposed can be widely used in smart home, smart grid, industrial monitoring, environment monitoring etc.

In [7], authors present the requirements for distributing IoT gateways, and propose architecture for the corresponding system to be distributed in the gateways. A Zigbee and GPRS protocols based IoT gateway system helps partly in dealing with the heterogeneity problem and therefore enables the WSNs to communicate with the mobile telecommunication network [8].

Authors in [9], presented the subtleties of integrating WSNs into the Internet in order to control the appliances of electric. They presented simple architecture which can be easily adopted for similar deployments. In this architecture, authors highlighted relevant constraints mainly IPv4 to IPv6 gatewaying.

Authors in [10] presented an overview of the premise of IoT concept, its enabling technologies, protocols, applications, and the recent research addressing different aspects of the IoT, and they cited some of the challenges and issues that pertain to the design and deployment of IoT implementations have been proposed. Moreover, in this paper, the interplay between the big data analytics, IoT, cloud and fog computing has been discussed.

3 Proposed Approach

We consider our network model as WSN in which sensors gather sensed data of interest from the environment, and send these data to the selected gateways which in turn convey the locally gathered and compressed sensed data to BS_WiMax. So, in a WSN sensors are of two types: normal sensors and gateway sensors. In model of WSN, the locals of sensors and gateways are random, whereas that of BS_WiMax is predetermined based on application. Additionally, the Zigbee communication system is used as a mean of data transmission between sensors within WSN, while the WiMax communication system is used as mean of data transmission between gateways and Internet.

Let N set of sensors and G is the set of gateways, where, $N = \{ n_i | n_i \in N \wedge 1 \leq n_i \leq n_{imax} \}$, and $G = \{ g_i | g_i \in G \wedge 1 \leq g_i \leq g_{imax} \}$. Each sensor can establish a wireless link within its communication range. In WSN, gateways which have functionality and establish the connectivity to Internet. Therefore, we can model establishment wireless link as follow:

$$C_{(x,y)} = \begin{cases} 1 & \text{if x connect to y} \\ 0 & \text{Otherwise} \end{cases} \quad (1)$$

In this way, we can conclude from Eq. (1) two cases:

- **Case 1**: if $x = n_i$ then $y = n_j \; \forall n_j \in N$ (and $n_i = n_j$) or $(y = g_i)$
- **Case 2**: if $x = g_i$ then $y = BS\_WiMax$

Operation of our proposed approach is divided into rounds. Each round is divided into three phase; (1) network configuration, (2) selection gateways, (3) data scheduling and transmission.

3.1 Network Configuration

Initialization phase begins at the start of round. In this phase after the sensors are distributed in network area, each sensor has information about their relative coordinates, position ID and received signal strength. HELLO packet exchange mechanism (Fig. 1) has been used to inform each sensor with the position ID, received signal strength, energy level status, status of links, and the coordinates of all other sensors along with BS_WiMax.

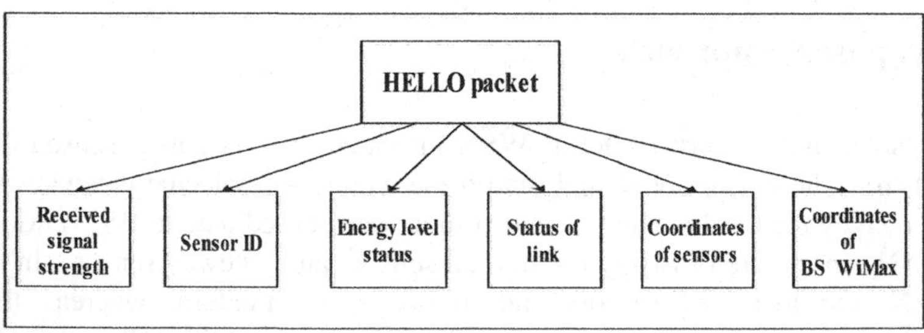

Fig. 1 HELLO packet format

3.2 Selection of Gateways

In this section, we present selection criteria for a sensor to become gateway, in order to enhance energy efficiency and throughput maximization. To connect the WSN to the Internet, our proposed selects new gateways in each round. BS_WiMax knows the position ID, distance and energy level status of the sensors, status of links. BS_WiMax computes the link cost of all sensors and transmit this link cost to all sensors. On the basis of this link cost, each sensor decides whether to become gateway or not. If n_i is ID of sensor than link cost of sensor n_i given by:

$$L \cdot C(n_i) = \frac{d(n_i)}{E_L(n_i)} \quad (2)$$

where $d(n_i)$ is the distance between sensor n_i and BS_WiMax, $E_L(n_i)$ is the energy level of sensor n_i and is calculated by subtracting the current energy of sensor from initial total energy. A sensor with minimum link cost is selected as a gateway for current round. All the neighbor sensors stick together with gateway sensor and send their sensed data to its gateway. Gateway sensor aggregates sensed data and convey to BS_WiMax. In addition, gateway sensor has high energy level and minimum distance to BS_WiMax; therefore, it consumes minimum energy to transmit sensed data to BS_WiMax.

3.3 Scheduling and Transmission Data

After that the gateways are selected, the gateways receive HELLO packets from others sensors intended to associate with them. Similarly, receives messages format from gateways selected intended to communicate directly with it. These message format receptions are followed by data scheduling, where the gateways assign Time Division Multiple Access (TDMA) schedules to their respective sensors and the also BS_WiMax then creates TDMA schedules to gateways, telling them when to forward. Data transmissions, with the assumption that gateways always have sensed data to send, begin soon after the fixed assignment of TDMA based schedules. These transmissions minimize the energy consumption due to the gateways selection mechanisms. The radio of each sensor is turned off until the nodes' allocated TDMA based schedules, thus the energy consumption is further decreased. The receiver of gateway is kept on to receive all the gathered information. The gateway, after receiving locally data, perform signal processing functions to compress these sensed data into a single packet. These high energy composite signals are then transmitted to the BS_WiMax. When all these data transmissions end, then next round begins with the network configuration deciding the next gateways or non-gateways based on the proposed operation.

4 Simulation Results and Discussion

Our objective in conducting simulations is to evaluate the performance of our proposed approach by comparing it with grids based proposed approach (Proposed-4 Grids and Proposed-6 Grids). Tests are carried out using immobile sensors and plane coordinates. 100 sensors are randomly distributed in the network are of 100 m × 100 m. As sensors are equipped with limited energy, when they drain their energy during of simulation, they cease to send or receive information. For the simulations, energy is consumed whenever a sensor receives or sends or aggregates sensed data. Average results with 85 % confidence interval presented after running the simulation 6 times. Moreover, the sensors in our proposed approaches are initially supplied with 2 J and the distance between WSN and BS_WiMax is 50 m.

When sensed packet travel from sensor to gateway or from gateway to BS_WiMax across a wireless channel, some of sensed packet fail to reach receiver (called dropped packets). To offset for dropped packets, Random Uniformed Model [11] has been used with the assumption that dropped packet is related to link status through which it is propagating. If a given wireless link is in good status, packet is successfully received at the BS_WiMax, otherwise it is dropped.

- **Stability period**: is the time interval from the establishment of network until of the death of first alive sensor.
- **Network lifetime**: is the time interval from the establishment of network until of the death of last alive sensor.
- **Throughput**: Number of packets successfully received at the BS_WiMax.

Figure 2 shows the impact of varying sensors density on stability period in three proposed approach. The stability period increases when the number of sensors in three proposed are varied. It is clear that downsizing the network in terms of the number of sensors results in increased stability period in our proposed approach. In other hand, stability period in our proposed based on grids prolong the stability period further due to balance the energy consumption in grids which means that the workload on gateways is decreased i.e., the gateway in grid now forward relatively less data to BS_WiMax.

Fig. 2 Stability period with varying number of sensors

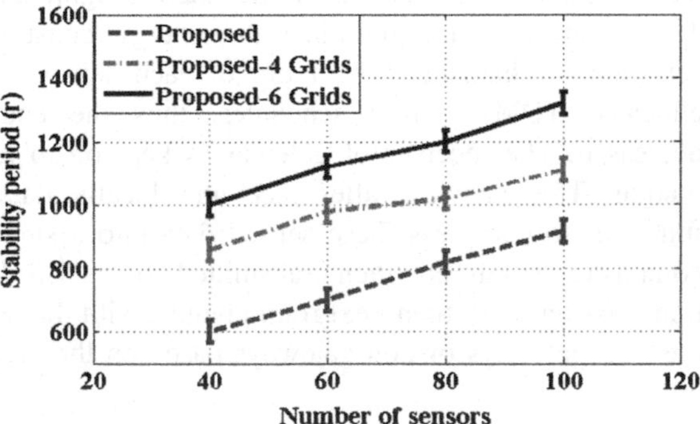

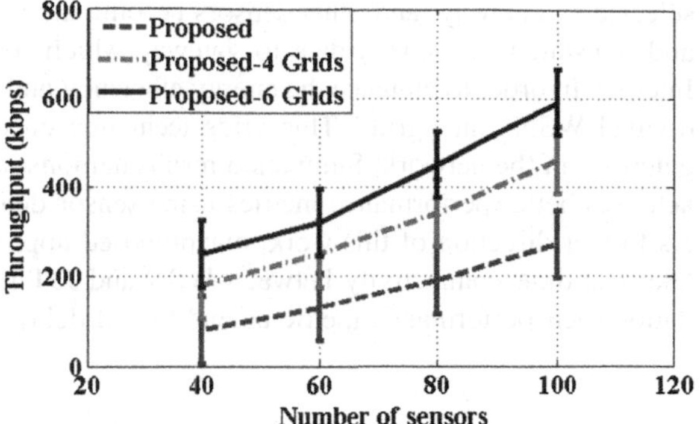

Fig. 3 Throughput of WSN with varying number of sensors

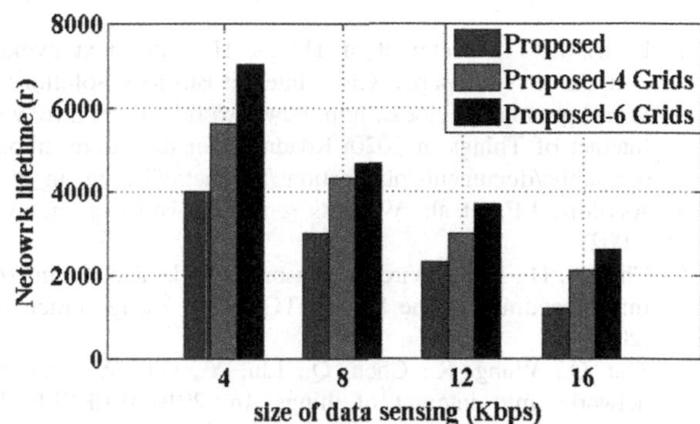

Fig. 4 Network lifetime with varying size of data sensing

From the Fig. 3, it is obvious that our proposed based on grids can achieve better throughput than proposed approach. Additionally, according on [12] in grids based networks area, provide efficient data delivery in WSNs, in that each sensor only maintains a simple transmitter candidate set for it to send sensed data. Figure 4 shows the impact of varying size of sensed data on network lifetime in the three proposed approaches. Reason for this interesting result is clear i.e., in grids based our proposed, sensors which remain alive for longer for a longer time transmit for longer time too whereas more number of contending sensors maximizes the chances of collision also leads to decreased network lifetime. Moreover, the relative network lifetime and throughput efficiencies of our proposed approach based on grids in comparison to our proposed.

5 Conclusion and Future Work

Integration of WSNs into IoT is key component in IoT application systems. In this paper, we proposed a novel approach which used the link cost to select appropriate gateway to integrate WSNs into IoT. Sensors with minimum value of link cost are selected as gateway, and other sensors become the normal sensors of that gateway and transmit their sensed data to gateway which forwards them to user through Internet. In order to enhance the energy efficiency and throughput of WSNs-IoT, we divided WSNs into grids. This grids technique encourages better distribution of gateways in the network. Simulation results demonstrated that our routing approach achieves better performance metrics using sensor density and size of sensing data. As further direction of this work, our proposed approach routing can be extend to take real-time connectivity between WSN and IoT into account, and further validation such performance metric as end to end delay.

References

1. Evans, D.: The Internet of Things. How the next evolution of the internet is changing everything, Whitepaper. Cisco Internet Business Solutions Group (IBSG) (2011)
2. World population clock., http://www.worldometers.info/world-population/
3. Internet of Things in 2020: Roadmap for the future. http://www.smart-systems-integration.org/public/documents/publications/Internet-ofThings_in_2020
4. Akyildiz, I.F., et al.: Wireless sensor networks: a survey. Comput. Netw **38**(4), 393–422 (2002)
5. Christin, D., et al.: Wireless sensor networks and the internet of things: selected challenges. In: Proceedings of the 8th GI/ITG KuVS Fachgespräch Drahtlose Sensornetze, pp. 31–34 (2009)
6. Zhu, Q., Wang, R., Chen, Q., Liu, Y., Qin, W.: IoT gateway: Bridgingwireless sensor networks into internet of things. In: 2010 IEEE/IFIP 8th International Conference on Embedded and Ubiquitous Computing (EUC), pp. 347–352. IEEE (2010)
7. Jia, X., Feng, Q., Fan, T., Lei, Q.: RFID technology and its applications in Internet of Things (IoT). In: 2012 2nd International Conference on Consumer Electronics, Communications and Networks (CECNet), pp. 1282–1285. IEEE (2012)
8. Li, L., Hu, X., Chen, K., He, K.: The applications of WiFi-based wireless sensor network in internet of things and smart grid. In: 2011 6th IEEE Conference on Industrial Electronics and Applications (ICIEA), pp. 789–793. IEEE (2011)
9. Khalil, N., Abid, M.R., Benhaddou, D., Gerndt, M.: Wireless sensors networks for Internet of Things. In: 2014 IEEE Ninth International Conference on Intelligent Sensors, Sensor Networks and Information Processing (ISSNIP), pp. 1–6. IEEE (2014)
10. Al-Fuqaha, A., Guizani, M., Mohammadi, M., Aledhari, M., Ayyash, M.: Internet of things: a survey on enabling technologies, protocols, and applications. IEEE Commun. Surv. Tutor. **17** (4), 2347–2376 (2015)
11. Zhou, Q., Cao, X., Chen, S., Lin, G.: A solution to error and loss in wireless network transfer. In: International Conference on Wireless Networks and Information Systems, 2009. WNIS'09, pp. 312–315. IEEE (2009)
12. Liu, X.: Atypical hierarchical routing protocols for wireless sensor networks: a review. IEEE Sens. J. **15**(10), 5372–5383 (2015)

Cognitive Radio Spectrum Sensing Based on Energy

Oualid Khatbi, Ahmed Mouhsen and Zakaria Hachkar

Abstract The surveys that have been done on the use spectrum shows that some frequency bands are less busy compared with others. Cognitive radio has come to give the possibility to use the free band of a primary user for a secondary user. Confirmation of the presence of the primary user is typically performed by a spectrum sensing. A function of cognitive radio based on the ability to find unoccupied spectrum without interference. The cognitive radio spectrum sensing method considered in this work is Energy detection method based on the central limit theorem and the sensing performance of this scheme is quantified by the receiver operating characteristic (ROC), such as between the probability of detection versus the probability of false alarm. A simulation is carried out in the Matlab environment to show the relation between Pd and Pf with various SNR values.

Keywords Cognitive radio • Spectrum sensing • Energy detection method • The probability of detection and probability of false alarm

1 Introduction

The Wireless communication saw a very fast growth in these last years. The growth of the systems and the wireless services showed that the availability of spectrum became severely limited; it becomes obvious that the increasing number of data cannot be satisfied by the current statistical plans of frequency assignment. As a

---

O. Khatbi (✉) · A. Mouhsen · Z. Hachkar
Faculté des Sciences et Techniques, Ingénierie Mécanique, Management Industriel et Innovation (IMMII), Université Hassan 1er, BP 577 26000 Settat, Morocco
e-mail: o.khatbi@gmail.com

A. Mouhsen
e-mail: mouhsen.ahmed@gmail.com

Z. Hachkar
e-mail: zhachkar2000ster@gmail.com

Á. Rocha et al. (eds.), *Europe and MENA Cooperation Advances in Information and Communication Technologies*, Advances in Intelligent Systems and Computing 520, DOI 10.1007/978-3-319-46568-5_44

consequence, the need to have a new way of exploitation of the spectrum became essential. The cognitive radio appears to be a solution of temptation of the problem of spectral congestion by presenting the opportunist use of the frequency bands which are not heavily occupied by authorized users [1]. Whereas there is no agreement on the formal definition of cognitive radio from now on, the concept developed recently to include diverse meanings in several contexts [2]. In this paper, we use the definition adopted by Federal Communications Commission (FCC): "Cognitive radio: A radio or system that senses its operational electromagnetic environment and can dynamically and autonomously adjust its radio operating parameters to modify system operation, such as maximize throughput, mitigate interference, facilitate interoperability, access secondary markets." [3]. In the terminology of the cognitive radio, primary users PU can be defined as the users who have the capacity and the priority to use a specific party of the spectrum. Examples of licensed technology are global system for mobile communications (GSM) [4], worldwide interoperability for microwave access (WiMax) [5, 6], and long term evolution (LTE) [6]. On the other hand, secondary users SU, who have a lower priority to exploit this spectrum so that they do not cause interference to the primary users. Thus, the secondary users must be capable of detecting the unused specters by the primary users in a reliable way and of exploiting them. The reliability of the spectral detection is the most important factor which determines the performance of the cognitive radio, and it is at present based on three techniques different [7]: matched filter detection, energy detection and cyclo-stationary detection.

The matched filter based detection requires a perfect knowledge of the PU signal for synchronization and a dedicated receiver for each kind of PU signal but it is very good at low SNR region. In cyclo-stationary detection, we require exact information about the cyclic frequencies of the PU signal which is not an easy task. Energy detection (also denoted as non-coherent detection), is the signal detection mechanism using an energy detector (also known as radiometer) to specify the presence or absence of signal in the band. The most often used approaches in the energy detection are based on the Neyman-Pearson (NP) lemma. The NP detection criterion enlarges the probability of detection Pd for a given probability of false alarm fa [8] It is an essential and a common approach to spectrum sensing since it has moderate computational complexities, and can be implemented in both time domain and frequency domain [9, 10]. To adjust the threshold of detection, energy detector requires knowledge of the power of noise in the band to be sensed [11]. The signal is detected by comparing the output of energy detector with threshold which depends on the noise floor.

In this paper with the help of Central Limit Theorem we have proposed new threshold based energy detector to improve the performance of spectrum sensing for CR. Simulation results shows that our proposed scheme based on this threshold depend the SNR values and better for higher SNR. The rest of the paper is organized as follows. System model is presented in Sect. 2. Spectrum sensing has been done in Sect. 3. Energy Detection, Central limit theorem approach, in Sects. 4 and 5 respectively, Simulation results are shown in Sect. 6 and finally our conclusion is drawn in Sect. 7.

2 System Model

The cognitive radio is the system of wireless communication where a transmitter-receiver can intelligently detect the communication channels which are not used and to reach it without creating interferences. This property gives the possibility to use the frequency spectrum. The availability of the idle spectrum of radio is generally varied according to time, the frequency and the location. The spectral detection allows the secondary users Known how to have a dynamic spectral access and get such idle spectrum in a opportunist way. It also allows knowing if the primary user decided to take back his band once again.

2.1 Types of CR

There are two types of Cognitive Radios:

- Full Cognitive Radio: Full Cognitive Radio (CR) considers all parameters. A wireless Node or network can be conscious of every possible parameter observable [12].
- Spectrum Sensing Cognitive Radio: Detects channels in the radio frequency spectrum. Fundamental requirement in cognitive radio network is spectrum sensing. To enhance the detection probability [13] many signal detection techniques are used in spectrum sensing.

2.2 Characteristics of CR

There are two main characteristics [14] of the cognitive radio and can be defined

- Cognitive capability: Cognitive Capability defines the ability to capture or sense the information from its radio environment of the radio technology. Joseph Mitola first explained the cognitive capability in term of the cognitive cycle "a cognitive radio continually observes the environment, orients itself, creates plans, decides, and then acts"
- Reconfigurability: Cognitive capability offers the spectrum awareness, Reconfigurability refers to radio capability to change the functions, enables the cognitive radio to be programmed dynamically in accordance with radio environment (frequency, transmission power, modulation scheme, communication protocol).

2.3 Functions of CR

There are four major functions of Cognitive Radio:

- Spectrum Sensing: The goal of spectrum sensing is to find and determine the presence of primary users on a band.
- Spectrum management: Supply and choose by the cognitive radio the just spectrum immediately after its detection.
- Spectrum Sharing: Cognitive Radio assigns the unused spectrum (spectrum hole) to the secondary user (SU) as long as primary user (PU) does not use it. This property of cognitive radio is described as spectrum sharing.
- Spectrum Mobility: When a licensed (Primary) user is detected the Cognitive Radio (CR) vacates the channel.

3 Spectrum Sensing

Spectrum sensing is the most important function in the cognitive radio; it is always in the study shift and development. Secondary user (SU) should be able to detect spectrum in a continuous and real time way and must be equipped with highly reliable spectrum sensing functions [15]. Many different methods are proposed to identify the presence of signal transmission and can be used to enhance the detection probability.

3.1 System Model

The problem of detecting the presence or absence of the PU transmission is formulated as a binary hypothesis testing problem. The null hypothesis denoted by H0 corresponds to the received signal being only noise. On the other hand, the alternative hypothesis denoted by H1 indicates that the received signal contains the PU signal along with noise. As an example, a simple binary hypothesis test for detecting the PU transmission in an AWGN channel is given by:

$$\begin{aligned} &\text{H0: } x(n) = w(n) \\ &\text{H1: } x(n) = s(n) + w(n) \end{aligned} \tag{1}$$

where n = 1, 2, 3, ……. N is the number of samples of received signal, x(n) is the received sample signal by the SU, w(n) is the white Gaussian noise with mean zero and variance σ_n^2, s(n) is the received PU signal with mean zero and variance σ_s^2.

3.2 Performance Criteria

Performance of spectrum sensing algorithms may differ in different scenarios. It is therefore important to compare and choose the best scheme for a given scenario. At the same time, it is necessary to choose proper performance criteria for a fair comparison. In this section, we briefly present important performance parameters which can be used to evaluate the sensing algorithms:

- False alarm probability: It is defined as the probability that the detector declares the presence of PU, when the PU is actually absent.
- Missed detection probability: It is defined as the probability that the detector declares the absence of PU, when the PU is actually present.
- Sensing time: If the receiver chain is time-duplexed for reception and sensing, it is desirable that the sensing durations are shorter and the data transmission durations are longer.
- SNR: The SNR of the received PU signal at the sensor depends on the PU transmitted power and the propagation environment. The detection performance improves with an increase in the SNR.

3.3 Detection Techniques

Fundamental to the theory of detecting the signal in noise is the theory of statistical decision, where the decision making depends on the hypothesis testing. In binary hypothesis testing, the problem resides in defining a decision rule that indicates which of two hypotheses should be chosen: the null hypothesis (H0) or the alternative hypothesis (H1). If the null and alternative hypotheses are defined in terms of signal(s), hypothesis H0 (signal absent) and hypothesis H1 (signal present). The decision rule can be represented as:

$$\begin{aligned} \text{H1}&: \Lambda(y) > \lambda \\ \text{H0}&: \Lambda(y) < \lambda \end{aligned} \quad (2)$$

where λ is the threshold and Λ(y) is a function that depends on the measurements. If it exceeds the threshold, then H1 is selected; otherwise, H0 is decided. The aim of the detection theory is, hence, to design the most effective detector by definition Λ(y) and λ. Let $y = [y_0, \ldots, y_{N-1}]$ be the observation vector and $P(y/H_i)$, i = 0, 1, denote the joint probability density function (PDF) of these N elements of observing y given that H_i was true, is often referred to as the likelihood function of the observation vector y. Thus, we can define the Λ(y) is the likelihood ratio test (LRT) as

$$\Lambda(y) = \frac{p(y/H_1)}{p(y/H_0)} \tag{3}$$

4 Energy Detection

Energy detection is a signal detection mechanism in which the presence or absence of primary user is determined by measuring the radio frequency energy in the channel [16]. Then this calculated energy of the received signal over specified time duration is compared with the threshold value (chosen) and decision about the occupancy of channel is made accordingly. The energy detection technique is quite simple as it does not require any prior knowledge about the primary user. The test statistics for energy detection is given as:

$$T(x) = \frac{1}{N}\sum_{n=1}^{N}|x(n)|^2 \tag{4}$$

where T(x) is received signal energy and N is sampling, If the threshold value is less than the calculated energy, decision is made that primary user is present and if the calculated energy value comes out to be less than threshold value than the decision is made that primary user is absent and the band is free and can be allotted to secondary user [17]. The performance of the detection can be characterized with two probabilities: probability of detection P_d and probability of false alarm P_f. It can be formulated as:

$$\begin{aligned} P_d &= P\{T(x) > \lambda/H_1\} \\ P_d &= Q\left(\frac{\lambda - (\sigma_S^2 + \sigma_n^2)}{(\sigma_S^2 + \sigma_n^2)/(\sqrt{N/2})}\right) \end{aligned} \tag{5}$$

$$\begin{aligned} P_f &= P\{T(x) > \lambda/H_0\} \\ P_f &= Q\left(\frac{\lambda - \sigma_n^2}{\sigma_n^2/(\sqrt{N/2})}\right) \end{aligned} \tag{6}$$

where

$$Q(x) = \frac{1}{\sqrt{2\pi}}\int_x^{\infty} e^{\frac{t^2}{2}}dt \tag{7}$$

The detection threshold for a fixed P_f can be given as:

$$\lambda = \sigma_n^2 \sqrt{(N/2)Q^{-1}(P_f)} + \sigma_n^2 \tag{8}$$

5 Central Limit Theorem Approach

The CLT shows that the sum of N random variable N with a finite mean and variance approaches a normal distribution when N is large enough. Using the CLT, the distribution of the test statistic (4) can be accurately approximated with a normal distribution for a sufficiently large number of samples as:

$$\Lambda \sim \begin{cases} N\left(\mathrm{N}(2\sigma_n^2), \mathrm{N}(2\sigma_n^2)^2\right) : H_0 \\ N\left(\mathrm{N}(2\sigma_n^2)(1+\gamma), \mathrm{N}(2\sigma_n^2)^2(1+\gamma)^2\right) : H_1 \end{cases} \tag{9}$$

where γ is the SNR, By using each mean and variance in (9), an approximated false alarm probability is:

$$P_f = Q\left(\frac{\lambda - \mathrm{N}(2\sigma_n^2)}{(\sqrt{\mathrm{N}}(2\sigma_n^2)}\right) \tag{10}$$

Similarly, approximated detection probabilities are:

$$P_d = Q\left(\frac{\lambda - \mathrm{N}(2\sigma_n^2)(1+\gamma)}{\sqrt{\mathrm{N}}(1+\gamma)(2\sigma_n^2)}\right) \tag{11}$$

6 Simulations Results

To observe the detection performance of the CR under the CLT theorem, simulation results are shown in this section to evaluate the CLT approach in energy detection scheme.

Figure 1 shows the probability of detection versus SNR curves for simulated case. We have taken 1000 samples and probability of false alarm is fixed to 0.01. As it can be easily observed from the Fig. 1 that the probability of detection is better from SNR = −10 dB.

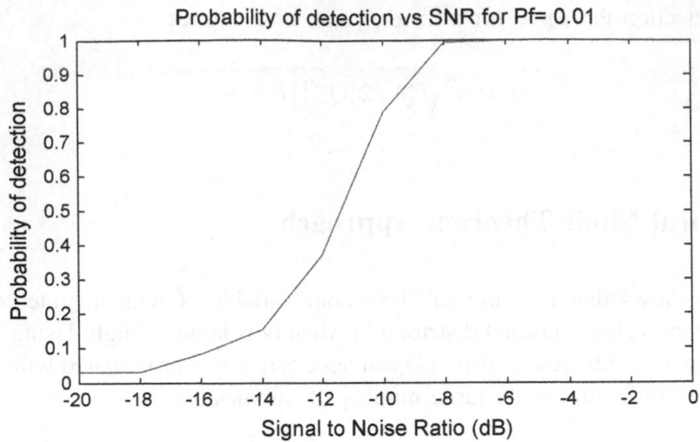

Fig. 1 Probability of detection vs SNR for $P_f = 0.01$

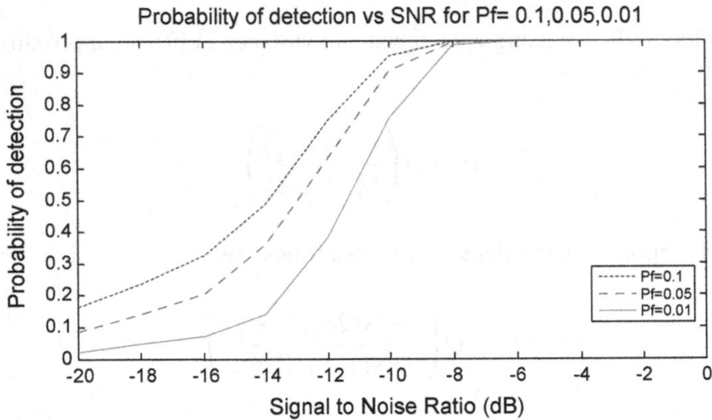

Fig. 2 Probability of detection vs SNR for $P_f = 0.1, 0.05, 0.01$

The graphs are drawn between probability of detection and SNR under various probability of false alarm. From Fig. 2 it is clear that for low SNR region probability of detection increases with increase in the probability of false alarm and for higher SNR graphs are converged.

It clear that with the increasing of the SNR (from –20 dB to 0) the detections we get also increased. It indicates that with the increasing of the SNR, the more spectrums which are occupied we can detect. By changing the value of the SNR, we get the relationship between the SNR and the detections, from the diagram, we can see from −10 dB to 0, SNR makes the energy detector performs best.

7 Conclusion

This paper proposes a threshold based on the central limit theorem and the sensing performance of this scheme is quantified by the receiver operating characteristic (ROC), such as between the probability of detection versus the probability of false alarm. Simulation results demonstrates that the increasing of the SNR increase the detections and the more spectrums which are occupied is detected, and also the relationship between the SNR and the detections, from SNR = −10 dB, the energy detector performs best.

References

1. Mitola, J., Maguire, G.Q.: Cognitive radio: making software radios more personal. IEEE Pers. Commun. Mag. **6**(4), 13–18 (1999)
2. Neel, J.O.: Analysis and design of cognitive radio networks and distributed radio resource management algorithms, Ph.D. dissertation, Virginia Polytechnic Institute and State University (2006)
3. Federal Communications Commission. Notice of proposed rulemaking and order: Facilitating opportunities for flexible, efficient, and reliable spectrum use employing cognitive radio technologies, ET Docket No. 03–108 (2005)
4. Association for global system for mobile communications. http://www.gsm.org. Accessed 19 Dec 2011. (Cited on page 2.). Mouly, M., Pautet, M.: The GSM system for mobile communications. Telecom Publishing, 1992, 704 pp. (Cited on p. 2)
5. IEEE 802.16 working group on broadband wireless access standards. http://www.ieee802.org/16/. Accessed 1 Dec 2011. (Cited on p. 2, 25)
6. Korowajczuk, L.: LTE, WiMAX and WLAN Network Design, Optimization and Performance Analysis, 782 p. Wiley, Chicester, UK (2011). (Cited on p. 2.)
7. Zhang, W., Mallik, R., Letaief, K.: Optimization of cooperative spectrum sensing with energy detection in cognitive radio networks. IEEE Trans. Wirel. Commun. **8**(12), 5761–5766 (2009)
8. Energy detector with baseband sampling for cognitive radio: real-time implementation. http://www.dx.doi.org/10.4236/wet.2012.34033. Accessed Oct 2012
9. Cabric, D., Mishra, S.M., Brodersen, B.W.: Implementation issues in spectrum sensing for cognitive radios. Berkeley Wireless Research Center, University of California, Berkeley (2004)
10. Pawetczak, P., Janssen, G.J.M., Prasad, R.V.: Performance measures of dynamic spectrum access networks. In: IEEE Global Telecommunications Conference, San Francisco, 27 Nov – 1 Dec 2006, pp. 1–6
11. Urkowitz, H.: Energy detection of unknown deterministic signals. Proc. IEEE **55**(4), 523–531 (1967). doi:10.1109/PROC.1967.5573
12. Tabaković, Ž.: A survey of cognitive radio systems. Post and Electronic Communications Agency, Jurišićeva 13, Zagreb, Croatia
13. Haykin, S.: Cognitive radio: brain-empowered wireless communications. IEEE J. Sel. Areas Commun. **23**(2), 201–220 (2005)
14. Akyildiz, I.F., Lee, W.-Y., Vuran, M.C., Mohanty, S.: Next generation/dynamic spectrum access/cognitive radio wireless networks: a survey. In: Broadband and Wireless Networking Laboratory, School of Electrical and Computer Engineering, Georgia Institute of Technology, Atlanta, GA 30332, United States. Accessed 2 Jan 2006. Accessed 2 May 2006

15. Lakshmi, M., Saravanan, R., Muthaiah, R.: A study on spectrum sensing methods for cognitive radio. Int. J. Eng. Technol. (IJET) **5**(2), Apr–May 2013
16. Subhedar, M., Birajdar, G.: Spectrum sensing techniques in cognitive radio networks: a survey. Int. J. Next Gener. Netw. (IJNGN) **3**(2), 37–51, Jun 2011
17. Adulsattar, M.A., Hussein, Z.A.: Energy detection technique for spectrum sensing in cognitive radio: a survey. Int. J. Comput. Netw. Commun. (IJCNC) **4**(5), 223–245 (2012)

Implementation of Clustering Metrics in Vehicular Ad-Hoc Networks

Abdelali Touil and Fattehallah Ghadi

Abstract The clustering in Ad-hoc networks, which has interested a large community of scientists, is one of solutions used to ensure the stability of the network and to reduce packets loss. The challenge is to find the best metric to generate fewer clusters. In this paper, we compare between four metrics used in Mobile Ad hoc Network (MANet) for choose the cluster-heads, namely Unique ID, Degree (Connectivity), Distance and Mobility. The intent of this study is to deduce the best metric suited to the characteristics of VANet such as high velocity and organized mobility.

Keywords VANet · Clustering · Ad-hoc · Clusters · Cluster-head · Metrics · Hierarchical routing

1 Introduction

The increase in the worldwide transport networks inducted to several serious problems, such as congestion in the urban environment, pollution and accidents. Despite the efforts made by the authorities to reduce the damage, the number of accidents is still very high. Vehicular networks are a projection of intelligent transport systems (ITS). Those systems have recently attracted many researchers and manufacturers attention in the automotive sector. Vehicles communicate with each other through inter-vehicular communication (V2V) as well as with the equipment of the road through the vehicle-to-infrastructure communication (V2I). The objective of these networks is to make vehicular networks contribute to design safer and more efficient roads in the future by providing real time information for

A. Touil (✉) · F. Ghadi
Sciences Engineering Laboratory, Faculty of Science, IBN ZOHR University, Agadir, Morocco
e-mail: abdelali.touil@edu.uiz.ac.ma

F. Ghadi
e-mail: f.ghadi@uiz.ac.ma

Á. Rocha et al. (eds.), *Europe and MENA Cooperation Advances in Information and Communication Technologies*, Advances in Intelligent Systems and Computing 520, DOI 10.1007/978-3-319-46568-5_45

drivers and concerned authorities [1–3]. Moreover each vehicle must be able to reinforce its protection after each external attack. They must respect the principles of security: authentication, integrity, confidentiality and non-repudiation [7, 8].

Vehicular Ad-hoc networks are considered as a special case of Mobile Ad-hoc Networks with few characteristics [2, 8, 9]: Rather than arbitrary moves, vehicles tend to travel in an organized way; VANet are faster than MANet Networks; Energetic constraints are no longer a concern for this type of networks.

This paper is organized as follows: Sect. 2 presents the vehicular Ad-hoc wireless access environment standards. Section 3 shows the architectures of vehicular Ad-hoc networks. Section 4 presents the clustering approach and the cluster-heads selection metrics. Section 5 describes the clustering process and the structure of the algorithm. Section 6 shows the results of the simulation. Section 7 concludes the document and presented the future works.

2 Standards for Wireless Access in VANet

In 1992, American researchers have begun research in dedicated short-range communications (DSRC) [17], specially designed for intelligent transport systems, which means all of protocols and standards involved in communications between vehicles and infrastructure or between vehicles. In October 1999, the USA Federal Communications Commission has allocated 75 MHz of the electromagnetic spectrum in the band 5.9 GHz. Similarly, in August 2008, European Telecommunications Standards Institute has allocated 20 MHz of spectrum in the band 5.9 GHz. One of the DSRC applications in roads networks, is the automatic toll on the highway, it uses the DSRC communication standard. The embedded system in the vehicle consists of a fixed device on the vehicle windscreen, sensed by a receiver installed in the toll. DSRC have been migrated to the standards of IEEE 802.11, early based on the 802.11a standard for wireless transmission link data at rates up to 54 Mbit/s and the frequency band 5 GHz. To solve the problem of scaling to cover wider networks, IEEE group spends more effort on new amendments called Wireless Access for Vehicular Environments (WAVE) [16, 17], this last one is a modification made to the IEEE 802.11p norm. This amendment is designed for transport intelligent system applications based on the short-range communications. A frequency band of 5.9 GHz (5.85–5.925 GHz) is reserved. WAVE is defined to enable communication between the vehicles in one hand and between vehicles and infrastructure in another hand [5]. This helps exchanging data in real time, improves driver safety and resolves the congestion issues in urban networks.

3 Vehicular Ad-Hoc Network Architectures

3.1 Vehicle-to-Vehicle Communication

Vehicle to vehicle (V2V) refers to a decentralized architecture based on inter-vehicular communications [2, 3, 6]. The V2V paradigm requires no infrastructure. Indeed, each vehicle takes the role of the sender, receiver and the router at the same time [16]. A vehicle can communicate directly with another vehicle equipped with an on-Board Unit (OBU), or by a multi-hop protocol which is responsible to send the data from source to destination by using nearby vehicles as relay [17].

3.2 Vehicle-to-Infrastructure Communication

Vehicle to infrastructure communication allows sharing of resources and services available for all vehicles in the vicinity, such as internet through the deployment of RoadSide Unit (RSU) placed every kilometer or less on the roads [4]. The RSU provides very high bandwidth useful for exchanging large data in a reduced transmission time. It also serves to inform the authorities in case of accidents and speed violations [2, 17] (Fig. 1).

4 Clustering Operation

We used the clustering in this paper because of its numerous advantages. It allows dividing a physical network of vehicles into a set of virtual groups geographically close. These groups are called clusters. They are generally managed by particular vehicles named the cluster-heads. The objective of this operation is to optimize the size of the routing tables and the sharing of available resources for more stability [15].

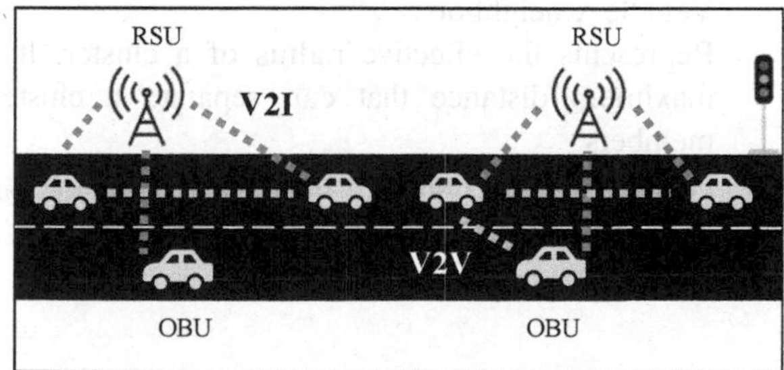

Fig. 1 Vehicle-to-infrastructure communication (V2I) and Inter-vehicle communication (V2V)

4.1 Different Status of a Vehicle in Clustering Network

In clustering network a vehicle could have three roles:

Cluster-head or organizer vehicle: Selected on an election process, only the vehicles that reach some selective criteria (depending on metrics) may take this role. It helps to build clusters, and maintain the structure of clusters after each re-affiliation or connection loss.

Cluster members: are vehicles that constitute the cluster, also called the ordinary vehicles by some authors.

Gateway has a feature on the cluster member's vehicles. This one is connected at least to two different cluster-heads. The main role of the gateway vehicle is to exchange data between the connected clusters.

4.2 Cluster-Head Selection Criteria

Unique Identifier

It is assumed that each vehicle has a unique identifier to distinguish it from others. Clustering algorithms compare the identifier of each vehicle with its neighbors, the vehicle that has the smallest or the biggest ID depending on the algorithm is elected as the cluster-head [9, 18, 19]. In our work, we have simulate a case when the cluster-head is the vehicle that has the biggest Identifier, noted that the identifier is an integer.

Degree

In graph theory, the degree of a node x is the number of edges adjacent to this node. The algorithms based on this criterion, calculates the number of connection of each vehicle, the vehicle that has the biggest number of connections is a candidate to become a cluster-head [10, 18].

$$D_x = |N(x)| = \sum_{y \in V,\ y \neq x} dist\,(x, y) < tx_{range}$$

$N(x)$	Vehicle x neighbors.
tx_{range}	Represents the effective radius of a cluster. It expresses the maximum distance that can separate a cluster-head of its members.
$dist\,(x, y) < tx_{range}$	Neighbor in the radius of the vehicle signal tx_{range}.

Mobility

The mobility of a vehicle x is calculated based on an application of the Pythagorean Theorem Euclidean geometry ($AC^2 = AB^2 + BC^2$), it can calculate the hypothesis distance (AC) in a time t and time t + 1 [11].

Using the basic speed formula d(m)/T(s), we deduce mobility of vehicle x as follows.

$$M_x = \frac{1}{T} \sum_{t=1}^{T} \sqrt{(X_t - X_{t-1})^2 + (Y_t - Y_{t-1})^2}$$

M_x Vehicle x mobility;
T Travel time between two points $dist\,(A, C)$.
The vehicle that has less mobility is the best candidate to be cluster-head; this allows an extended life time to clusters.

Distance

The metric of the distance calculating distances of all neighborhood members of a vehicle. The vehicle with less distance is considered as cluster-head. This metric is calculated using location-based, GPS equipment for example.

$$d_x = \sum_{y \in N(x)} dist\,(x, y)$$

d_x The sum of the distances between a vehicle x and its neighbors.
$N(x)$ Neighborhood members of x.

5 Clustering Algorithm Description

The objective of the clustering algorithm is to select fewer vehicles to be cluster-heads based on the election criteria (Identifier, Degree, Distance and Mobility) [15]. The clusters are made of these cluster heads and their 1-hop neighbor vehicles. They cover the whole network.

Our protocol consists of two main layers, the synchronization layer and the clustering layer:

5.1 *Synchronization Layer*

The synchronization layer allows vehicles to fill and update the neighborhood tables automatically, to establish the connectivity between vehicles. It periodically sends messages in a fixed time interval. Each time the synchronization message is received by vehicle, it initializes a variable called TTL (Time-To-Leave). The TTL variable contains the time of the last message received, if the receiver detects that the TTL has not been updated in a predefined time, it means that the transmitter is not in its neighborhood and that he did not receive any synchronization message from him, and then automatically transmitter vehicle is removed from the receiver's neighbor table. We have modified wave short message used in Wireless Access in Vehicular Environment (WAVE) to send information between vehicles, because it is very useful for broadcast applications and particularly suitable for safety messages between devices on vehicles and roads. Designed primarily to support almost instantaneous exchange of security information, the WSM message is adapted to our clustering approach by adding five data fields. The first field represents the status; it is an integer type that allows receivers to know the decision of the transmitter. This field can contain three values: zero for undefined status which explains that the transmitter is not included in the clustering process; one means that the transmitter is a cluster-head and two indicates that the transmitter is a cluster member. The second field corresponds to the transmission time of the message. The third one is the value of the election's metric. The fourth field contains the network address of the cluster-head. The fifth one holds in the weight of the transmitter.

5.2 Clustering Layer

The clustering layer includes all the features of the clustering process; this layer is arranged on two sub-layer creation and maintenance of clusters:

The cluster creation layer is used to build clusters. At the beginning of this operation, each vehicle queries its neighbors table to check if there is at least one vehicle with a cluster-head status and with biggest weight than his. If there is a vehicle that meets these criteria, it automatically sends a JOIN message to it. Otherwise if there is only a vehicle z that is not cluster-head but with biggest weight than his, he waits until the vehicle z executes the process of clustering to confirm their status. If z it becomes a cluster head, the vehicle running this process joins his cluster. If not it becomes cluster-head to build its own cluster, and sends a CH message to every vehicle in its neighbor table.

The maintenance layer is called when an event has occurred in the network after cluster formation. In other words, this layer is called after each new or broken connection in the topology.

A vehicle has joined a cluster, if the cluster-head has bigger weight than his. When a vehicle leaves the cluster, we have two cases: If it is a cluster-head: All the cluster-members delete his entry from their tables. If they are no longer able to receive periodic messages from the leaving vehicle, the members choose a new cluster-head. Otherwise if the leaving vehicle is a member, the cluster-head removes the entry corresponding to this vehicle from the cluster and the neighbor tables.

6 Simulation

The simulation is conducted to evaluate the performance of our algorithm. We apply the simulation platform composed by the OMNet++ simulator [12] and the SUMO [13] tools. In order to connect these two tools, we resorted to use the framework Veins [14]. OMNet++ is a network simulator representing the network features such as number of vehicles, topography, time steps and velocity. We implement the functionalities of our clustering algorithm and modeled the wireless communication among the vehicles by using OMNet++. SUMO supports both micro-mobility and macro-mobility features and through it, the mobility patterns for different vehicle densities are generated as well.

To measure the performance of each metric, we simulated those data for a period of 120 s using a clustering algorithm presented in Sect. 5. The simulations are performed in a geographical area of 2500 m^2 with transmission range of 300 m. A metric is considered relatively better, if it improve the stability of connection. We will test the stability of these metrics using the following three factors: The average of re-affiliation or average of cluster changes per vehicle, the number of lost packets and the average number of clusters generated throughout the simulation period.

Figure 2, shows the change in the average number of clusters generated by each metric with respect to the unit time in an urban environment with a speed between 15 and 80 km/h. In this result we observe that the degree is the metric that generated less clusters in the simulations, this metric considers that the vehicle that has a greater degree means in its neighborhood is a candidate to become the Cluster-head.

Figures 3 and 4, shows respectively the average of changes and the number of packets lost during the simulation.

In the first figure (Fig. 3), we can obviously see that the unique identifier and the degree have the same re-affiliation average, followed by the mobility. However, the average number of re-affiliation in the simulation and the distance are greater than the other metrics.

Each metric clustering incurs some packet loss during the exchange of message construction and maintenance of clusters that primarily affects the efficiency of the algorithm and the stability of clusters. We consider the number of lost packets

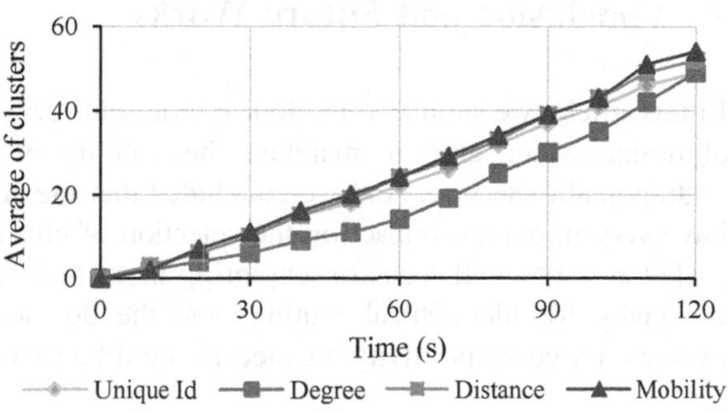

Fig. 2 Average number of generated clusters

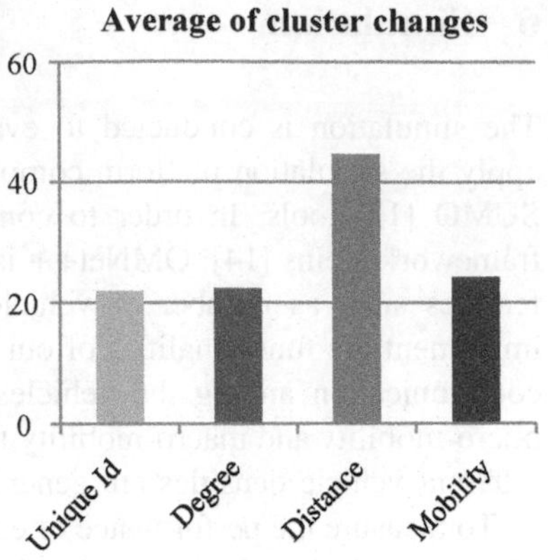

Fig. 3 Average cluster change by vehicle

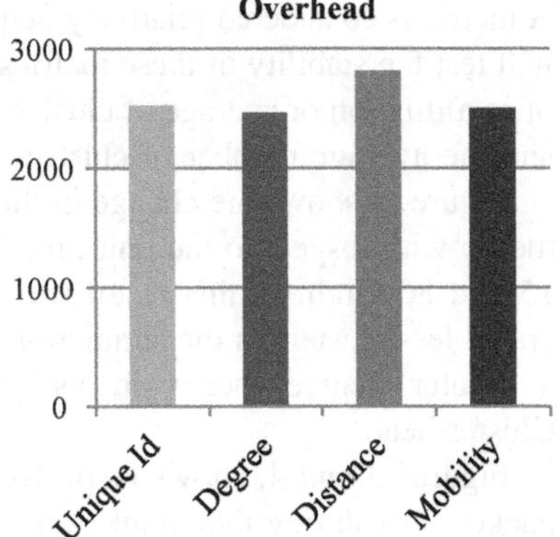

Fig. 4 Packets loss

caused by the shaping and cluster-head selection as the overhead in the second figure (Fig. 4). After comparing the overhead generated by each metrics, we observe that the degree is the metric that generates less loss during the packets exchanges.

7 Conclusion and Future Works

In this article, we simulated the four metrics cluster-head selection with a clustering algorithm which aims to maintain the stability of the vehicular network while getting quality service. And we concluded that the connectivity factor (the degree) have very important impact on the selection of cluster-head.

Future work will focus on exploring more results obtained from this research, especially for hierarchical routing, and the development of a new hierarchical protocol based connectivity adapted to the urban environment.

References

1. Martinez, F.J., Toh, C.K., Cano, J.C., Calafate, C.T., Manzoni, P.: A survey and comparative study of simulators for vehicular ad hoc networks (VANETs). Wirel. Commun. Mob. Comput. **11**(7), 813–828 (2011)
2. Liang, W., Li, Z., Zhang, H., Wang, S., Bie, R.: Vehicular ad hoc networks: architectures, research issues, methodologies, challenges, and trends. Int. J. Distrib. Sensor Netw. (2014)
3. Jiménez, F.: Connected vehicles, V2V communications, and VANET. Electronics **4**(3), 538–540 (2015)
4. Sou, S.I., Tonguz, O.K.: Enhancing VANET connectivity through roadside units on highways. IEEE Trans. Veh. Technol. **60**(8), 3586–3602 (2011)
5. Eichler, S.: Performance evaluation of the IEEE 802.11p WAVE communication standard. In: Proceedings of Vehicular Technology Conference, pp. 2199–2203 (2007)
6. Sichitiu, M.L., Kihl, M.: Inter-vehicle communication systems: a survey. IEEE Commun. Surv. Tutorials **10**(2), 88–105 (2008)
7. Yadav, V.S., Misra, S., Afaque, M.: Security of Self-Organizing Networks (2010)
8. Stampoulis, A., Chai, Z.: A Survey of Security in Vehicular Networks (2007)
9. Gela, M., Tsai, J.T.: Multiuser, mobile, multimedia radio networks. Wirel. Netw. **1**, 255–265 (1995)
10. Chen, G., Nocetti, F., Gonzalez, J., Stojmenovic, I.: Connectivity based k-hop clustering in wireless networks. In: Proceedings of International Conference on System sciences, vol. 7, pp. 188.3 (2002)
11. Shea, C., Hassanabadi, B., Valaee, S.: Mobility-based clustering in VANETs using affinity propagation. In: Global Telecommunications Conference, 2009. GLOBECOM 2009, pp. 1–6. IEEE (2009)
12. Varga, A.: OMNET++ discrete event simulation system user manual, 4.2.2 2011
13. Michael Behrisch JEDK Laura Bieker. SUMO—simulation of urban mobility: an overview. In: The Third International Conference on Advances in System Simulation, SIMUL, pp. 63–68 (2011)
14. Sommer, C., German, R., Dressler, F.: Bidirectionally coupled network and road traffic simulation for improved IVC analysis. IEEE Trans. Mob. Comput. **10**(1), 3–15 (2011)
15. Rawashdeh, Z.Y., Mahmud, S.M.: A novel algorithm to form stable clusters in vehicular ad hoc networks on highways. EURASIP J. Wirel. Commun. Netw. **2012**(1), 1–13 (2012)
16. Ahmed, S.A., Ariffin, S.H., Fisal, N.: Overview of wireless access in vehicular environment (WAVE) protocols and standards. Indian J. Sci. Technol. **6**(7), 4994–5001 (2013)
17. Zeadally, S., Hunt, R., Chen, Y.S., Irwin, A., Hassan, A.: Vehicular ad hoc networks (VANETS): status, results, and challenges. Telecommun. Syst. **50**(4), 217–241 (2012)
18. Fan, P., Haran, J., Dillenburg, J., Nelson, P.C.: Traffic model for clustering algorithms in vehicular ad-hoc networks. In: Proceedings of CCNC, pp. 168–172 (2006)
19. Wolny, G.: Modified DMAC clustering algorithm for VANETs. In: 2008 Third International Conference on Systems and Networks Communications, pp. 268–273. IEEE (2008)

Assessing Cost and Response Time of a Web Application Hosted in a Cloud Environment

Karim Abouelmehdi, Abderrahim Beni-hssane, Zakaria Benlalia, Abdellah Ezzati and Abdelmajid Moutaouikkil

Abstract Cloud computing offers a possibility of web applications hosting more cheaply and easy compared to traditional hosting in servers that have very limited capacities. The main objective of our study is to show the importance of assessing cost, and response time of a web application which will be hosted in a cloud environment, in several geographical areas before choosing the optimum configuration as cloud solution provider.

Keywords Cloud computing · Hosting cost · Datacenter response time · Web application

1 Introduction

Cloud computing progress open up many new opportunities for developers of Internet applications. Previously, the main concern of developers was the deployment and application hosting because it required acquisition of a server which has a fixed capacity capable, in peak demand, to manage expected applications, installation and maintenance of the entire software infrastructure of the platform supporting the application.

K. Abouelmehdi (✉) · A. Beni-hssane · Z. Benlalia · A. Moutaouikkil
LAROSERI laboratory, Sciences Faculty, Computer Science Department,
Chouaïb Doukkali University, El Jadida, Morocco
e-mail: karim.abouelmehdi1@gmail.com

A. Beni-hssane
e-mail: abenihssane@yahoo.fr

Z. Benlalia
e-mail: benlalia.zakaria@gmail.com

A. Ezzati
LAVETE laboratory, Sciences and Technics Faculty, Hassan 1st University,
Settat, Morocco
e-mail: abdezzati@gmail.com

Á. Rocha et al. (eds.), *Europe and MENA Cooperation Advances in Information and Communication Technologies*, Advances in Intelligent Systems and Computing 520, DOI 10.1007/978-3-319-46568-5_46

In addition, the server was underused because the peak traffic is occurred only at specific times. With the advent of cloud, deployment and hosting has become cheaper and easier. Faced with the ongoing rise in costs of implementation and maintenance of IT systems, companies are increasingly outsourcing their IT services by entrusting them to specialized enterprises such as cloud providers. In this context, a number of challenges including isolation (resources, failures, performances and user spaces), optimal management of resources and security arise to allow a real development of cloud [1].

In this study, we show the importance of assessing cost, and response time of a web application, which will be hosted in a cloud environment in several geographical areas, before choosing cloud provider.

2 Related Works

Many studies using simulation techniques to study the behavior of large-scale distributed systems and tools to support such researches. Among these simulators: Micro-grids [2], Christophe Gaultier [3], SimGrid [4], GridSim [5] and CloudANALYST [6]. While the first three simulators focus on Grid computing systems, the Cloudanalyst is the only simulation Framework designed for Cloud computing systems study. However, grid simulators were used to evaluate the costs of running distributed applications in cloud infrastructures [7, 8]. In 2012, Dhaval and all [9] proposed an algorithm that selects the most cost-effective datacenter. This reduces the overall cost, but increases the response time. This algorithm is tested by cloud analyst. In 2012, Sarfaraz [10] also proposed a solution that selects the data center with a high-performance configuration. This reduces the response time, but increases the overall cost. It is also tested by Cloud analyst.

The objective of our study is to show the simplicity of using cloud technology to host a large-scale web application in several geographic areas, by assessing, with CloudAnalyst simulator, the hosting cost and response time. And then make decision of the convenient configuration.

3 Case Study: Case Simulation of a Social Network Application

In this case study that is made by the simulator CloudAnalyst we worked with data from social network application “Facebook” on which we will calculate the transfer time and cost of deploying Datacenter. For each region, we used several scenarios. We will work with the data found in 2012 with a scale of 1/1000. This application is accessed by users from all around the world. Asia (2678100 users), North America (2365200 users), South America (1444000 users), Europe (2499900 users), Oceania (146200 users) and Africa (504,300 users).

Table 1 User bases used in the experiment

User base	Region	Peak hours	Online users during peak hrs	Online users during off-peak hrs
UB1	N. Ameria	3:00–9:00	236520	23652
UB2	S. America	3:00–9:00	144400	14440
UB3	Europe	3:00–9:00	249990	24999
UB4	Asia	3:00–9:00	267810	26781
UB5	Africa	3:00–9:00	50430	5043
UB6	Oceania	3:00–9:00	14620	1462

3.1 *Configuring the Simulation*

Table 1 Based on the above information, we set for the number of users each user group was approximately 10 % of all users during peak hours, the number of users accounting for nonpeak hours to 10 % of peak time users. The detailed settings are showed in Table 1. And we assume the Americas, Europe, Oceania user requests once every 5 min; Asia users request once every 3 min; Africa once the user requests every 10 min. The user requests data size is 100 b each time [4]. Cloud resource services agency policy is the closed data center policies. Data center contains 100 identically configured servers, each with the x86 architecture, Linux operating system, 2 G memory, 100 G storage spaces, and four processors. And the processor speed is 10,000 MIPS; Bandwidth is 1000 Mbps; Virtual machine resource scheduling policy is the time-sharing. Applications in the experiments virtual local storage size is 100 MB, the virtual machine has 1 GB of RAM and 10 MB of available bandwidth. Users are grouped by a factor of 1000, and requests are grouped by a factor of 100. Each user request requires 250 instructions to be executed. In terms of the cost of hosting applications in a Cloud, we assume a pricing plan which closely follows the actual pricing plan of Amazon EC2. The assumed plan is: Cost per VM per hour (1024 Mb, 100 MIPS): $ 0.10; Cost per 1 Gb of data transfer (from/to Internet):$0.10 [11].

3.2 *Simulated Scenarios*

Several scenarios are considered in our case study. The easiest one is to model the case in which only one centralized data center is used to host the application of social network “Facebook”. In this model, all requests of all users across the globe are processed by this single data center. A data center has 40 virtual machines assigned to the application is located in North America.

The second scenario consists to use two data centers, each one of them has 20 virtual machines dedicated to the application and they are located in Europe and North America. **The third scenario** is similar to the second one; the only difference

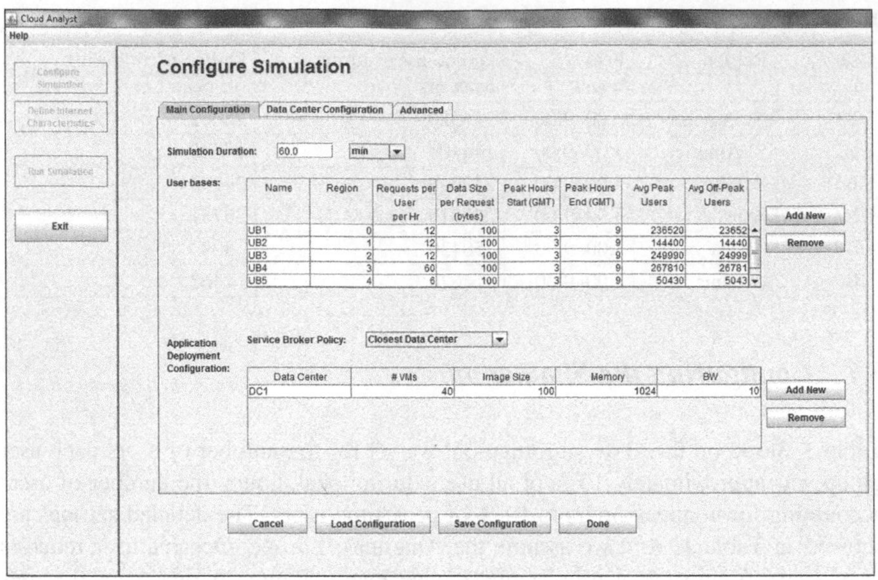

Fig. 1 Data entry

is that each data center has 40 virtual machines without load sharing between them. **The fourth and fifth scenarios** are similar to the third one. We just changed the load balancing algorithm. In **the sixth scenario** three data centers, each one has 40 virtual machines, are used. In this case, data centers are located in North America, Europe and Asia. Finally, in the last scenario, three data centers each contains the following number of virtual machines (60, 40, and 20) are also located in North America, Europe and Asia (Figs. 1 and 2).

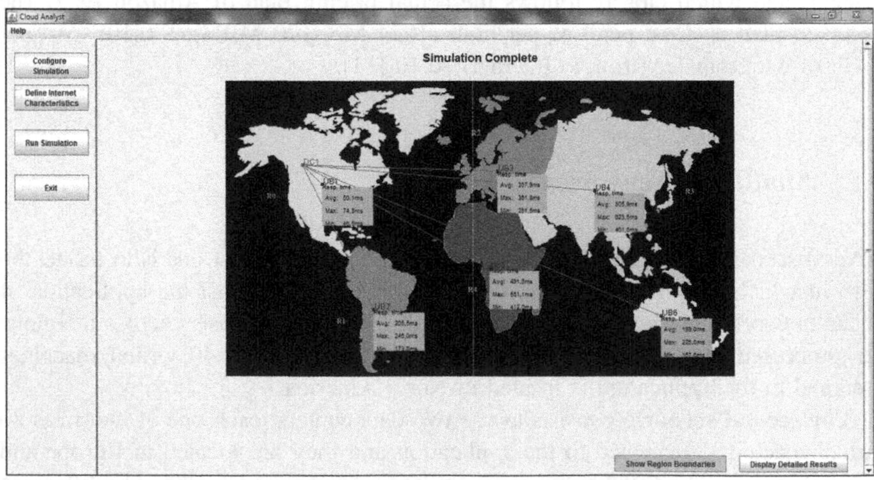

Fig. 2 Result of the first scenario

Table 2 Simulation results

	Configuration	Load balancing policy across VM	Overall average response time (ms)	Average data centre processing time (ms)	VM cost ($/hrs)	Data transfert cost ($/hrs)	Total cost ()
1	1 data center with 40 VMs (N. America)	Round robin	401,34	3,75	4,02	23,33	27,35
2	2 data centers with 20 VMs each (N. America, Europe)	Round robin	242,52	9,49	4,02	23,33	27,35
3	2 data centers with 40 VMs each (N. America, Europe)	Round robin	237,72	4,61	8,03	23,33	31,37
4	2 data centers with 40 VMs each (N. America, Europe)	Equally spread current execution load	237,74	4,56	8,03	23,33	31,37
5	2 data centers with 40 VMs each (N. America, Europe)	Throttled	236,82	3,69	8,03	23,33	31,37
6	3 data centers with 40 VMs each (N. America, Europe, Asia)	Throttled	73,57	5,93	12,05	23,33	35,38
7	3 data centers with 60, 40, 20 VMs (N. America, Europe, Asia)	Throttled	78,13	10,65	12,05	23,33	35,38

3.3 Results

(Table 2).

3.4 Discussion of Results

For a simple and visible debate, the results are shown in Figs. 3 and 4

- When more than Datacenter are used, the average response time is reduced
- The higher the number of virtual machines in the data center, the higher the average time of datacenter treatment is minimal even in the case of a single Datacenter. Therefore the most important is the number of VMs and not the number of Datacenter

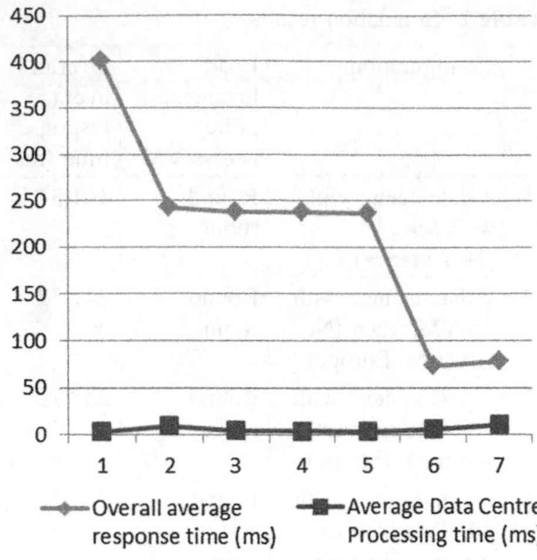

Fig. 3 Performance comparison with recent data center services agency policy

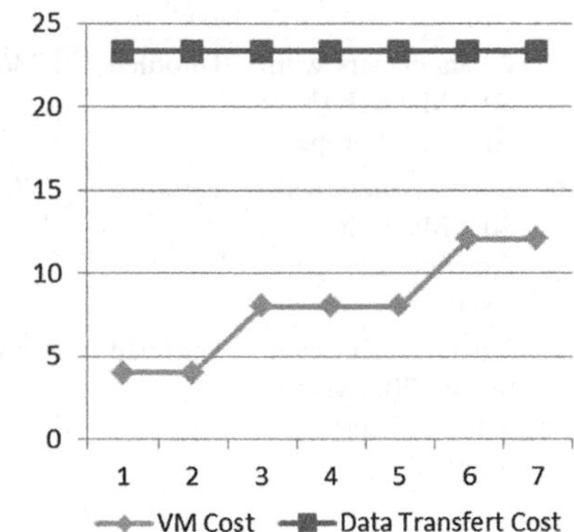

Fig. 4 The variation in costs

- The greater the number of virtual machines, the more the cost is high
- The data transfer costs remains constant because we worked with the same amount of information on the 7 scenarios.

4 Conclusion

In this work, we showed the importance of assessing cost, and response time of a web application to be hosted in a cloud environment in several geographic areas before choosing cloud provider. In fact, it can provide an idea about the optimum configuration depending on what suits us with an optimum cost before committing to a Cloud solution provider. This type of simulation experiment has a great potential to help testers to identify new features in order to model and develop new mechanisms and algorithms for managing cloud resources.

References

1. Weiss, A.: Computing in the clouds. netWorker **11**, 16–25 (2007)
2. Liu, X., Song, H., Jakobsen, D., Bhagwan, R., Zhang, X., Taura, K., Chien, A.: The MicroGrid: a scientific tool for modeling computational Grids. In: Proceedings of the ACM/IEEE Supercomputing Conference, IEEE Computer Society, Nov. 2001
3. Dumitrescu, C., Foster, I.: GangSim: a simulator for grid scheduling studies. In: Proceedings of the 5th International Symposium on Cluster Computing and the Grid (CCGrid 05), IEEE Computer Society, May 2005
4. Legrand, A., Marchal, L., Casanova, H.: Scheduling distributed applications: the SimGrid simulation framework. In: Proceedings of the 3rd IEEE/ACM International Symposium on Cluster Computing and the Grid (CCGrid 07), pp. 138–145, May 2001
5. Buyya, R., Murshed, M.: GridSim: a toolkit for the modelin and simulation of distributed resource management and scheduling for Grid computing. Concurr. Comput. Pract. Exp. **14** (13), 1175–1220 (2002)
6. Buyya, R., Ranjan, R., Calheiros, R.N.: Modeling and simulation of scalable cloud computing environments and the cloudsim toolkit: challenges and opportunities. In: Proceedings of the 7th High Performance Computing and Simulation Conference (HPCS 09), IEEE Computer Society, June 2009
7. Deelman, E., Singh, G., Livny, M., Berriman, B., Good, J.: The cost of doing science on the Cloud: the Montage example. In: Proceedings of the 2008 ACM/IEEE Conference on Supercomputing, IEEE, Nov. 2008
8. Assunção, M., di Costanzo, A., Buyya, R.: Evaluating the cost benefit of using cloud computing to extend the capacity of clusters. In: Proceedings of the 18th International Symposium on High Performance Distributed Computing. ACM Press, June 2009
9. Limbani, D., Oza, B.: A proposed service broker strategy in cloudanalyst for cost-effective data center selection. Int. J. Eng. Res. Appl. India **2**(1), 793–797 (2012)
10. Ahmed, A.S.: Enhanced proximity-based routing policyfor service brokering in cloud computing. Int. J. Eng. Res. Appl. India **2**(2), 1453–1455 (2012)
11. Gustedt, J., Jeannot, E., Quinson, M.: Experimental methodologies for large-scale systems: a survey. Parallel Process. Lett. **19**, 399–418 (2009)

For Formed Entrepreneurial Culture

Mouad EL Omari, Mohammed Erramdani and Rachid Hajbi

Abstract Culture greases the wheels of entrepreneurial spirit. Without it, even the rich and business plan would remain nothing more than a document. It creates values to have really clear-cut visionary maps and principles to discover the realm of entrepreneurship. Entrepreneurs use their mindsets and essential entrepreneurial thinking to build successful family businesses and share their experiences to help get good change of economical system to be more attractive for the futuristic entrepreneurs which can pave the ground building up startups and the transition of innovative minds towards successful firms. The aim of this paper, however is to describe this special attitude, which manifests itself in entrepreneurial thinking, and offers a solution to help successors in family businesses and the whole society to not only revamp but also improve their entrepreneurial vitality. Thereupon, providing a programmed platform modeling by MDA-model driven architecture–to help anyone know his weakness and strength in entrepreneurial landmark. Hence, improve the potential weaknesses and strength too for the purpose of having statistics enable those countries by using big Data to usher the state of entrepreneurial values in the community culture into forward-thinking.

Keywords Culture · MDA · Entrepreneurs · Clear visionary · Startups · Innovation · Family businesses · Programs · Big data

---

M. EL Omari (✉) · M. Erramdani · R. Hajbi
Système d'information et Science humaine, ESTO, Université Mohammed Ist,
Oujda, Morocco
e-mail: elomari.mouad@gmail.com

M. Erramdani
e-mail: m.eramdani@gmail.com

R. Hajbi
e-mail: r.hajbi@gmail.com

Á. Rocha et al. (eds.), *Europe and MENA Cooperation Advances in Information and Communication Technologies*, Advances in Intelligent Systems and Computing 520, DOI 10.1007/978-3-319-46568-5_47

1 Introduction

The Culture and especially family education has a primordial role to play in order to obtain this concept. The descendant of family business steps back often to their founding-fathers history and bring an overview available to the upcoming generation. The specification and pursuit of opportunities leading to acquiring ventures revolve around a founder-centered culture. The first generation profit from their leadership spirit to try to module the way for their descendant. The values of this organizational culture can be adopted by countries to establish a strategic plan which can motivate anyone who had idea to turn it to real business.

While studies highlight the relevance of founders in the buildup of entrepreneurial cultures [1] there is limited understanding of how entrepreneurial cultures are transmitted and sustained. The transmission of entrepreneurial values to potential successors is often neglected by founders or incumbents [2]. Moreover, while entrepreneurial cultures may be continued by one or more individuals, a collective approach by family members to opportunity identification and pursuit is often ignored [3].

Despite the widely acknowledged importance of entrepreneurial orientation in the establishment of successful ventures, the literature lacks evidence regarding the way individuals can be classified in terms of their individual and social entrepreneurial orientation, as a means of responding to the expectations of business partners. The identification of EO–entrepreneurial orientation–among individuals is considered a complex undertaking, given the vast number of intuition–and subjectivity-based factors which underlie entrepreneurial decision processes. Indeed, individual-level orientation has typically been considered an under researched area relative to firmlevel EO [4] and the entrepreneur is often seen as no more than "an independent business owner [5]. Yet, for individual entrepreneurship orientation to result in practical advantages in the current complex and volatile environment, it needs to be properly measured and managed [6].

Continuity of knowledge and expertise to any person to obtain the culture entrepreneurial is our target, as people aim to promote this aspect but it does not find platforms that meet this cognitive need.

So we must implement a cognitive map collecting big factors which influence EO, and we will start by discussing the huge role of Culture in the build-up of EO, those factors will help us to implement a program, which can test and identify strengths and weak talents and give some advice to enhance them. Another program, above all as well as the public state, based upon Big Data concept offers a golden-key to survey the state of entrepreneurial values among people.

This article stresses the entrepreneurial cognitive map based on culture. Yet different sections will be, otherwise, the ground upon which the Sect. 2 section is to be devoted to present some of culture effect in EO, or at any rate family education and the organizational culture. In the same vein, Sects. 3, 4 and 5 approach our solution we table to test the entrepreneurial talents. However the conclusion will take a glance at the future along with the works coming with, all these will be appended in the conclusion of the article.

2 Culture and Entrepreneurial Orientation

The Experience with groups in general, in particular strategic planning work groups, underlines the importance of establishing norms for group behavior with respect each group norms as the strategic planning work must be coupled and accustomed with the variety of norms each group behavior show.

The importance of entrepreneurship to society has been identified and discussed since at least the fifteenth century [7]. The questions of whether and how entrepreneurial skills and competences can be fostered during education and later followed up by Cantillon [8]. From these historical roots, Entrepreneurship Education (EE) has evolved to become a prominent field. This field is born of diverse disciplines, which include economics, management, education, and technical studies [9].

All Personal and organizational decisions and orientations are driven by the values of people involved in those decisions and actions, values are the underlying principles or standards that guide all human actions.

2.1 Family Business

Family present our first source of values. The foundation stories of family firms are typical entrepreneur stories. Actions like seeking the market gap, deciding on the business branch, taking risks, and following innovations are roles all entrepreneurs undertake actively in the start-up step. This step is also the step in which family values start to form [10]. Family businesses are driven by entrepreneurs who found them, set up the corporate culture, and transform visions into values. Nevertheless, the founder decides what the subject of the business will be, who the customers will be, and which products or services the business will provide. The founders set enough goals and opportunities for their descendent so that to plunge into business world and get more real concrete touch with their goals thereof. lots of values are, nevertheless taken from the past and therefore the planning education comes into the surface to play the basics to manipulate the closure and the gap step by step pending heavily upon the dimensions of where the strategic thinking reach a business and economic bubble to the utmost. The organization leadership will make decisions that shape their future, increase or decrease risk, choose to move into new business or not, decide whether the current business is part of the future, and make other decisions that impact the future of the firm, but unformed culture isn't enough because Worldwide statistics indicate that approximately 70 % of family-owned businesses do not survive into the 2nd generation and 90 % are no longer controlled by the 3rd generation of the family [11].

The family business fail [12]: (1) markets and technology change, (2) competitors quickly copy successful strategies, (3) overtaking with outside buyer willing to pay more to acquire the company than it is worth and owners are unable to resist the premium to sell out, (4) lack of financial capabilities, and (5) lack of staff skills. But beyond these typical family business pitfalls. The issues lies, I see, behind the incapacity of how would the next generations set their plans to funnel the opportunities suiting their capacity and plans as well. And that's perhaps significantly clarify their need to take abstract values in the prior by establishing a Design thinking which can be defined as a creative concept to generate innovative competitive strategies.

Creative thinking is used by designers to stimulate out-of-the-box thinking and contribute to innovation in an organization [13]. Design thinking is the form of thought that enables one to gain a nearly inexhaustible, long-term business advantage, but without committed leadership, no business can realize the structural, process, and cultural adjustments needed to become a design thinking organization. Design thinking cannot help to manage enterprises but, when combined with other entrepreneurial competencies, provides a framework for creating new visions and ideas [14].

2.2 *Organizational Culture*

Organizational culture can be defined the specific collection of values and norms that are shared by people and groups in an organizational environment and that control the way they interact witch each other. Organizational values are beliefs and ideas about what kinds of goals members of an organization should pursue and ideas about the appropriate kinds or standards of behavior organizational members should use to achieve these goals. A somewhat different approach to understanding organizational culture.

Organizational culture is the "residue of success" in that those behaviors and beliefs that have led the entrepreneurs to success in the past become institutionalized as its culture, with continuous reinforcements the organization becomes less conscious of this culture and treats its beliefs and values as non-negotiable or taken for–granted.

As a result, business founders shape the organization's culture as their view of doing things becomes shared or commonly held among members of the organization, experiences prove that we can influence population like publicity do to customers.

3 Big Data Role to Obtain Formed Entrepreneurial Culture

Surveys are a systematic way of asking volunteering people to provide information about their attitudes, behaviors, opinions and beliefs. The success of a survey research rests on how closely the answers that people give to survey questions matches reality, in additions to how people really think and act. The first problem that a survey researcher has to tackle are the following. How to design the survey in order to get the right information from people? How accurate does the survey have to be? Is this a one-time survey or can the researcher repeat the survey on different occasions and with different settings? How will the results be used?

All these problems make us thinking about the necessity of new and alternative ways to know public values and norms on entrepreneurship. A way that will be easy, efficient, permanent, significant using an area abundant with people and content: the web content.

Obtaining those pieces of information allow to discover the attitudes of a large number of informants that could give us a huge behavior background to choose the perfect strategic plan which can propagate their entrepreneurial values.

To assure the good working of this mechanism, we will verify three past words before the value which we want to know how people look at, our database will contain adjectives and by them we can define how this writer look to this value.

3.1 *Using Big Data to Get Public Entrepreneurial Values*

Our first method consists of the use of Big Data to get public values and their vision of entrepreneurship, based on content derived from the Web-Social Media, articles-. For that we will use an analytics workflow for Big Data. (Figure 1 illustrates the different steps of our approach). In fact, our approach is divided into 4 major

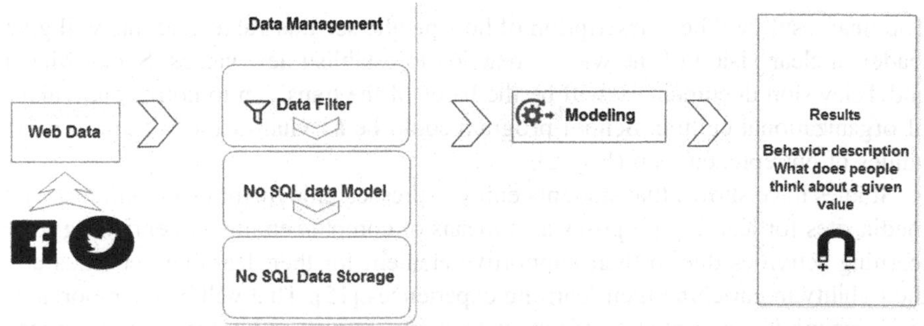

Fig. 1 Our approach of the analytics workflow for Big Data to define public opinion about entrepreneurship value based on content from the web-facebook or twitter articles for example-

Steps:

- Database resource: will get related information from different web source which contain some key words like entrepreneurship, venture and risks and three past words….
- Data management: consists of modeling and dividing data for preparing it to be stored on an NoSQL storage.
- Modeling: consists of the implementation of an algorithm which extract data by mining "Web mining" to analyze data, for example we can find someone who share in his social media profile that successful human shouldn't take risks and here we will note that this person had a bad vision to this value due to the existence of negation word and a specified value.
- Result: consists of visualizing the results by showing clearly the positive or negative opinion of people about our value.

3.2 Analyzing Data: Web Mining

So we will semantically analyze each rows of data, entrepreneurship values, to get more detail on organizational culture, by using the Text mining techniques [15]. This processing is known as "Web mining". It's contains the following steps:

Get different opinion terms about a given value after pre-processing each row of data from NoSQL storage.

Determine if the value is positive or negative by a semantic analyze.

If the number of positive value exceeds the number of negative value, that mean the public acquired with the given value. If it's the opposite, that mean the public disagree with the given value like determination or confidence [16].

3.3 Exploiting Result

The final result will be a description of how people see this value, and this will give leaders a clear vision of the way of reaction to establish new values. Social Media and Television documentary will be the bases of the transition to norms and values of organizational culture. School program could be an vital factor to acquire basic values of entrepreneurship (Fig. 2).

Studies have shown that students enjoy expressed enjoyment using online social media sites for learning purposes as a means of complementing and enriching their learning activities due to their supportive element for their learning activities and their ability in enriching their learning experiences [15]. That will be of importance in extracting the current state of values then government or big firms can intervene

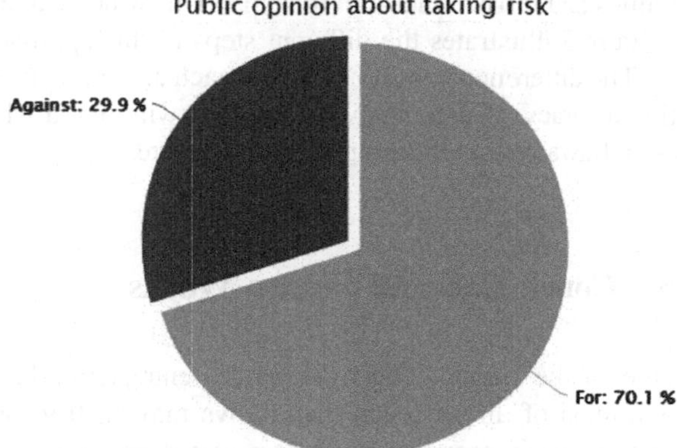

Fig. 2 Public opinion about a the value of taking risks

by including some strategic plans which can be like some web documentary solution to strengthen entrepreneurial values or other ways like venture games for boys who had problems with taking risk opportunities.

4 Using Questionnaires About User Values

Answering a question requires that respondents interpret the question to understand what is meant and retrieve relevant information from memory to form an answer. In most cases, they cannot provide their answer in their own words but need to map it onto a set of response alternatives provided by the researcher.

By answering of those questions we will have a clear entrepreneurial values map to this user, Each answer has a coefficient and in the end an algorithm will calculate the rate of each value and will give some advice to this user to improve the weak

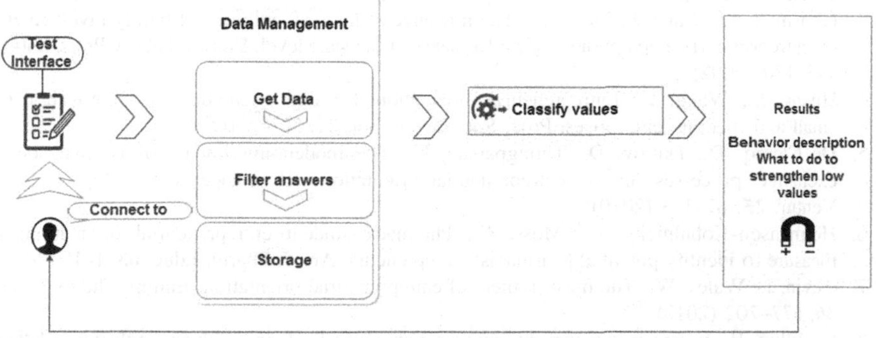

Fig. 3 Our approach of the analytics data obtained from forms based on MDA to define self values about entrepreneurship

values and reorient her/him to some skills which can help him to acquire this value. (Figure 3 illustrates the different steps of this approach).

The difference between this approach and the other based on Big Data appears in the accuracy of data analyzed and that will give a whole vision about the self and social awareness of entrepreneurial culture.

5 Conclusion and Future Works

One of the major challenges in the entrepreneurial development process is the definition of an approach that allows moving from weakness to strengths values, our approach provides a solution to the problem of self awareness and transformation. In this paper, we presented a various kinds of entrepreneurial culture sources and some procedures to discover if persons do conform with the best values of entrepreneurial culture which allows to easily to implement strategic plan to fill insufficiency norms.

We are now working on modeling other factors which build good entrepreneur personality and that will give us the CIM layer and after we will use an MDA. The ongoing work is extended to improve the role of the construction of the CIM level obtained from business model process to transform it to PIM layer via the QVT language generator to obtain PSM for user Test.

All models of the PIM are obtained through an automatic transformation of CIM level realized, a business process model is used in CIM level.

References

1. Chirico, F., Nordqvist, M.: Dynamic capabilities and trans-generational value creation in family firms: the role of organizational culture. Int. Small Bus. J. **25**(5), 487–504 (2010)
2. Brockhaus, R.: Family business succession: suggestions for future research. Fam. Bus. Rev. **17**(2), 165–177 (2004)
3. Chrisman, J.J., Chua, J., Steier, L.: The influence of national culture and family involvement on entrepreneurial perceptions and performance at the state level. Entrep. Theory Pract. **26**(4), 113–130 (2002)
4. Huang, S., Wang, Y.: Entrepreneurial orientation, learning orientation, and innovation in small and medium enterprises. Proc. Soc. Behav. Sci. **24**, 563–570 (2011)
5. De Clercq, D., Dimov, D., Thongpapanl, N.: The moderating impact of internal social exchange processes on the entrepreneurial orientation-performance relationship. J. Bus. Ventur. **25**, 87–103 (2010)
6. Hermansen-Kobulnicky, C., Moss, C.: Pharmacy student entrepreneurial orientation: a measure to identify potential pharmacist entrepreneurs. Am. J. Pharm. Educ. **68**, 1–10 (2004)
7. Covin, J., Wales, W.: The measurement of entrepreneurial orientation. Entrep. Theory Pract. **36**, 677–702 (2012)
8. Cantillon, R.: Essai sur la nature du commerce en général, ed. and with an English translation by Henry Higgs for the Royal Economic Society. Macmillan & Co. London (1931)
9. Davidsson, P.: The Entrepreneurship Research Challenge (2008)

10. Erdem, F., Başer, G.G.: Family and business values of regional family firms: a qualitative research. Int. J. Islam. Middle East. Financ. Manag. **3**(1), 47–64 (2010)
11. Leach, P.: Family Business the Essentials, pp. 146. Profile Books, London (2011). ISBN 978-1-86197-861 5
12. Ward, J.L.: Keeping the Family Business Healthy: How to Plan for Continuing Growth Profitability, and Family Leadership, pp. 3–15. Palgrave Macmilalan, New York (2011). ISBN 978-0-230 11121-9
13. Wattanasupachoke, T.: Design thinking, innovativeness and performance: an empirical examination. Int. J. Manag. Innov. **4**(1) (2012)
14. Roger, M.: The design of business: why design thinking is the next competitive advantage, pp. 18–28. Harvard Business Press (2009). ISBN 978-1-4221-7780-8
15. Khan, K., Baharudin, B., Khan, A., Ullah, A.: Mining opinion components from unstructured reviews: a review. J. King Saud Univ. Comput. Inf. Sci. **26**(3), pp. 258–275 (2014). ISSN 1319-1578
16. Veletsianos, G., Navarrete, C.: Online social networks as formal learning environments: learner experiences and activities. Int. Rev. Res. Open Distance Learn. **13**(1), 144–166 (2012)

Part IV
Security, Privacy & Data Forensic

Cloud-RSA: An Enhanced Homomorphic Encryption Scheme

Khalid El Makkaoui, Abdellah Ezzati and Abderrahim Beni-Hssane

Abstract Homomorphic encryption technique is a form of encryption which allows specific types of computations to be performed on encrypted data and generates an encrypted result. The decrypted result of any operation is the same such as working directly on raw data. The ability to perform computations on ciphertexts makes homomorphic encryption a widely adopted technique for ensuring confidentiality of both storage and treatment of outsourced data into a third party. Indeed, the RSA is the first asymmetric-key encryption scheme with a homomorphic property. Unfortunately, many attacks can threaten the encrypted data confidentiality of this scheme. In this paper, we will address the standard RSA drawbacks by suggesting an enhanced encryption scheme of it, which keeps a homomorphic property and resists more the well-known attacks against RSA.

Keywords Cloud-RSA · RSA · Homomorphic encryption · Confidentiality · Third party

1 Introduction

Currently, the majority of companies and organizations choose to move their data-centers into a third party (e.g., Cloud Provider) in order to reduce data storage and management costs, and to benefit from the advantages offered by the selected

K. El Makkaoui (✉) · A. Ezzati
LAVETE laboratory, FST, Univ Hassan 1, B.P.: 577, 26000 Settat, Morocco
e-mail: kh.elmakkaoui@gmail.com

A. Ezzati
e-mail: abdezzati@gmail.com

A. Beni-Hssane
LAROSERI laboratory, Sciences Faculty, Department of Computer Science,
Chouaïb Doukkali University, El Jadida, Morocco
e-mail: abenihssane@yahoo.fr

Á. Rocha et al. (eds.), *Europe and MENA Cooperation Advances in Information and Communication Technologies*, Advances in Intelligent Systems and Computing 520, DOI 10.1007/978-3-319-46568-5_48

third party. However, the concerns over the confidentiality of sensitive data is become a major barrier to outsourcing data-center adoption.

Regarding the storage service, data can be encrypted before sending them to the third party, using one of the most efficient secret-key encryption schemes. But, to give a third party the ability to perform any treatments, these data must be decrypted. It is this step that can be considered as a breach of confidentiality. Thus, researchers stressed a useful technique called "*Homomorphic Encryption*" which provides to a third party the ability to perform specific operations on encrypted data, without knowing any information about the data itself [1]. These operations generate an encrypted result. The result of any operation is the same as working directly on raw data. Indeed, the ability to perform operations on ciphertexts makes this technique popular for ensuring the confidentiality of both storage and treatment of outsourced data into a third party.

Indeed, the RSA is the first asymmetric-key encryption scheme with a homomorphic property [1]. The RSA was invented by Rivest, Shamir and Adleman in 1977 [2]. It uses two separate keys for encrypting and decrypting data. The public key pk = (N, e) is used to encrypt the plaintexts, where the modulus $N = pq$ is the product of two large distinct primes of the same bit-size and the public exponent e is an integer such that $1 \leq e < \phi(n)$ and $gcd\ (e, \phi(n)) = 1$ where $\phi(n) = (p\text{-}1)(q\text{-}1)$. Whereas, the private key sk = (N, d) is used to decrypt the ciphertexts, where the integer d is the private exponent such that $ed = 1(mod\ \phi(n))$.

The RSA is a deterministic encryption algorithm and its security is based on two computational problems: the factoring problem and the eth root problem [1]. Indeed, the RSA is one of the most popular public-key encryption schemes and frequently used to ensure the security of digital data [3]. Unfortunately, many attacks make the RSA insecure and threaten also encrypted data confidentiality. More specifically, these attacks achieve results when using weak modulus or/and low value of the exponents. Furthermore, the outsourced encrypted data will remain stored for a long time by the same encryption key in a third party environment; thus, the RSA can be easily broken by a brute-force attack.

To strengthen RSA security, the researchers have proposed to pad messages with random bits before encryption by using OAEP (Optimal Asymmetric Encryption Padding) scheme in order to convert the RSA into a probabilistic encryption scheme and to achieve semantic security [4]. However, the padded RSA loses the homomorphic property [1]. To address the RSA drawbacks, in this work we will suggest an enhanced encryption scheme of the standard RSA which keeps a homomorphic property and resists more the well-known attacks against RSA.

The rest of the paper is organized as follows: In Sect. 2, we will give a formal definition of homomorphic encryption. We will first present in Sect. 3, the standard RSA encryption scheme and see in Sect. 4, our proposal encryption scheme. Finally, in Sect. 5, we will present our conclusions and future works.

2 Homomorphic Encryption

Homomorphic encryption can be considered as group homomorphism [1]. In mathematics, a group is a set **G** equipped with a composition law $\circ$. (***G***, $\circ$) must satisfy the following four axioms [5]:

1. $\forall$ *a*,*b* $\in$ ***G***, $a \circ b$ is also in ***G*** ($\circ$ is an internal composition law).
2. $\forall$ *a*,*b*,*c* $\in$ ***G***, $a \circ (b \circ c) = (a \circ b) \circ c$ (i.e., $\circ$ is associative).
3. $\exists$ *e* $\in$ ***G*** such that $\forall$ *a* $\in$ ***G***, $a \circ e = a$ and $e \circ a = a$ (*e* is the **identity element**).
4. $\forall$ *a* $\in$ ***G*** there exists an element *b* in ***G*** such that $a \circ b = b \circ a = e$ (*b* is the **inverse** of *a*)

A group homomorphism is a function between groups which preserves the algebraic structure [5]. Let us give two groups, ***G*** together with an operation $\circ$, and ***H*** together with an operation $\diamondsuit$. A group homomorphism from (***G***, $\circ$) to (***H***, $\diamondsuit$) is a function

$$f: G \rightarrow H$$

such that

$$f(a \circ b) = f(a) \diamondsuit f(b)$$

for all *a*, *b* $\in$ ***G***.

Let the five-tuple (*P*, *C*, *K*, *E*, *D*) be an encryption scheme, where *P* is a finite set of possible plaintexts (called the plaintext space), *C* is a finite set of ciphertexts (called the ciphertext space) and *K* is a finite set of possible of keys (called the key space). For each $k \in K$, there is an encryption rule $e_k \in E$ and a corresponding decryption rule $d_k \in D$. Let us assume that the plaintexts form a group (*P*, $\circ$) and that the ciphertexts form a group (*C*, $\diamondsuit$), then e_k is a map from the group *P* to the group *C* and d_k is a map from the group *C* to the group *P*, i.e., e_k: $P \rightarrow C$ and d_k: $C \rightarrow P$, where $k \in K$ [1, 6].

For all *a*, *b* in ***P***, their corresponding ciphertexts c_a, c_b in ***C***, and *k* in ***K***, if

$$e_k(a) \diamondsuit e_k(b) = e_k(a \circ b)$$

and

$$d_k(c_a \diamondsuit c_b) = d_k(c_a) \circ d_k(c_a)$$

Therefore, the encryption scheme is homomorphic.

Since homomorphic encryption schemes are numerous, in this paper we will focus on the standard RSA encryption scheme. We will present its algorithm in Sect. 3, and some well-known attacks used against this encryption scheme.

3 RSA Encryption Scheme

In 1977, Ron Rivest, Adi Shamir, and Leonard Adleman proposed for the first time a practical public-key encryption scheme with a homomorphic property, named RSA [2]. RSA is a deterministic encryption algorithm that uses two separate keys: the public key (is published to the public) is used to encrypt the plaintexts; whereas the private key (is known to its owner only) is used to decrypt the ciphertexts (Fig. 1).

The RSA encryption scheme is composed of key generation, encryption, and decryption algorithms as follows [2]:

3.1 *Homomorphic Property*

Let us assume that $\boldsymbol{K} = \{pk = (N, e)$ and $sk = (N, d)\}$ is the key space, the plaintexts form a group $(\boldsymbol{P}, \bullet)$ and the ciphertexts form a group $(\boldsymbol{C}, \bullet)$, where $\bullet$ is the modular multiplication. Given two plaintexts m_1, m_2 in $\boldsymbol{P}$ and corresponding

---

RSA Key Generation Algorithm

Input: Two large primes *p* and *q* randomly and independenly of each other.

- Compute $N = pq$ and $\phi(N) = (p\text{-}1)(q\text{-}1)$ where $\phi(N)$ is the Euler totient function.
- Choose randomly an integer $\boldsymbol{e}$ such that $1 \leq e < \phi(N)$ and $gcd\,(e, \phi(N)) = 1$.
- Determine the private exponent $\boldsymbol{d}$ the multiplicative inverse of the public exponent e (mod $\phi(N)$) such that $ed = 1(mod\ \phi(N))$.

Output: (pk, sk)

The public key is **pk** = $(N, \boldsymbol{e})$ and the private key is **sk** = $(N, \boldsymbol{d})$.

RSA Encryption Algorithm

Input: message *m*, where $m \in Z_N$.

- Compute the ciphertext *c* as: $c = m^e\ (mod\ N)$

Output: $c = E\ (pk, m)$

RSA Decryption Algorithm

Input: ciphertext *c*, where $c \in Z_N$.

- Recover the plaintext message as: $m = c^d\ (mod\ N)$

Output: $m = D\ (sk, c)$

Fig. 1 RSA encryption scheme

ciphertexts c_1, c_2 in $\boldsymbol{C}$, such that, $c_1 = E(pk, m_1)$ and $c_2 = E(pk, m_2)$ [2]. We have that

$$\begin{aligned} E(pk, m_1) \bullet E(pk, m_2) &= m_1^e \bullet m_2^e \ (mod\ N) \\ &= (m_1 \bullet m_2)^e \ (mod\ N) \\ &= E(pk,\ m_1 \bullet m_2) \end{aligned}$$

and

$$\begin{aligned} D(sk,\ c_1 \bullet c_2) &= (c_1 \bullet c_2)^d \ (mod\ N) \\ &= (m_1^e \bullet m_2^e)^d \ (mod\ N) \\ &= m_1^{ed} \bullet m_2^{ed} \ (mod\ N) \\ &= m_1 \bullet m_2 \ (mod\ N) \\ &= D(sk,\ c_1) \bullet D(sk,\ c_2) \end{aligned}$$

Therefore, RSA encryption scheme has a homomorphic property.

3.2 RSA Security

The problem of determining the primes (i.e., p and q) factorization of $N = pq$, specially when N is large and the eth root problem form the security basis of the RSA encryption scheme [1]. Although the RSA is most commonly used both for providing privacy and for ensuring the authenticity of digital data [3], many powerful attacks make improper use of this encryption scheme. More specifically, when using weak modulus or/and low exponents. These attacks can be divided into three categories:

Factorization attacks: Since the initial publication of the RSA, many algorithms for factoring the modulus N have been proposed. Currently, the best known algorithms are the Number Field Sieve [7] and the Elliptic Curve Method [8]. The Number Field Sieve is the most powerful algorithm for factoring large modulus N [9]. Its running time depends on the bit-length of N. Moreover, the factorization attacks can efficiently achieve the results when the modulus being is not generated properly.

Low private exponent attacks: In RSA, the decryption or signature-generation time is proportional to the private exponent bit-size [2]. The usage of low private exponent seems to be an effective way to reduce this time (see [3, 9]). Unfortunately, many powerful low private exponent attacks such as, the attack of M. Wiener [10] and the attack of Boneh and Durfee [11] can easily recover the private exponent d.

Low public exponent attacks: In RSA, the encryption or signature-verification time is also proportional to the public exponent bit-size [2]. The reduction of this time by using low public exponent has been proven to be insecure against some low public exponent attacks [9]. The most powerful low public exponent attacks are based on Coppersmith's theorem [12] (see also [3]).

Indeed, the most noticeable way to attack RSA encryption scheme is to attempt to factor the modulus N [6]. Thus, an attacker can easily compute the secret exponent d from the public key(N, e) since $ed = 1(mod\ \phi(N))$ where $\phi(N) = (p\text{-}1)(q\text{-}1)$. Besides, if an attacker succeeds to recover the private exponent d, he can efficiently factor the modulus N [3]. These attacks can efficiently achieve the results when this encryption scheme is adopted to ensure the confidentiality of both storage and treatment of outsourced data into a third party due to the long time of storing the encrypted data by the same public key. Currently, the most obvious way to solve RSA security problem is to pad the plaintexts with random bits before encryption by using OAEP (Optimal Asymmetric Encryption Padding) scheme [4]. Unfortunately, this method makes the RSA lose homomorphic property [1]. In order to address the standard RSA drawbacks, in the next section we will suggest an enhanced encryption scheme of it, which keeps a homomorphic property and resists more the well-known attacks against RSA.

4 Proposal Encryption Scheme

To overcome the standard RSA drawbacks (in terms of security), in this section, we suggest an enhanced encryption scheme of it, which keeps a homomorphic property and resists more the well-known attacks against RSA. The enhanced encryption scheme is called "Cloud-RSA". The Cloud-RSA uses two separate keys, an evaluation key and a private key. But, the key generation algorithm strategy is somewhat different from the standard RSA algorithm. The evaluation key is used by a third party to perform operations on encrypted data. Whereas, the private key is known only to the data owner and used to encrypt and decrypt data (Fig. 2).

The Cloud-RSA encryption scheme is also composed of key generation, encryption and decryption algorithms, as follows:

Indeed, the Cloud-RSA loses some main functions of the standard RSA such as keys exchange confidentiality of symmetric-key encryption schemes and digital signatures. However, it keeps a homomorphic property and it is most suitable for ensuring the confidentiality of outsourced data into a third party.

4.1 The Communication System Architecture of the Cloud-RSA

The communication system architecture of the Cloud-RSA encryption scheme can be illustrated as in Fig. 3.

Cloud-RSA Key Generation Algorithm

Input: Two independenly large primes of the same bit size p and q.

- Compute $N = pq$ and $\phi(N) = (p-1)(q-1)$ where $\phi(N)$ is the Euler totient function.
- Choose randomly an integer $\boldsymbol{e}$ such that $1 \leq e < \phi(N)$ and $gcd(e, \phi(N)) = 1$.
- Determine the exponent $\boldsymbol{d}$ the multiplicative inverse of the exponent $e \pmod{\phi(N)}$ such that $ed = 1(mod\ \phi(N))$.

Output: (ek, pk)

The evaluation key is $\boldsymbol{ek} = (N)$ and the private key is $\boldsymbol{pk} = (\boldsymbol{N}, \boldsymbol{e}, \boldsymbol{d})$.

Cloud-RSA Encryption Algorithm

Input: message m, where $m \in Z_N$.

- Compute the ciphertext c as: $c = m^e (\ mod\ N)$

Output: $c = E\ (pk, m)$

Cloud-RSA Decryption Algorithm

Input: ciphertext c, where $c \in Z_N$.

- Recover the plaintext message as: $m = c^d (\ mod\ N)$

Output: $m = D\ (pk, c)$

Fig. 2 Cloud-RSA encryption scheme

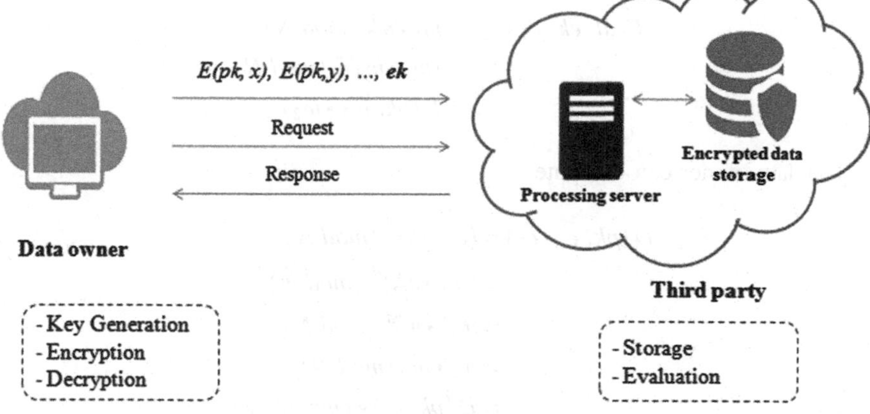

Fig. 3 Communication system architecture of Cloud-RSA encryption scheme

Where,

- **Key generation**: The data owner generates both the evaluation key (***ek***) and the private key (***pk***).
- **Encryption**: The data owner encrypts data under the private key. Then he sends the encrypted data and ***ek*** to a third party.
- **Storage**: The encrypted data and ***ek*** are stored into the third party database.
- **Request**: The data owner sends a request to the server to perform operations on encrypted data.
- **Evaluation**: The processing server processes the request and performs the operations requested by the data owner, using ***ek***.
- **Response**: The third party returns to the data owner the processed result (in encryption form).
- **Decryption**: The data owner decrypts the returned result, using ***pk***.

4.2 Homomorphic Property

The Cloud-RSA encryption scheme keeps a homomorphic property of the standard RSA. Let us assume that the evaluation key $ek = (N)$, the private key $pk = (N, e, d)$, the plaintexts form a group $(\boldsymbol{P}, \bullet)$ and the ciphertexts form a group $(\boldsymbol{C}, \bullet)$, where $\bullet$ is the modular multiplication. Given two plaintexts m_1 and m_2 in $\boldsymbol{P}$ and corresponding ciphertexts c_1, c_2 in $\boldsymbol{C}$, that computed by data owner under its private key such that $c_1 = E\,(pk, m_1) = m_1^e\ (mod\ N)$ and $c_2 = E\,(pk, m_2) = m_2^e\ (mod\ N)$. The third party can compute

$$\begin{aligned} Eval\,(ek,\, c_1 \bullet c_2) &= m_1^e \bullet m_2^e\ (mod\ N) \\ &= (m_1 \bullet m_2)^e\ (mod\ N) \\ &= E\,(pk,\, m_1 \bullet m_2) \end{aligned}$$

The data owner can compute

$$\begin{aligned} D\,(pk,\, c_1 \bullet c_2) &= (c_1 \bullet c_2)^d\ (mod\ N) \\ &= (m_1^e \bullet m_2^e)^d\ (mod\ N) \\ &= m_1^{ed} \bullet m_2^{ed}\ (mod\ N) \\ &= m_1 \bullet m_2\ (mod\ N) \\ &= D\,(pk,\, c_1) \bullet Dec\,(pk,\, c_2) \end{aligned}$$

Hence, Cloud-RSA encryption scheme has a homomorphic property.

4.3 Cloud-RSA Security

Recall that the Cloud-RSA encryption scheme uses two separate keys, the evaluation key $ek = (N)$ that allows a third party to perform some operations on encrypted data, and the private key $pk = (N, e, d)$ is known only to the data owner and used to encrypt and decrypt data. Its security is based on two computational problems, the problem of determining the primes factorization of the modulus $N = pq$ and the Cloud-RSA e^{th} root problem. To decrypt a ciphertext encrypted under the Cloud-RSA encryption scheme is equivalent to taking e^{th} root of $c \bmod N$ such that $c = m^e \pmod N$. When the factorization of N and the private exponents (i.e., e and d) are unknown, this task is intractable.

The RSA function can be defined as $x \rightarrow x^e \bmod N$; this function is easily computed, but it is difficult to invert without using the trapdoor d [3]. However, if an attacker succeeds to factorize the modulus N, he can easily recover the private key d as $ed = 1(\bmod \phi(N))$ where $\phi(N) = (p\text{-}1)(q\text{-}1)$. In the case of Cloud-RSA it is hard to recover the private exponent d even if the factorization of N is given since the exponent e is also private. Therefore, if an attacker succeeds to factorize the modulus N, he must try to find two exponents: e and d such that $ed = 1(\bmod \phi(N))$ and that decrypted messages are semantically correct.

5 Conclusion and Future Works

In this paper, we have suggested an enhanced encryption scheme of the standard RSA, Cloud-RSA. The Cloud-RSA uses two separate keys, evaluation and private keys. The evaluation key $ek = (N)$ is used by a third party to perform operations on encrypted data; whereas, the private key $pk = (N, e, d)$ is known only to the data owner and is used to encrypt and decrypt data. Its security is based on the problem of determining the primes factorization of the modulus N and on the Cloud-RSA e^{th} root problem. Even if the factorization of the modulus N is given, the decryption of ciphertexts encrypted under the Cloud-RSA encryption scheme is an intractable way since the exponents e and d are privates. Indeed, the Cloud-RSA loses some main functions of the standard RSA (i.e., the confidentiality of keys exchange of symmetric key schemes and digital signatures). However, it keeps a homomorphic property and is most suitable for ensuring the confidentiality of outsourced data into a third party.

In our future works we will implement the Cloud-RSA encryption scheme and we will analyze its performance (in term of response time). In our future implementation we will work on a virtual platform and use Python programming language.

References

1. Yi, X., Paulet, R., Bertino, Elisa: Homomorphic Encryption and Applications. Springer Briefs in Computer Science, 1st edn. Springer International Publishing, (2014)
2. Rivest, R., Shamir, A., Adleman, L.: A method for obtaining digital signatures and public-key cryptosystems. Commun. ACM **21**(2), 120–126 (1978)
3. Boneh, D.: Twenty years of attacks on the RSA cryptosystem. Not. Am. Math. Soc. **46**, 203–213 (1999)
4. Bellare, M., Rogaway, P.: Optimal asymmetric encryption. In: EUROCRYPT'94. Lecture Notes in Computer Science, vol. 950, pp. 92–111. Springer-Verlag (1994)
5. Feil, T., Anderson, Marlow: A First Course in Abstract Algebra: Rings, Groups, and Fields, 3rd edn. CRC Press, (2014)
6. Stinson, D.R.: Cryptography: Theory and Practice, Discrete Mathematics and Its Applications, 3rd edn. Chapman Hall, (2006)
7. Lenstra, Arjen K., Lenstra, Hendrik W.: The Development of the Number Field Sieve. Lecture Notes in Mathematics, vol. 1554, 1st edn. Springer-Verlag, Berlin Heidelberg (1993)
8. Lenstra, H.: Factoring integers with elliptic curves. Ann. Math. **126**, 649–673 (1987)
9. Nitaj, A., Rachidi, T.: Factoring RSA moduli with weak prime factors. In: First International Conference on Codes, Cryptology, and Information Security. Lecture Notes in Computer Science, vol. 9084, pp. 361–374 (2015)
10. Wiener, M.: Cryptanalysis of short RSA secret exponents. IEEE Trans. Inf. Theory **36**, 553–558 (1990)
11. Boneh, D., Durfee, G.: Cryptanalysis of RSA with private keydless than N0:292. In: Advances in Cryptology-Eurocrypt'99. Lecture Notes in Computer Science, vol. 1592, pp. 1–11. Springer-Verlag (1999)
12. Coppersmith, D.: Small solutions to polynomial equations, and low exponent RSA vulnerabilities. J. Cryptol. **10**, 233–260 (1997)

A Clustering Algorithm for Detecting and Handling Black Hole Attack in Vehicular Ad Hoc Networks

Badreddine Cherkaoui, Abderrahim Beni-hssane and Mohammed Erritali

Abstract A vehicular ad hoc network (VANET) basically consists of a group of vehicles that communicate with each other through a wireless transmission and requires no pre-existing management infrastructure. This communication, as the main objective, streamlining traffic for drivers. This exchange of information is not always reliable because of several constraints such as the existence of malicious users aimed falsifying information to serve their own interests. In this paper, we will simulate the Black Hole attack in a VANET environment with a generated real world mobility model using MOVE Tool and SUMO and analyse the performance of this communication under this attack. And then we propose a clustering algorithm to detect and react against the black hole attacker node.

Keywords VANET · Black hole attack · Mobility model · NS2 · Clustering algorithm

1 Introduction

VANETs (Vehicular Ad hoc Networks) are a new form of Mobile Ad hoc Networks (MANETs) and it is considered as a subclass of MANETs [1]. The main difference between MANETs and VANETs is node mobility [1]. VANETs are characterized by a frequent and rapid change in the topology which makes it

B. Cherkaoui (✉) · A. Beni-hssane
LAROSERI laboratory, Sciences Faculty, Computer Science Department,
Chouaïb Doukkali University, El Jadida, Morocco
e-mail: b.cherkaoui@ucd.ac.ma

A. Beni-hssane
e-mail: abenihssane@yahoo.fr

M. Erritali
TIAD Laboratory, Sciences and Technics Faculty, Computer Sciences Department,
University of Sultan Moulay Slimane, Beni Mellal, Morocco
e-mail: m.erritali@usms.ma

Á. Rocha et al. (eds.), *Europe and MENA Cooperation Advances in Information and Communication Technologies*, Advances in Intelligent Systems and Computing 520, DOI 10.1007/978-3-319-46568-5_49

difficult to manage. There is also a limited coverage network which had a direct impact on V2 V communications. VANETs are used in several applications, such as collision warning systems used to notify other drivers to change direction to avoid congestion. Like all communication systems, there's always malicious attacks from a kind of users that pretends malicious.

Among these attacks, we find the Black Hole attack that already exists in MANETs, so it is necessarily replicated in VANETs. Security constraints in VANETs are very complex to manage in an ad hoc environment. These constraints are due to the permanent changing of the topology and the high-speed of vehicles.

Vimal Bibhu and Kumar Roshan [2] have achieved a performance analysis of Black Hole attack under VANET using OPNET Simulator to compare between AODV and OLSR routing protocols. Sonia and Padmavati [3] made a comparison study between Dynamic Source Routing (DSR), Ad hoc On-demand Distance Vector (AODV) and Ad hoc On-demand Multipath Distance Vector (AOMDV) under Black Hole attack with 20 vehicular nodes moving with a constant speed of 10 m/s simulated with NS-2.35. We find also the works achieved by Mahesh KUMAR and Kuldeep BHARDWAJ which made an analysis of the impact of Black Hole attack on AODV by introducing many attackers in a VANET network composed of 50 vehicular nodes moving with 125 m/s. Besides that, Vaishali D. Khairnar and S.N. Pradhan [4] made a mobility model generated with SUMO and MOVE Tool, and which used by NS-2.

Concerning prevention, the black hole, Ming-Yang Su [5] has performed an IDS to prevent a selective Black Hole attack on MANETs by setting a sniffing mode to detect an abnormal difference between the routing messages transmitted from the node. In [6] the authors proposed a DMN to detect malicious nodes and improve the network performance. Jaskaran Preet Singha, Rasmeet S. Balib [7] proposed a hybrid backbone based clustering algorithm for VANETs to limit the number of nodes witch could be a cluster-head and improve the re-organization of the cluster mechanism.

Our work is about designing a mobility model to simulate continuous road traffic with SUMO and MOVE Tool to generate a real world simulation. Then, we implemented a Black Hole attack inside this model to give a real aspect to the attack, and then we analysed the results to see the impact of this attack on the network communications. Thereafter, we proposed a cluster-based algorithm to detect and isolate the malicious node from the network with an algorithm complexity analysis.

2 Our Proposed Mobility Model

With the MOVE tool [8], implemented in Java and runs on top of SUMO simulator [9], we can generate a "real-world mobility model" for VANETs. The generation of a "roadmap" can be done with 3 different ways: manual creation, automatic creation, and import of a real world map. In our case, we will manually create a map,

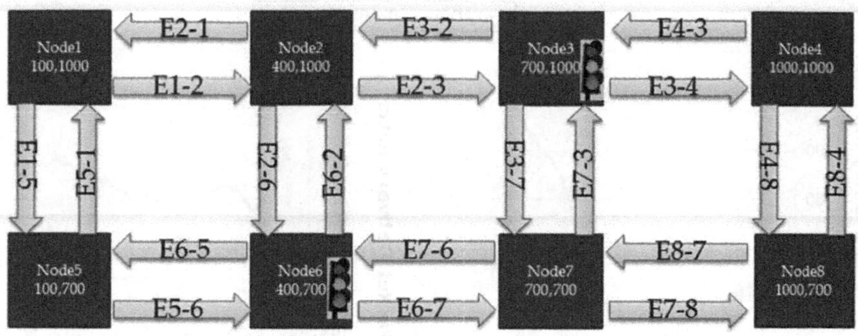

Fig. 1 Road topology

and define the movements of vehicles, traffic lights and roads that a vehicle can take in moving on the map as shown in the Fig. 1. This model will allow us to generate traffic by defining the points of departure and points of arrival. We defined also vehicles direction changing probabilities for each junction. All these parameters will give us a realistic scenario for our simulation. Now, we implement the attack in NS2 by introducing one single attacker circulating on the map that generates the attack.

After the deployment of this model, we use the Network Simulator NS-2.35 [10] to simulate the data exchange between the vehicles (Fig. 2).

Parameters	Values
Network size	952 m * 352 m
Number of vehicles	60
Max speed	40 m/s
Traffic model	CBR
Routing protocol	AODV
Number of attackers	1
Simulation time	500 s

We notify that the PDR decrease under a Black Hole attack and the E2E Delay increase during the same attack:

Packet End-to-End Delay depends on the routing protocol, AODV in our case, the number of nodes and the distance between the transmitter and the receiver in a given area. So AODV takes more time, generating RREQs and RREPs, to reach the destination in a topology that is constantly changing. However, during a Black Hole attack, the malicious node does not need to find new routes to forward packets. It responds to RREQs it receives from transmitters to take control of the road and then intercept and destroy the packets.

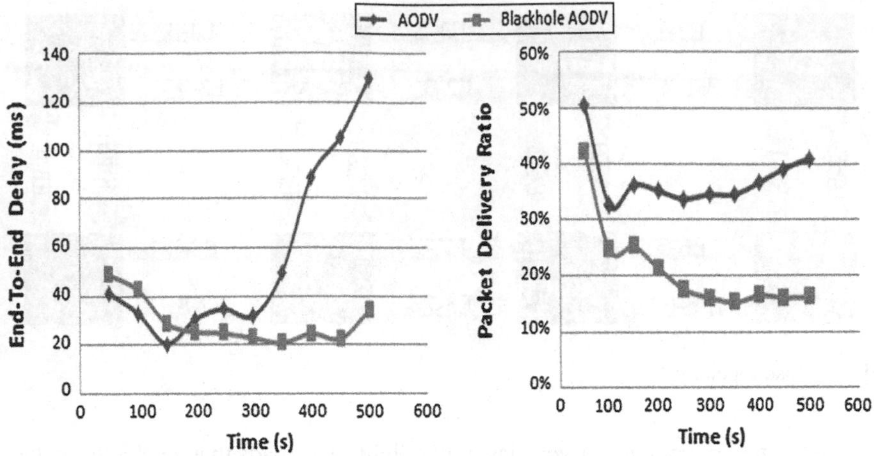

Fig. 2 End-to-End Delay and packet delivery ratio metrics

Packet Delivery Ratio depends on the number of packets successfully delivered to their destination compared to the number of the packets sent by the transmitter. The Black Hole node intercepts a good amount of data packets and drops them, so they can't arrive at their destination. That's why we notify the decrease of the PDR under Black Hole attack comparing of a normal AODV routing environment.

3 Our Proposed Algorithm

3.1 Basic Idea

As we mentioned earlier, a vehicle that generates a black hole attack reserves the road, then it starts receiving packets to be transmitted to the destination, as this removes vehicle instead of transferred. A vehicle is considered as malicious when the number of packets received is abnormally low compared to the received packets. Our detection strategy is based on the monitoring of vehicles that have acquired the road. For this, it must designate a vehicle as an honest node to oversee the data packet transfer process. The changing topology problem persists in this step, so to manage this problem we opted for clustering. Clustering is an effective method to easily manage the topology. Clustering in VANETs topology allows splitting the network into a group of clusters. Each cluster defines his Cluster-Head (CH) who can monitor the communication between vehicles inside the cluster. In our case, the CH is the node that will oversee the cluster to detect any Black Hole attacks. It is also responsible for broadcasting warning messages to other members of the cluster. Our idea is to take the clustering approach of [7] and concatenate with a mechanism to detect and isolate the malicious node.

To select a CH, each cluster has to select his leaderships, and then one of these could be the CH. The concept of this process is the fact that each member of cluster leaders calculates relative speed to calculate a parameter known as the ARV (Aggregate Relative Velocity).

Cluster leaders are trained on the basis of the degree of connectivity (δ) and ARVs. All nodes begin with an unknown initial status (UN). After the neighbours discovery phase, the nodes count the number of nodes in their neighbour table (δ). Each one then calculates its ARV relative to its neighbours.

$$V_i^{rel}(j) = |Vi - Vj|$$
$$V_i = var(Vi(jn)$$
$$\Omega_i = \delta_i - V_i$$

where:

- V_i: Velocity of node i.
- V_j: Velocity of node j.
- n: Number of leaders.
- δ: Degree of connectivity.
- Ω: Value obtained by subtracting ARV from δ.

If the value of Ω is higher than the threshold predefined value then the node tagged as a leader (LE), else the node is tagged as an ordinary cluster member (MN). Now the leaders can proceed to the election process to select a CH. The leader with a minimum ARV is tagged as a CH.

Hereafter, the CH start to verify the number of packets received and packets transferred for a given route in a predetermined time interval.

3.2 Proposed Architecture

The general strategy of our algorithm is illustrated in Fig. 3.

3.3 Proposed Algorithm

Step 1: Initialize Status to all vehicles in the environment
Step 2: Sum of *N* vehicles in the neighbours list of node *i*
Step 3: Compute relative velocity of vehicle *i* with regard to *j* and ARV for node *i*
Step 4: Find out the vehicle with the lowest for each cluster to be the CH
Step 5: Update Status for all Vehicles
Step 6: Allocate all CH vehicles obtained from **Step 4** to be verifiers

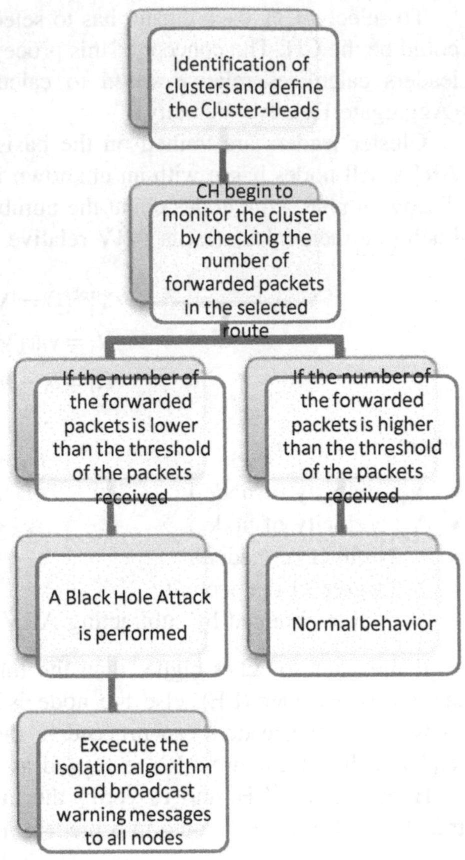

Fig. 3 The proposed strategy

Step 7: CH starts to monitor the vehicles who obtained routes to deliver packets to their destinations

Step 8: CH calculate the average Threshold of packets transferred to the destination.

Step 9:

If ($P_T < T$) // Where P_T is the Packets transferred and T is the average Threshold of packets transferred

$C_p = C_p - 1$ // Where C_p is the confidence Parameter

If(C_p ==0)

Vehicle Status is tagged as malicious

Goto **Step 10**

Else Goto **Step 7**

Else

$C_p = C_p + 1$

Goto **Step 7**

Step 10: Broadcast warning messages to all vehicles
Step 11: Update Blacklist Table to all vehicles
Step 12: Isolate the Vehicle from the network

4 Algorithm Complexity Analysis

The complexity of an algorithm is a function that describes the efficiency of this algorithm based on the amount of data to be processed by the algorithm. Mostly, there are two main measures of complexity of the efficiency of an algorithm are Time complexity and Space complexity:

- Time complexity describes the amount of time taken by an algorithm to run a given function. The time complexity of an algorithm is generally expressed using Big O notation. This latter eliminates the coefficients and lower order terms.
- Space complexity is the total amount of memory space needed to run a given program.

In our case, we have studied the time complexity with the Big O notation to express the worst-case time complexity. This allows us to have an idea about the efficiency and the feasibility of our algorithm.

In **Step1**, it starts with a status initialization for all vehicles in the environment. We suppose that we have n vehicle in the environment. To initialize the status for all vehicles we need to execute this instruction "STATUS (i) ← CN" n times. If we want to dissect this instruction, we will write:

```
        N = n; // N is the number of the vehicles in the
    environment
        for i = 0 to i = N
                STATUS (i) = CN;
        Endfor
```

So the number of operations that we need in this step is: $f_1(n) = n + n = 2n$.

In **Step2**, we need to sum of nodes in the neighbor list of vehicle i.

```
for i = 1 to i = N
        δi = Σ^N i=0 Neighbor - List(i)
endfor
```

The number of operations in the **Step2** is $f_2(n) = n$.

In **Step3**, we need to compute the relative velocity of vehicle *i* with regard to vehicle *j*.

```
for i = 1 to i = N
        for j = 1 to j = N
                Vi^rel (j)= | Vi – Vj|
                Vi = Vi^rel (j)
        endfor
        Ωi = δi - Vi
endfor
```

The number of operations needed to calculate the relative velocity for each vehicle is:

$$f_3(n) = (n * 1) * (n * 2) = 2n^2.$$

In **Step4**, we have to find out the vehicle with the lowest Ω for each vehicle to participate in the CH elections and update status for all vehicles in the environment.

```
for i = 1 to i = N
        If Ωi ≥ ST
                set STATUS (i) ← LE
        else
                set STATUS (i) ← CN
        endif
endfor
```

The number of operations needed to calculate and find the vehicle with lowest Ω and update status is: $f_4(n) = n\ (1 + 1 + 1) = 3n$.

In **Step5**, we can proceed to the election process by choosing the vehicle which has the lowest velocity with regard to his neighboring to be the CH. This process is for each cluster.

```
for i = 1 to i = n
        for j = 1 to j = n
                if (j € Leaderships (LE)) then
                        if (Vi < Vj)
                                set STATUS ← CH
                        endif
                endif
        endfor
endfor
```

The number of operations needed to select the CH is: $f_5(n) = n * (1 + 1 + 1)\ n = 3n^2$.

In **Step6**, **Step7**, **Step8,** and **Step9** CHs start to verify the packets in their respective clusters. CHs calculate the average threshold of packets transferred to the destination. We suppose that we have x cluster, and each cluster contained n vehicle.

```
for i =1 to i = x
        for j = 1 to j = n
                if (STATUS (j) == CH) then
                        for j =1 to j = n
                           PT = Σ packet received (j) / Σ packet send (j)
                        endfor
                endif
        endfor
endfor
```

The number of operations needed to achieve these steps is:

$$f_6(n) = x*(n*(1 + n*(1))) = x*\left(n + n^2\right) = n^2 + xn$$

In the rest of steps, the complexity is linear. Thus allowing to deduct that is negligible. So the complexity of the totality of the algorithm is:

$$f(n) = f_1(n) + f_2(n) + f_3(n) + f_4(n) + f_5(n) + f_6(n) = 6n^2 + n(6 + x)$$

To estimate the global complexity of the algorithm, we focused to the asymptotic performance. The common notation used if the Big O. In our case we have:

$$f(n) = 6n^2 + n(6 + x) = O(n^2)$$

After this analysis, we can say that the running time of our algorithm is polynomial.

5 Conclusion and Perspectives

VANETs networks, which have the objective to provide solutions for security and traffic management, need several security measures to ensure reliable communication. After this analysis, we find that the quality of service decreases at a Black Hole attack on a routed environment by AODV. We proposed a cluster-based algorithm to detect the attack and isolate the malicious node. Our future works are about implementing this algorithm and analyze the results.

References

1. Chouhan, P., Kaushal, G., Prajapati, U.: Comparative study MANET and VANET. Int. J. Eng. Comput. Sci. 16079–16083 (2016)
2. Bibhu, V., Kumar, R., Kumar, B.S., Singh, D.K.: Performance Analysis of Black Hole Attack in Vanet, Int. J. Comput. Netw. Inf. Secur. **11**, 47–54 (2012)
3. Sonia, S., Padmavati, P.:Performance analysis of Black Hole Attack on Vanet's Reactive Routing Protocols. Int. J. Comput. Appl. **73**(9), 0975–8887 (2013)
4. Khairnar, V.D., Pradhan, S.N.: Mobility models for vehicular ad-hoc network simulation. Int. J. Comput. Appl. **11** (4), 0975–8887 (2010)
5. Su, M.-Y.: Prevention of selective black hole attacks on mobile ad hoc networks through intrusion detection systems. Comput. Commun. **34**, 107–117 (2011)
6. Khana, U., Agrawala, S., Silakaria, S.: Detection of Malicious Nodes (DMN) in Vehicular Ad-Hoc Networks. Proc. Comput. Sci. **46**, 965–972 (2015)
7. Singha, J.P., Balib, R.S.: A hybrid backbone based clustering algorithm for vehicular ad-hoc networks. Proc. Comput. Sci. **46**, 1005–1013 (2015)
8. MOVE: http://www.cs.unsw.edu.au/klan/move/
9. SUMO: http://sumo.sourceforge.net/
10. The Network Simulator NS-2 version 2.35: http://www.isi.edu/nsnam/ns/index.html

A Multi Agent System for Service Restoration Based on Resilience Policy in Critical Infrastructure

Yaou Hamida, Baina Amine and Bellafkih Mostafa

Abstract In the information age, critical infrastructures (CI) have become largely computerized and tightly interconnected. Indeed, a failure in one critical infrastructure could lead to serious consequences on national security, economic well-being, public health, safety, or any combination thereof, generating cascading effects because of their synergies. Consequently, the reliability, performance, continuous operation, safety, and protection of these critical infrastructures are essential toward society and its economy. These mutual interdependencies strengthen the systems to be more resilient in case of failure. In this paper we propose resilience framework that aims to manage disruptions in critical infrastructure using multi agent system. This allows each CI to have a global view of resilience plan of the others.

Keywords Critical infrastructure · Resilience · Resilience strategy · Multi agent system · Service restoration · Resilience policy

1 Introduction

Critical infrastructure [1] constitute a core of the modern society by providing it with services that are essential for its functioning (electrical power, telecommunication system, water supply, banking…). They are often described as large-scale, spatially distributed systems with high degrees of complexity [2]. Hence, these characteristics strengthen the interdependencies [3] that exist among the systems, which enable by

Y. Hamida (✉) · B. Amine · B. Mostafa
STRS Lab, National Institute of Post and Telecommunication Rabat, Rabat, Morocco
e-mail: yaou@inpt.ac.ma; yaou.hamida@gmail.com

B. Amine
e-mail: baina@inpt.ac.ma

B. Mostafa
e-mail: bellafkih@inpt.ac.ma

Á. Rocha et al. (eds.), *Europe and MENA Cooperation Advances in Information and Communication Technologies*, Advances in Intelligent Systems and Computing 520, DOI 10.1007/978-3-319-46568-5_50

then cascading failures from a system to another. These infrastructures can suffer from faults, malicious attacks and disruptions coming from malfunctioning operations or mishaps, so is of paramount significance that they are reliable and robust. These characteristics can be applied by implementing resilience strategy on the infrastructure. The concept of resilience is recognized as the capacity of the system to tolerate disruptions and sustain a minimum of service level after this interruption.

From the state-of-art regarding resilience in critical infrastructure, it is recognized that there is no effective management skills that conduct the system in case of failure or reconfiguration of the infrastructure. Thus, the primary focus of this paper is directed toward employing multi agent system architecture to achieve resilience through independent management of collaborative organizations. The restoration plan of each service in critical infrastructure must be recognized by the others in order to achieve a unified agreement in case of disruption.

The paper is laid out as follows: Sect. 2 presents related works that have studied some aspects of resilience in critical infrastructures. Section 3 presents resilience strategy, which describe the basis of resilience systems relied on ResiliNet strategy. In Sect. 4, we discuss the comparison between some interdependency models. Section 6 describes the methodology of resilience framework based on multi agent system. Section 7 provides a case study applied on financial followed by a concluding comments and future work.

2 Related Work

Several projects have been interested on resilience of critical infrastructure. In the European project CRUTIAL (Critical Utility Infrastructural Resilience) [4], resilience holds great importance next to the protection of critical infrastructure that is dedicated specifically to electric power systems. The project focuses on technologies such as fault and intrusion tolerance as well as self-healing mechanisms through developed device CIS (CRUTIAL Information Switch) [5]; it then describes the operations of the system in case of failure and proceeds through modeling. INSPIRE (INcreasing Security and Protection through Infrastructure Resilience) project aims to ensure protection of critical information infrastructure through the identification of their vulnerabilities and the development of innovative techniques for securing networked process control systems and thus by configuring, managing and securing the underlying communication network.

3 Resilience Strategy

ResiliNet [6] is a resilience network initiative that has developed framework for resilient network serving as the basis of the ResumeNet (Resilience and Survivability for Future Networking: Framework, Mechanisms, and Experimental

Evaluation) project. It describes a comprehensive strategy for network resilience engineering. ResiliNet framework is based on a set of axioms: inevitable, understand, expect, and respond. These four axioms motivate the ResiliNet Strategy supported by the ResiliNet Principles and implemented by the ResiliNet Mechanisms.

Resilience strategy is formalized on six steps two phases D^2R^2 + DR referring to: Defend, Detect, Remediate, Recover, Diagnose, Refine, which in turn support the four axioms presented before. The first phase D^2R^2 describes a real-time control loop to allow dynamic adaptation of networks in response to challenges and the second phase a non-real time control loop that aims to improve the design of the network, including the real-time loop operation, reflecting on past operational experience.

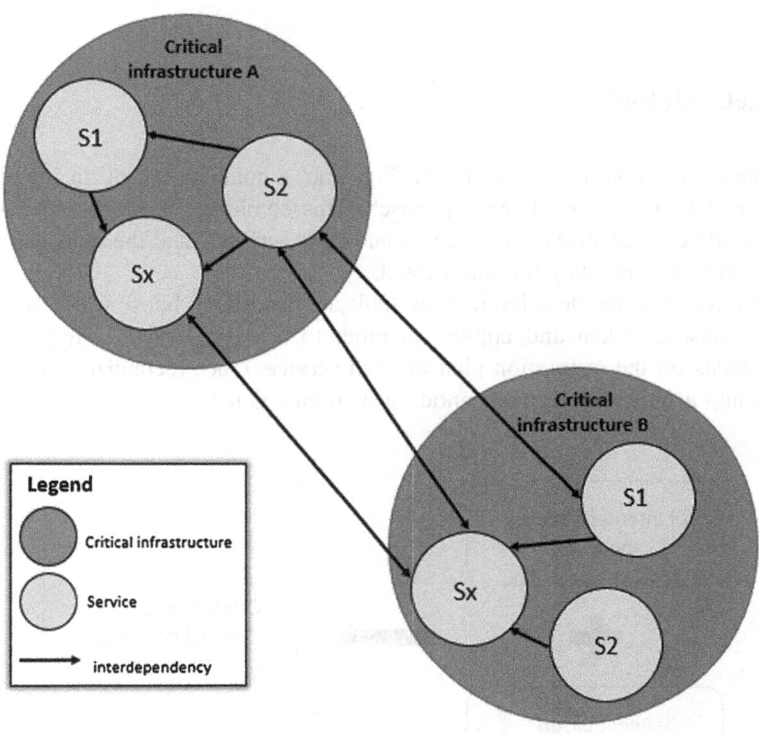

Fig. 1 Interdependency between services [9]

4 Interdependency Modeling

Critical infrastructures are known as complex adaptive systems (CAS) involving the interaction and collaboration of their different organizations, with different objectives and goals. There is several approaches that study interdependencies between CIs which is a fundamental property of its. In [7] the author compare the existing approaches according to the notion of resilience defined on three properties: resistant capacity, absorptive capacity and restorative capacity. From this comparison, the author mentioned three modeling approaches relied on agent based model (ABM), flow-based methods and HLA-based model that support effectively resilience. Among the various existing modeling approaches, ABM can properly represent the autonomous behavior of these infrastructures in terms of resilience and observe the general response of the system as a result of individual action [8]. And this requires also to integrate the others approaches into a single one. It is supposed that critical infrastructure studied encompasses collaborative organizations, each one of them provides critical services essential for the others.

The Fig. 1 demonstrates the dependencies and interdependencies that exist between services and CIs.

5 Methodology

The methodology is illustrated on the Fig. 2 as a combination of an interdependency models and the resilience approach. This combination allows the management of interdependencies between CIs and their services, and the management of their resilience when they are interrupted.

In this architecture, decision making tools are needed to choose which resilience policies must be taken and applied in order to achieve service restoration. The agents focus on the restoration plan of each service. Once recognized, the system can go into a unified agreement under system emergency.

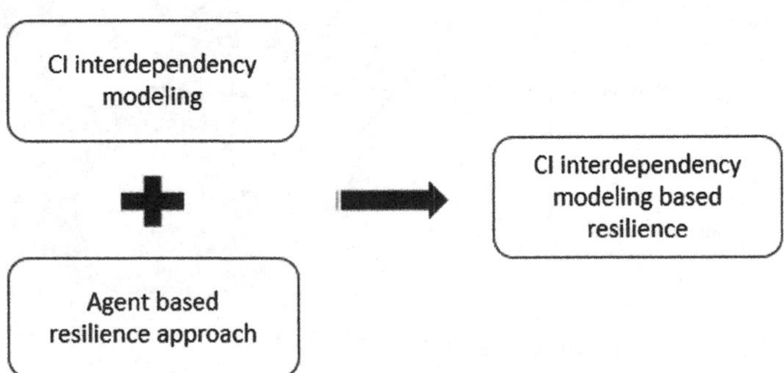

Fig. 2 Resilience modeling approach

In our context this autonomous interactions will be implemented by JADE platform. (Java Agent Development framework) known for their ability to communicate by exchanging messages and the reaction components integration in simplified way. It is an open source platform for development of peer-to-peer agent based applications. This choice is made based on [10] which argue that this agent platform responds intimately to the expectations in terms of agents' functionalities, security, performance, standardization, and secure communication between agents.

5.1 Resilience Policy Management

Critical infrastructures are frequently exposed to many changes during their functioning. That is why they are supposed to be more resilient. Adding or removing nodes are one of these changes. The use of policies to ensure communication between the entities have a potential benefits to configure and coordinate the interaction between mechanisms implemented on each entity. It defines how the operation of the several components in network should be modified in response to pre-specified events. The challenges and changes represented in the events may be used as a trigger for the reconfiguration of the system. Then, the policy interpreter evaluates the events and makes decisions subject to the current set of policies. The applicability of these policies mitigate the challenges in particular permitting modifications of the strategy during run-time by means of adding and removing policies depending on the challenge required such as a malicious attack or a new network configuration.

5.2 Framework Scenario

The main goal of our framework is to insure resilience management between these organizations using MAS that respond intimately to the requirement of these infrastructures especially in a distributed way. Furthermore, to improve resilience it is necessary to design a comprehensive approach for the system with reference to ResiliNet strategy discussed previously. It assists the system to be more interactive to the changes exhibited on its environment.

Each agent communicates with the nodes of services situated on the organization. In case of disruption on one node, the agent detects these changes and tries to manage the situation first on the node itself then between the other ones. The agents have some intelligence to decide on which appropriate way to search about the missed services.

This architecture is highly decentralized in terms of management, which make the system fundamentally resilient.

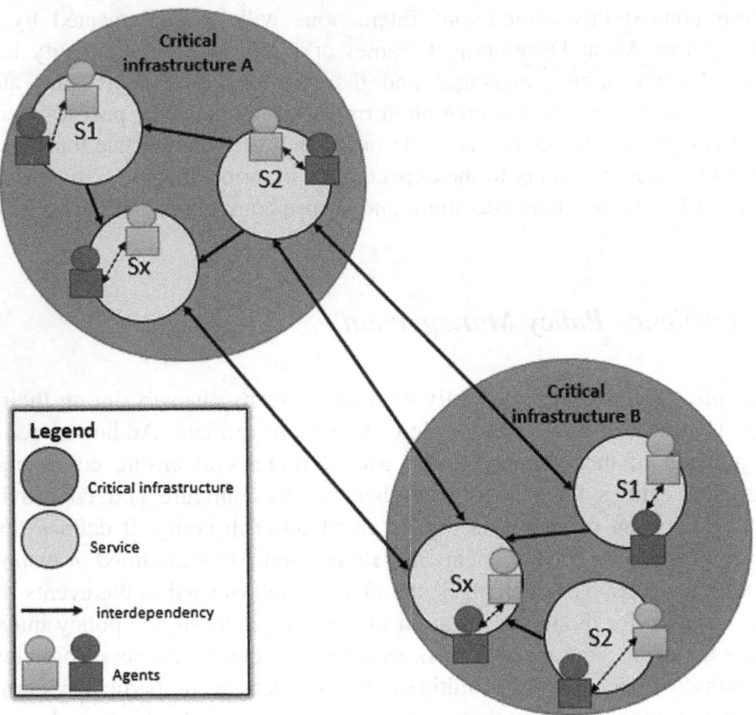

Fig. 3 Interdependent infrastructure based MAS

Agents have to be implemented on each service in CI (Fig. 3). Once a disruption occurs, the agent (Decision agent) makes his own decision about the current state of the service itself by charging the appropriate policies. Then, it transfers this decision to deployment agent in order to react on the service provided. The policies here focus on the restoration plan of each service.

The interaction (Fig. 4) between agents on one service starts by detecting event which interrupt it. Decision agent analyses this event and try to choose the

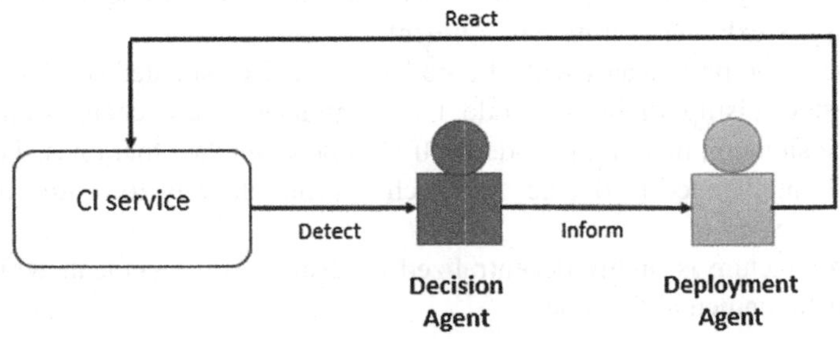

Fig. 4 Agents flow

appropriate resilience policy for the current situation by means of a decision making tools. These decisions are transferred to deployment agent who is responsible of applying those policies on the service in question.

6 Case Study: Financial Infrastructure

In order to implement this approach on a critical infrastructure, we have proposed a case study applied on financial system. It concerns the interaction between Acquiring companies and Issuing ones for payment card transaction. In general, the transaction flows are established as mentioned in Fig. 5:

There are two processes that must occur relating to a card transaction:

- Authorization, the process by which the customer's bank (the issuer) approves the transaction;
- Clearing and settlement, the process by which merchants and financial institutions are paid for their services.

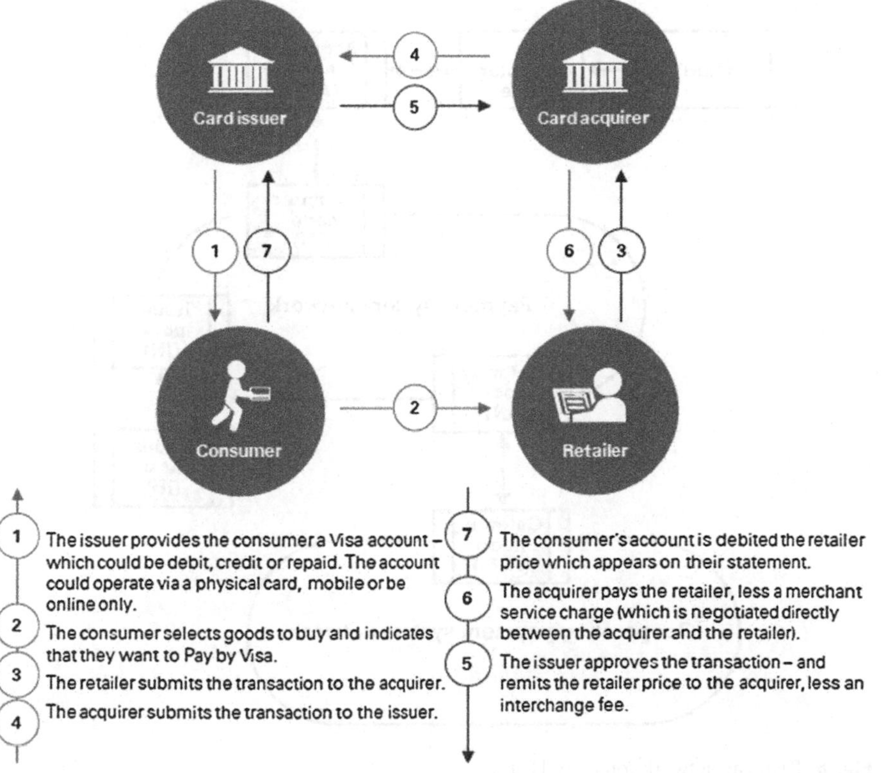

Fig. 5 Simplified processing flow in financial transaction for card payment [11]

The payment system network is responsible for managing the network and for settling funds between the issuer and the acquirer following a payment transaction. Each acquirer host (AH) and issuer host (IH) is connected to separate nodes of the payment network, referred to as the acquirer node (AN) and issuer node (IN), respectively. In order to increase the availability of the issuer's service, an issuer can duplicate the functionality of an IH through a second computer acting as an active reserve. In a complex scenario (Fig. 6), the acquirer managing the terminal at the point of service and the issuer of the card involved in the payment transaction are subscribers to the services offered by different payment system networks, which have established mutual supporting agreements. In order to guarantee the compatibility between these two different networks, two gateway nodes, GN1 and GN2 (one on each payment network), must provide the message translation between the two heterogeneous environments.

Supposing the interdependencies between these payment systems, in case of disruption of this complex scenario we deploy agent in each node that have to share their resilience plan with each other: between acquirer/issuer institutions and also between different payment systems. This can be integrated on the mutual supporting agreement already established.

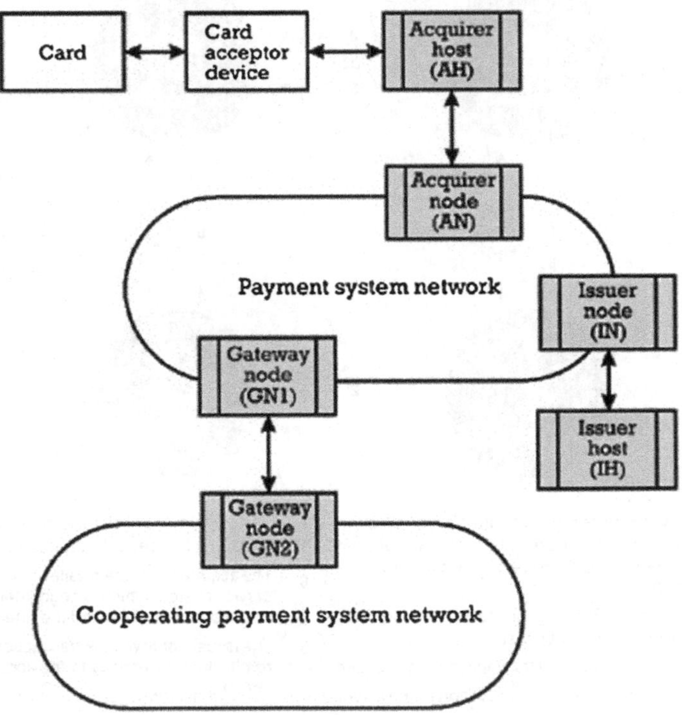

Fig. 6 Payment network topology [12]

7 Conclusion and Future Work

Critical infrastructures are becoming increasingly complex and difficult to manage, in particular when considering the failure of one component or service. Using multi agent system is envisioned as an opportunity to improve the management and interpretation of the interdependent system to increased efficiency in the configuration of resources and in doing so maintain resilience behavior on it. Also, to support decision making on these agents it is needed to integrate some mechanism and algorithms for this purpose. An open modeling framework to capture both short-term and long-term change and evolution of CIs is also more desired for applications. As a future work, it is envisaged to:

- Give a modeling language for the ongoing framework
- Integrating an appropriate decision making tool
- Implementing this approach with JADE platform

References

1. Alcaraz, C., Zeadally, S.: Critical infrastructure protection: requirements and challenges for the 21st century. Int. J. Crit. Infrastruct. Prot. **8**, 53–66 (2015)
2. Johansson, J., Hassel, H.: An approach for modelling interdependent infrastructures in the context of vulnerability analysis. Reliab. Eng. Syst. Saf. **95**, 1335–1344 (2010)
3. Rinaldi, S.M., Peerenboom, J.P., Kelly, T.K.: Identifying, understanding, and analyzing critical infrastructure interdependencies. IEEE Control Syst. **21** (2001)
4. Giovanna Dondossola, G.D.: Critical Utility Infrastructural Resilience, 2012, pp. 11–25
5. Bessani, A.N., Sousa, P., Correia, M., Neves, N.F., Verissimo, P.: The CRUTIAL way of critical infrastructure protection. IEEE Secur. Priv. **6**, 44–51 (2008)
6. Sterbenz, J.P.G., Hutchison, D.: ResiliNets: multilevel resilient and survivable networking initiative (2008)
8. Jones, A., Merabti, M., Randles, M.: Resilience framework for critical infrastructure using autonomics (2012)
9. Aubert, J., Schaberreiter, T., Incoul, C., Khadraoui, D., Gâteau, B.: Risk-based methodology for real-time security monitoring of interdependent services in critical infrastructures. In: International Conference on Availability, Reliability, and Security, pp. 262–267 (2010)
10. Bulut, E., Khadraoui, D., Marquet, B.: Multi-agent based security assurance monitoring system for telecommunication infrastructures. In: International Conference on Communication, Network and Information Security, pp. 90–95 (2007)
11. UK Payments Infrastructure: Exploring Opportunities (2014)
12. Radu, C.: Implementing electronic card payment systems (2002)

Improvement of SPEKE Protocol Using ECC and HMAC for Applications in Telecare Medicine Information Systems

Taoufik Serraj, Moulay Chrif Ismaili and Abdelmalek Azizi

Abstract To ensure patient's privacy and protect data exchanged in Telecare Medicine Information Systems (TMIS), several authenticated key agreement schemes were proposed. In this paper, we propose an elliptic curve instantiation of Abdalla and Pointcheval's Simple Password-Based Encrypted Key Exchange Protocol (SPEKE) including an additional key confirmation step. The proposal is based on short passwords without requiring a complex (PKI). So, it is more suitable to authenticate medical devices and secure medical data exchanged through (RFID) technology, or to ensure remote authenticated key exchanges between patients and servers in TMIS. In addition, we discuss the different security aspects of the proposed protocol, including resistance against side channel attacks in real-world implementations.

Keywords Authentication · Elliptic curve · Key agreement · RFID · Password · Side channel attacks · Telecare medicine information system

1 Introduction

During the last decades, information technology (IT) has changed our lives at a rapid rate. Nowadays, new services like e-Government, e-Business, e-Learning or e-Health use modern IT to facilitate administrative and commercial transactions or to provide remote educational and medical services. In the health domain, health care is one of the biggest economic and social challenges around the world, especially in the aging countries, it has a high cost for governments in terms of finance and human resources. Generally, chronic diseases (e.g., heart and respiratory diseases) require continuous health monitoring, health information processing and sharing to enhance the future disease diagnosis. Especially, in the case of any emergency. To overcome these problems TMIS services are provided. Recently, many electronic medical devices have

T. Serraj (✉) · M.C. Ismaili · A. Azizi
ACSA Laboratory, Faculty of Sciences,
Mohammed First University, 60000 Oujda, Morocco
e-mail: taoufik.serraj@gmail.com

Á. Rocha et al. (eds.), *Europe and MENA Cooperation Advances in Information and Communication Technologies*, Advances in Intelligent Systems and Computing 520, DOI 10.1007/978-3-319-46568-5_51

been used to continuously measure the patient's physiology information, the data collected by medical sensors will be sent to a remote server in the Telecare medicine information system or locally processed (for example, with smart phone applications).

Related work. In the last few years, various authenticated key agreement schemes were proposed to secure the sensitive medical data exchanged between a patient and a server in the Telecare medicine information system. Srivastava et al. [29] and Xu and Wu [31] present schemes based on symmetric cryptographic keys only. Indeed, symmetric cryptographic primitives are secure and fast for the encrypting/decrypting of big data, but suffer from the key sharing problems. On the other hand, Wen and Guo [30] use RSA while Xu et al. [32], Islam and Khan [14] and Zhang and Zhu [33] propose schemes based on Elliptic Curve Cryptography (ECC). The security of these last schemes is based on public/secret key cryptography, which requires additional cryptographic devices to store and manage long keys and certificates in Public Key Infrastructure (PKI). In addition, their security analysis does not take into account side channel attacks, which present a serious threat in practice.

Contributions. In this paper, we propose a simple password-authenticated key agreement protocol for Telecare medicine information systems, the proposed protocol is an ECC instantiation of Abdalla and Pointcheval's Simple Password-Based Encrypted protocol [2] with an additional key confirmation step. The proposal enables two parties which share a small size password *pw* (e.g., a PIN) to generate a strong and authenticated session key, which will be used to secure subsequent communications over a public network without requiring complex public key infrastructures. The proposed protocol is efficient and achieves security goals even in the presence of side channel attacks.

Organization. The rest of the paper is organized as follows: Sect. 2 recalls some notions on elliptic curves and briefly reviews known mathematical and physical attacks on elliptic curve based cryptosystems. The proposed protocol and the corresponding security analysis are presented in Sects. 3 and 4, the design choices and performance are discussed in Sect. 5. Finally, a conclusion is provided.

2 Mathematical and Cryptological Background

This section recalls notions related to elliptic curves over finite fields and provides a brief survey on various attacks on cryptosystems based on elliptic curves.

2.1 *Elliptic Curve Cryptography (ECC)*

Elliptic Curves Over Finite Fields

Definition 1 An elliptic curve E over a field $\mathbb{K}$ of characteristic >3, denoted by $E/\mathbb{K}$ is given by the reduced Weierstrass equation

$$y^2 = x^3 + ax + b \ , \tag{1}$$

where $a,\ b \in \mathbb{K}$ and $4a^3 + 27b^2 \neq 0$. If $\overline{\mathbb{K}}$ is an extension field of $\mathbb{K}$, we denote by $E(\overline{\mathbb{K}})$ the set of points $(x, y) \in \overline{\mathbb{K}}^2$ satisfying (1) together with the point at infinity $\mathcal{O}$, $E(\overline{\mathbb{K}})$ is an additive Abelian group [27].

We can also define the point addition law in this case by:
if $P = (x_1, y_1)$, $Q = (x_2, y_2) \in E$ and $P \neq -Q$ we can get
$R = (x_3, y_3) = P + Q$ where

$$x_3 = \lambda^2 - x_1 - x_2, \quad y_3 = \lambda(x_1 - x_3) - y_1$$

$$\text{and} \quad \lambda = \begin{cases} \frac{y_2 - y_1}{x_2 - x_1} & \text{if} \quad P \neq Q \\ \frac{3x_1^2 + a}{2y_1} & \text{if} \quad P = Q \end{cases} . \tag{2}$$

The most important operation in ECC is the scalar multiplication

$$k.P = P + P + \cdots + P \ \ (k\,\text{times}) \ . \tag{3}$$

There are various algorithms to do this operation, most of these algorithms use the double and addition operations.

Elliptic curves used in cryptographic applications are either defined over prime fields or binary fields. In this paper, we will use elliptic curves over prime fields $\mathbb{F}_p$ where $p > 3$ for security purposes [9].

The Discrete Logarithm Problem in Elliptic Curves

Definition 2 (*Elliptic Curve Discrete Logarithm Problem*) Let E be an elliptic curve over a finite field $\mathbb{F}_p$. Let $P, Q \in E$, such that Q is in the subgroup of E generated by P. The elliptic curve discrete logarithm problem (ECDLP) is the problem of determining an integer $k \geq 1$ such that $k.P = Q$.

The security of elliptic curve cryptosystems is mainly based on the presumed hardness of the ECDLP.

2.2 Attacks on Elliptic Curve Cryptosystems

Elliptic curve cryptosystems are vulnerable to a wide range of mathematical and physical attacks.

Mathematical attacks. If the elliptic curve is not carefully chosen, the ECDLP can be solved mathematically by using: Shanks [25], Pollard [20, 21], Pohlig and Hellman [19] algorithms in general cases, and Menezes et al. [16], Frey and Rück [11] and Satoh and Araki [23], Semaev [24], Smart [28] algorithms in very special cases. In traditional security models, cryptographic primitives are considered only as mathematical objects. The current mathematical knowledge enables us to build secure cryptographic protocols from this point of view.

Side channel attacks. Unfortunately, in the real-world applications, implementation of cryptographic primitives is very complex and depends on the physical properties of the hardware. Thus, new forms of attacks arise from this fact. In this setting, an attacker exploits not only the mathematical structures of cryptographic primitive, but also the inherent properties of the implementations. The attacker can analyze the information leaked by the cryptographic devices, during the normal or abnormal execution of the cryptographic protocol, and recovers secret keys without solving ECDLP, these last attacks are called Side Channel Attacks (SCAs). An attacker can analyze the time required to perform the scalar multiplication to deduce the secret key k bit by bit through Sato-Schepers-Takagi timing attack [22]. Also, he can exploit the traces of power consumptions or electromagnetic emanations of the cryptographic devices to deduce k using Coron attack [7], Goubin attack [12] or Heyszl attack [13]. The previous attacks are passive, an attacker can also influence the cryptosystem operations by injecting faults. The main idea in fault attacks on ECC [5, 10] is to move computations from the secure curve to a weak one by injecting faults, then solving the ECDLP in the weak curve.

Side channel attacks are serious threats in practice, especially, in smart cards or on unprotected medical devices. Unfortunately, such attacks are not well covered by current security models or security standards.

Quantum attacks. If we succeed one day in the future to build a quantum computer, cryptosystems based on RSA and elliptic curves will be broken due to the Shor's algorithm [26]. Until this happens, quantum attacks remain theoretical.

3 The Description of the ECC-SPEKE Protocol

In this section, we give a description of the proposed SPEKE protocol using elliptic curves, key derivation and HMAC functions.

3.1 Notations

Table 1 illustrates the notations used in the rest of this paper.

Table 1 The notations used in this paper

Notation	Descriptions
U	Patient in TMIS
S	Telecare server in TMIS
pw	Patient's password randomly chosen from a dictionary D
p, q	Two large primes
$E/\mathbb{F}_p$	Elliptic curve over a finite prime field $\mathbb{F}_p$ of order q
$\mathcal{G} = (a, b, p)$	The parameters of the curve $E/\mathbb{F}_p$
$\mathcal{K}$	The session key
G, P and Q	Points in the curve $E/\mathbb{F}_p$
H(.)	Secure cryptographic hash function
KDF(.)	Key derivation function
HMAC(.)	Hash-based message authentication code
RNG(.)	(Pseudo)-random number generator
sid	Session identification
$\|\|$	String concatenation operation

3.2 Description of the Proposed Scheme

The complete protocol is depicted in Table 2. The patient and the server share a secret password pw, and they agree on the following public parameters: $\mathcal{G}$, G, P and Q. The proposed protocol can be also used to secure the contactless communication between medical devices. Roughly, the ECC-SPEKE protocol can be divided into four steps. **Randomization**. In this step, the patient (resp. the server) chooses randomly x (resp. y) from $[1, q-1]$, computes $X = x.G$, $k_{pw} = H(0\|pw)$ and $X^\star = X + k_{pw}.P$ (resp. $Y = y.G$, $k_{pw} = H(0\|pw)$ and $Y^\star = Y + k_{pw}.Q$);

Key exchange. The patient and the server exchange the values $X^\star$ and $Y^\star$. On receiving $Y^\star$ (resp. $X^\star$), the patient (resp. the server) recovers Y (resp. X) and computes $x.(Y^\star - k_{pw}.Q)$ (resp. $y.(X^\star - k_{pw}.P)$);

Key derivation. Both parties have the same value $xy.G$, they use the key derivation function KDF to compute

$$\mathcal{K} = \mathrm{KDF}\left(X^\star \| Y^\star \| xy.G\right) = k_{enc} \| k_{mac} \| sid.$$

From this key, the patient and the server extract two sub-keys k_{enc}, k_{mac} and a session identifier sid. The key k_{enc} will be used to secure subsequent communications, while k_{mac} is used in the next step to confirm knowledge of the session key by the communicating parties.

Table 2 The ECC-SPEKE

Patient		Server
$x \leftarrow_R [1, q-1]$		$y \leftarrow_R [1, q-1]$
$X = x.G$		$Y = y.G$
$k_{pw} = H(0\|pw)$		$k_{pw} = H(0\|pw)$
$X^\star = X + k_{pw}.P$		$Y^\star = Y + k_{pw}.Q$
	$\overset{X^\star}{\longrightarrow}$	
	$\overset{Y^\star}{\longleftarrow}$	
$k_U = x.(Y^\star - k_{pw}.Q)$		$k_S = y.(X^\star - k_{pw}.P)$
	$k_U = k_S = xy.G$	
	$\mathcal{K} = k_{enc}\|k_{mac}\|sid$	
$u =$HMAC$(k_{mac}, (Y^\star, \mathcal{G}))$		$v =$HMAC$(k_{mac}, (X^\star, \mathcal{G}))$
	$\overset{u}{\longrightarrow}$	
	$\overset{v}{\longleftarrow}$	
Abort if v invalid		Abort if u invalid

Key confirmation. In this step, the patient (resp. the server) computes $u =$ HMAC $(k_{mac}, (Y^*, \mathcal{G}))$ (resp. $v =$ HMAC$(k_{mac}, (X^*, \mathcal{G}))$ and sends u (resp. v) to the server (resp. the patient). Each party checks the other HMAC and reports a failure in case of a mismatch.

4 Security Analysis

4.1 Security of Cryptographic Primitives

A good choice of the elliptic curve parameters and the method of performing the scalar multiplication can avoid mathematical attacks and a wide range of side channel attacks.

Mathematical attacks. According to the NIST [18] or the ANSSI [4] recommendations, we should select a curve $E/\mathbb{F}_p$ with prime order q of size at least 224 bits to avoid Shanks, Pollard, Pohlig-Hellman and Index-calculus attacks. In addition, we must have $q \neq p$ to prevent Sato-Araki-Semaev-Smart attacks. Finally, MOV and Frey-Rück attacks can be avoided if:

$$p^r \not\equiv 1 \pmod{q}, \forall 1 \leq r \leq 10^4 \ . \tag{4}$$

Side channel attacks. To prevent simple power, timing or electromagnetic analysis attacks, the computations performed should not depend on the secret data and branching should be avoided. We can use regular scalar multiplication algorithms (e.g., [15]) in the implementation. Additionally, we can prevent Goubin-type attacks by choosing a non-square coefficient b in $\mathbb{F}_p$ for the second curve parameter, and twist fault attacks if the curve $E/\mathbb{F}_p$ has a twist of prime order.

Additional requirements. The message authentication code HMAC must be unforgeable against adaptively chosen message attacks. RNG, KDF and the hash function H should ensure a random uniformly distributed values in the output.

4.2 Security of the Protocol Construction

The original protocol [2] is proved to be secure as a password-based authenticated key exchange scheme in the security model of Abdala Fouque Pointcheval (AFP) [1]. The proposed protocol remains secure in this model under the assumptions mentioned in the previous subsections since H, RNG, KDF and HMAC still behave as random oracles.

In the following, we provide an heuristic security analysis:

The session key confidentiality. It means that each run of the protocol should produce a unique secret key named session key which is not known to anyone except the patient U and the server S. Under the assumptions that the elliptic curve and the scalar multiplication algorithms are chosen to be secure against mathematical and side channel attacks, both parties U and S succeed to generate a secure key, since an attacker cannot solve the ECDLP.

The resistance against dictionary attacks. It means that an attacker is not able to find the correct session key by guessing passwords. Dictionary attacks can be divided into on-line and off-line attacks. We can avoid on-line dictionary attacks by limiting the use of passwords and aborting the protocol after a certain number of failed attempts. In addition, an attacker tries to find the password pw by guessing passwords pw^* in an off-line manner and comparing results to X^* (or Y^*), since q and the output of the hash function have sizes of at least 224 bits, the off-line attack becomes a brute force attack, which is computationally infeasible. Therefore, the proposed scheme is secure against dictionary attacks.

The secure mutual authentication. In the proposed protocol, only *U* and *S* know the password *pw* and can generate a secure session key, the sub key k_{mac} will be used in the authentication process by HMAC. Then, both parties explicitly confirm knowledge of the session key. If verification fails, parties abort the session.

The security against men-in-the-middle attack. In the proposed protocol, the session key is not shared if the HMAC in the confirmation step is not valid. So, an attacker cannot pretend to be *U* or *S* since he does not know the password. Therefore, the proposed protocol can resist the man-in-the-middle attack.

The perfect forward secrecy. It means that a loss of the password *pw* should not compromise already-distributed session keys. Indeed, the random choice of *x* and *y* in randomization step ensures that each session generates its own random key (with a session identification *sid*), which is independent of other session keys. Then, the proposed protocol achieves perfect forward secrecy.

5 Design Choices and Performance Analysis

5.1 Design Choices

Recently, many elliptic curves were proposed by international security organizations for current implementations $p \approx 224$ bits (or future implementations $p \approx 256$ bits). For instance, P-256 proposed by NIST in FIPS-184 [17], FRP256*v*1 introduced by ANSSI in [3], secp256*r*1 recommended by Certicom Research [6] and the brainpoolP256*r*1 curve [8]. By analyzing these curves, the brainpool curve meets all the security requirements discussed above, with the following parameters:

Curve-ID: brainpoolP256*r*1:

p = A9FB57DBA1EEA9BC3E660A909D838D726E3BF623D52620282013481D 1F6E5377
a = 7D5A0975FC2C3057EEF67530417AFFE7FB8055C126DC5C6CE94A4B44 F330B5D9
b = 26DC5C6CE94A4B44F330B5D9BBD77CBF958416295CF7E1CE6BCCDC 18FF8C07B6

For the hash function H, and the underlying hash functions used in KDF or in HMAC, we proposed the use of SHA-256 standardized by the National Institute of Standards and Technology (NIST).

5.2 Performance Analysis

The use of ECC ensures the same security level as RSA but with short key size (e.g., 224 bits instead 2048 bits for RSA [4]), which give to elliptic curve cryptography more advantages in terms of efficiency for embedded systems. In protocols using ECC, we need to store or transmit an elliptic curve point $P = (x, y)$. Instead of transmitting the two values x and y, we use point compression technique, we transmit only the x-coordinate along with an additional bit to say which value of the y-coordinate we should take. Since hash functions, KDF, HMAC and negation of an elliptic curve point are almost free, the time of the protocol execution is dominated by scalar multiplications. In the proposed protocol, each party needs four scalar multiplications. Using PARI/GP system in the environment (CPU: 2.16 GHz, RAM: 2.0 G), we obtain that a scalar multiplication with the brainpoolP256r1 curve parameters takes an average 16 ms. For more environments we can use the Multiprecision Integer and Rational Arithmetic Cryptographic Library (MIRACL).

6 Conclusion

In this paper, we presented an ECC instantiation of the general SPEKE Protocol proposed by Abdala and Pointcheval. We have added a key confirmation step using KDF and HMAC to enhance mutual authentication. We propose the use of hash, KDF and HMAC functions standardized by NIST due to their security and efficiency, also to preserve the origin security proof in the random oracle model since the proposed functions are still behaving as random oracles. Since the proposed protocol uses short keys (compared to RSA) and do not require a complex PKI, it will be more suitable to secure contactless communications between electronic medical devices. Finally, if the security parameters and the implementation methods are well chosen, a wide range of mathematical and physical attacks can be avoided.

References

1. Abdalla, M., Fouque, P.A., Pointcheval, D.: Password-based authenticated key exchange in the three-party setting. In: Public Key Cryptography—PKC 2005. LNCS, vol. 3386, pp. 65–84. Springer (2005)
2. Abdalla, M., Pointcheval, D.: Simple password-based encrypted key exchange protocols. In: Topics in Cryptology—CT-RSA 2005. LNCS, vol. 3376, pp. 191–208. Springer (2005)
3. ANSSI: Publication d'un paramétrage de courbe elliptique visant des applications de passeport électronique et de l'administration électronique française (2011)
4. ANSSI: Mécanismes cryptographiques règles et recommandations concernant le choix et le dimensionnement des mécanismes cryptographiques. Technical report (2014)
5. Biehl, I., Meyer, B., Müller, V.: Differential fault attacks on elliptic curve cryptosystems. In: Advances in Cryptology—CRYPTO 2000. LNCS, vol. 1880, pp. 131–146. Springer (2000)
6. Certicom Research: Sec 2: Recommended elliptic curve domain parameters (2010)
7. Coron, J.S.: Resistance against differential power analysis for elliptic curve cryptosystems. In: Cryptographic Hardware and Embedded Systems. LNCS, vol. 1717, pp. 292–302. Springer (1999)
8. ECC Brainpool: ECC brainpool standard curves and curve generation (2005)

9. Faugère, J.C., Perret, L., Petit, C., Renault, G.: Improving the complexity of index calculus algorithms in elliptic curves over binary fields. In: Advances in Cryptology—EUROCRYPT 2012. LNCS, vol. 7237, pp. 27–44. Springer (2012)
10. Fouque, P.A., Lercier, R., Réal, D., Valette, F.: Fault attack on elliptic curve montgomery ladder implementation. In: 5th Workshop on Fault Diagnosis and Tolerance in Cryptography, 2008. FDTC'08, pp. 92–98. IEEE (2008)
11. Frey, G., Rück, H.G.: A remark concerning m-divisibility and the discrete logarithm in the divisor class group of curves. Math. Comput. **62**(206), 865–874 (1994)
12. Goubin, L.: A refined power-analysis attack on elliptic curve cryptosystems. In: Public Key Cryptography PKC 2003. LNCS, vol. 2567, pp. 199–211. Springer (2003)
13. Heyszl, J., Mangard, S., Heinz, B., Stumpf, F., Sigl, G.: Localized electromagnetic analysis of cryptographic implementations. In: Topics in Cryptology–CT-RSA 2012. LNCS, vol. 7178, pp. 231–244. Springer (2012)
14. Islam, S.H., Khan, M.K.: Cryptanalysis and improvement of authentication and key agreement protocols for telecare medicine information systems. J. Med. Syst. **38**(10), 1–16 (2014)
15. Joye, M.: Highly regular right–to–left algorithms for scalar multiplication. In: Cryptographic Hardware and Embedded Systems—CHES 2007. LNCS, vol. 4727, pp. 135–147. Springer (2007)
16. Menezes, A.J., Okamoto, T., Vanstone, S.A.: Reducing elliptic curve logarithms to logarithms in a finite field. IEEE Trans. Inf. Theory **39**(5), 1639–1646 (1993)
17. NIST: FIPS PUB 186-4: Digital signature standard, DSS (2013)
18. NIST: Transitions: Recommendation for transitioning the use of cryptographic algorithms and key lengths. NIST Special Publication 800-131A Revision 1 (2015)
19. Pohlig, S.C., Hellman, M.E.: An improved algorithm for computing logarithms over $GF(p)$ and its cryptographic significance (Corresp.). IEEE Trans. Inf. Theory **24**(1), 106–110 (1978)
20. Pollard, J.M.: Monte Carlo methods for index computation (mod p). Math. Comput. **32**(143), 918–924 (1978)
21. Pollard, J.M.: Kangaroos, monopoly and discrete logarithms. J. Cryptol. **13**(4), 437–447 (2000)
22. Sato, H., Schepers, D., Takagi, T.: Exact analysis of montgomery multiplication. In: Progress in Cryptology—INDOCRYPT 2004. LNCS, vol. 3348, pp. 290–304. Springer (2004)
23. Satoh, T., Araki, K.: Fermat quotients and the polynomial time discrete log algorithm for anomalous elliptic curves. Commentarii Mathematici Universitatis Sancti Pauli **47**(1), 81–92 (1998)
24. Semaev, I.: Evaluation of discrete logarithms in a group of p-torsion points of an elliptic curve in characteristic p. Math. Comput. Am. Math. Soc. **67**(221), 353–356 (1998)
25. Shanks, D.: Class number, a theory of factorization, and genera. Proc. Symp. Pure Math. **20**, 415–440 (1971)
26. Shor, P.W.: Polynomial-time algorithms for prime factorization and discrete logarithms on a quantum computer. SIAM Rev. **41**(2), 303–332 (1999)
27. Silverman, J.H.: The Arithmetic of Elliptic Curves. Springer, New York (2009)
28. Smart, N.P.: The discrete logarithm problem on elliptic curves of trace one. J. Cryptol. **12**(3), 193–196 (1999)
29. Srivastava, K., Awasthi, A.K., Kaul, S.D., Mittal, R.: A hash based mutual RFID tag authentication protocol in telecare medicine information system. J. Med. Syst. **39**(1), 1–5 (2015)
30. Wen, F., Guo, D.: An improved anonymous authentication scheme for telecare medical information systems. J. Med. Syst. **38**(5), 1–11 (2014)
31. Xu, L., Wu, F.: Cryptanalysis and improvement of a user authentication scheme preserving uniqueness and anonymity for connected health care. J. Med. Syst. **39**(2), 1–9 (2015)
32. Xu, X., Zhu, P., Wen, Q., Jin, Z., Zhang, H., He, L.: A secure and efficient authentication and key agreement scheme based on ECC for telecare medicine information systems. J. Med. Syst. **38**(1), 1–7 (2013)
33. Zhang, L., Zhu, S.: Robust ECC-based authenticated key agreement scheme with privacy protection for telecare medicine information systems. J. Med. Syst. **39**(5), 1–11 (2015)

Performance Analysis of an Intrusion Detection Systems Based of Artificial Neural Network

Mohammed Saber, Ilhame El Farissi, Sara Chadli, Mohamed Emharraf and Mohammed Ghaouth Belkasmi

Abstract The Artificial Neural Network (ANN) enables systems to think and act intelligently. In recent years, ANNs are applied in security of network. Therefore, there are several researches in this area, particularly in Intrusion Detection System which are based on ANN. The objective of this paper is to select the most important and crucial parameters in order to provide an optimized ANN for Pattern Recognition which is able to detect attacks including the recently developed ones. First of all, we have taken some and all of the basic attributes to aliment the networks input and to verify the dependence between these parameters and attacks. Then, we have added the parameters relating to content and time-based ones in order to demonstrate their utility and performance and also to present in which case they are crucial.

Keywords Intrusion detection system · Artificial neural network for pattern recognition · KDD data · KDD parameters · Attack categories

M. Saber (✉) · I. El Farissi · M. Emharraf · M.G. Belkasmi
Laboratory LSE2I, National School of Applied Sciences,
First Mohammed University, Oujda, Morocco
e-mail: mosaber@gmail.com
URL: http://wwwensa.ump.ma

I. El Farissi
e-mail: ilhame.elfarissi@gmail.com

M. Emharraf
e-mail: m.emharraf@gmail.com

M.G. Belkasmi
e-mail: ghaouth@gmail.com

S. Chadli
Laboratory Electronics and Systems, Sciences Faculty,
First Mohammed University, Oujda, Morocco
e-mail: chad.saraa@gmail.com

Á. Rocha et al. (eds.), *Europe and MENA Cooperation Advances in Information and Communication Technologies*, Advances in Intelligent Systems and Computing 520, DOI 10.1007/978-3-319-46568-5_52

1 Introduction

In recent years, researches aim to develop the most performant and powerful Intrusion Detection Systems(IDS). Particularly the IDSs which are based on neural network. Indeed, this type of IDS has the ability to detect also the attacks recently developed [1, 2].

The ultimate objective of any IDS is to minimize the attack success and increase the detection rate. Indeed, many of the IDSs which are based on the Artificial Neural Network employ the KDD dataset. This dataset contains recorded attacks depending on 41 features. These last ones belong on four categories attributes. In the realized researches, all these categories are exploited to develop a neural network. However, some of these properties can be unnecessary in attack detection, which can lead to bad-functioning system. Thus, it seems interesting to promote an optimum neural network-based IDS with restricted proprietes.

This set of work aims to analyze the utility of each category. Indeed, there are four categories in KDD dataset; basic attributes, attributes which are related to content, attributes which are based on the time using windows of two-second time and time-based attributes using windows of 100 connections time. In order to develop an optimum IDS basing on neural network, we have designed four scenarios; the first one consists on alimenting the neural network by the basic attributes. In the second scenario we have taken into account also the attributes which are based to content. Then, we have powered the neural network, in the third scenario, by the basic attributes and attributes which are based on the time using windows of two-second time. Concerning the last scenario, we have taken into consideration the basic attributes and features belonging to time-based attributes using windows of 100 connections time category.

This article is structured as follows: As far as the first section, the main concepts are presented. It includes KDD dataset, Intrusion Detection Systems and Artificial Neural Network. The second part is dedicated to description of the issue consisting on presenting the previous works realized in this area and also to introduce the proposed solution as an optimized system of attack detection. As far as the third section of this article, it aims to present the conception and realization of an optimum IDS which is based on neural network and KDD dataset and also to analyse the obtained results from the different scenarios.

2 Basic Concepts in Neural Network-Based IDS

The current work aims to design an IDS based on neural network. For that, its necessary to employ a dataset attack, explore the existing IDS and define the neural network concept. This is the purpose of this section of the article.

2.1 Description of KDD Dataset

KDD [3] is a publically available dataset of attacks for IDS. It contains 41 features and one more attribute for class. In fact, there are 41 parameters in KDD set, their values depend on the attacks type which is indicated in the 42nd feature.

First of all, before getting into the crux of the discussion it is important to describe the KDD content. In one hand, the exploited parameters are divided into four categories:

3 Problem Statement and Resolution Methodology

3.1 Problem Statement

Detection precision and detection stability are two key indicators which allow evaluating intrusion detection systems (IDS) [4]. In order to enhance the detection precision and detection stability, many researches have been realized (e.g., [5]). In the early stage, the research focus lies in using rule-based expert systems and statistical approaches [6]. But when encountering larger datasets, the results of rule-based expert systems and statistical approaches become worse. Thus, a lot of data mining techniques have been introduced to solve the problem (e.g., [7, 8]). Among these techniques, Artificial Neural Network (ANN) is one of the widely used techniques and has been successful in solving many complex practical problems. Furthermore, ANN has been successfully applied into IDS.

However, the main drawbacks of ANN-based IDS exist in two aspects: (1) lower detection precision, especially for low-frequent attacks, e.g., Remote to Local (R2L), User to Root (U2R), and (2) weaker detection stability [1]. For the above two aspects, the main reason is that the distribution of different types of attacks is imbalanced. For low-frequent attacks, the learning sample size is too small compared to high-frequent attacks. It makes ANN not easy to learn the characters of these attacks and therefore detection precision is much lower. In practice, low-frequent attacks do not mean they are unimportant. Instead, serious consequence will be caused if these attacks succeeded. For example, if the U2R attacks succeeded, the attacker can get the authority of root user and make everything he wants to the targeted computer systems or network device. Furthermore, in IDS the low-frequent attacks are often outliers. Thus, ANN is unstable as it often converges to the local minimum [9]. Although prior research has proposed some approaches, when encountering large datasets, these approaches become not effective [5, 10].

Table 1 Number of the employed patterns per attacks class

Dataset1		Dataset2	
Class	Patterns	Class	Patterns
Normal	60593	Normal	3883370
DOS	223298	DOS	972781
Probe	2377,00	Probe	41120
R2L	5993	R2L	1126
U2R	39	U2R	52
Total	292300	Total	4898449

3.2 *Proposed Approach: An Optimum Neural Network System Such as an Efficient IDS*

In recent researches, the IDSs are developed through ANNs. Most of them are alimented by KDD dataset. KDD dataset contains 41 features belonging on four categories. In previous works, the four categories are exploited. In order to increase the detection precision, it seems interesting, for us, to specify the employed categories of features in the networks input layer.

Moreover, we have designed four scenarios; In the first scenario, we have taken into account uniquely the basic attributes which are presented in the Table 1. The second scenario consists on using the basic attributes and the attributes which are based on content. As far as the third scenario, we have alimented the network by the basic attributes and attributes which are based on the time using windows of two-second time. The last scenario consists on using the basic attributes and features belonging to time-based attributes using windows of 100 connections time category.

4 Conception, Realization and Diagnostic of an Optimum IDS Based on Neural Network and KDD Dataset

As noted before, the KDD features are divided into four categories; basic attributes, attributes which are related to content, attributes based on the time using windows of two-second time and time-based attributes using windows of 100 connections time. All the neural network-based IDS which are already developed depends on these four categories. In this work, we aim to specify the role of each one by adopting multiple scenarios.

Thus, in order to improve the IDSs performance by avoiding the redundant and ineffective parameters and increase rate detection. we have developed the following scenarios.

Furthermore, it is important to note that we have used MATLAB tool to put into practice our approach. The model employed consists on Neural Pattern Recognition, it allows to import input and output data and specify the percentage of patterns used in Training, Validation and Testing steps. The patterns employed in training phase are presented to the network during training, and the network is adjusted according to its error. As far as validation patterns, they are used to measure network generalization, and to halt training when generalization stops improving. Finally, the testing patterns have no effect on training and so provide an independent measure of network performance during and after training. In our case, we have applied respectively 55 %, 15 %, 30 % of patterns in training, validation and testing phases.

As a result, we will present the confusion matrix of each scenario. It contains information about the obtained and estimated classifications performed by a classification system. The target and output classes are hosted by a number. The normal execution, U2R, R2L, DOS and Probe correspond respectively to 1, 2, 3, 4 and 5 which are indicated in the confusion matrix below.

Furthermore, according to [2, 11] the attributes A9, A20 and A21 have no role in attack detection, the attributes A15, A17, A19, A32, A40 have minimum role and the features A7, A8, A11 and A14 have almost all zero values in dataset. Consequently, in the first, second and fourth scenarios we have taken due note of this.

4.1 *First Scenario: Basic Attributes as the Neural Networks Input*

This scenario is divided into two parts. In one hand, we have included the nine basic attributes (already presented in Table 1) such as input network. By selecting distinct recordings from the database Dataset1, we have obtained 317594 patterns. 55 % of them have been used in training, 15 % in validation and 30 % in test phase. The following confusion matrix (Fig. 1) presents the general result.

This Fig. 1 demonstrates that nearly 100 % of normal cases have been well classified. As far as attacks, only 70.8 % of Probe have been detected.

In the other hand, the attributes A7, A8 and A9 have been omitted in order to verify their effectiveness. In the rest of the article we will call the basic attributes without A7, A8 and A9 the optimum basic attributes. The confusion matrix that has been generated in this case is as follows.

According to the Fig. 2, the obtained results in this scenario demonstrate that the basic attributes are essential to detect Probe attacks but they are not sufficient and they also allow us to confirm that the attributes A7, A8 and A9 do not impact the detection rate. Consequently, in the rest of work we have not employed these three attributes.

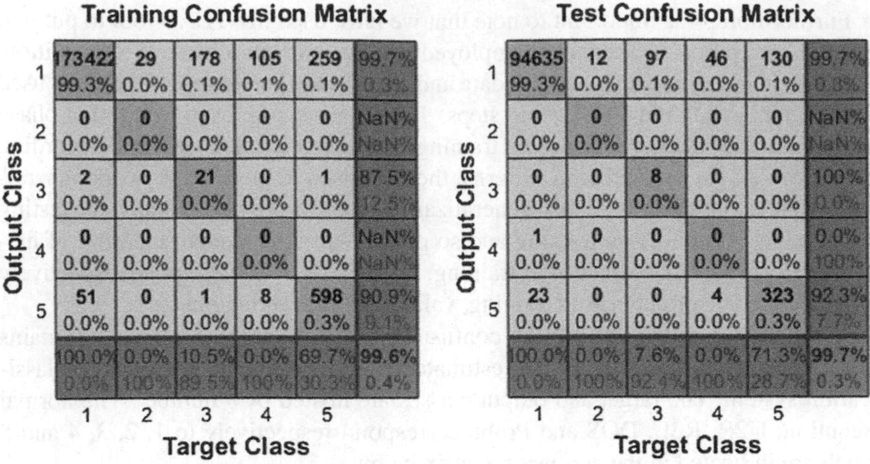

Fig. 1 Confusion matrix of neural network alimented by the KDD basic attributes

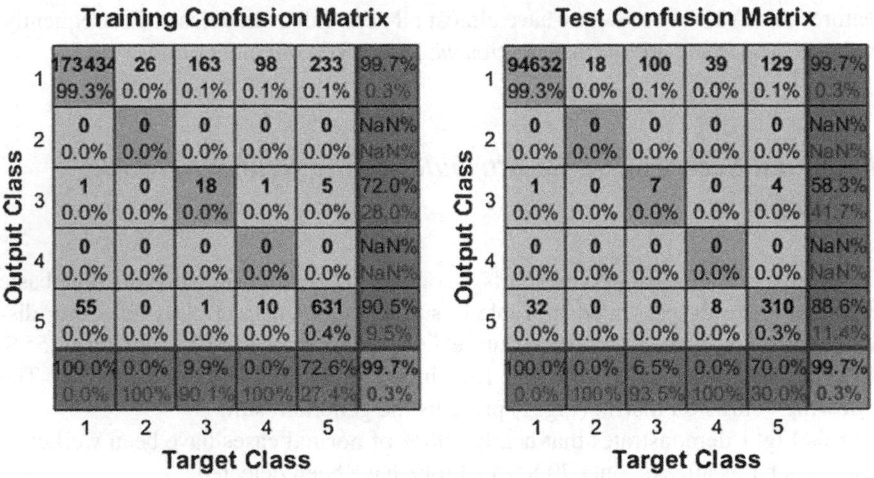

Fig. 2 Confusion matrix of neural network alimented by the optimum basic attributes

4.2 Second Scenario: The Optimum Basic Attributes + Attributes Related on the Content as Networks Input

First of all, we initiate this part by presenting the results obtained through a neural network alimented by 19 attributes including the six first basic attributes and all of the attributes which are related on the content. The generated confusion matrix is presented in Fig. 3.

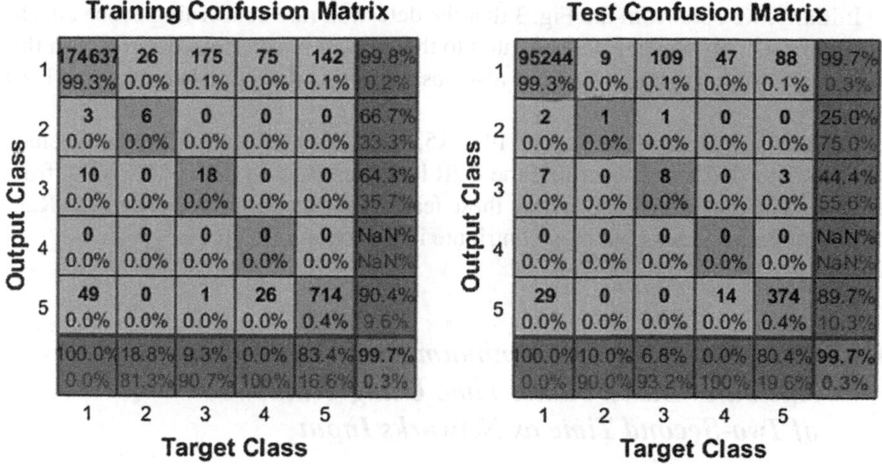

Fig. 3 Confusion matrix of neural network alimented by the optimum basic attributes + attributes related on the content as networks input

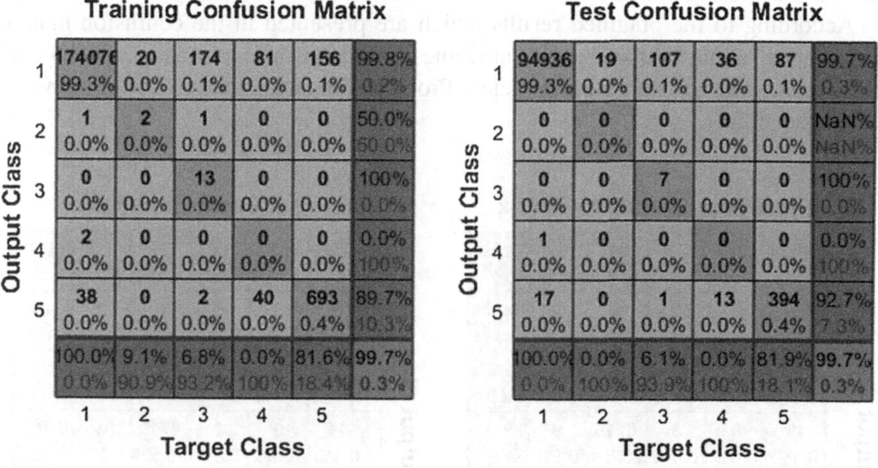

Fig. 4 Confusion matrix of neural network alimented by the optimum basic attributes + A10, A12, A13, A16, A18, A22 as networks input

The second part of this scenario consists on alimenting a neural network with the optimum basic attributes + Attributes related on the content without the attributes A11, A14, A15, A17, A19, A20 and A21. The corresponding confusion matrix is depicted in Fig. 4.

Initially, we note from the Fig. 3 that the detection rate concerning Probe attacks has increased. So, the attributes related to the content have a significant role in this scenario. On the same figure, we are also observing that detection percentage of U2R is 16.3 %.

Neglecting the attributes A11, A14, A15, A17, A19, A20 and A21, we conclude that the detection rate concerning the U2R had decreased by 12 %. Therefore, from the second scenario we confirm that these features have no role concerning the R2L, DOS and Probe attacks but they contribute in detection of U2R category.

4.3 *Third Scenario: The Optimum Basic Attributes + Attributes Based on the Time Using Windows of Two-Second Time as Networks Input*

In this Scenario, we have extracted 61754 distinct patterns from the database Dataset2. As mentioned above, 55 % are employed in training, 15 % in validation and 30 % in testing phase.

According to the obtained results which are presented in the confusion matrix (Fig. 5), we acknowledge that the attributes based on the time using windows of two-second time are necessary to detect Probe attack and sufficient and effective in DOS detection.

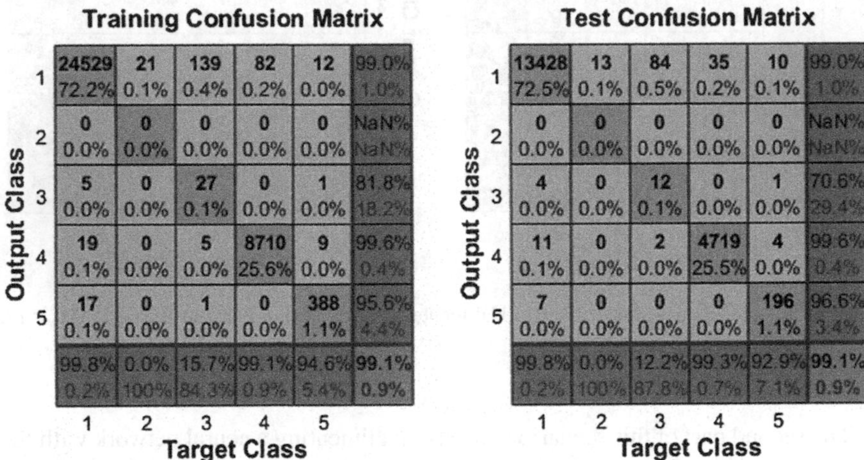

Fig. 5 Confusion matrix of neural network alimented by the optimum basic attributes + attributes based on the time using windows of two-second time

4.4 *Fourth Scenario: The Optimum Basic Attributes + Time-Based Attributes Using Windows of 100 Connections Time*

By adding the time-based attributes using windows of 100 connections time to the optimum basic attributes, and as presented in Fig. 6 the detection rates of R2L, DOS and Probe attack categories are respectively 55.4 %, 96.7 %, 90.5 %. Therefore, these

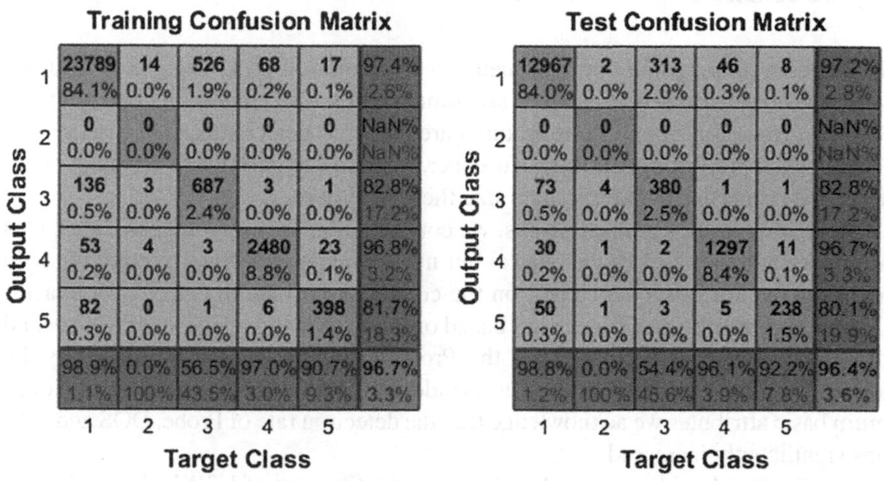

Fig. 6 Confusion matrix of neural network alimented by the optimum basic attributes + time-based attributes using windows of 100 connections time

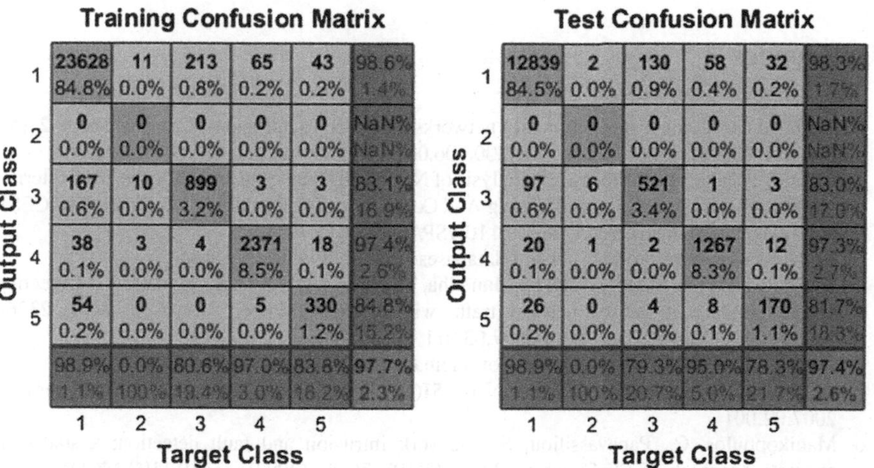

Fig. 7 Confusion matrix of neural network alimented by the optimum basic attributes + time-based attributes using windows of 100 connections time without A32 and A40

attributes are largely sufficient to detect DOS and Probe attacks, but they do not contribute in U2R case and they should be improved to detect R2L attacks.

By eliminating A32 and A40 features and as presented in Fig. 7, we observe that the detection rate of R2L category attack has increased to 80.5 %. However, it really proves that some parameters in some cases are unnecessary and can decrease the IDSs performance.

5 Conclusion

This article proposes an optimum neural network-based IDS. Thus, most realized IDSs which rely on neural network are alimented by KDD dataset including the 41 features. However, not all these features are useful to detect attacks. Consequently, the detection precision decreases. Moreover, we have designed a neural network with specified properties in order to optimize the IDS system.

According to the obtained results, we conclude that the optimum basic attributes are necessary to detect Probe attacks but not sufficient to recognize other ones. In addition, the attributes which rely on the content contribute to detect U2R attacks. Concerning the attributes which are based on the time using windows of two-second time, they are crucial to recognize the Probe and DOS attacks. Subsequently, by adding the time-based attributes using windows of 100 connections time to the optimum basic attributes we acknowledge that the detection rate of Probe, DOS and R2L has significantly increased.

Finally, we should mention that the low detection rate of U2R is due to the fact that it is obligatory to carry out another type of attack before launching an U2R one.

References

1. Beghdad, R.: Critical study of neural networks in detecting intrusions. Comput. Secur. **27**(5), 168–175 (2008). doi:10.1016/j.cose.2008.06.001
2. Ingre, B.; Yadav, A.: Performance analysis of NSL-KDD dataset using ANN. In: 2015 International Conference on Signal Processing And Communication Engineering Systems (SPACES), pp. 92–96, 2–3 January 2015. doi:10.1109/SPACES.2015.7058223
3. KDD data set. http://kdd.ics.uci.edu/databases/kddcup99/kddcup99.html
4. de S Silva, L., dos Santos, A. C. F., Mancilha, T. D., da Silva, J. D. S., Montes, A.,: Detecting attack signatures in the real network traffic with ANNIDA. Expert Syst. Appl. **34**(4), 2326–2333 (2008). doi:10.1016/j.eswa.2007.03.011
5. Patcha, A., Park, J.M.: An overview of anomaly detection techniques: existing solutions and latest technological trends. Comput. Netw. **51**(12), 3448–3470 (2007). doi:10.1016/j.comnet.2007.02.001
6. Manikopoulos, C., Papavassiliou, S.: Network intrusion and fault detection: a statistical anomaly approach. IEEE Commun. Mag. **40**(10), 76–82 (2002). doi:10.1109/MCOM.2002.1039860. Oct
7. Wu, S.Y., Yen, E.: Data mining-based intrusion detectors. Expert Syst. Appl. **36**(3), 5605–5612 (2009). doi:10.1016/j.eswa.2008.06.138

8. Salvatore J.S., Lee, W., Chan, P.K., Fan, W., Eskin, E.: Data mining-based intrusion detectors: an overview of the columbia IDS project. SIGMOD Rec. 30, 4 (December 2001), 5–14. doi:10.1145/604264.604267
9. Haykin, Simon: Neural Networks: A Comprehensive Foundation, 2nd edn. Prentice Hall PTR, Upper Saddle River (1998)
10. Joo, D., Hong, T., Han, I.: The neural network models for IDS based on the asymmetric costs of false negative errors and false positive errors. Expert Syst. Appl. **25**(1), 69–75 (2003). doi:10.1016/S0957-4174(03)00007-1
11. Bajaj, K., Arora. A.: Improving the intrusion detection using discriminative machine learning approach and improve the time complexity by data mining feature selection methods. Int. J. Comput. Appl. (0975-8887) **76**(1), 5–11. doi:10.5120/13209-0587

Towards a Novel Privacy-Preserving Access Control Model Based on Blockchain Technology in IoT

Aafaf Ouaddah, Anas Abou Elkalam and Abdellah Ait Ouahman

Abstract Access control face big challenges in IoT. Unfortunately, it is hard to implement current access control standards on smart object due to its constrained nature while the introduction of powerful and trusted third party to handle access control logic could harm user privacy. In this work we show how blockchain, the promising technology behind Bitcoin, can be very attractive to face those arising challenges. We therefore propose FairAccess as a new decentralized pseudonymous and privacy preserving authorization management framework that leverages the consistency of blockchain technology to manage access control on behalf of constrained devices.

Keywords Internet of things · Security · Privacy · Access control · Blockchain · Bitcoin · Cryptocurrency

1 Introduction

We believe that the concept of a distributed IoT is a promising approach to release [1]. As devices increase their computational capacity, there are more opportunities to bring intelligence, mainly security and access control logic, on devices themselves. Actually, with this edge intelligence principle, users have more control over the granularity of the data they produce. However, as side effect of this approach, end-users are not expected to be experts to use security mechanisms. A simple mistake or a misconfiguration can lead to huge breaches in their privacy. For this reason, access control, mainly within the decentralized approach, have to be enough usable for ordinary people. Furthermore, the decentralized approach faces the following challenges: implementing current security standards and access control solutions on the device's side is more complicated. It requires intensive and

A. Ouaddah (✉) · A.A. Elkalam · A.A. Ouahman
OSCARS Laboratory, ENSA of Marrakesh, Cadi Ayyad University,
BP 575, 40000 Marrakesh, Morocco
e-mail: aafafouaddah@gmail.com

Á. Rocha et al. (eds.), *Europe and MENA Cooperation Advances in Information and Communication Technologies*, Advances in Intelligent Systems and Computing 520, DOI 10.1007/978-3-319-46568-5_53

computational capabilities which is not always available, especially in devices like sensors, actuators or RFID tags etc. While, relieving those devices from the burden of handling a vast amount of access control-related information by outsourcing these functionalities to a powerful entity prevents end-to-end security to be achieved. In addition, delegating the authorization logic to an external service requires a strong trust relationship between the delegated entity and the device. Moreover, all communications between them must be secured and mutually authenticated, so that the delegated entity security level is at least as high as if the authorization logic were implemented internally. Hence, we believe that IoT needs a new access control framework suitable to its distributed nature, where users may control their own privacy and, rather than being controlled by a centralized authority, and at the same time, the need arises for centralized entity handling authorization function to hardly constrained IoT devices. Then, the goal of this paper is to introduce FairAccess framework as a balance solution and equilibrium that solve the dilemma of centralized and decentralized access control management challenges highlighted above by leveraging the block chain technology.

Contribution: we introduce FairAccess as a novel Distributed Privacy Preserving Access Control framework in IoT scenario that combines, for the first time, access control models and cryptocurrency blockchain mechanisms. In FairAccess, we propose the use of SmartContract [2] to express fine-grained and contextual access control policies to make authorization decisions. We opt for authorization tokens as access control mechanism, delivered through emergent cryptocurrency solutions. We use blockchain firstly to ensure evaluating access policies in distributed environments where there is no central authority/administrator, and guarantee that policies will be properly enforced by all interacting entities and secondly to ensure token reuse detection.

Organization: The rest of this paper is structured as follows: in Sect. 2, we review related work and discuss the benefit of a decentralized peer-to- peer architecture. In Sect. 3, we show how blockchain can be used in distributed and transparent access control. We then introduce our proposed framework in Sect. 4 and finally Sect. 5 concludes the paper.

2 Related Work

In one hand, numerous efforts have emerged in adapting traditional access control model such as The Role Based Access Control (RBAC) model [3] that was extended to a new model named context based access control by the introduction of context which is provided by the web service. In this model the permission is assigned to the role according to the characteristics and contextual information collected from the environment of the physical object however its feasibility in constrained devices has not been demonstrated. The Capability-based access control model (CapBAC) was also chosen in [4] where it was directly implemented on resource-constrained devices, within a fully distributed security approach but the model introduced was coarse grained and not user-driven. Another generic Authorization Framework for the Internet-of-Things is proposed in [5]. It supports fine-grained and flexible access control for any objects with low power and memory resources. Based on current Internet standards and access control solutions such as XACML and Security Assertion Markup Language (SAML). But it introduces a Trusted Third Party as an authorization engine to handle access control logics.

In other hand, across the industry, many companies implement their own proprietary authorization software based on the OAuth protocol [6], in which they serve as centralized trusted authorities. For instance, EU project CALIPSO [7] adopts a centralized approach where the authorization logic is outsourced from the smart and constrained device to a more powerful server called IoT-OAS. However, it has demonstrate in [8] the impossibility to run all OAuth logic in a constrained device due to its heavy communication and processing overheads.

Unfortunately, those typical security and access control standards today are built around the notion of trust where a centralized trusted entity is always introduced. However, significant drawbacks arise when centralized approaches are considered on a real IoT deployment. On one hand, the inclusion of a central entity for each access request clearly compromises end-to-end security properties. On the other hand, the dynamic nature of IoT scenarios with a potential huge amount of devices complicates the trust management with the central entity, affecting scalability. In addition, they are built around a single logical server and multiple clients. As a consequence, access control is often done within the server side application, once the client has been authenticated. IoT reverses this paradigm by having many devices serving as servers and possibly many clients, taking part in the same application. More importantly, servers are significantly resource-constrained, which results in the minimization of the server side functionality. Subsequently, access control becomes a distributed problem.

We therefore turn our attention to blockchain, the technology behind Bitcoin protocol, to conceive our new FairAccess authorization framework as ultimate solution and equilibrium that solve all IoT authorization challenges previously highlighted above. Actually, the blockchain is the first technology that has successfully overcome the problem related to how consensus can be reached in distributed anonymous participants, some of whom may be behaving with malicious intent without the intervention of any centralized party. It is a universal digital ledger that functions at the heart of decentralized financial systems such as Bitcoin, and increasingly, many other decentralized systems such as Storj,[1] a decentralized peer-to-peer cloud storage network. Onename,[2] a distributed and secured identity platform. IBM's Adept, an Internet of things architecture [9], Enigma [10] to

[1]http://www.storj.io.

[2]http://www.onename.com/.

enhance user privacy and many others. However, to our knowledge, the use of a blockchain in access control filed has never been explored yet. In the next section the characteristics of the blockchain that will help provide a decentralized privacy preserving access control model are described.

3 How Blockchain Can Be Used in Distributed and Transparent Access Control

3.1 Background

Cryptocurrency and blockchain: Cryptocurrencies are a new form of virtual currency, first introduced with creation of Bitcoin, developed by Nakamoto [10]. A cryptocurrency is a decentralized digital currency built on cryptographic protocols providing an open, self-regulating alternative to classical currencies managed by central authorities such as banks. It is the first technology to successfully overcome the problem related to how consensus can be reached in a group of anonymous participants, some of whom may be behaving with malicious intent without the intervention of any centralized party. Actually, specific nodes known as miners are responsible for collecting transactions, solving challenging computational puzzles (proof-of-work) to reach consensus, and adding the transactions in form of blocks to a distributed public ledger known as the blockchain.

The blockchain: The blockchain technology provides everyone with a working proof of a decentralized trust. All cryptocurrencies utilize what can best be described as a public ledger that is impossible to corrupt. Every user or node has the exact same ledger as all of the other users or nodes in the network. This ensures a complete consensus from all users or nodes in the corresponding currencies blockchain.

Transactions: A transaction records the transfer of a value (altcoin) from some input address to output addresses. Transactions are generated by the sender and distributed amongst the peers in the network. Transactions are only valid once they have been accepted into the public history of transactions, the blockchain. Actually, the fundamental building block of a cryptocurrency transaction is an unspent transaction output, or UTXO. UTXO are a value of the currency locked to a specific owner, recorded on the blockchain, and recognized as currency units by the entire network. The UTXO consumed by a transaction are called transaction inputs, and the UTXO created by a transaction are called transaction outputs. The recipient is identified through their public key, so cryptocurrency transactions can be traced throughout the blockchain, to the beginning of the creation of the cryptocurrency. This forms the mechanism for checking the ownership of cryptocurrency bitcoins. Publicly verifiable transactions by any node avoids double spending and provides a high degree of certainty to the participants of the cryptocurrency ecosystem.

Scripting language and smart contract: Each UTXO has to specify a person (or several persons) eligible for spending virtual money associated with it. To accomplish this, the Bitcoin protocol introduces a scripting language. The language describes the execution of a certain program on a stack machine. Each transaction output contains a script which locks, or encumbers, the money associated with the *UTXO*. This script is commonly referred to as **scriptPubKey**. To spend this money, a user of the Bitcoin network must demonstrate the proof of ownership in the form of an unlocking script **scriptSig**. Hence, a script is a part of each input and each output of a transaction. When we generalize this scripting language computation to arbitrary Turing complete logic, we obtain an expressive smart contract system. The interest in smart contract applications steadily risen since 2014 due to the appearance of Bitcoin-like technologies, such as Etherum [2] and many other works designed specifically to decentralized smart contract system.

3.2 Blockchain in FairAccess

Most cryptocurrencies solutions are designed with a currency in mind. In our FairAccess framework, we define an Authorization Token rather than a bitcoin. This token is simply a digital signature that represents the access right or the entitlement defined by the creator of the transaction to its receiver in order to access a specific resource designed by its address. Blockchain specifications vary from cryptocurrency to cryptocurrency to meet the purpose of specific applications. Our FairAccess Framework uses a custom blockchain transaction specification that includes additional fields tailored to the requirements of a granular access control model. FairAccess provides several useful mechanisms using the blockchain. In fact, in FairAccess, the blockchain is considered as a database that stores all access control policies for each pair (resource, requester) in form of transactions, it serves also as logging databases that ensures auditing functions. Furthermore, it prevents forgery of token through transactions integrity checks and detects token reuse through the double spending detection mechanism.

4 FairAccess: A Token-Based Access Control Model Enforced by the Blockchain Technology

4.1 Technical Description

Preliminaries: We will denote key pairs using the capital letters (e.g. A), and refer to the private key and the public key of A by: A.Sk and A.pk, respectively then: $A = (A.sk, A.pk)$. In addition, we will use the following convention: if A = (A.sk, A.pk) then $\text{sig}_{\text{A}}(m)$ denote a signature on a message m computed with A.Sk and let

Table 1 FairAccess main interacting entities

Acronym	Its meaning
IDx	The index of the current transaction Tx where x = H(Tx), H is a hash function
rs	The address of requested resource
rq	The address of the requester who is the receiver of the current transaction
πx	Locking script (access control policies written in scripting language)
$TKN_{rq,rs}$	Encrypted access token associated to couple (rs, rq)
ref	Point to the previous transaction output

$\text{check}_A(m, \sigma)$ denote the result (true or false) of the verification of the signature σ on the message *m* with respect to the public key A.pk and finally we note H as a hash instantiated by a SHA-256 implementation.

The acronyms we use to describe the functionalities of our proposed framework and their meanings are summarized in Table 1.

Our authorization Framework is based on the following main authorization functionalities: (1) Registering a new resource with a corresponding address. (2) Grant access. (3) Request access. (4) Delegate access. (5) Revoke access. We describe each function in this paragraph.

FairAccess work flow: Typical FairAccess based access control works as follows. A subject (e.g., a device A, identified with the address rq) wants to perform an action (e.g., modify) on a protected resource (e.g., Device B temperature, identified with address rs). The subject submits this request to the authorization management point (AMP = wallet) acting as a policy Enforcement Point (PEP) that manages the protected resource. The PEP formulates such a request to a GetAcess transaction. Then, the PEP broadcasts this transaction to the network nodes till it reach miners, those later act as distributed Policy Decision Point, and evaluate the transaction. The PDP checks the request with the defined policy, by comparing the unlocking script of this transaction to the locking script of the previous GrantAccess transaction. Then determines whether the request should be permitted or denied. Finally, if it is permitted the transaction is valid and it will be recorded in the blockchain, else the transaction will be rejected and a notification will be sent to its sender.

Phase 1: Reload access control policy to the blockchain trough: Grant access transaction

Before requesting access to device B, the Smart device A needs to obtain an access token. For this purpose, it sends a request to device B owner (RO) indicating the address of the target resource rs and the action to be performed on. Then, the device B owner defines his access control policy and reloads it to the blockchain through a GrantAccess transaction. This later is created by the RO wallet. The GrantAccess transaction encapsulates the defined access control policy in form of locking script in its output, the address rq of device A as receiver and the access token signed as UTXO. Then the wallet broadcasts the GrantAccess transaction to the peer to peer nodes. The peer to peer nodes verify the transaction and record it in the block chain in case of success validation. At this stage, the network witnesses that the device B owner has entitled the device A to get access to the specific

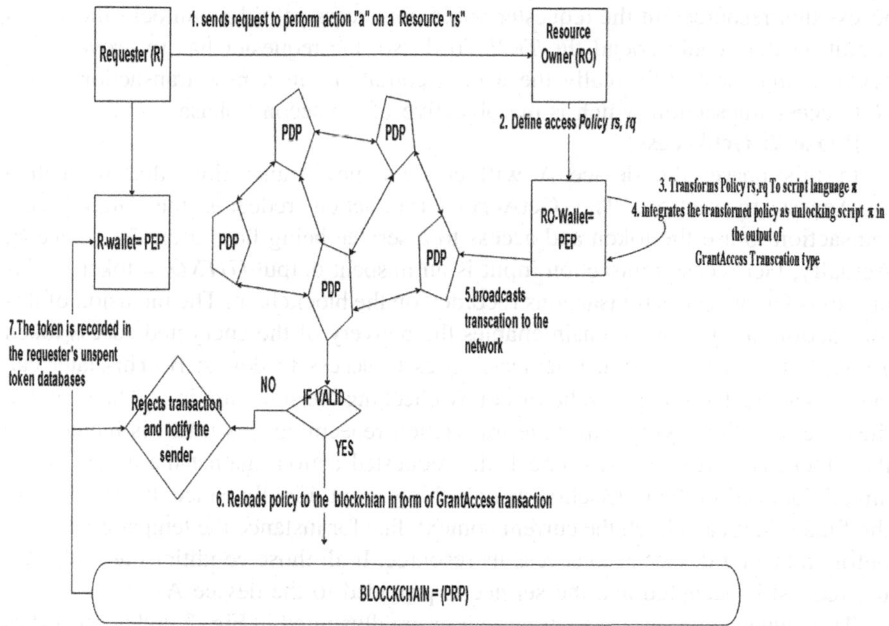

Fig. 1 GrantAccess transaction process

service provided by device B. But device A could not get yet access to the target service till he meets the access control policy and unlock the output of the transaction which is the token.

The sequence of GrantAccess transactions are illustrated in Fig. 1 and described as follows:

1. The RO defines for the couple (Resource *rs*, Requester *rq*) an access control policy $POLICY_{rs,\,rq}$
2. The wallet transforms this access control policy to a scripting language $POLICY_{rs,\,rq} \rightarrow \pi_x$
3. The RO, generates a Token $TKN_{rs,\,rq}$ encrypted with the requester public key.
4. The wallet generates a **GrantAccess** Transaction in the following form:

$$T_x = (m, sig_{rs}(m)) \ where \ m = (ID_x, input(rs), \ output(rq, \pi_x, TKN_{rs,\,rq})$$

5. Each node verifies the transaction within the transaction validation process.
6. If the transaction is valid the unspent transaction output:$TKN_{rs,\,rq}$ is recorded in the blockckain and shown in the requester's wallet as part of the available $TKN_{rs,\,rq}$. Else, the transaction will be rejected.

At the end of this phase, if the transaction appears in the blockchain, it means that a new *TKN* is added to the requester *available unspent TKN* database. Meaning that the network witnesses that the Resource owner had entitled the requester to access that resource but the requester could not access yet till he unlocks the access condition then could spend the *TKN* To do so, the requester has to prove to the network that he fulfills really the access conditions in a new transaction called **GetAccess** transaction which is the objective of the second phase.

Phase 2: GetAccess

In this phase, the device A will create a new transaction, that we call a GetAccess transaction. The GetAccess transaction redeems the GrantAccess transaction to use the token and access to a service being hosted on the device B. Actually, GetAccess transaction input is an unspent output ($UTXO =$ token) of its previous GrantAccess transactions recorded on the blockchain. The inclusion of this transaction into the block chain enables the delivery of the encrypted access token $TKN_{rs,rq}$ to device A. When device A tries to access to device B. This later can check whether the token is valid or not by checking the signature in one hand and in the other hand checking either the transaction redeeming this token is included in the blockchain. It could also check the requested action against the access rights already defined in the transaction as unlocking script. Finally, since the device have the final say, it can check the current context, like for instance the temperature level, before allowing device A to access its resource. If all those conditions are fulfilled, the request is accepted and the service is provided to the device A.

The sequence of GetAccess transactions are illustrated in Fig. 2 and described as follows:

1. The requester will, first, scan his available TKN database $ScanTKN(rq) \rightarrow TKN_{rs,rq}$
2. The wallet decrypts the token $decrypt(TKN_{rs,rq})$
3. The wallet gets the locking script $GetLockingscript(TKN) \rightarrow \pi_x^{'}$
   Where: $\pi_x^{'}$ is the locking script in the corresponding GrantAcces transaction.
4. The requester fulfills access control condition placed in $\pi_x^{'}$ and generate an unlocking script ψ

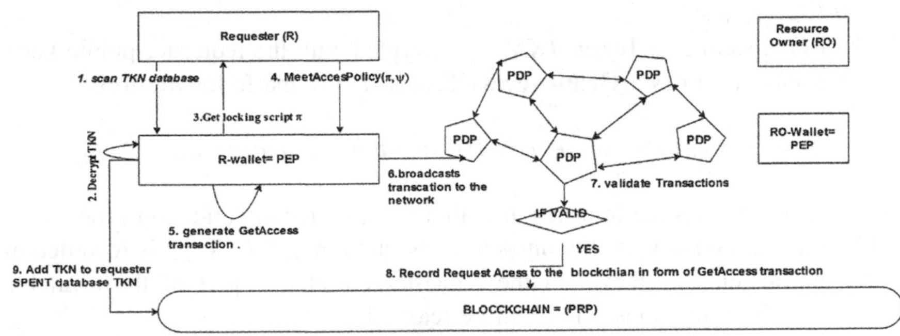

Fig. 2 GetAccess transaction process

$$MeetAccessControlPolicy(\pi_x^{'}) \rightarrow \psi$$

5. The wallet generates a GetAccess transaction type in the following form:

$$T_x = (ID_x, input(ref, rs, \psi), output(rq, TKN_{rq,rs})$$

6. The wallet broadcasts the transaction to the network
7. The network nodes verify and validate the transaction if it was valid it will be included in the blockchain else it will be rejected and a notification is sent to its sender.
8. Once the transaction appears in the blockchain. It means the network witnesses that the requester has full filled the access condition (unlocking script) then the Token is now valid and could be spent.
9. The requester device sends the token to the target device
10. The target device checks the validity of the token by checking the inclusion of GetAcess transaction in the blockchain. If it was valid, the access is allowed else the access is denied.

Delegate access through *DelegateAccess Transaction type*:
Consider Device A again as example. This later can delegate access rights or part of his granted rights over the service *rs* provided by Device B to another Device C identified with *C.pk* to access resource *rs* through this transaction.

1. Device A owner's wallet generate the following transaction:

$$T_x = (m, sig_A(m)) \ where \ m = (ID_x, input(ref, rs, \psi), \ output(C.pk, \pi_x, TKN_{C.pk,rs})$$

2. The wallet broadcasts the transaction
3. The network nodes validate transaction
4. If the transaction is valid, the unspent transaction output: $TKN_{C.pk,rs}$ is recorded in the blockchain and showed in the device C owner's wallet as part of the available $TKN_{C.pk,rs}$
5. When device C wants to access to that resource *rs*, it creates a GetAccess transaction that releases the encumbrance, unlocking the output by providing an unlocking script meeting the access conditions.

Revoke/Update access through a new GrantAccess Transaction type:
The Resource Owner could revoke or update the permissions granted to a requester at any time by simply issuing a GrantAccess transaction with a new set of permissions, including revoking access to previously designed resource. This new transaction will override all rights granted by all previous transactions. Since transaction are recorded in a chronological way in the blockchain.

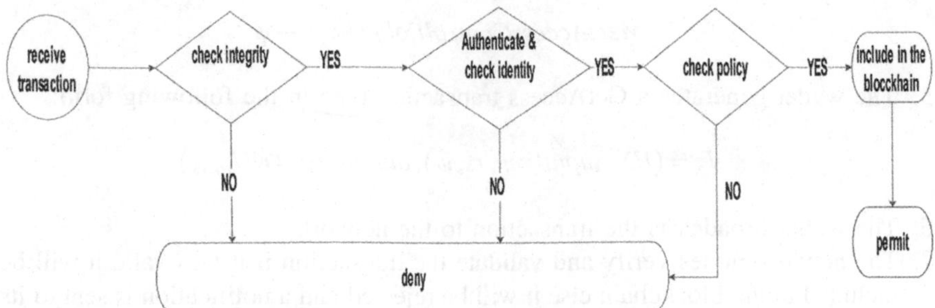

Fig. 3 Authorization evaluation process

Evaluation the access control policy by validation transaction:
Each peer to peer node receives a signed transaction in the following form:

$$(T_x, sig_A(T_x))\ where\ T_x = (ID_x,\ input1(ref, A.pk, \psi),\ output(B.pk, \pi_x, TKN_{A.pk,\,B.pk})$$

To validate the transaction and evaluate the access control policy, the node executes the following functions as illustrated in Fig. 3:

1. **CheckIdentity**: checking the signature of the owner prove the following properties: (1) Authenticate the owner. (2) Prove his ownership to the resource (3) prove his non repudiation. This function is ensured by executing this procedure: $Check_A(T_x, \sigma) = True.$
2. **CheckIntegriy** (T_x): Hash the transaction and compare to its ID_x to ensure that the transaction has not been altered during its propagation in the network. This function is ensured by executing this procedure: Compare $(\mathcal{H}(T_x).ID_x) = True$
3. **CheckPolicy**: check if the sender fulfill access control policy which is simply the unlocking scripts for each input must validate against the corresponding output locking scripts. To do so, for each input in the transaction, the validation function CheckPolicy will first retrieves the output of the previous transaction referenced by the input of the current transaction. This output contains a locking script defining the conditions required to get the TKN. The validation function will then take the unlocking script contained in the input that is attempting to get this TKN and execute the two scripts. We note that the $\pi_x^{'}$ is permanently recorded in the blockchain, and therefore is invariable and is unaffected by failed attempts to spend it by reference in a new transaction. This function is ensured by executing these two procedures:

    a. GetOutputFromRef(ref) $\rightarrow \pi_x^{'}$
    b. Compare$(\psi, \pi_x^{'}) = $True

If any result other than "True" remains after execution of described function, then transaction is considered as invalid. The access is denied and a notification is sent to its sender.

Hence we have shown that access control process can be easily achieved using the blockchain, and through the same mechanism, additional integrity and security protections can be added. Furthermore, blockchain scripts allows for intelligent programming of actions within a transaction. This enables FairAccess to implement more granular access control policies, expressed by any access control model as soon as this model could be transcoded to a script language.

5 Conclusion

In this paper, we have inaugurated a new applicability domain of blockchain that is access control through our FairAccess framework. Our framework leverages the consistency offered by blockchain-based cryptocurrencies to solve the problem of centralized and decentralized access control in IoT highlighted in the beginning of this paper. In FairAccess, we provide a stronger and transparent access control tool. We have explained, the main building blocks and functionalities of our framework. In future work, we will implement FairAccess with RaspberryPI IoT device and bitcoin blockchain as example.

References

1. Vermesan, P., Friess, P., Guillemin, S., Gusmeroli, H., Sundmaeker, A., Bassi, I.S., Jubert, M., Mazura, M., Harrison, M.D.: Internet of things strategic research roadmap. In: Cluster of European Research Projects on the Internet of Things, CERP-IoT (2011)
2. SZABO, Nick: Formalizing and securing relationships on public networks. First Monday, **2** (9) (1997)
3. Zhang, G., Tian, J.: An extended role based access control model for the Internet of Things. In: 2010 International Conference on Information Networking and Automation (ICINA), pp. V1-319–V1-323. IEEE, (2010)
4. Hernández-Ramos, J.L., Jara, A.J., Leandro, M., et al.: Dcapbac: embedding authorization logic into smart things through ecc optimizations. Int. J. Comput. Math. no ahead-of-print, 1–22 (2014)
5. Seitz, L., Selander, G., Gehrmann, C.: Authorization framework for the internet-of-things. In: 2013 IEEE 14th International Symposium and Workshops on a World of Wireless, Mobile and Multimedia Networks (WoWMoM), pp. 1–6. IEEE (2013)
6. Hardt, D. (ed.): The OAuth 2.0 authorization framework. In: IETF, RFC6749, October 2012
7. Connect All IP-Based Smart Objects (CALIPSO)—FP7 EU Project [Online]. http://www.ict-calipso.eu/. Accessed 15 Oct 2014
8. Cirani, S., Picone, M., Gonizzi, P., Veltri, L., Ferrari, G.: Iot-oas: an OAuth-based authorization service architecture for secure services in IoT scenarios. IEEE Sens. J. **15**(2), 1224–1234 (2015)
9. Sanjay, P., Sumabala, N., Paul, B., Pureswaran, V.: ADEPT: an IoT practitioner perspective, Draft copy for advance review. IBM (2015)
10. Zyskind, G., Nathan, O.: Decentralizing privacy: using blockchain to protect personal data. In: Security and Privacy Workshops (SPW), 2015 IEEE, pp. 180–184. IEEE (2015). Satoshi Nakamoto. Bitcoin: A peer-to-peer electronic cash system (2008)

A Fair Comparison Between Several Ciphers in Characteristics, Safety and Speed Test

Youssef Harmouch and Rachid El Kouch

Abstract As networks applications are growing fast, data are more generated and transmitted, leaving it vulnerable to modification. Besides that, the significance, sensitivity and preciseness of that information cause a big security issue, increasing the needs to keep it safe. One proper solution for this problem is cryptography, which is a technique used to transform data to unrecognizable information and useless to any unauthorized person. By being the main core of network security and since new attacks techniques are daily invented, ciphers are constantly under test by cryptanalysis attacks to harden their safety. This paper contains a fair comparison between serval ciphers in speed test under different platforms, also in performance and risk to globalize the vision about the most fast safe algorithm.

Keywords Cipher · Speed-test · Cipher-time · Performance and structure · Cryptanalysis attacks

1 Introduction

Because information privacy has become a major concern for all—users and companies, the cryptography is considered as a standard for providing information trust, security, electronic financial transactions, controlling access to resources, prevents eavesdroppers from obtaining critical or private information.

Cryptosystem is the name of a system providing cryptography. A Cryptosystem uses an encryption/decryption algorithm that determines how simplicity or complexity this process will be. In this process, key is information used under conversion of plaintext to cipher text or vice versa. The big key space used the more

Y. Harmouch (✉) · R. El Kouch
Department of Mathematics Computing and Networks,
National Institute for Post and Telecommunication, Rabat, Morocco
e-mail: harmouch@inpt.ac.ma

R. El Kouch
e-mail: elkouch@inpt.ac.ma

Á. Rocha et al. (eds.), *Europe and MENA Cooperation Advances in Information and Communication Technologies*, Advances in Intelligent Systems and Computing 520, DOI 10.1007/978-3-319-46568-5_54

possible keys can be created. The strength of the encryption algorithm depends on the length of the key, secrecy of the key, the complexity of the process and how they all work together [1, 2].

However, executing encryption algorithms consumes both time and energy. While using strong algorithms may result in severely reduced lifetimes for the battery-powered devices, certain encryption algorithms may be less vulnerable to compromise than others. In other words, utilizing stronger encryption algorithms may consume more energy and drain the device battery faster than using less secure algorithms. As result of processing requirements and the limited computing power, using strong algorithms might also increase the delay of data transmissions. Thus, users, software and system designers need to be conscious of the benefits and costs of using various encryption algorithms [3].

This research clarifies the answer regarding both—safety and execution time for various encryption algorithms to help their comparison by experimental measurements. The studies cipher algorithms are AES, Blowfish, Camellia, CAST128/256, DES/3DES, GOST, IDEA, MARS, RC2/5/6, Serpent, SHACAL2, SHARK, SKIPJACK, Three-way, Two-fish, and finally XTEA.

The rest of the paper is organized as follows. The Sect. 2 gives a global definition about the ciphers using into this research with a cipher characteristic comparison, while Sect. 3 focuses on related work. As for Sect. 4, it proposes a security comparison by presenting the best-known cryptanalysis attacks and theirs success rate. Finally, the Sect. 5 studies the cipher-time, which is the encryption and decryption time that cipher uses during the process, it gives a few simulations and result discusses while conclusion is left after.

2 Main Cipher Algorithms

Before starting evaluation, a short definition of the most common encryption algorithms must be mention. As so:

AES (called Rijndael) is a block cipher. It has variable key length of 128, 192, or 256 bits. This key helps to encrypt data blocks of 128 bits in 10, 12 and 14 round relying on the key size. AES seems faster and flexible and has been tested for many security applications [4–6]. AES is a substitution permutation network (SPN).

Blowfish is a block cipher uses a variable-length key, starting from 32 bits to 448 bits, and it has nearly variants of 14 rounds. Blowfish is available free for all clients [5].

Camellia cipher has key lengths of 128, 192, or 256 hits, using to processes data blocks of 128-bits. Camellia algorithm has the same interface as the AES algorithm [6].

CAST-128 is a Feistel cipher and is similar in structure to Data Encryption Standard (DES) [9] algorithm, but differs from it in term of round key and S-box size. DES's round key is a permutation of secret key while CAST-128 round key is generated by using multiple layers of non-linear element (S-box) [7].

CAST-256 is "Feistel" network (**FN**) private-key block cipher, it uses a 128-bit block size and a 256-bit primary key that is used in the algorithm's key schedule scheme to generate two sets of sub-keys, each of which is used per round. There are 48 rounds of encryption [8].

DES (Data Encryption Standard) was the first encryption standard to be used by National Institute of Standards and Technology. DES uses 64 bits as key size and 64 bits in block size [9].

3DES is an improvement of DES, it has 64-bit block size with 192 bits key size. The encryption method is similar to the one in the original DES but applied three times to increase the encryption level and the average safe time [9].

GOST is a for symmetric cipher operating on 64-bit blocks using 256-bit key, it is a classic Feistel net with 32 rounds, developed in 1989 as symmetric block ciphers standard in Russian Federation [10].

IDEA is a block cipher that consists of a cascade of eight identical blocks known as rounds, followed by a half-round or output transformation. In each round, XOR, addition and modular multiplication operations are applied [11].

MARS is a 128 bit-block cipher with a variable key size ranging from 128 to 444 bits, it is a symmetric-key block ciphers. MARS is based on a "Feistel" network structure and a "round function". The strength of the algorithm is directly related to the degree of diffusion and non-linearity properties of the function [12].

RC2 can be seen as a replacement for DES; it is a 64-bit block code and can have a key size from 40 bits to 128-bits (in increments of 8 bits) [13].

RC5 is a fast block cipher that has a parameterized algorithm with a variable block size (32, 64 or 128 bits), a variable key size (0–2048 bits) and a variable number of rounds (0–255). It has a heavy use of data dependent rotations, and the mixture of different operations as data-dependent rotations, modular addition and XOR operation, which assures that RC5 is secure [13].

RC6 is a cipher operates with block size of 128 bits and supports key sizes of 128, 192 and 256 bits. RC6 is very similar to RC5 in structure and defers only in use of an extra multiplication operation in order to make the rotation dependent on every bit in a word, and just not the least significant few bits. RC6 can be viewed as interweaving two parallel RC5 encryption processes [14].

Serpent is a block size cipher of 128 bits and 32-round substitution permutation network (SPN) operating on four 32-bit words, it can encrypts a 128-bit plaintext to a 128-bit cipher-text in 32 rounds with 33 sub-keys. The cipher consists first in initial permutation IP and then 32 rounds, each round involves a key mixing operation, a pass through S-boxes, and a linear transformation, and the last round, the linear transformation is replaced by an additional key mixing operation. In the end, a final permutation FP is applied [15].

SHACAL-2 is a 256-bit block cipher based on the compression function of SHA-256 [47]. It inputs the plaintext to the compression function as the chaining variable, and inputs the key to the compression function as the message block. First, a 256-bit plaintext is divided into eight 32-bit words. Then, a state update function updates these eight 32-bit variables in 64 steps [16].

Table 1 Cipher characteristic comparison

Factors / ciphers	Cipher type	Key length (bits)	Possible key	Block size (bytes)	Structure	Rounds	Secure services
AES	Symmetric block cipher	128, 192, 256	2^{128}, 2^{192}, 2^{265}	16	SPN	10, 12, 14	Confidentiality
Blowfish	Symmetric block cipher	32–447	2^{32} to 2^{447}	8	FN	16	Confidentiality
Camellia	Symmetric algorithm	128, 192, 256	2^{128}, 2^{192}, 2^{265}	16	FN	18, 24	Confidentiality
CAST128	Symmetric block cipher	40–128	2^{40} to 2^{128}	8	FN	12, 16	Confidentiality
CAST256	Symmetric block cipher	138–256	2^{136} to 2^{256}	16	FN	48	Confidentiality
DES	Symmetric block cipher	64	2^{64}	64	FN	16	Confidentially
3DES	Symmetric block cipher	192	2^{192}	64	FN	48	Confidentiality
GOST	Symmetric block cipher	256	2^{256}	8	FN	32	Confidentiality
IDEA	Symmetric block cipher	128	2^{128}	8	Lai-Massey scheme	8.5	confidentiality
MARS	Symmetric block cipher	128–448	2^{128} to 2^{448}	16	FN	32	Confidentiality
RC2	Symmetric block cipher	0–1024	1 to 2^{1024}	8	FN	18	Confidentiality
RC5	Symmetric block cipher	0–2048	1 to 2^{2048}	4, 8, 16	FN	0 to 255	Confidentiality
RC6	Symmetric block cipher	128, 192, 256	2^{128}, 2^{192}, 2^{256}	16	FN	20	Confidentiality
Serpent	Symmetric block cipher	0–256	1 to 2^{256}	16	SPN	32	Confidentiality
SHACAL2	Symmetric block cipher	128–512	2^{128} to 2^{512}	32	Hash function	80	Confidentiality
SHARK	Symmetric block cipher	128	2^{128}	8	SPN	6	Confidentiality
SKIPJACK	Symmetric block cipher	80	2^{80}	8	Unbalanced FN	32	Confidentiality
Three-way	Symmetric block cipher	96	2^{96}	12	SPN	11	Confidentiality
Twofish	Symmetric block cipher	128, 192, 256	2^{128}, 2^{192}, 2^{256}	16	FN	16	Confidentiality
XTEA	Symmetric block cipher	128	2^{128}	8	FN	64	Confidentiality

SHARK cipher combines highly nonlinear substitution boxes and maximum distance separable error correcting codes (MDS-codes) to guarantee a good diffusion. It is resistant against differential and linear cryptanalysis after a small number of rounds [17].

Skipjack is a block cipher with 64-bit block size and an 80-bit key size designed by the US National Security Agency (NSA). After designing Skipjack, it was classified to be used in Clipper chips and tamper-resistant Capstone for US government purposes, e.g., voice, mobile and wireless communications [18].

Three-way is an iterated block with 96 bits block uses and 96 bits in key length. It repeats some relatively simple operations a specified number of rounds [19].

Twofish is a symmetric key block cipher with 128 bits block size and 128, 192, 256 bits as key size, it uses precomputed key-dependent S-boxes build it by building four 8 × 8 bit key using permutation to keep the attacker unknowing what actually the S-boxes are, and a complex key schedule. Twofish key is divided into two halves, one-half is use as actual key and the other half is used to modify the algorithm. It is fast, flexible and conservative design and it is derive from blowfish [9].

TEA is a Feistel type cipher that uses operations from mixed (orthogonal) algebraic groups. It is designed to minimize memory footprint and maximize speed. TEA was extended to XTEA (Extended TEA) and like TEA; XTEA makes use of arithmetic and logic operations. XTEA proposed to fix the two minor weaknesses pointed, the first enhancement is to adjust the key schedule and the second is to introduce the key material slowly [20].

By offering each cipher description, this section presents the first look before any test or comparison between ciphers. In main fact, it gives a global idea about the study ciphers characteristics, which can lead in addition with different tests to deduce the safety level required for system by users. The following Table 1 presents a characteristic comparison between cipher already mention:

3 Related Work

To give forthcoming about the performance of the compared algorithms, this section discusses the results obtained from other resources.

The author in [21] presents a speed test and comparison between different open-source cryptography libraries and compiler flags at two different test programs. In the first, the author verifies each cipher implementations against each other, as for the second one, the author accomplishes a speed tests based on time for both a buffer encryption and decryption (The buffer length is varied from 16 bytes to 1 MB in size) different ciphers exported from different libraries.

As for this paper [22], it provides an evaluation of five encryption algorithms: AES, DES, 3DES, Blowfish, and author proposed algorithm. This evaluation examines a method for analyzing trade-off between efficiency and security for WIFI uses.

In addition, this author [23] evaluates in his paper six different encryption algorithms: AES, DES, 3DES, RC2, Blowfish, and RC6 to make obvious the effectiveness of each algorithm. This evaluation is based on different settings for each algorithm such as different data types, different sizes of data blocks, different key size, battery power consumption and finally encryption and decryption speed. Simulation results are given.

As for this paper [24], the author present an experimental results of comparing block and stream ciphers for securing VoIP in terms of end-to-end delay and subjective quality of perceived voice. He also, proposed a new technique that provides automatic synchronization of stream ciphers on a per packet basis, without the overhead of an initialization vector in packet headers or without maintaining any state of past-encrypted data. He shows that this technique mitigates the trade-off between subjective quality and confidentiality.

Those presented papers process a few discipline to cipher comparison, either time execution or power consumption or characteristic and resistance to attacks, while it should not been split. This article study cipher comparison in many approach to generalize a global vision to users in order to simplify the image to make decision easier and successful.

4 Cipher Weakness to Cryptanalysis Attacks

The main goal for ciphers developing is to secure communication, as so many researchers are daily verifying their strangeness based on cryptanalysis study [25, 26]. This later expresses a test for both solidness and weakness for the cipher through various theoretical or developed attacks. Therefore, in this section, we illustrate for each cipher in the following Table 2, the best successful knowing attack in order to allow a risk comparison between them. This comparison is based on presenting the danger level for using each cipher, because successful attack rate and safety are linearly correlated. In main fact, this paragraph present the main interest information for every user who look for safety comparison to deduce the safety level required for their own system.

5 Speed Test and Simulation

5.1 Experimental Environment

The experimental environment for development and test used to compare ciphers performances speed was C++ code application, with as tools:

- Visual studio C++ expresses 2010 for Windows 7 desktop.
- GCC 4.8.2 for Centos 7 machine.

Table 2 The best-known cryptanalysis attack on multiple ciphers with **BRV** as Attack only breaks a reduced version of the cipher, **TB** as Theoretical break—attack breaks all rounds and has lower complexity than security claim and **ADP** as Attack demonstrated in practice

	Cipher	Security claim	Best attack	Publish date	Attack name
TB	AES128	2^{128}	$2^{126.1}$ time, 2^{88} data, 2^{8} memory	2011-08-17	Biclique attack [27]
TB	AES192	2^{192}	$2^{189.7}$ time, 2^{80} data, 2^{8} memory		
TB	AES256	2^{256}	$2^{254.4}$ time, 2^{40} data, 2^{8} memory		
BRV	Blowfish	2^{448}	4 of 16 rounds	1997	Differential cryptanalysis [28]
BRV	Camellia128	2^{128}	11 of 18 rounds ($2^{123.6}$ time, $2^{120.5}$ data)	2011	Impossible differential attacks [39]
BRV	Camellia 192	2^{192}	12 of 24 rounds ($2^{171.4}$ time, $2^{120.6}$ data)		
BRV	Camellia 256	2^{256}	14 of 24 rounds ($2^{238.2}$ time, $2^{121.2}$ data)		
TB	CAST (not CAST-128)	2^{64}	2^{48} time, 2^{17} chosen plaintexts	1997-11-11	Related-key attack [34]
BRV	CAST-128	2^{128}	6 of 16 rounds ($2^{88.51}$ time, $2^{53.96}$ data)	2009-08-23	Known-plaintext linear cryptanalysis [35]
BRV	CAST-256	2^{256}	24 of 48 rounds ($2^{156.2}$ time, $2^{124.1}$ data)		
ADP	DES	2^{56}	2^{39}–2^{43} time, 2^{43} known plaintexts	2001	Linear cryptanalysis [29]
BRV	3DES	2^{168}	2^{113} time, 2^{32} data, 2^{88} memory	1998-03-23	Meet-in-the-middle attack [30]
TB	GOST	2^{256}	2^{101} time, 2^{64} data, 2^{47} memory	2012	Guess-then-determine attacks [40]
TB	IDEA	2^{128}	$2^{126.1}$ time	2012-04-15	Narrow-biclique attack [36]
BRV	MARS	2^{400}	21 of 32 rounds (2^{50} plaintext, 2^{197} memory)	2004	Meet-in-the-middle attack [30]
TB	RC2	2^{64}–2^{128}	Unknown [clarification needed] time, 2^{34} chosen plaintexts	1997-11-11	Related-key attack [34]
ADP	RC4	Up to 2^{2048}	2^{20} time, $2^{16.4}$ related keys (95 % success probability)	2007	PTW attack [31]

(continued)

Table 2 (continued)

	Cipher	Security claim	Best attack	Publish date	Attack name
BRV	RC5	2^{128}	12 of 12-255 rounds with 64 bit-block (2^{44} plaintexts)	1998	Differential attack [28]
BRV	RC6	2^{128}	12 of 20 rounds (2^{31} data, 2^{52} memory)	2007	Chi-square Test Attack [41]
BRV	RC6	2^{192}	16 of 20 rounds (2^{31} data, 2^{52} memory)		
BRV	RC6	2^{256}	16 of 20 rounds (2^{31} data, 2^{52} memory)		
BRV	Serpent-128	2^{128}	10 of 32 rounds (2^{89} time, 2^{118} data)	2002-02-04	Linear cryptanalysis [32]
BRV	Serpent-192	2^{192}	11 of 32 rounds (2^{187} time, 2^{118} data)		
BRV	Serpent-256	2^{256}			
BRV	SHACAL-1	2^{512}	59 of 80 rounds ($2^{489.3}$ time, $2^{149.72}$ plaintext)	2006	Related-key rectangle attack [42, 43]
BRV	SHACAL-2	2^{512}	42 of 64 rounds ($2^{488.37}$ time, $2^{243.38}$ plaintext)	2008	
BRV	Shark	2^{128}	5 of 6 rounds (2^{83} time, 2^{52} plaintext)	1997	The Interpolation Attack [44]
TB	Skipjack	2^{80}	31 of 32 rounds (2^{41} plaintexts)	1999	Impossible differential attacks [39] ECRYPT II recommendations note that, as of 2012, 80-bit ciphers provide only "Very short-term protection against agencies". [37] NIST recommends not to use Skipjack after 2010 [38]
TB	Three-way	2^{96}	(2^{22} chosen-plaintexts)	1997	Related-key attack [34]
BRV	Twofish	2^{128}–2^{256}	6 of 16 rounds (2^{256} time)	1999-10-05	Impossible differential attack [33]
TB	TEA	2^{128}	2^{32} time, 2^{23} chosen plaintexts	1997-11-11	Related-key attack [34]
BRV	XTEA	2^{128}	36 of 64 rounds ($2^{126.44}$ time, $2^{64.98}$ data)	2009	Related-key rectangle attack [45]

The ciphers used in this study are token from an open source library called Crypto++ [46], the program was developed and tested under two machines with different OS (Windows 7 desktop ×32 and Centos 7) and the same configurations: Intel i3 processor with 4 GB in RAM storage.

5.2 Experimental Results

The speed test was measuring through computing the encryption and decryption time that cipher took for different packet size, i.e. the size of packet data belong to

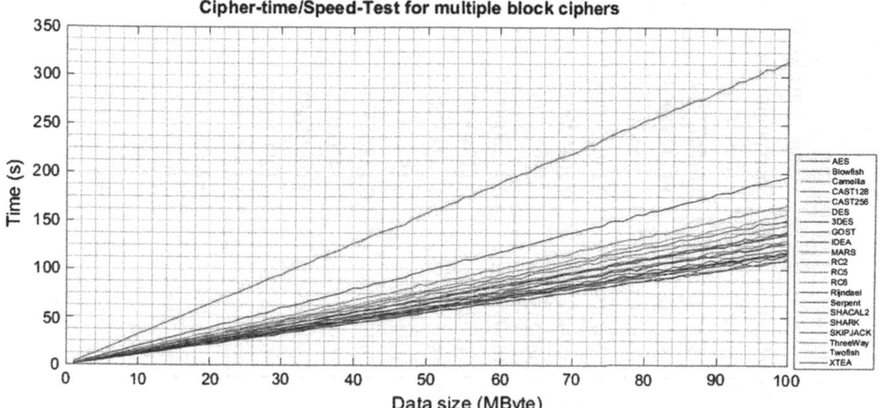

Fig. 1 Windows 7 cipher-time computing

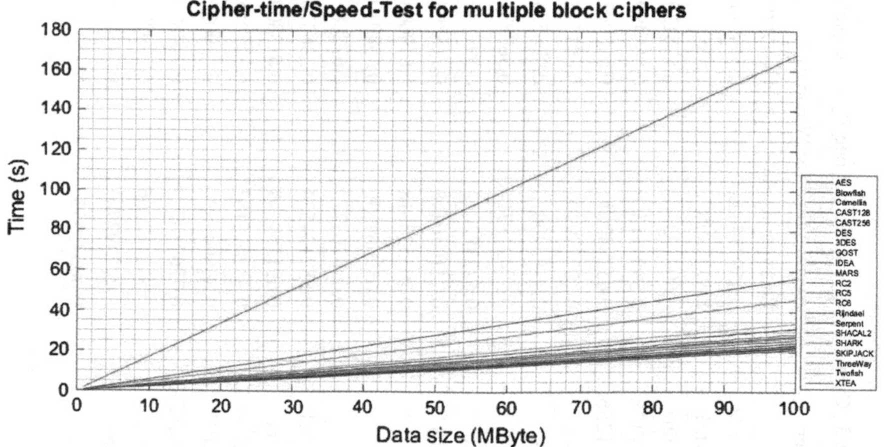

Fig. 2 Centos 7 cipher-time computing

Table 3 Cipher average speed per one Byte comparison

Cipher	Average time for Windows 7 × 32			Average time for Centos 7		
	Encryption time	Decryption time	Cipher-time	Encryption time	Decryption time	Cipher-time
AES 128	0.559775423	0.534911812	1.09468724	0.109624007	0.10194227	0.2115663
Blowfish	1.594330551	1.563594158	3.15792471	0.845677787	0.834793157	1.6804709
Camellia	0.60419046	0.578179894	1.18237035	0.114251903	0.105298873	0.2195508
CAST128	0.592311306	0.568499234	1.16081054	0.122301535	0.109387031	0.2316886
CAST256	0.652847014	0.623903457	1.27675047	0.146336308	0.132513827	0.2788501
DES	0.707779205	0.681713536	1.38949274	0.164695972	0.152701126	0.3173971
3DES	0.998690016	0.970555094	1.96924511	0.289438125	0.275600499	0.5650386
GOST	0.584218714	0.560881192	1.14509991	0.115808083	0.109081734	0.2248898
IDEA	0.622382052	0.59724628	1.21962833	0.113400073	0.107016148	0.2204162
MARS	0.741046648	0.718398548	1.4594452	0.142322866	0.128525141	0.270848
RC2	0.703530818	0.679608448	1.38313927	0.150419422	0.138945717	0.2893651
RC5	0.659964339	0.638010938	1.29797528	0.132863284	0.126148586	0.2590119
RC6	0.695767774	0.672883797	1.36865157	0.14260191	0.131415531	0.2740174
Rijndael	0.557134574	0.538570335	1.09570491	0.108676624	0.101753314	0.2104299
Serpent	0.691569349	0.673809425	1.36537877	0.129072923	0.115028303	0.2441012
SHACAL2	0.647515183	0.625245585	1.27276077	0.119130896	0.105340085	0.224471
SHARK	0.848945521	0.829030881	1.6779764	0.233960764	0.226545519	0.4605063
SKIPJACK	0.776778739	0.744824904	1.52160364	0.161402234	0.1540664	0.3154686
Three-way	0.602462306	0.579373637	1.18183594	0.109082294	0.102659514	0.2117418
Twofish	0.797397707	0.771283916	1.56868162	0.177186012	0.163901056	0.3410871
XTEA	0.60573553	0.57611449	1.18185002	0.119579337	0.113687194	0.2332665

the interval from one to 100 Mbyte and the speed was counting in second's number. Thus, for windows 7 × 32 machine resp. centos 7, we have (Figs. 1, 2):

5.3 Results and Discussion

From graphs and simulation, we can assume the average time per one Mbyte for each cipher to encrypt or decrypt data. Thus, the result is figured in the Table 3.

The results from the table above suggest using in preference (with order) AES-128, GOST, CAST128, Three-way, Camellia, XTEA, IDEA and SHACAL2 for real time application or high throughput network communication in any environment (same or heterogeneous OS). This result does not take any security claim in consideration. However, this result cannot be split away without the carefully study for the chosen cipher characteristic shown in the previous section. Still, any decision by user should take an earlier cryptanalysis study, which help to express the weaknesses for each cipher and the most successful attack developed as well.

In this case, while studying each cipher complexity, characteristic, key length with the speed test (linear dependence with power consumption in reverse mode) the earlier order change into AES128, Camellia128, GOST, CAST128, IDEA, XTEA, SHACAL2 and Three-way.

This result presents the general case to help any user to make a decision about the appropriate cipher for its own system security. Besides, user should define the security level accepted, the tolerance level for attacks, the quality of service required and how ciphers can affect that… Those requirements are essential and they will allow setting a barrier for a scale comparison in order to filter and choose the most convenient cipher to which the earlier order can change through considering each user recommendations.

6 Conclusion

Encryption is the main core for any security solution. However, this technique consumes a significant amount of computing resources such as CPU time, memory, and battery power and packet size. Those computing resources are linearly depended and linked to speed test, which makes it an important test to measure and compare the most optimal cipher for the desirable service or solution. In fact, this paper presents a fair performance comparison by computing the cipher-time and giving the average time required per one Byte for multiple OS. This cipher-time is affected by different settings for each cipher, such as different sizes of data blocks, different data types, and different key size. This comparison allows user to choose the best cipher for each required service. Besides that, the paper traits the security concept by comparing and showing the best-known successful cryptanalysis attack for some of them to help proving the protection level offered via each one. In addition, it must be mentioned that "speed, power consumption, safety…" cannot be achieved to their complete potential, due to their conflicting nature, which forces us to compromise between requirements. According to this and by considering those requirements in normal use, the paper suggests an order of the best eight convenient ciphers to use in general case.

References

1. Tornea, E.G.: Contributions to DNA cryptography: applications to text and image secure transmission (2003)
2. Mathur, M., Kesarwani, A.: Comparison between DES, 3DES, RC2, RC6, BLOWFISH and AES. In: Proceedings of National Conference on New Horizons in IT-NCNHIT (2013)
3. Creighton, T., Hager, R., Midkiff, S.F., Park, J.-M., Martin, T.L.: Performance and energy efficiency of block ciphers in personal digital assistants
4. Lu, C.C., Tseng, S.Y.: Integrated design of AES encrypter and decrypter
5. Nie, T., Zhang, T.: A study of DES and blowfish encryption algorithm
6. Denning, D., Imine, J., Devlin, M.: Compact iterative FPGA Camellia algorithm implementations
7. Boey, K.H., Lu, Y., O'Neill, M., Woods, R.: Differential power analysis of CAST-128. In: 2010 IEEE Annual Symposium on VLSI
8. Riaz, M., Heys, H.M.: The FPGA implementation of the RC6 and CAST-256 encryption algorithms. In: Proceedings of the 1999 IEEE Canadian Conference on Electrical and Computer Engineering
9. Mushtaque, M.A., Hussain, S., Dhiman, H., Maheshwari, S.: Evaluation of DES, TDES, AES, blowfish and two fish encryption algorithm: based on space complexity. IJERT **3**(4) (2014)
10. Korobitsin, V., Ilyin, S.: GOST-28147 encryption implementation on graphics processing units. In: The 3th International Conference on Availability, Reliability and Security
11. Leong, M.P., Cheung, O.Y.H., Tsoi, K.H., Leong, P.H.W.: A bit-serial implementation of the international data encryption algorithm IDEA
12. Wunnava, S.V., Rassi, E.: Data encryption performance and evaluation schemes. In: Proceedings IEEE Southeastcon (2002)
13. Buchanan, W.J.: RC2 encryption and decryption in microsoft.NET
14. Mohamed, A.B., Zaibi, G., Kachouri, A.: Implementation of RC5 and RC6 block ciphers on digital images. In: 8th International Multi-Conference on Systems, Signals and Devices (2011)
15. Amiri, M.A., Mahdavi, M., Atani, R.E., Mirzakuchaki, S.: QCA implementation of serpent block cipher. In: 2009 Second International Conference on Advances in Circuits, Electronics and Micro-Electronics
16. Isobe, T., Shibutani, K.: All subkeys recovery attack on block ciphers: extending meet-in-the-middle approach. Sony Corporation
17. Rijmen, V., Daemen, J., Preneel, B., Bosselaers, A., De Win, E.: The cipher SHARK
18. Kim, J., Phan, R.C.-W.: A cryptanalytic view of the NSA's skipjack block cipher design. In: 3-th International Conference and Workshops Proceedings, ISA 2009 Seoul, Korea, June 25–27, 2009
19. Daemen, J., Govaerts, R., Vandewalle, J.: A New Approach to Block Cipher Design, pp. 18–32. Springer (1994)
20. Andem, V.R., Tuscaloosa, A.: A cryptanalysis of the tiny encryption algorithm (2003)
21. Bingmann, T.: Speedtest and comparison of open-source cryptography libraries and compiler flags **14**, 53 (2008) Permlink. Tags: cryptography crypto-speedtest

22. Masadeh, S.R., Aljawarneh, S., Turab, N., Abuerrub, A.M.: A comparison of data encryption algorithms with the proposed algorithm: wireless security
23. Elminaam, D.S.A., Abdul-Kader, H.M., Hadhoud, M.M.: Performance evaluation of symmetric encryption algorithms. Commun. IBIMA **8** (2009). ISSN: 1943-7765
24. Elbayoumy, A.D., Shepherd, S.J.: Stream or block cipher for securing VoIP? Int. J. Netw. Secur. **5**(2), 128–133 (2007)
25. Heys, H.M.: A tutorial on linear and differential cryptanalysis. In: Electrical and Computer Engineering Faculty of Engineering and Applied Science Memorial University of Newfoundland
26. Sinkov: Elementary cryptanalysis. Math. Assoc. Am. (1966)
27. Bogdanov, A., Khovratovich, D., Rechberger, C.: Biclique cryptanalysis of the full AE (2011)
28. Rijmen, V.: Cryptanalysis and Design of Iterated Block Ciphers. Ph.D thesis (1997)
29. Pascal, J.: On the complexity of Matsui's attack. Sel. Areas Cryptogr. 199–211 (2004). Archived from the original on 27 May 2009
30. Lucks, S.: Attacking triple encryption. Fast Softw. Encryp. **1372**, 239–253 (1998)
31. Tews, E., Ralf-Philipp, W., Pyshkin, A.: Breaking 104 bit WEP in less than 60 seconds. WISA (2007)
32. Biham, E., Dunkelman, O., Keller, N.: Linear cryptanalysis of reduced round serpent. FSE (2002)
33. Ferguson, N.: Impossible differentials in twofish (1999)
34. Kelsey, J., Schneier, B., Wagner, D.: Related-key cryptanalysis of 3-WAY, Biham-DES, CAST, DES-X NewDES, RC2, and TEA. In: Lecture Notes in Computer Science (1997)
35. Wang, M., Wang, X., Hu, C.: New linear cryptanalytic results of reduced-round of CAST-128 and CAST-256 (2009)
36. Khovratovich, D., Leurent, G., Rechberger, C.: Narrow-bicliques: cryptanalysis of full IDEA
37. Yearly report on algorithms and key sizes, D.SPA.20 Rev. 1.0, ICT-2007-216676 ECRYPT II, 09/2012
38. Transitions: recommendation for transitioning the use of cryptographic algorithms and key lengths, NIST
39. Bai, D., Li, L.: New impossible differential attacks on camellia. In: 8th International Conference, ISPEC 2012, Hangzhou, China, April 9–12, 2012 Proceedings, pp. 80–96. Springer (2012)
40. Courtois, N.T., Gawinecki, J.A., Song, G.: Contradiction immunity and guess-then-determine attack on GOST. Versita (2012). Retrieved 2014
41. Hinoue, T., Miyaji, A., Wada, T.: The security of RC6 against asymmetric chi-square test attack. IPSJ J. **48**(9), 1–10 (2007)
42. Kim, J., Kim, G., Hong, S., Lee, S., Hong, D.: The related-key rectangle attack application to SHACAL-1. In: Information Security and Privacy on 9th Australasian Conference, ACISP 2004 Sydney, Australia, July 13–15, 2004 Proceedings. Springer
43. Kim, J., Lu, J.: Related-key rectangle attack on 42-round SHACAL-2. In: Information Security on 9th International Conference, ISC 2006, Samos Island, Greece, August 30–September 2, 2006. Proceedings pp. 85–100. Springer (2006)
44. Jakobsen, T., Knudsen, L.R.: The interpolation attack on block ciphers. In: Fast Software Encryption on 4th International Workshop, FSE'97 Haifa, Israel, January 20–22 1997 Proceedings, pp 28–40. Springer (1997)
45. Lu, J.: Related-key rectangle attack on 36 rounds of the XTEA block cipher. Int. J. Inf. Secur. **8**(1), 1–11 (2008)
46. Wei: Free C++ library for cryptographic schemes. http://www.cryptopp.com
47. NSA: On the secure hash algorithm family

Multi-layer Neural Network for EMV Evaluation

Noura Ouerdi, Tarik Hajji, Abdelmalek Azizi and Amina Yahia

Abstract In this work, we based on EMV specifications to implement a multi-layer artificial neural network. Indeed, we used the machine state diagram that has already been achieved in our previous work [4] where we model an EMV transaction between terminal and payment card using UML language. In order to detect potential attacks or vulnerabilities, we chose to use a neural network multilayer with back propagation. We made a comparative study to select the architecture of the best performing ANN. In this paper, we discuss in detail our methodology.

Keywords EMV · Artificial neural network · Multi-layer · Back propagation

1 Introduction

EMV is the international standard of security of payment cards (smart cards). It takes its name from three organizations: Europay, MasterCard and Visa International. EMV specifies the interoperability between EMV cards and EMV payment terminals worldwide. The transition to EMV Standard payment cards provides benefits to improving transaction security. Therefore, EMV cards have become the most currently used, which leads to the attackers who always look for ways to

N. Ouerdi (✉) · A. Azizi · A. Yahia
Laboratory of ACSA, FSO, Mohammed First University, Oujda, Morocco
e-mail: noura.ouerdi@gmail.com

A. Azizi
e-mail: abdelmalekazizi@yahoo.fr

A. Yahia
e-mail: aminayahia@yahoo.fr

T. Hajji
Lab CRDEI, Private University of Fez, Fez, Morocco
e-mail: hajji-tarik@hotmail.com

Á. Rocha et al. (eds.), *Europe and MENA Cooperation Advances in Information and Communication Technologies*, Advances in Intelligent Systems and Computing 520, DOI 10.1007/978-3-319-46568-5_55

attack them by exploiting potential vulnerabilities. To detect possible vulnerabilities in an EMV transaction, we thought to use neural networks and more particularly a multi-layer neural network with back-propagation.

The first neural networks were not able to solve nonlinear problems; this limitation was removed through back propagation of the error gradient in multilayer systems. In this work, we took advantage of this benefit.

The paper is organized as follows: the second section presents the related works including the previous work on the neural network implementation for a portion of the EMV specifications (without loops) [2]. The third section details our methodology of multi layer neural network (with loops). In the fourth section, we present obtained results.

2 ANN and Related Works

2.1 Artificial Neural Network

This is a problem resolution mechanism, inspired by the operation of the human brain. The basic unit of ANN is the artificial neuron. An ANN is made up of collections of neurons organized in layers: The first layer is called the input layer, the last one is called the output layer and the middle layers are respectively called hidden layer 1, hidden layer 2, etc. (Fig. 1) [3].

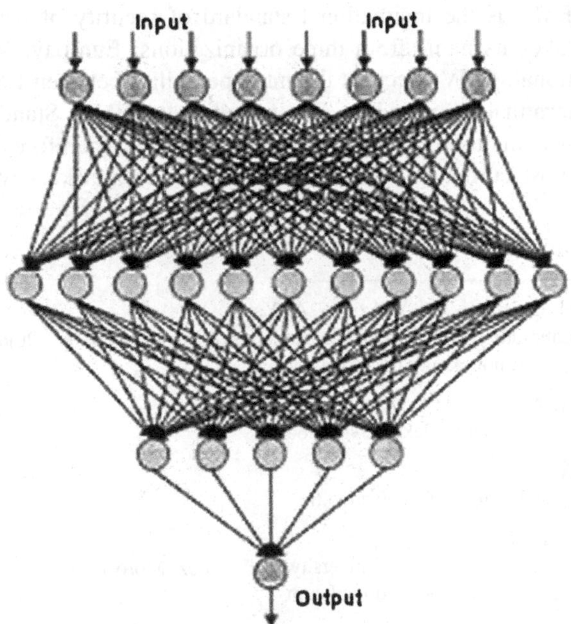

Fig. 1 Layers of an ANN

An ANN is characterized by a specific architecture to define its neural structure. In this work we propose a notation to define this architecture, for example the notation ANN (64, 16, 64) indicates the use of a multilayer ANN compound by:

- An input layer composed by 64 artificial neurons,
- One hidden layer composed by 16 artificial neurons,
- An output layer composed by 64 artificial neurons.

We distinguish two ways of using ANN:

Use with learning: we have a database composed by a partial solution of the problem. Then, we use an algorithm to teach ANN the resolving mechanism of the different instances of the problem. The gradient back propagation is a famous algorithm able to comprehend ANN comprehension mechanisms. Finally, we can use the ANN after learning to solve new instances of the problem.

Use without learning: in some cases, we have no information about the solution space for the problem studied. For example, the problem of classification which we do not have classes to classify elements. So we can use an ANN in unsupervised mode to find the classes firstly.

2.2 *Artificial Neuron*

It is a mathematical operator that can perform an unweighted sum of its inputs (see Fig. 2) [3]. It represents the atomic unit that composes ANN.

It uses non-linear functions to calculate its output as the sigmoid function (Fig. 3) or the hyperbolic tangent (Fig. 4).

2.3 *Artificial Neural Network Applied to EMV Specifications*

In our previous work [2], we implemented an artificial network to detect potential attacks or vulnerabilities in EMV specifications of terminal-Card transaction. Indeed, we used state transition diagram illustrating the modeling of an EMV

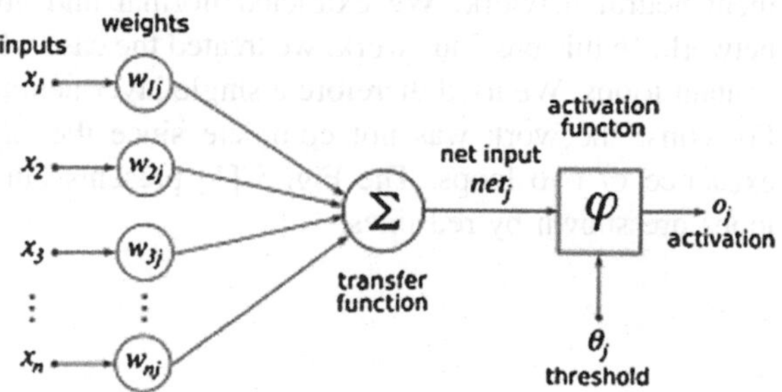

Fig. 2 The atomic unit composing an ANN

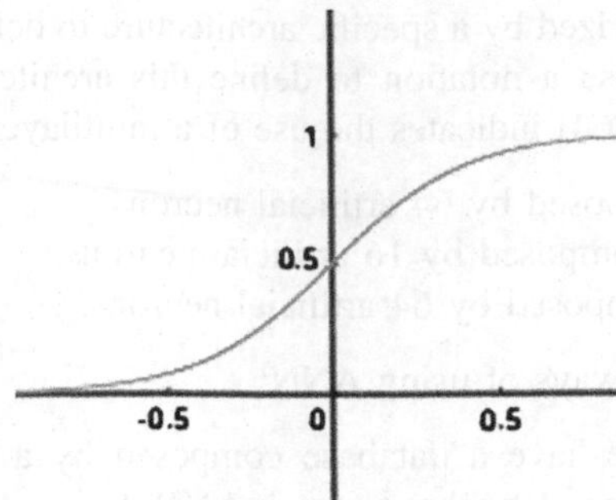

Fig. 3 Sigmoid function

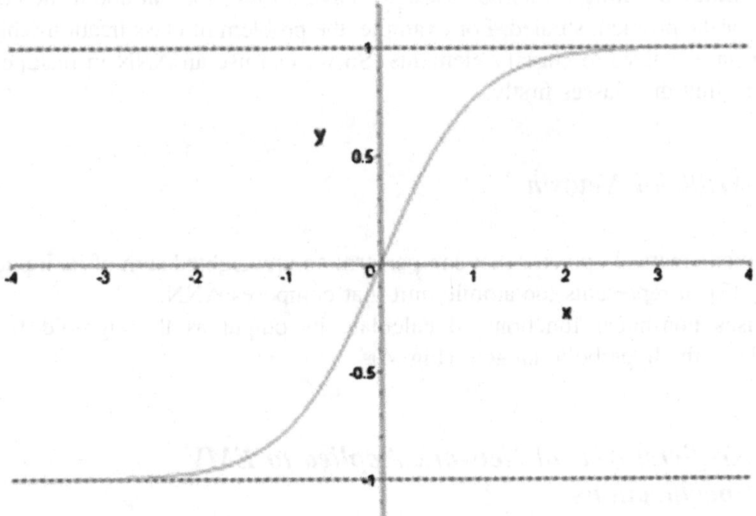

Fig. 4 Hyperbolic tangent

transaction. Then, we deducted the graph on which we based our order to implement neural network. We extracted normal and abnormal paths to aliment our network. In this previous work, we treated the case where the state diagram does not contain loops. We used therefore a single layer neural network and it was effective. For cons, the work was not complete since the EMV specifications require the existence of two loops. The Fig. 5 [1] presents our complete state diagram. The loops are shown by red lines.

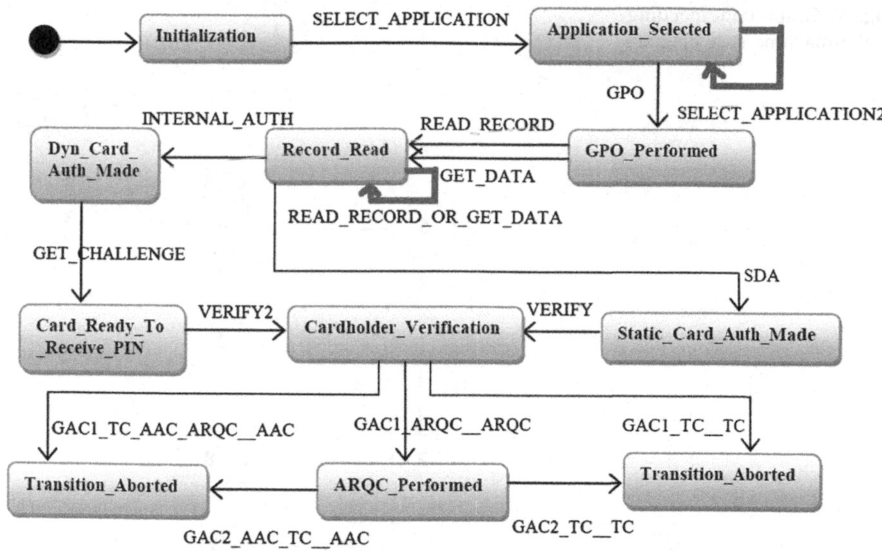

Fig. 5 Machine state diagram modeling EMV transaction

3 Our Methodology

First, we started with matching an integer value to each state of the machine state diagram of the Fig. 6.

We made a comparative study to select the architecture of the best performing ANN. We unified the basis of training data and we used our automatic generator of ANN by a meta-model description to generate several types of ANN with distinct architecture. The description of this generator is published in the article "Generating an Artificial Neural Network from a Meta-Model Description" [4].

Learning algorithm:

The algorithm is inspired by the back-propagation gradient [3] and it is as follows:

- Step 1: Initialize the connection weights (weights are taken randomly).
- Step 2: Propagation entries entered the E_i are presented to the input layer:

$$X_i = E_i$$

- The spread to the hidden layer is made using the following formula:

$$Y_i = f\left(\sum_{i=1}^{7} X_i * W_{ji} + X_0\right)$$

Fig. 6 Graph corresponding to the machine state diagram

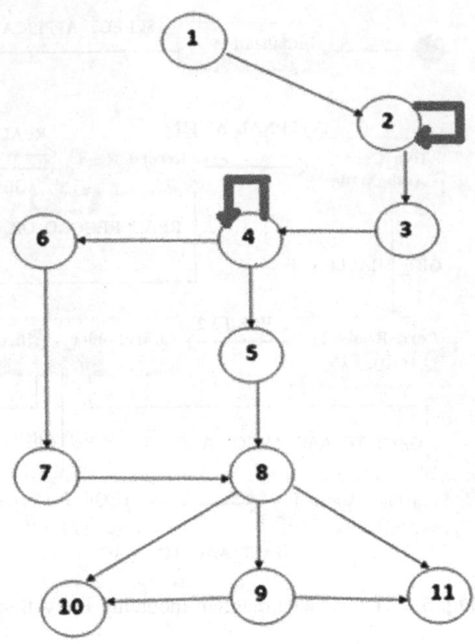

Then from the hidden layer to the output layer, we adopt:

$$Z_k = f\left(\sum_{i=1}^{3} Y_i * W_{kj} + Y_0\right)$$

X_0 and Y_0 are scalar;
f(x) is the activation function: $f(x) = 1/(1 + e^{-x})$

- Step 3: Back propagation of error at the output layer, the error between the desired output S_k and Z_k output is calculated by: $E_k = Z_k (Z_k - 1)(S_k - Z_k)$
- The error calculation is propagated on the hidden layer using the following formula:

$$Y_j = f_j(1 - Y_j) \sum_{k=1}^{7} W_{kj} E_k$$

- Step 4: Fixed connection weights: the connection weights between the input layer and hidden layer is corrected by:

$$DW_{ji} = n\, X_i F_j$$
$$DY_0 = n\, F_j$$

Table 1 Comparative study between different architectures of ANN

ANN architecture			Tests of stops		Results
Input	Hidden layers	Output	Square error	Loops	Recognition rate
8	4	2	0.0001	10^9	80
8	6, 4	2	0.0001	10^9	88
8	6, 4, 3	2	0.0001	10^9	78
8	6, 5, 4, 3	2	0.0001	10^9	70
9	5	2	0.0001	10^9	85
9	6, 4	2	0.0001	10^9	90
9	8, 6, 4	2	0.0001	10^9	88
9	8, 6, 4, 3	2	0.0001	10^9	63
10	5	2	0.0001	10^9	85
10	8, 4	2	0.0001	10^9	92
10	8, 6, 4	2	0.0001	10^9	88
10	8, 6, 4, 3	2	0.0001	10^9	53
11	5	2	0.0001	10^9	90
11	8, 4	2	0.0001	10^9	92
11	9, 6, 3	2	0.0001	10^9	83
11	9, 7, 5, 3	2	0.0001	10^9	52
12	6	2	0.0001	10^9	91
12	8, 4	2	0.0001	10^9	91
12	9, 7, 4	2	0.0001	10^9	84
12	10, 8, 6, 4	2	0.0001	10^9	53
13	6	2	0.0001	10^9	91
13	8, 4	2	0.0001	10^9	94
13	9, 7, 5	2	0.0001	10^9	83
13	10, 8, 6, 4	2	0.0001	10^9	50
14	7	2	0.0001	10^9	92
14	8, 4	2	0.0001	10^9	93
14	12, 8, 4	2	0.0001	10^9	86
14	12, 8, 6, 4	2	0.0001	10^9	51
15	7	2	0.0001	10^9	91
15	8, 4	2	0.0001	10^9	91
15	12, 8, 4	2	0.0001	10^9	83
15	12, 8, 6, 4	2	0.0001	10^9	51
16	**8**	**2**	**0.0001**	$\mathbf{10^9}$	**95**
16	8, 4	2	0.0001	10^9	90
16	12, 8, 4	2	0.0001	10^9	86
16	14, 12, 8, 4	2	0.0001	10^9	53

Then, we change the connections between the input layer and the output layer by: $DW_{kj} = n\ Y_j\ E_k$ (N is a parameter to be determined empirically)

- Step 5 loops: Loop to step 2 until a stop to define criterion (Error threshold, the number of iterations)

4 Obtained Results

To feed our learning base, we used 512 paths including 250 valid paths. The following table contains the results depending on the architecture of the ANN used:

By analyzing the recognition rate given by Table 1, it is concluded that the ANN, composed of an input layer with 16 neurons, a hidden layer composed by 8 neurons, and an output layer composed of 2 neurons, is the most efficient because it has the highest rate of recognition (95 %).

As synthesis, we can say that an ANN with a reduced architecture is unable to store a sufficient amount of knowledge to solve the problem. By against a wide architecture for network allows to dispersing this knowledge in the synapses of the ANN, as a result we obtain a lower recognition rate.

5 Conclusion

To verify an EMV card and detect possible vulnerabilities, we proposed in this paper a methodology based on neural networks. Indeed, we based on the state machine diagram modeling an EMV transaction between a terminal and an EMV card. Our model has a number of states with two possible loops. We opted for the choice of a multi-layer neural network with back propagation because it is able to solve our nonlinear problem and thus treat loops. Therefore, we conducted a comparative study between the different architectures in order to find the best one. The best recognition rate found was 95 % for the neural network architecture (16, 8, 2) (see Table 1).

As future work, we aim to compare this methodology with other vulnerability detection methodologies like [1, 5].

References

1. Ouerdi, N., Azizi, M., Ziane, M., Azizi, A., Lanet, J.L., Savary, A.: Security vulnerabilitics tests generation from SysML and Event-B models for EMV cards. Int. J. Secur. Appl. **8**(1), 373–388 (2014). http://dx.doi.org/10.14257/ijsia.2014.8.1.35
2. Ouerdi, N., Elfarissi, I., Azizi, M., Azizi, A.: Artificial neural network-based methodology for vulnerabilities detection in EMV. In: 11th International Conference en Information Assurance and Security, Marrakesh, Morocco, pp. 14–16 December 2015. IEEE Xplore. doi:10.1109/ISIAS.2015.7492750
3. Hajji, T., Jaara, E.M., Esbai, R., Erramdani, M.: Generating an artificial neural network from a metamodel description. Open. Res. Socio-econ. World. 10, 11 and 12 April 2014
4. Hajji, T., Jaara, E.M.: Digital watermaking and signing by artificial neural networks. Am. J. Intell. Syst. **4**(2), 21–31 (2014)
5. Ouerdi, N., Azizi, M., Ziane, M., Azizi, A., Lanet, J.L., Savary, A.: Generation of test cases based on SysML models. 2013 International Conference on Electronic Engineering and Computer Science. IERI Procedia **4**, 133–138 (2013)

Access Control System in Campus Combining RFID and Biometric Based Smart Card Technologies

Mohamed El Beqqal, Mohammed Amine Kasmi and Mostafa Azizi

Abstract For universities where security is primordial and accessing to certain areas must be checked, an access control system should be used in order to enhance security in general and to reduce time-consuming in access control of an important number of candidates in the same time. RFID and smart cards are widely used technologies; RFID guarantees a simultaneous reading of the identified objects and the smartcards offer a storage capacity and enable processing information. Several problems of security remain still unresolved for both RFID and smart card used separately. We present in this paper a solution of access control, which combines specific identification and authentication technologies with synchronous and asynchronous data processing. In our case, we target the academic context. We demonstrate by two use cases how much the system security could be improved by combining RFID and Smartcards in a complementary way. For this purpose, the design of our system focuses in the first scenario on performing the verification of the presence of students in exams by using both synchronous and asynchronous techniques based on the coupling of RFID and smart card technologies. The second scenario focuses on the security of accessing sensitive areas by providing a synchronous verification method requiring immediate validation of the access based on RFID technology and the fingerprint of staff. Further technologies are also used to ensure accurate authentication, such as cameras and liveness sensors.

Keywords RFID · Smartcard · Biometric test · Access control · Identification · Authentication · Validation · Academic context

M. El Beqqal (✉) · M.A. Kasmi · M. Azizi
University Mohamed First, B.P. 524, 60000 Oujda, Morocco
e-mail: elbeqqal.mohamed@gmail.com

M.A. Kasmi
e-mail: makasmi.ensao@gmail.com

M. Azizi
e-mail: azizi.mos@gmail.com

Á. Rocha et al. (eds.), *Europe and MENA Cooperation Advances in Information and Communication Technologies*, Advances in Intelligent Systems and Computing 520, DOI 10.1007/978-3-319-46568-5_56

1 Introduction

Using the latest contactless technologies in many areas has become necessary because of the great benefits offered by them. Among the most known domains which make use of these contactless solutions there is logistics that uses RFID technology to identify and track product of the company. Another area is the payment transactions where smartcards are massively used. Each technology is efficient in a specific context of use. For example in the domain of logistics, companies prioritize speedy processing of items and simultaneous identification while in the domain of banking, operations security is strongly favored.

Our interest is directed to the university area and campus management. Many works upon security and optimization have been conducted in this regard. Attendance management system is one of the subjects that have been widely studied as mentioned in [1, 2]; Nowadays managing library system is considered among the most RFID application in institutes [3]. This paper will focus specifically on the identification and authentication processes that will give authorization to access the most important campus resources.

The design of our system is divided into two parts; first part will be devoted to verification of students in halls of examinations, on this purpose, the RFID technology will help us do a first identification of students present in the exams. To avoid student blocking at the door, a biometric authentication with smartcards will be processed synchronously by matching the captured fingerprint directly on smartcard, combined with an asynchronous verification in the server side to ensure the true identity of students. The second part will deal with the control of access to sensitive areas, the system is different from the first one, we will use also RFID identification, and a synchronous and asynchronous biometric authentication based on matching respectively the captured fingerprint with stored one in smartcard and with another stored remotely on a fingerprint server.

The remaining of this paper is organized as follows: Sect. 1 presents the concept on an RFID system, Smart card technology and a brief review on current challenges and some frauds on access control in campus where we focus on the problem of security and authentication. Then, we present related works in Sect. 2. Section 3 presents our contribution of an enhanced access control system. Section 4 concludes this paper.

2 Technologies

In the access control field, we often use smart secure technologies like Bluetooth, NFC or RFID for identification, to verify identified people smart cards or biometrics are used for authentication. In this section we give more details on each of these technologies which we will use in our work.

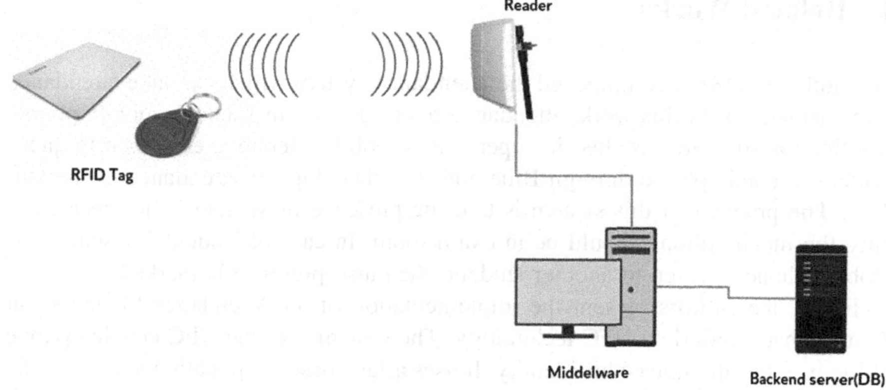

Fig. 1 General architecture of RFID based system

Radio frequency identification (RFID) [4] is a fast automatic identification technology that uses radio frequency to identify objects carrying tags when they pass near an RFID reader. The Fig. 1 shows the general architecture of an RFID based system.

An RFID system comprises:

1. Tag, consist of a chip, antenna and a certain amount of computational and storage capabilities
2. Reader, interrogates tag to obtain tag information and forwards the encrypted information gathered from the tag to the backend server for verification purpose.
3. Software responsible to gather data from reader, and dispatch it to the backend servers.
4. Backend server contains a local database and some processors.

A smart card is an object made in plastic containing a microelectronic circuit providing the majority of features of a computer. It is produced as a small card, incorporating an electronic circuit performing a set of operations (storage, calculation or verification) in a secure way. Any smart card has its own computing power and its own storage capacity. Everything concerning the standardization of the smart card is defined in the ISO 7816 standard with the International Organization for Standardization (ISO) [5]. Among the relevant applications of smart cards we find: financial applications, Government programs, information security, physical access, health card, campus identification, e-ticketing and others [6].

Biometric technology allows to check identity through characteristics such as fingerprint, face, voice, iris, and so on. The physical aspect of data considered as unique is receiving attention as a personal authentication method that is more reliable than standard methods such as a password or ID cards. [7] Biometric is defined as the "automated identification or verification of human identity through the measurement of repeatable physiological and behavioral characteristics" (Association of Biometric 2004).

3 Related Works

The authors in [8] have proposed the attendance system which can take attendance using Bluetooth. In this work, attendance realized using instructor's mobile phone. Application software installed in supervisor's mobile telephone enables it to query student's mobile phone through Bluetooth for checking the attendance of the student. The problem of this system is that the presence of student is not necessary, only the mobile phone should be in exam room. In case of students' absent if his mobile phone is given to another student, then also presence is marked.

In [9], the authors present the implementation of an Attendance Management System that is based on NFC technology. The system uses the NFC enabled phone of the user to authenticate his identity. It uses a Java based application to receive the NFC tag IDs. The problem in this case is that the student must have an NFC enabled phone to mark presence in the class room, which cannot be possible for every student.

The authors in [10] proposed a fingerprint based attendance system. The students validate their presence by placing their finger on the fingerprint reader. In 2013, [11] proposed also a system for employee attendance using fingerprint. The main problem in those cases is it is very time consuming as it checks one fingerprint template with all templates stored in the database. Also, every student needs to stand in long line to mark attendance.

In 2010, [12] has proposed a wireless iris recognition attendance management system. This system takes a long time to check the student who has to wait a considerable time to scan his iris.

In [13], Student attendance is being taken using the Face Recognition method. A camera is fixed at the center of the classroom near the blackboard, so that the view of the camera covers the entire classroom. The system requires a person to be at a medium distance from the camera, the proposed system uses the Viola Jones algorithm to detect the human faces. The major problem in this system is that the captured face has to be compared with all the records stored in the database which is time consuming.

In [14], authors present a new system based on the combination of RFID and the Facial Recognition to verify the student authentication. It would simply need a camera in the class and as the lecture starts it identifies all present students, however, the technology available today is still not robust enough hence the need to continue research and development in this area. Some problems to consider are lighting as a very big problem, there is many attempts to find a solution that most are not satisfactory as they degrade the image too extensively. A more hardware issue would be camera quality.

Even though there are many technologies available for identification taking like Bluetooth, Infrared, Wi-Fi, and NFC, then we select RFID technology. The main reason is that the RFID has a capability of reading of several labels simultaneously and can contain information like student ID.

There are different technologies for authentication but we select fingerprint recognition technology as it has the lowest cost. In addition, all the above technologies having many drawbacks which will be cover by fingerprint recognition. Furthermore, the fingerprint result is much higher than those of camera or iris. So fingerprint recognition is chosen for authentication.

In our proposed system, identification will be made using RFID technology and authentication will be done by fingerprint recognition.

4 Issues Raised in Campus

Based on the current state of the majority of our open access institutions, we identify the basic areas that require protection within the campus:

- General Campus access,
- Access to the administration office,
- Access to the library,
- Access to professor's offices,
- Access to classrooms,
- Access to laboratories,
- Access to exam rooms.

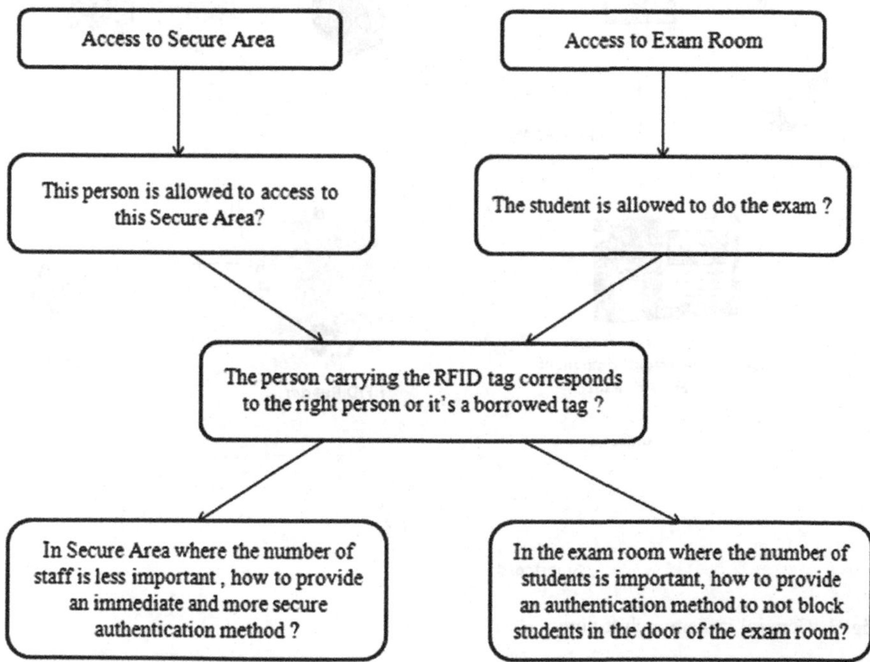

Fig. 2 Challenges in our access control system

Our study here focuses on access control to the laboratories, restricted-access rooms, and exam rooms. In Fig. 2, we list the challenges that will be taken into account in our access control system.

5 Our Proposed System for Access Control

In our system, we assume a few rules and prerequisites that are considered by the institute:

- Each student is provided with an RFID tag which reveals his unique identity within the campus.
- Each student is also provided with a smartcard containing the id tag and biometric fingerprint (later we will see the utility of this coupling).
- The institute has a record of data concerning the associated biometric fingerprint and RFID tag of the student stored in a central database server.

The Fig. 3 presents the general architecture of the target system.

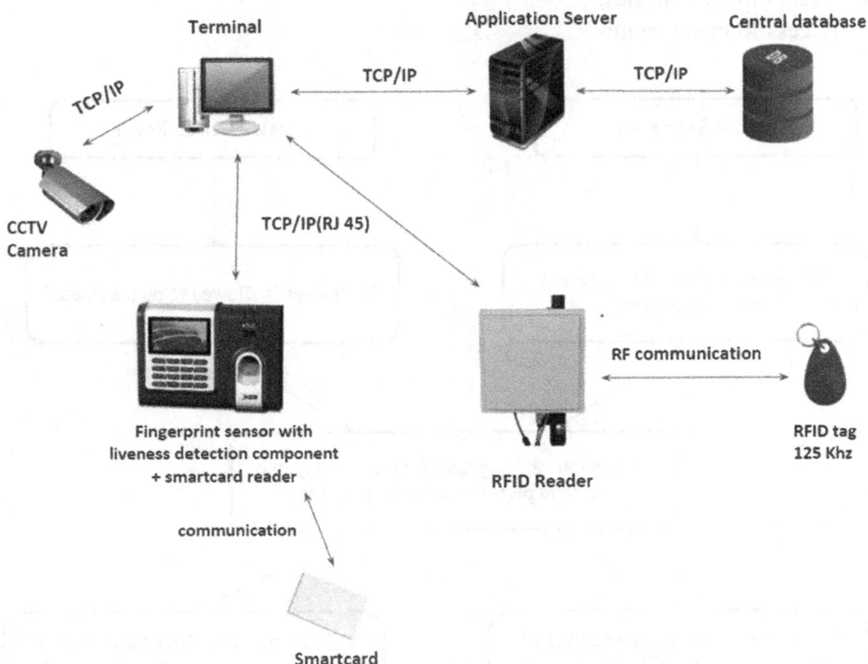

Fig. 3 General system architecture

6 Flowchart of the Exam Attendance System

Our contribution focuses on the security aspect and specifically the access control of student to the exam rooms.

To ensure performance and processing speed, we decided to load data from central server in a local database. We also planned to make an intranet application designed primarily for managing the exams. For each exam room, there is a specific list of students authorized to access in.

The cinematic of our system can be recognized easily through the flowchart shown in Fig. 4.

(1) When students approach the exam room where a stationary RFID reader is installed, they will declare their tags (here, passive tags are used to reduce scope (<1 m) transmission and also avoid the problem interference [10]) near the reader to validate their right to access to the room.
(2) A first synchronous identification is made by checking the captured tag with the list of students allowed to do the exam.
(3) The dashboard (RFID tag column) is updated with success if the student is authorized. However if the student does not meet the conditions of this exam, the reader will transmit a sound of failure and the dashboard will be updated with this result.
(4) The student who validated the RFID identification, must insert his smart card in reader and puts his finger on the fingerprint sensor to validate the authentication synchronously with the smart card reader. The reader will send the captured fingerprint to the smartcard for matching with the template stored in smartcard [15]. The matching decision is computed inside the smartcard and consequently the entire original template is never released from the smartcard. Once the validation is done, we get the RFID tag ID from the smart card, this id allow us to match directly the student's captured fingerprint with the corresponding record in the database, without performing a one to many comparison, which will reduce the response time.
    To enhance the security mechanism, our fingerprint recognition system will have the ability to detect if a presented fingerprint is from a live person or an artificial finger [16]. Nowadays, fingerprint readers are being upgraded to include liveness detection solutions [17] that will be able to detect if the submitted fingerprint is a spoof or live finger through the use of blood pressure, electrocardiogram, temperature or other new methods. Our proposed system will also add a camera to capture a picture for the student who has put his finger on the fingerprint sensor, this picture will be stored for an additional verification if desired by the teacher.
    The Smart Card Biometric reader will send the pair (tag id, biometric fingerprint) to the middleware for asynchronous verification with the fingerprint in the database.

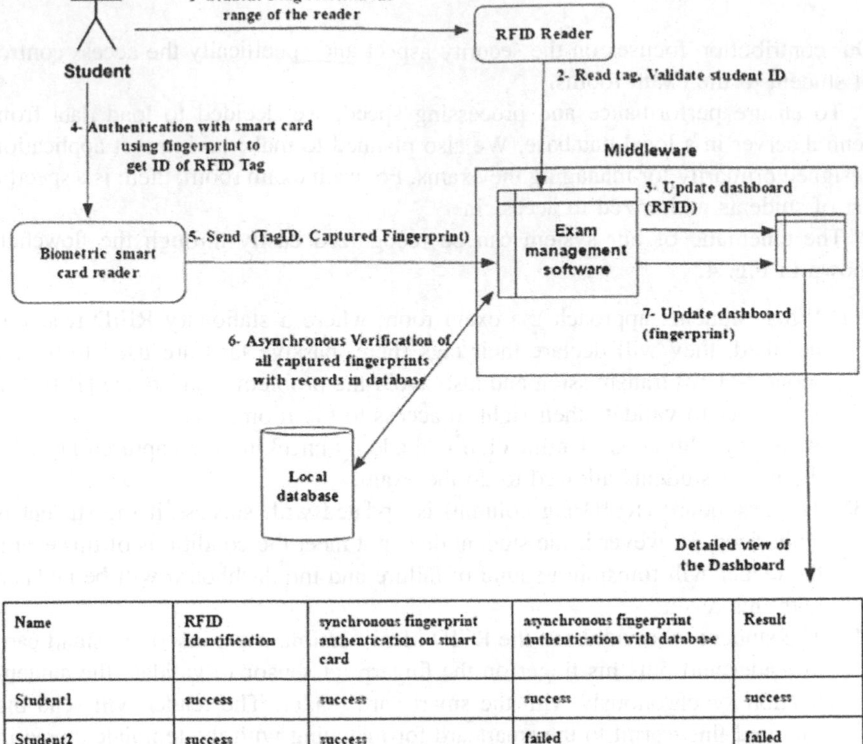

Name	RFID Identification	synchronous fingerprint authentication on smart card	asynchronous fingerprint authentication with database	Result
Student1	success	success	success	success
Student2	success	success	failed	failed
Student3	success	failed	success	failed
Student4	success	failed	failed	failed
Student5	failed	success	success	failed
Student6	failed	success	failed	failed
Student7	failed	failed	success	failed
Student8	failed	failed	failed	failed

Fig. 4 Flowchart of the exam management system

(6) The middleware will store the captured fingerprint temporarily in the computer existing in the exam room. This choice is based on the fact a large number of students will be susceptible to pass the same exam, we decided to treat this third identity control step asynchronously to avoid blocking the students at the door of the exam room. We took advantage of the rule that the student is not allowed to leave the exam room during the first 30 min to treat the comparison between the collected fingerprints and those already existing in the local database.

(7) Once the validation done, the system updates the status of the following columns of students in the software dashboard.

- RFID Identification,
- Fingerprint synchronous authentication,
- Fingerprint asynchronous authentication.

7 Access to Secure Areas

In this section we present another use case of our proposed system which is designed also for securing access to sensitive areas including university residences, research laboratories, and administration block. These areas can be reached by students, teachers or administrative staff according to their predefined permissions.

As described in the case of examination room and to optimize the execution time of requests processing, the data is stored in the local database. Also we will develop an access control feature which is designed for the human resources manager, and this feature will aim to define the authorized list for a specific area. The rule of the manager is to associate a list of authorized persons with the concerned area. Once the task validated, the selected persons are loaded into the local database which is attached to the corresponding secure area.

When the user wants to access a secure area he must identify himself with his identity card using RFID technology, for that a proximity RFID reader will be installed beside each secure area to capture the user tag information. A secure authentication with the smart card will be established using fingerprint. After a

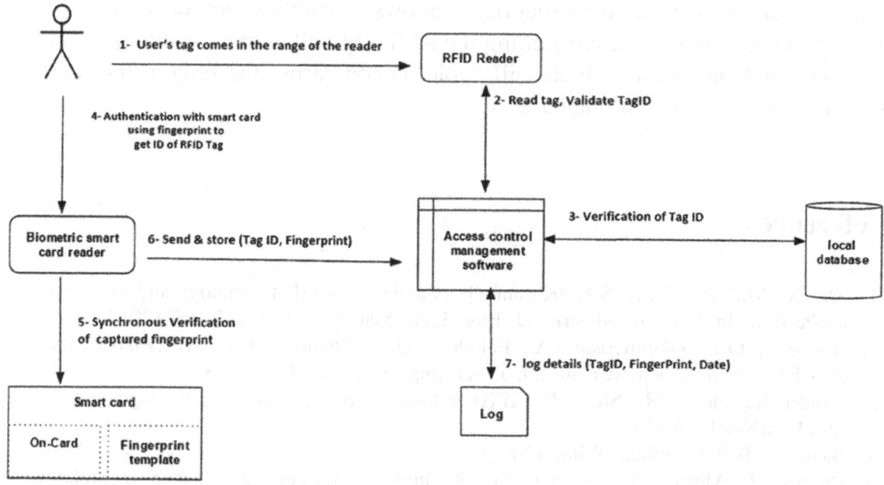

Fig. 5 Flowchart of the access control system in sensitive areas

successful authentication, the door will be unlocked, and a picture of the authorized person will be taken.

Compared to the case of the examination rooms we noticed that the number of staff supposed to access to the secure area is limited, moreover, these private areas needs an immediate verification for that our system will match synchronously every captured fingerprint with the original template of stored fingerprint in smart card, if the system detects incorrect or unauthorized users then the access is denied.

Finally the system logs all the details about the entry including the tag id, the fingerprint and the timestamp. The Fig. 5 shows easily the sequence of required steps to access a secure area and also interactions between objects of the system.

8 Conclusion

In this paper, we proposed a design of an access control system to augment the security level in university; precisely, we have focused on access to examination rooms and private areas that requires a high security. This solution is based on coupling RFID technology with smart cards based on fingerprint authentication.

Our proposed system provides:

- An immediate RFID identification for access control to exam rooms and secure areas.
- A synchronous verification of captured fingerprint of students with the stored fingerprint in smart card by using the match-on-card algorithm.
- An asynchronous verification approach of captured fingerprint with the database which is more appropriate to validate identities of students in the exam rooms.

Finally, the utility of coupling RFID and smart card-based biometrics has a double advantages, in term of security it allows a double check of identity of the student and in term of time consuming the RFID tag will help us to directly seek for the concerned fingerprint which will avoid us comparing the fingerprint gathered with all available ones in database.

References

1. Zhu, X., Mukhopadhyay, S.K., Kurata, H.: A review of RFID technology and its managerial applications in different industries. J. Eng. Tech. Manage. **29**(1), 152–167 (2012)
2. Arulogun, O.T., Olatunbosun, A., Fakolujo, O.A., Olaniyi, O.M.: RFID-based students attendance management system. Int. J. Sci. Eng. Res. **4**(2), 1–9 (2013)
3. Nainan, S., Parekh, R., Shah, T.: RFID technology based attendance management system (2013). arXiv:1306.5381
4. Banks, J.: RFID Applied. Wiley (2007)
5. Dreifus, H., Monk, J.T.: Smart Cards: A Guide to Building and Managing Smart Card Applications. Wiley (1998)

6. Online courses. http://www.smartcardbasics.com/smart-card-types.html, http://cedric.cnam.fr/~bouzefra/cours/CoursNFC_Bouzefrane_Decembre2013.pdf
7. Kadry, S., Smaili, M.: Wireless attendance management system based on iris recognition. Sci. Res. Essays **5**(12), 1428–1435 (2013)
8. Bhalla, V., Singla, T., Gahlot, A., Gupta, V.: Bluetooth based attendance management system. Int. J. Innovations Eng. Technol. (IJIET) **3**, 2319–2320 (2013)
9. Ahmad, B.I.: TouchIn: an NFC supported attendance system in a university environment. Int. J. Inf. Edu. Technol. **4**(5), 448 (2014)
10. Basheer, M.K.P., Raghu, C.V.: Fingerprint attendance system for classroom needs. In: India Conference (INDICON), 2012 Annual IEEE, pp. 433–438. IEEE (2012)
11. Rao, S., Satoa, K.J.: An attendance monitoring system using biometrics authentication. Int. J. Adv. Res. Comput. Sci. Softw. Eng. **3**(4), 379–383 (2013)
12. Kadry, S., Smaili, M.: Wireless attendance management system based on iris recognition. Sci. Res. Essays **5**(12), 1428–1435 (2013)
13. Behara, A., Raghunadh, M.V.: Real time face recognition system for time and attendance applications. Int. J. Electr. Electron. Data Commun. **1**(4). ISSN: 2320-2084
14. Patel, U.A.: Development of a student attendance management system using RFID and face recognition: a review (2014)
15. Chen, T., Yau, W., Jiang, X.D.: On-card matching. In: Encyclopedia of Biometrics, pp. 1014–1021. Springer (2009). ISBN: 978-0-387-73004-2
16. Buciu, I., Gacsadi, A.: Biometrics systems and technologies: a survey. Int. J. Comput. Commun. Control **11**(3), 315–330 (2016)
17. Mura, V., Ghiani, L., Marcialis, G.L., et al.: LivDet 2015 fingerprint liveness detection competition 2015. In: IEEE 7th International Conference on Biometrics Theory, Applications and Systems (BTAS), pp. 1–6. IEEE (2015)

Printed in the United States
By Bookmasters